Satellite Encryption

Satellite Encryption

John R. Vacca

ACADEMIC PRESS

A Harcourt Science and Technology Company

San Diego   London   Boston

New York   Syndey   Tokyo   Toronto

This book is printed on acid-free paper. ⓒ

ACADEMIC PRESS
A division of Harcourt Brace & Company
525 B Street, Suite 1900, San Diego, CA 92101-4495, USA
http://www.apnet.com

Academic Press
24–28 Oval Road, London NW1 7DX, UK
http://www.hbuk.co.uk/ap/

Library of Congress Catalog Card Number: 98-83124

International Standard Book Number: 0-12-710011-3

Printed in the United States of America
99 00 01 02 03 IP 9 8 7 6 5 4 3 2 1

This book is dedicated to all of the women and men that have sacrificed their

lives to fight terrorism around the world. They have served their planet well.

Contents

Author Preface

The purpose of this book is to show governments and organizations around the world how satellite encryption helps to preserve vital national secrets, limit attacks on a nation's information infrastructure, and eliminate security and authentication obstacles to electronic commerce. In addition, the book will show how in the wrong hands, however, it can be used to plan or cover up domestic and international crimes or overseas military operations. Satellite encryption software could make it easy and cheap for adversaries or terrorists to conceal deadly messages (like the surprise embassy bombings in Kenya and Tanzania, killing scores of people, including 12 Americans in the summary of 1998).

Next, I will discuss how the Federal Bureau of Investigation and the National Security Agency seek to preserve their ability to intercept and decode domestic and international communications. In other words, the book discusses how the FBI and NSA would like to inhibit the use of Public Key Exchange (PKE) in satellite communications to generate unbreakable codes; and, how stopping it altogether may be technically impossible and may raise constitutional issues.

I will also cover the *electronic battlefield* of the future. United States military strategists increasingly believe that high-speed encrypted satellite information technology (more than tanks, ships, or fighter jets) is the key to remaining the world's dominant force during the next 25 years. The ability to automatically identify friend or foe and transmit encrypted (and compressed) targeting data to the right weapon (via a satellite uplink and downlink) at the speed of light could produce a combat edge comparable to the blitzkrieg (like the U.S. cruise missile strikes on terrorists in Afghanistan and Sudan). At the very least, satellite encryption should help resolve one glaring problem for the battlefield of the future that occurred during the Persian Gulf war: friendly fire deaths, which accounted for 35 of the 146 Americans killed in action.

Finally, I will show how establishing a high confidence level in the security of electronic commerce is impossible without the protected satellite transmission of intellectual property. I will discuss how satellite encryption will make Peripheral Component Interconnect (PCI)-based computers, modems, web browsers and set-top-boxes safer for intellectual property distribution and electronic commerce through the hardware implementation of PCI bus-compatible real-time data encryption/decryption chip solutions.

Now, let us take a brief journey of the chapters contained in this book. See you there!

Part I

We begin our journey with the first part of this book. Chapter 1 provides an overview of current satellite encryption policies; the threat from the Internet; encrypted satellite data transmissions (downlink) and receiving (uplink); and encryption cracking. In Chapter 2, I outline the problem of growing information vulnerability and the need for technology and policy to mitigate this problem. In Chapter 3, I examine possible roles for satellite encryption in reducing information vulnerability and place cryptography into context as one element of an overall approach to ensuring information security. Finally, in Chapter 4, I discuss nongovernment needs for access to encrypted satellite information and related public policy issues, specifically those related to information gathering for law enforcement and national security purposes. I also discusses the right to satellite encryption by the financial community, international law firms, international management consulting firms, and, CEOs and Corporate Senior Managers.

Part II

The second part of our journey brings us to the instruments and goals of the current U.S. satellite encryption policy and some of the issues raised by current policy. Here, Chapter 5 is concerned primarily with export controls on satellite encryption, a powerful tool that has long been used in support of national security objectives but whose legitimacy has come under increasing fire in the last several years. In Chapter 6, I address the following: Sales of individual images from satellites owned and operated by U.S. firms; sales of real-time bit streams from such satellites; satellites owned by third parties (who could, unless otherwise limited, sell to whomever they chose), but operated by U.S. firms; satellites owned and operated by third parties that rely on support from U.S. firms; and, satellites owned and operated autonomously by third parties.

Chapter 7 discusses why many commercial users (who constitute the bulk of the Global Positioning System (GPS) system's beneficiaries), have expressed nervousness about relying on a DOD-controlled system. In Chapter 8, I examine the following: Issues concerning the fall-out from possible widespread deployment of Public-key Encryption (PKE) technologies; how encryption helps to preserve vital national secrets, limit attacks on the nation's information infrastructure, and eliminate security and authentication obstacles to electronic commerce, and export controls. Chapter 9 addresses the following: How the history of commercial space launch is a story of continually increasing competition; how the U.S. has taken advantage of the fact that it supplies over two-thirds of all commercial payloads — essentially communications satellites — to persuade China and Russia to restrict their launch rates and limit their price discounts;

how the U.S. has taken steps to inhibit Russia's sale of rocket equipment to India; how the idea of using export controls over satellite payloads to stabilize markets for launchers may be counterproductive in the long run by casting doubt on the reliability of U.S. suppliers of satellites; and, if technology comes to favor small satellites for some missions (as may be the case for land-mobile communications and surveillance) then small, quick-turnaround launch vehicles may be preferred over large ones.

In Chapter 10, I cover how Direct Broadcast Satellites (DBS) raise potential national security issues because they challenge state sovereignty over communications by bypassing national communications monopolies. Chapter 11 covers the growing tendency for U.S. forces to engage in multinational coalitions, coupled with the transition to computer-based information systems

Chapter 12 shows how the exploitation and control of space will enable U.S. forces to establish information dominance over an area of operations. In Chapter 13, I address escrowed satellite encryption: an approach aggressively promoted by the federal government as a technique for balancing national needs for information security with those of law enforcement and national security. Finally, in Chapter 14, I discuss other dimensions of the national satellite encryption policy, including the Digital Telephony Act of 1995 (aka the Communications Assistance for Law Enforcement Act) and a variety of other levers used in national cryptography policy that do not often receive much attention in the debate.

Part III

This next part of our journey covers development, implementation and management of advanced satellite encryption options and strategies that will forever change how organizations do business now and in the foreseeable future. Chapter 15 discusses the following: the planning process, details, implementing the plan, and, the commanding uplink (receiving) process. In Chapter 16, I analyze the following: how we listen; Telemetry Delivery System (TDS); downlink (transmitting) process; and, how to serve a world of data consumers.

Chapter 17 covers the following: geosynchronous satellites; low-earth orbit satellites; multiple access methods; geosynchronous systems; low-earth orbit systems; and, advantages and disadvantages of each. In Chapter 18, I focus on the following: Channel Control System and AMS software intricacies; DEC's MicroVax II computers and VMS computer operating system; the exact, technical nature of video and audio baseband systems; the idiosyncrasies of satellite uplink equipment; the technical characteristics of current generation satellites; the wide variety of satellite downlink systems; the technical liaison with General Instruments and DEC; the interface with the uplink operator to ensure transmission quality control; support of cable headend technicians that have intranet reception troubles; loading of DBS/IPPV program epochs from printed schedules; creation of all DBS tier-addressed messages; routine tape backups of DEC computer system hard disks; spot monitoring of signal quality with our test equipment; software development to enhance encryption systems; full-time and occasional leasing of Video-Cipher encryption systems; immediate utilization of all features; absolutely no learning curve; quick identification and resolution of operational problems; very low cost to achieve and maintain complete operational competence; security of the Videocrypt system; ISO card protocol; Videocrypt protocol; and, VCL File Format.

In Chapter 19, I present a framework for dealing with the following: satellite encryption policy; protection of personal data and privacy; security of information systems; intellectual property protection; international instruments; and, rising demand for hardware-based data security. Chapter 20 discusses the following: Mobile tracking; satellite encryption systems planning, design, and operation at C and Ku Bands; principles of satellite encryption systems; mobile satellite communications (MSAT); land mobile satellite encryption systems; MSAT propagation; modulation and coding for MSAT; space segment technology; mobile segment; vehicle antenna technology; aeronautical mobile satellite encryption systems; maritime mobile communications and surveillance; radio determination satellite encryption services; Inmarsat systems; mobile satellite encryption integration; current and future mobile systems; and, trends in mobile satellite encryption systems. Finally, in Chapter 21, I discusses the following: Getting ready for the next war; putting the electronic battlefield to the test; the new terror fear: biological weapons; advances in encrypted signal processing technology for electronic warfare; radar target identification; advanced electronic warfare principles; theater missile defense; soldier identification encryption system utilizing low probability of intercept (LPI) techniques; battle management systems; and, IRCM and IRCCM (Infrared Countermeasures and Counter-Countermeasures).

Part IV

Next, we continue on with the fourth part of our journey. Chapter 22, covers privacy; the Clipper Chip; and, banning satellite encryption for private citizens, while the government uses the same technology to listen in on the conversations of its own citizens. In Chapter 23, I examine the growing threat from our allies to steal encryption technology; and, the intensified efforts by the FBI to prevent such theft.

Chapter 24 discusses the political backlash and emotional fallout of the bombing of the federal building in Oklahoma City; and, how the FBI has begun to wage their own private war on the use of private encryption schemes. Finally, in Chapter 25, I continue the discussion started in Chapter 24 with an examination of how drug cartels are buying sophisticated encryption technology; how drug cartels will use telephones and other communications media with impunity knowing that their conversations are immune from our most valued investigative technique; why encryption capabilities available to criminals, both now and in days to come, must be dealt with promptly; and, the rise of an off-shore, black market satellite encryption trade. The chapter will also look at crackers who can take down encrypted military sites, nuclear weapons labs, Fortune 100 companies, and scores of other institutions.

Part V

The final part of our journey through this book evaluates enlarging the space of possible satellite encryption policy options, and offers findings and recommendations. Chapter 26 discusses a variety of options for the satellite encryption policy, some of which have been suggested or mentioned in different forums (in public and/or private input). In Chapter 27, I cover the following: Tightly-coupled secure GPS/INS systems; military leads in secure GPS/INS integration; an optimized secure GPS carrier phase ambiguity search method focusing on high speed and reliability; secure intelligent vehicle highway systems; and, preoperational testing of encrypted data link-based air traffic management systems in Magadan, Far East Russia.

In Chapter 28, I examine why ones and zeros, not bombs and bullets, may win tomorrow's battles; why within 25 years, accurate, up-to-date encrypted information from satellites about what's happening on a battlefield will provide more of an advantage than any weapon system; why a satellite computer picture will show where all U.S. forces are and where many of the enemy forces are; and, satellites collecting strategic battlefield information. Chapter 29 concludes with a discussion on how attacks on satellites could threaten national security as we approach the Year 2000; and, how DOD and businesses face significant challenges in countering attacks on satellites. Finally, Chapter 30 wraps up with a discussion on the findings and recommendations with regards to: The problem of information vulnerability; cryptographic solutions to satellite information vulnerabilities; the policy dilemma posed by satellite encryption; national satellite encryption policy for the information age; and, how satellite encryption is essential to freedom.

Part VI

Oops! Almost forgot! There are two appendices that provide additional resources that are available for satellite encryption security. Finally, there's a glossary of satellite encryption technology terms and acronyms.

Overall, this book will leave little doubt that a new world infrastructure in the area of satellite communications and encryption is about to be constructed. No question, it will benefit organizations and governments, as well as their advanced citizens. For the disadvantaged regions of the world, however, the coming satellite communications revolution could be one of those rare technological events that enable traditional societies to leap ahead and long-dormant economies to flourish in security.

Acknowledgments

There are many people whose efforts on this book have contributed to its successful completion. I owe each a debt of gratitude and want to take this opportunity to offer my sincere thanks.

A very special thanks to my editors Zvi Ruder and Chuck Glaser, without whose initial interest and support would not have made this book possible. And editorial assistant Della Grayson, who provided staunch support and encouragement when it was most needed. Thanks to my production editor, Vanessa Gerhard, whose fine editorial work has been invaluable. Thanks also to my marketing manager Bob Donegan; and, marketing assistant Sam Libby whose efforts on this book have been greatly appreciated.

Thanks to my wife, Bee Vacca, for her love, her help, and her understanding of my long work hours.

I wish to thank the organizations and individuals who granted me permission to use the research material and information necessary for the completion of this book.

As always, thanks to my agent, Margot Malley, for guidance and encouragement over and above the business of being an agent.

Foreword

Tim Matthews
RSA Data Security, Inc.

The night of May 19, 1998 was much like any other night, except that it was unusually quiet. Urgent pages did not interrupt dinner. Doctors on call were not called. Off-duty policemen went undisturbed by emergency calls. But this Tuesday night was unusually quiet because something was amiss. Salespeople, doctors, policemen, as well as thousands of other people, became unwittingly incommunicado.

Orbiting more than 22,000 miles above the equator, something went wrong on the Galaxy 4 satellite. Run by PanAmSat Corp., the Galaxy 4 is a critical link in the U.S. paging network. A glitch in the positioning system caused the satellite to spin out of alignment and turn away from Earth, where it is normally pointed. In addition to 90% of the 45 million pagers in the U.S., credit card authorization systems, radio signals, and television broadcasts were disrupted.

While most of us are blasé about these high-flying marvels, they are a critical link in our modern lives. The technology inside satellites is some of the most sophisticated available, and companies like PanAmSat spend hundreds of millions of dollars to build them. The tricky launches—and especially launch failures—are widely covered by the media. Yet not even the combination of high technology, high cost, and high risk causes the average person to give satellites a second thought — until the link breaks.

Perhaps the most amazing thing about satellites is that we all take for granted. Few technologies touch so many aspects of our daily lives. We notice when telephone lines are down, because they carry our conversations and connect our data networks. Satellites carry both, and much more. Paging and mobile phones rely on satellites. Live news and a large portion of radio and television broadcasts are transmitted via satellite. Phone networks use satellites for backup. Even gas stations depend on satellites. Our experiences using credit cards at the pump, and buying snacks at the station market would not be as convenient without them.

It seems strange to think that something most of us have never seen — could not go to see — is so critical to our lives. We can barely comprehend where they exist (more than 22,000 miles above the earth) and even the technically savvy among us would be hard pressed to explain how one works.

Galaxy 4 reminded all of us just how inconvenient life could be without functioning satellites. We only realize the impact of a technology when we are without it. Galaxy 4 is now back in position, and transmissions are working normally. But what of the security of satellite transmissions? The safety of bouncing signals across the world is perhaps a worthy question. How secure are these links that are connecting us all?

Taking a look at a technology closer to home may help provide an answer. Similar questions have been asked about the Internet. We are all, by now, familiar with the security problems of the Internet. Break-ins of every kind have appeared in the mainstream media. No type of organization — commercial, government, military or otherwise — has been spared. The hard lessons we have learned with the Internet can help prevent the same mistakes being made with satellites.

The irony of the Internet—which could be true of the growing satellite network—is that if offers the same promise to the mischievous and the criminal as it does to the rest of us. Think of it: a massive, interconnected pathway to most of the computers on the planet. All conveniently hooked together. All able to communicate using the same protocol. And most without someone always at the watch. We are doing a tremendous service to crackers and criminals by tying all of our systems together for them via the Internet.

Hackers breaking into systems are not the only threat on the Internet. No, every day stealing information becomes easier because data passes unprotected through a matrix of way stations not under our control. In an effort to make the Internet useful, we bridge the gaps. We connect offices. We communicate with suppliers and customers on-line. We read e-mail from hotel rooms. With each step we take in the name of convenience and utility, we potentially take a step backward in security. Why should anyone break in when the data is bound to pass by through network way stations?

Satellites may only broaden the risks. Unlike on the Internet, information transmitted via satellite is out there—literally. It is whizzing by us all the time. Sending data via satellite—in any wireless form—is the ultimate exposure. There is not even so much as a wire to contain the signals carrying our information.

The risk with satellites is multiplied by the amount of data that is sent. Because satellites handle pager, voice, data, radio and television signals, there is significantly more exposed information to be stolen. Since we rely on satellites to carry more and more personal and sensitive information, the protection should be accordingly better.

It took years of break-ins, and uncountable financial loss, for us to wake up to the security issues on the Internet. The complacency is finally gone. We should not allow ourselves to be lulled into complacency again with satellites. Raising awareness of the security risks posed by satellites is critical. The exposure of wireless transmissions, combined with a multiple of in the amount and type of information transmitted, cries out for vigilance.

Reading *Satellite Encryption* is a good first step. If you are new to satellite technology, then presumably you bought this book to find out more about the myriad vulnerabilities involved in beaming data from land to space and back again. With nearly 3000 satellites scheduled to launch in the next 10 years, these vulnerabilities will only grow in number and magnitude. As we rely more heavily on satellites, we must pay heed to security.

For the cynics, *Satellite Encryption* may be an eye opener. After learning how exposed transmissions are, you may wonder how data can ever be safe. Cryptography is an extremely powerful technology that can be a potent defense against the theft and tampering of valuable information—be it intellectual property, personal information, or military secrets. Military intelligence agencies spawned and nurtured both satellite technology and cryptography. Just as satellites are becoming more prevalent in the commercial sector, cryptography is being called upon to provide us all with military-grade security.

The opportunist has perhaps the most to gain from *Satellite Encryption*. New products and services can now be delivered anywhere, instantly. A physical infrastructure does not need to be built. Developing countries can afford the most modern of telecommunications and broadcasts facilities. Moreover, services can be made better, because they do not rely on an existing infrastructure. Through digital means, cryptography can provide the 'security infrastructure.' Cryptographic techniques can be used not only to personalize information. These techniques provide analogs for things we now do only with paper. Digital signatures, for example, can provide a substitute for handwritten signatures on business correspondence, legal documents, even contracts.

Vacca offers a comprehensive look at satellite and security technology. The role of satellite security in our society is critical. A review of essential policy and implementation issues is provided. Of course, details on the threats to our privacy are also included. Perhaps the most intriguing section is a look at the future and what it holds. *Satellite Encryption* offers something for the novice, cynic, and opportunist alike.

Everyone who reads this book should be aware of another threat to their privacy. Ironically, the threat comes from the government. The U.S. government has made efforts to limit the rights of citizens to protect their privacy with cryptography. New proposals from Washington threaten to weaken the security in products we all buy. In some cases, the government wants to mandate key recovery—tantamount to a government janitor having a master key to your house.

The U.S. government is not alone in efforts that limit privacy, nor is it the worst offender. The privacy rights of citizens throughout the world are the same. Think of it—the job of a government is to protect the well being of citizens. In an age when more aspects of our lives depend on technology, our ability to protect ourselves is diminishing. Our exposure is up—and our protection may be legislated down. By preventing privacy of consumers, governments are also limiting the growth of business. We must all stand up for our digital rights to privacy.

The issue of satellite security has risen in importance, and becomes more critical with every new satellite that goes up. We must not let security be a technology we take for granted, and miss only when our privacy is compromised. The Internet, barely out of its teens, holds valuable lessons about taking security technology for granted. If we ignore it, the next quiet Tuesday night could portend something disastrous for us all.

Part One

Satellite Encryption's Role in Securing the Information Society

1

Satellite Data Encryption Security Nightmare

Although the boom in satellites in the next five years will change the way we work and live, it will be a security nightmare for those organizations and governments whose survival depends on the protection of intellectual property distribution, electronic commerce, electronic battlefields, and national security. Secure exchange of information among millions of users around the globe, involving perhaps billions of transactions, will prove vital to the continued growth and usefulness of satellite communications as well as the Internet and intranet. Encryption—especially several layers of encryption on top of compressed data that is to be transmitted (via a highly directional microwave radio signal) to a satellite (uplink) from Earth and then transmitted down to Earth (downlink) and decrypted—can effectively solve the Internet's confidentiality and authentication concerns.

This chapter sets the stage for the rest of the book by showing how governments and organizations around the world can use satellite encryption to help preserve vital national secrets, limit attacks

3

on a nation's information infrastructure, and eliminate security and authentication obstacles to electronic commerce. Specifically, this chapter provides an overview of current satellite encryption technology, the threat from the Internet, encrypted satellite data transmissions (downlink) and receiving (uplink), and encryption cracking. In addition, this chapter describes a new on-line compressed encryption-based technology that uses seven levels of encryption and is capable of high-volume data compression.

The *midnight* of the new century is upon us. The launching of thousands of secure satellites during the next few years will ensure that there will be a *dawn*! With that in mind, before delving into satellite encryption itself, a presentation of current and future satellite technology is in order first.

Current and Future Satellite Technology

During the Persian Gulf war, CNN reporters made their dramatic broadcasts over portable satellite telephones while Iraq was being hammered. Advanced for their day, the phones were not for everyone. They required a separate antenna the size of an umbrella to send a signal, each one cost an arm and a leg to operate, and, it took a suitcase to carry one.

Calling from war zones (and many less dramatic but equally out-of-the-way places) is about to become a whole lot easier. Starting late next year, satellite phone users will be able to pull a tiny flip phone from their pocket and dial from literally anywhere on Earth to anywhere else. Their calls will zip to one of 77 Iridium satellites (See Table 1.1) whirling overhead and be passed along, satellite to satellite, to the receiving party. The cost: $2.50 a minute.

Developed by Motorola, Iridium is only the beginning of a new generation of satellite-based services that promise to change the way we live and work. Companies from start-ups to the biggest and most established are expected, by some estimates (between now and the turn of the century), to spend up to $60 billion to build and launch new satellites. An additional $180 billion is expected to be spent for antennas, phones, switches, and other gear to support their birds aloft.

For those with no access to telephone service (which is to say, half the world's population) the implications are revolutionary. The launch of new satellites may, for them, be akin to the laying of the rails to the American West—an epochal new infrastructure that creates the potential for rapid economic development. People living in remote Indonesian islands, in central China, and in villages in India that have not changed much in a thousand years, should be able in the not-so-distant future to phone a nearby village or a relative in Maine. Or, browse the World Wide Web (WWW).

Impact

The impact on advanced economies will be more immediate, if less grandiose. By 1999, for example, upscale hunters, hikers, boaters, and other wilderness enthusiasts will never again have an excuse to become lost or be out of touch with the rest of the world. With an $888 hand-held device from Magellan, a subsidiary of Orbital Sciences Corp. of Dulles, VA, they will be able to type a short message to an Internet address anywhere in the world. The message will be carried via a constellation of 39 tiny Orbcomm satellites (see Table 1.1) that will be lofted next year. *Help me! I've been wounded in a fall, and I think that both of my legs are broken. I'm lying at the bottom of a cliff on the north side of Mount Hood.* Determined by a global positioning satellite receiver in the gadget, the text will always include the exact location of the transmission, so that rescuers will know just where to look. Television reruns of *Gilligan's Island* or *Lost In Space* may never seem the same.

Cellular phone tycoon Craig McCaw and the world's richest man, Bill Gates, have proposed the most audacious satellite scheme of all for those who despair that terrestrial telephone companies will never provide more than sluggish connections to the Internet. The scheme, known as Teledesic—*Internet in the sky*, is shown in Table 1.1. Teledesic would use 950 satellites (as planned today) in order to provide the same kind of broadband, multimedia connections, such as videoconferencing (which works best over fiber-optic cables). This setup will put into orbit more than three times the 260 commercial satellites in orbit today. The $10 billion service will be aimed initially at linking far-flung international corporate offices. Individual speedaholics who choose to sign up will be able to download multimedia data from the Internet at a sizzling 56 megabits per second— 2000 times faster than a dial-up telephone connection.

Table 1.1 Comparison of GEO, MEO and LEO Satellites

	Type of Satellite		
	Geostationary Earth Orbit (GEO)	Medium-Earth Orbit (MEO)	
Company	SPACEWAY (1)	ICO GLOBAL COMMUNICATIONS (2)	ODYSSEY (3)
Satellites (est.)	9	21	23
Orbits	Geostationary, 33,400 miles above equator	Inclined 50° to equator, 7,545 miles up	Inclined 55° to equator, 7,545 miles up
Operational in (est.)	First Global Region in 2000, rest in 2001	2001	2002
Cost (est.)	$4 billion	$4.8 billion	$4.3 billion
Speed	32 kilobits per second to 12 megabits per second	N/A	N/A
Applications	Ordinary fixed telephone service in developing areas, fax, videoconferencing, data, broadband multimedia, such as from Internet	Hand-held dual-mode phones that talk to satellites and cellular systems; phones for cars, ships, aircraft; fixed phones in developing areas	Hand-held dual-mode phones, other personal communications services, largely in developing world
Investor(s)	Hughes Electronics	47 telecommunications investors including COMSAT and Hughes	N/A
Partner(s)	N/A	N/A	N/A

Legend:
(1) Subscribers access service with $900 terminal, 25-in dish antenna.
(2) A London-based private offshoot of INMARSAT.
(3) Founded by TRW Inc. and Teleglobe Inc. to offer worldwide mobile phone service.
(4) A simple beefy system for worldwide mobile and fixed-phone service.

Low-Earth Orbit (LEO)			
GLOBALSTAR (4)	IRIDIUM (5)	ORBCOMM (6)	TELEDESIC (7)
59 Inclined 57° to equator, 874 miles up	77 Near polar orbit at 533 miles up	39 Inclined 50 and 75° to equator, 590 miles up	950 Near polar orbits, 530 miles up
Winter 1999 (partial), full in 2000 $3.6 billion N/A	Oct. 1999 $6 billion N/A	1998 $440 million N/A	2003 $10 billion N/A
Hand-held dual-mode phones, fixed ordinary phones, paging, low-speed data	Hand-held dual-mode phones, paging, low-speed data, fax	Data messages from individuals; tracking of barges, truck trailers; remote monitoring of industrial installations, oil wells	Broadband multimedia for corporate intranets, Internet, videoconferencing, up to 39 megabits per second
N/A	Mororola, Raytheon, Lockheed Martin, Sprint, Khrunichev State Research, 12 others	N/A	Bill Gates, Craig McCaw
Loral Space & Communications, Qualcomm Inc.	N/A	Orbital Sciences Corp., Teleglobe Inc.	N/A

(5) Designed so business travelers can call anywhere from anywhere.
(6) 84-lb minisatellites to provide message services for industry, outdoors enthusiasts.
(7) *Internet in The Sky* aims to make broadband multimedia connections anywhere just like fiber-optic cables.

Launch Capacity

Teledesic must find enough rockets to launch all their satellites before any of this happens, however. This is going to be a challenging task for Teledesic. Nevertheless, the world launch capacity is growing very fast as new rockets are coming to market. The forecast calls for a brief launch capacity surplus for the turn of the century. Teledesic will be looking at this quite closely. But there is no guarantee that any of this will happen. A lot has to go right for Teledesic to get all their satellites up. It may not be until at least 2006 before Teledesic gets them all launched and working. However, Teledesic plans to begin operating their system with fewer than 400 satellites.

Although McCaw and Gates do have a terrific vision, Teledesic may be late to the party even if it is on schedule. In 2001, Hughes, the dominant satellite builder and a major service provider, plans to offer its own $4 billion global broadband system called Spaceway (as shown in Table 1.1). Hughes will utilize nine of its giant HS 702 satellites. Loral Space & Communications hopes to have its own version, called CyberStar, operating in 2000. Similar services are also being planned by Motorola and Alcatel.

Satellite services are already being used by early adopters to change the way they do business. For instance, today in Tennessee, Quality Transportation Services of Kingsport uses American Mobile Satellite Corporation (AMSC) to provide instant two-way voice communications with its 330 vehicles that carry medical patients around the state. Opting for the satellite rather than building ground relay towers throughout the state saved Quality Transportation Services over a million dollars. Nagle Line, a Toledo, Ohio, trucking company, uses AMSC to transmit data back and forth to its 60 refrigerated 18 wheelers. In the past, it took a Nagle driver an hour to make a 4-min call by the time he or she pulled off the road and found a phone. Now it takes less than 4 minutes for the driver to stop, type a note into a laptop computer, and send it to Toledo.

All the new satellite services are aimed primarily at business users and affluent individuals. Nonetheless, the 3 billion or more people who cannot reach out and touch anyone are likely to be the biggest long-term beneficiaries because they live in areas where phone service is not now available.

Satellites provide the same coverage to India as to San Francisco, whether we want them to or not. It is an inherently egalitarian technology. Commercial space telephony is the most cost-effective way to *wire* the planet, because satellites can take the place of thousands of miles of terrestrial phone lines.

Ease of Calling

Globalstar, for example (as shown in Table 1.1), forecasts that more than half its revenues will come from providing regular phones to areas not now served. Remote villages could be served by portable and mobile units that will cost about $860 apiece. Public phone booths, powered by solar panels and linked directly to satellites, could cost $3000–$4700, depending on how many are connected to a single antenna.

A recent historic agreement by 80 nations at the World Trade Organization to open up their telecommunications markets will speed the process dramatically. More than half the signatories are countries from the developing world where the phone systems typically are inefficient government monopolies. These countries have conceded that deregulation, competition, and admission of foreign companies will benefit their economies by providing cheaper communications via satellites, fiber-optic cables, and other means. The agreement was a victory for U.S. negotiators, who had good reason to play tough. United States companies are the most competitive providers in the world. They are in the best position to compete and win under this agreement.

The Coming Boom in Satellite Services

Responsibility for the coming boom in satellite services is due to dramatic changes in technology, politics, economics, and consumer demand. This technology could not have been dreamt of 11 years ago. For example, Loral Space & Communications is the lead partner in Globalstar (as shown in Table 1.1)—a 59-satellite space-phone system that should be operating in late 1999 and competing directly with Iridium (by, among other things, charging lower rates). The same advances in microelectronics that make personal computers so powerful and inexpensive also apply to satellites. An Iridium satellite, built

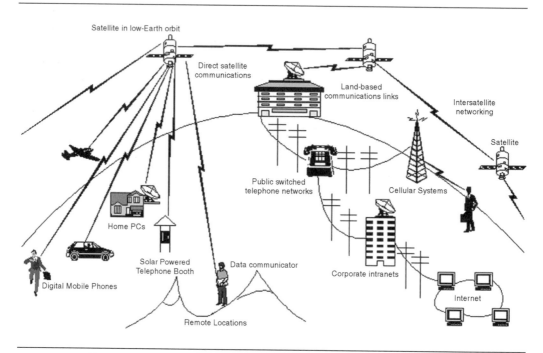

Figure 1.1 The Low-Earth Orbit (LEO) Network.

on a Motorola assembly line, is expected to cost less than $40 million — 15 percent the cost of past birds laboriously built one at a time by hand.

Huge rockets from Russia, Ukraine, and China now compete with French and American boosters to launch satellites. In the past 11 years, more than 2360 payloads, including manned missions, were launched worldwide; 50 percent were Russian spy satellites. In 1999 alone, industry analysts estimate that 87 purely commercial satellites will be orbited, up from 30 in 1997. An additional 232 are expected to go up in 2000. Over the next 11 years, roughly 2900 new satellites are proposed to be launched, 80 percent for commercial use. Lockheed Martin, in fact, is marketing Russian Proton launches. It plans to install Russian-designed engines on its latest Atlas booster. In order to increase its capacity, Lockheed Martin is also considering whether to invest in upgrading the Baikonur launch facilities in Kazakhstan. Boeing, meanwhile, has teamed up with Ukrainian and Russian rocket builders and a Norwegian shipbuilder to start launching from a modified offshore oil-drilling platform in the Pacific Ocean in July 1999.

How satellites will be deployed is the other big change. The new idea is to put up constellations of satellites that operate close to the Earth's surface. That translates from 500 to 1100 miles up, in low-Earth orbit (LEO) as shown in Figure 1.1 and discussed in the next Sidebar, "Low-Earth Orbit (LEO) Configuration." Most commercial satellites that today relay cable television, provide direct-broadcast television, and handle long-distance phone calls are in geosynchronous orbits 33,400 miles up. At that location and altitude, they move around the Earth at the same rate as does the Earth's surface. This makes the satellites appear to hover over the same point. It allows them to act like very tall transmission towers.

Low-Earth Orbit (LEO) Configuration

Eighty percent of all new satellites launched in the next 10 to 15 years will fly close to the Earth's surface, in low-Earth orbit, between 500 and 1100 miles up. The new satellites will carry calls from hand-held phones. Depending on the system, that could be both low and high-speed digital data. Customers include people in remote areas of the developing world, who today have no access to phones, international travelers, and far-flung corporate offices.

Direct Satellite Communications

This type of communication comprises direct links between digital mobile phones, pagers, home PCs, or other devices on the ground and satellites in orbit as shown in Figure 1.1.

Digital mobile phones: This type of communication includes hand-held units that can talk both to satellites and to local cellular phone systems. It would also include fixed units in cars, planes, and ships.

Home PCs: This type of communication is for systems such as Teledesic (see Table 1.1) that propose to offer broadband satellite connections that are as good as fiber-optic cable. Individuals could tune into the Internet from home with a dish antenna similar to those used by DirecTV.

Data communicator: Here, inexpensive low-power devices send and receive short data messages, including to and from Internet addresses. When combined with a Global Positioning System (GPS) receiver, they transmit their location (an aid to research and rescue operations anywhere). So that companies know where their assets are located, these communicators can be mounted on trucks and barges.

Solar-powered phone booth: This communication system could be located anywhere, such as in a Third World village square where it would be too costly to run wires and phone service does not exist today.

Land-Based Communication Links

Ground stations called *gateways* connect terrestrial networks with satellite systems as shown in Figure 1.1. Iridium (see Table 1.1) will switch calls aboard its satellites. The satellites will then talk to each other. Therefore, few gateways are needed. Globalstar's satellites (see Table 1.1) do not talk to each other and are much simpler. However, this type of system will require many more gateways.

Public switched telephone networks: These networks connect satellite systems to ordinary land-based telephone networks.

Intersatellite networking: Some of these systems *hand off* data from satellite to satellite and handle all switching in orbit. Simpler systems depend on ground stations to do this job.

Cellular systems: Most satellite phones will be dual mode. They can connect to the local cellular service, or they can talk directly to a satellite. The cellular provider could also deliver satellite calls as regular cellular calls and receive them as such.

Internet: Connecting to the Internet could be made from many parts of the world that now do not have access.

Corporate intranets: Data traffic inside companies is moving to intranets. Satellite systems will allow corporate offices in any part of the world to communicate easily.

Positioning

Few positions are available on the geosynchronous arc (stationary or fixed orbit) which is located precisely above the equator. This is a problem, however, because it takes powerful, costly rockets to boost heavyweight satellites into that position. Furthermore, it takes $\frac{1}{4}$ second for a signal to travel from Earth to a geosynch satellite (satellite in stationary orbit) and back to Earth because they are so far away (see Sidebar, "Advantages and Disadvantages of LEOs, MEOs, and GEOs"). The delay is irrelevant for data transmissions or television broadcasts. However, for voice phone calls, it can interfere with lively back-and-forth conversations.

Advantages and Disadvantages of LEOs, MEOs, and GEOs

There are many advantages as well as disadvantages to Geostationary Earth orbit (GEO), Medium-Earth orbit (MEO), and Low-Earth orbit (LEO) satellites. Table 1.1 presented a general comparison of the different types of satellites launched by various companies. The following lists some of the more specific comparisons:

Geostationary Earth Orbit (GEO)

The orbital arc directly above the equator (33,400 miles at the equator) is a scarce resource that cannot hold many more than the 260 satellites now located there.

Advantages: Satellites in GEO maintain a fixed position above the Earth's surface. Great for broadcasting direct to homes and sending television programs to cable operators.

Disadvantage(s): Two-way phone conversations are difficult because transmissions take $\frac{1}{4}$ second to send and return. Satellites are expensive and large.

Medium-Earth Orbit (MEO)

The orbital arc is inclined to the equator (about 8000 miles up). New data systems and mobile phone could put 36 or more large satellites in MEO in the next few years. The current proposed MEO has a constellation of 12 satellites.

> **Advantage(s):** Fewer satellites are needed to cover the Earth than in LEO orbits.
>
> **Disadvantage(s):** Phones need more power to reach satellites in MEO than in lower orbits.
>
> ### Low-Earth Orbit (LEO)
>
> The orbital arc is 500 to 1100 miles above the Earth's surface. The biggest area of growth over the next 10 to 15 years involves satellites that will skim close to the surface in low-Earth orbits. Iridium's LEO constellation will have 77 satellites (see Table 1.1). This constitutes about 11 each in 7 near polar orbits.
>
> **Advantage(s):** No apparent signal delay. Satellites are cheaper to launch and smaller. Hand-held phones have smaller batteries and need less power.
>
> **Disadvantage(s):** Moving data across constellations of many satellites will be complex. Technology is still untested.

Currently, there is practically no limit to the number of satellites in low-Earth orbit. Initially, some of the techniques for managing big constellations (dozens or hundreds of satellites working together) were hypothesized for Star Wars missile-defense schemes. As satellites in LEO are close to Earth, phone batteries can be smaller, there is essentially no delay to the signal, and less radio energy is required. The satellites themselves can be easier to launch, smaller, and less expensive. For instance, a big booster like Russia's Proton is able to carry up to 8 Iridium satellites on a single trip.

Most of the proposed and new satellite services (including Iridium, Globalstar, Orbcomm, and Teledesic) will operate in LEO as shown in Table 1.1. Nevertheless, at higher altitudes, there will also be plenty of action. For instance, ICO Global Communications (a private company spun off from INMARSAT) will operate its space phone system from satellites 7540 miles up in medium-Earth orbit, or MEO. At that altitude, only 11 satellites, similar to those used in geosynchronous orbit, are required. And, the signal delay is minimal. The first satellite will be launched in 1999.

Some of the players with big plans will no doubt be also-rans when the shootout in the skies is over 10 or 15 years from now. Furthermore, some immensely complex systems like Iridium and Teledesic may not work quite as expected. Certainly, there will be launch problems. The recent explosion of a Delta II rocket carrying an Air Force payload has caused other Delta II launches (including one carrying Iridium's first 3 satellites), to be delayed until at least January of 1999.

There is little doubt that a new world infrastructure is about to be constructed. It will benefit the most advanced communication companies like CNN. However, for the disadvantaged regions of the world, the coming satellite communications revolution could be one of those rare technological events that enable long-dormant economies to flourish and traditional societies to leap ahead.

Now that the foundation of current and future satellite technology is in place, a look at the potential threat to these orbiting systems from the Internet is of the utmost importance before covering how to use encryption to best protect them. As it stands today, the Internet is not secure, so the only option is to understand how attacks occur and how best to protect against them—the theme of this part of the chapter.

Internet Threat

With just a few keystrokes, a terrorist's strike may be just a modem away. Click, click.... That is all it might take, and then Wall Street takes a dive and throws financial markets into chaos. Banks lose their electronic records and explosive international tensions ignite. Airplanes plunge disastrously into mountainsides and oceans. Municipal water supplies shut down, leaving entire cities dry. These incidents could well be the norm a few years from now as 80 percent of the world's commercial and government organizations will depend on satellite communications to process their transactions via real time Internet access.

High-Tech Mayhem

Such incidents regularly fill Hollywood plot lines. For instance, in the recently released James Bond thriller, *Tomorrow Never Dies*, terrorists steal an encryption decoder from the CIA and use it to transmit false longitude and latitude coordinates from a U.S. military satellite to a U.S. ship carrying nuclear missiles to make it appear that it is in international waters. Instead, the ship is actually within Chinese waters. The Chinese send aircraft to the area to warn the ship. The terrorists then fire a torpedo at the U.S. ship. In response, the ship fires a missile at the Chinese aircraft, thinking that the plane fired the torpedo. The missile strikes one of the planes and destroys it. The threat of World War III now becomes imminent.

Incidents such as the one just described could be all too real in the near future because as satellite-connected networked computers expand their control over world governments, military, energy, power, water, finance, communications and emergency systems, the possibility of electronic attack and catastrophic terrorism becomes increasingly possible. A serious threat is sure to evolve if the international community does not take steps now to protect these systems in the future.

High-Tech Breed

Today's electronic highway is threatened by a new breed of *highwaymen*, called crackers, ranging from malicious pranksters to hardened terrorists. For the sake of public trust in the Internet, an infrastructure must be designed to support the safe use of land-based communication links or ground stations (called gateways, which connect terrestrial networks with satellite systems) as shown in Figure 1.1 and Table 1.1. Systematic mechanisms and protocols must be developed to prevent breaches of security while transmitting data to (uplink) a satellite, or receiving (downlink) data from it.

As the Internet is an international collection of independent networks owned and operated by many organizations, there is currently no uniform cultural, legal or legislative basis for addressing misconduct with regard to direct satellite communications or land-based communications links. Although most of the organizations connected to the Internet have their own security policies, because the Internet has no central authority through which it can regulate the behavior of those using it, these policies vary widely both in how they are put into effect and their objectives. See Part III, Satellite Encryption Policy Instruments, Chapters 5 through 14 for an in-depth discussion of this subject.

Today most companies have either an informal or formal information-security policy. This is an oral or written statement of objectives for ensuring that a system and the information in it meet with only appropriate treatment. In order to reach policy goals, this statement of objectives is associated with corporate and personal practices that must be implemented. Typical policy objectives include preventing unauthorized modification of satellite data transmissions (that is, ensuring data integrity), preserving the availability of system resources (such as computer time), and protecting the confidentiality of private information. All of this of course is in accordance with the needs and expectations of the system's users. The practices and policy can range from manuals of several volumes to a single page.

A security policy prevents the unacceptable use of an information system's resources and satellite data transmissions without impeding legitimate activity, just as a legal system is designed to stop wrong-doers from harming those who live within its jurisdiction. The policy must protect not only the data contained in the communications relayed by the network to a satellite (sent and received), but also the data stored on those company computers connected to the network as well. Electronic mail passed along by network routers must be as intact as personnel records stored on the corporate mainframe.

A formal security policy may consist of a set of constraints (when and how the constraints may exist), plus a mathematical model of the system as a collection of all of its possible states and operations. But just as it is hard to write security policies that formally and precisely express which activities are disallowed, it is difficult to write laws that precisely define unacceptable behavior.

Security policies are usually stated informally in current practice. In other words, they are stated in ordinary language—which hobbles the task of translating their intent into a computerized form that automates enforcement. Imprecise translation, however, is not the only problem. Automated security mechanisms also may be configured incorrectly. The system is open to malicious behavior in either case.

Denial of access to a system's data or resources by someone not cleared for such access is known as prevention of intrusion. This is basic to any security policy. Security can be violated even by unintentional intrusion.

Intrusion once it occurs is just the start of security problems. It is critical to determine what the intruder may have done once he or she gained access. For instance, a still-useful list of the types of mischievous actions an intruder can carry out may be summarized as:

- Masquerading (impersonating an authorized user or a system resource, such as an e-mail server).
- Unauthorized use of resources (running a lengthy program that eats up computing cycles and so keeps others from running programs).
- Denial of service (by, say, deliberately overloading a system with messages to keep others from gaining access to it).
- Unauthorized disclosure of information (illicitly reading or copying an individual's personal information, such as a credit card number, or sensitive corporate data, such as business plans).
- Unauthorized alteration of information (tinkering with file data being transmitted to and from a satellite).[1]

A single intrusion can result in a number of the problems listed here. What must be well thought out first is ways to detect an intrusion and assess what the intruder did. Intruders for the most part will rely upon the ability of each system on the Internet to keep a log of events. The logs are invaluable for analysis and intrusion detection. Indeed, the logs are basic to all post-attack analysis. Keeping in mind how the desired level of logging will affect system performance, authors of a security policy must determine what to log and how the logs should be analyzed. The logs should note what they have done as well as who has entered the system.

Before a detailed examination of security methods is made, the issues affecting security enforcement will be presented. After all, an ounce of prevention translates into a pound of detection or vice versa.

Prevention or Detection: That Is the Question

The use of either prevention or detection is a means of enforcing security policy. Prevention seeks to preclude the possibility of malicious behavior because it is prophylactic (guarding against or preventing). On the other hand, detection aims to discover and record any possibly malicious behavior as it occurs.

Among the protection mechanisms are: encryption for safeguarding the transmission of sensitive data (such as credit card numbers) to and from satellites; authentication by asking for a password; and, access controls such as permission to access files. All of these protection mechanisms are designed to ensure that only an authorized person can gain access to systems and alter information. On the other hand, audit mechanisms are investigative tools that detect and quantify malicious behavior. For instance, some tools check the records (called *audit logs*) of system behavior, while others examine user activities on the system as they occur.

One class need not be employed exclusively. Most systems, in fact, employ both. While audit logs usually have the highest levels of access control to prevent a cracker from covering his or her tracks by altering them, audit mechanisms can serve to review the effectiveness of access controls.

Because of the limitations inherent in translating policies stated in everyday language into the software that enforces its intent, policies cannot be enforced in an exact manner. A case in point: The file protection mechanism of the Unix operating system cannot prevent any user who has permission to read a file from making a copy of it, but it can limit access to a file.

Computers cannot distinguish between unintentional mistakes and malicious actions; but people usually can. There is even a gap between actual user behavior and policies. Thus, a system can be abused by careless authorized users.

Restriction of a user's activities to those allowed by the security policy is the goal of protection mechanisms. A security policy might forbid any external users' viewing of information on an internal Web server. It could permit only certain corporate users to add to or change information on the server, but allow all internal users to view the information.

The system's access control mechanism determines whether a person is authorized to access a protected object — be it a text file, a program, or some hardware resource, such as a server. If so, the person gains access. If not, the access is denied and an error message may be returned to the user. At the beginning of each access, the decision to grant access is usually made during run-time. Prior to making any

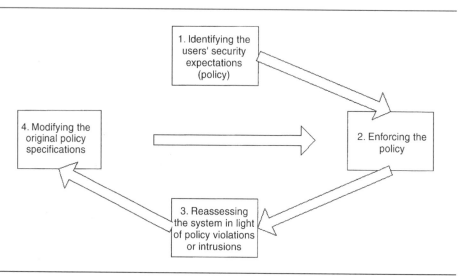

Figure 1.2 As with most software processes, the creation of a security system is cyclic. Once the policy developers determine what the ultimate users of a system expect and need in the way of security, the cycle of enforcement, reassessment, and modification that makes up its life begins.

accesses, users may alternatively be given an electronic token, which they turn in at system start-up.

As shown in Figure 1.2[2], creating and maintaining a security policy is an iterative process. During the process, the policy's authors must modify the original policy specification; identify the organization's and users' security expectations; set them forth in a policy; enforce the policy; and, reassess the system in light of policy violations or intrusions. During the reassessment step of each iteration, both policy and protection mechanisms update the policy to accommodate new user and organization requirements, refine it in order to address new attacks, and prevent vulnerabilities.

Vulnerability Protection

A cracker transgresses a security policy by exploiting the vulnerabilities in the system. All attacks would fail if there were none. When the system's designers, implementers, and administrators consider the problems would-be intruders might present, vulnerabilities exist. This is because the designers, implementers, and administrators make assumptions and tradeoffs about and the use to which the system will be put, the environment in which the system will be used, and, the quality of satellite data transmissions on which it will work. These assumptions proceed from the laws and cultural customs of the workplace, personal experience, and beliefs about the environment in which the system operates. Vulnerabilities can be extremely subtle, existing in systems for years before being noticed or exploited. Furthermore, the conditions under which systems can be exploited may be quite fleeting.

The development of the Unix operating system is a good example of how assumptions breed vulnerability. Unix was created by programmers in a friendly environment. Here, security mechanisms had to deal only with simple threats, such as one user accidentally deleting another's files. But, as Unix spread into commercial realms, the threats were very different.

For instance, the original design of the Unix system had only one all-powerful user (the *super-user*). The existence of this kind of user is a serious flaw in military and many other environments. In fact, most attackers attempt to gain access in the guise of this user. Attackers want to modify log-in programs, system libraries or even the Unix kernels so that they can return later. Thus the initial assumptions about security needs did not generalize well into other environments even though they were reasonable initially in the environment in which the Unix system was born.

Advances that were not foreseen when these systems were built also permit vulnerabilities to arise. Suppose, for example, that a company decided that data from external World Wide Web servers should be barred from the company's satellite connected network, say, to prevent unauthorized software from being sneaked into the system. To this end, the company could set up a firewall. This is software that can be configured to block specific types of communications between internal and external satellite connected networks. The firewall might be configured to block communications using server port 80 to prevent Web traffic. This is the default port used by the Web's hypertext transport protocol (http) to transfer data to and from a satellite. However, the firewall would let satellite communications to that server through on port 25 if someone outside the firewall purposely ran a World Wide Web server that accepted connections on port 25. Thus, the site is vulnerable because this kind of usage is not covered by the assumption that all http satellite communications will go through the firewall at port 80.

Any flaw in the software system's implementation is another source of weakness. Early server implementations that did not check input data is a good example. This allowed attackers to send messages to a Web server and have it execute any instructions in those messages.

Used to provide downloadable and executable programs called applets, the initial implementation of Sun Microsystems Computer Corp.'s Java programming language is another good example. The designers restricted the actions the applet could perform in order to limit the dangers to the system receiving an applet. Nevertheless, a number of implementation flaws allowed little programs to breach these restrictions.

Additionally, it was the intent to constrain each applet to connect back only to the system from which it was downloaded during the initial Java designing phase. To do this, an applet had to be written so that it identified the download system by its alias. Or, it had to be written so that the domain name (say, www.xfile.com) was not its absolute, or network address (that is its IP (Internet Protocol) address: 198.56.7.8)—which is the actual address the Internet uses to locate the download system.

The problem here was that when the applet asked to be reconnected to the download system, it had to request the download server's network address over the Internet by sending its domain name to a domain-name server (whose job it was to return the actual network address). If an attacker corrupted the domain-name-to-network-address translation tables in the domain server, it could *lie* and give a false network address.

Java's implementers trusted that the domain-name server's look-up table would be correct and reliable. However, under fire from an attacker, the look-up table would be corrupted. They later fixed this leak by not having applets refer to all systems by names but by addresses.

Other vulnerabilities can arise from errors made when configuring the security system. For instance, most World Wide Web servers allow their system administrators to use the address of the client asking for a page to control access to certain Web pages — such as those containing private company data. The company security policy can be violated should the system administrator mistype an address, or fail to restrict sensitive pages. A vulnerability exists any time a system administrator or user must configure a security-related program (as in the specific case noted) by typing in a list of allowed-user addresses.

Usually more subtle, hardware vulnerabilities can also be exploited. So that the cryptographic keys inserted by the card's issuer could be discovered by comparing good and bad satellite transmitted data, researchers have studied *artificially injecting faults* into smart cards by varying operating voltages or clock cycles. While the *burning* of keys into hardware is supposed to protect them, it may not protect them well enough.

Like servers, vulnerabilities are not confined to end systems. The computers, protocols, software, and hardware along the path to the server (that is, those that make up the Internet itself) have weak points, too. Consider the vulnerability of a router. This is a computer designed to forward satellite transmitted data packets to other routers as those packets traverse the Internet to their final destination.

In order to determine the path along which the packet will be forwarded, a router uses a routing table. Also, in order to make it possible to reconfigure the satellite connected network dynamically as more paths are added to it, routers update each other's tables periodically. If a router were to announce that it was the closest one to all other routers (through design or error), they would all send it their packets. The misconfigured router would try to reroute the packets. But, because all routes would lead back to it, the packets would never reach their destination. A perfect example of this is the denial-of-service attack — one that would bring the Internet to its knees.

It Is a Matter of Trust

The issue of what or whom do you trust is central to the problem of vulnerability. Designers and engineers trust that a system will be used in a certain way and under certain conditions; design teams trust that the other teams did their jobs correctly so that pieces fit together. Program designers trust that the coders did not introduce errors. And consumers trust that a system will perform as specified. Thus, in the chain of trust, vulnerabilities arise at every loose link.

Trust is demanded by the vast scope of the Internet. Suppose Lisa in Indianapolis wants to send a business letter via electronic mail to Jeremiah in Dublin. Lisa types the letter on a computer and uses a mail program to send it to Jeremiah, trusting that:

- The mail message contains the letter as typed, not some other letter.
- The mail program correctly sends the message from the local network to the next network.
- The message is sent on a path, chosen for efficiency by routers, over the Internet to Jeremiah's computer.
- The destination computer's mail-handling program will receive the message, store it, and notify Jeremiah that it has arrived.
- Jeremiah will be able to read the message successfully using a mail-reading program.[3]

Multiple pieces of hardware (including computers and dedicated routers) and the transport medium (be it twisted pair, fiber-optic cable, satellite link, or some combination thereof) must operate in the way intended for Lisa's confidence to be well-placed. In addition, numerous pieces of software (including the mail programs, the operating systems, and the software that implements message transportation) must work correctly. In fact, the number of components in the network can become quite large. They must all interact correctly to guarantee that electronic mail is delivered safely. If, however, one of the components acts in some other way, Lisa's trust has been misplaced.

The Odd Person Out!

Suppose that an attacker is competing with Lisa for Jeremiah's business, and wants to intercept their e-mail billet-doux. If the messages traveling over the Internet via satellite connection can be modified en route, the message Jeremiah receives need not be the one Lisa sent. To do this, the attacker must change the router tables so that all e-mail messages between Lisa's and Jeremiah's computers are forwarded to some intermediate satellite system to which the attacker has easy access. The attacker can then read the messages on this intermediate satellite site, change their contents, and forward them to the original destination as if the intermediate site were legitimately on the message's path — a so-called *odd person out* attack.

Using encryption to hide the contents of messages, while often seen as the ultimate answer to this problem, is merely a part of the solution, because of a simple yet fundamental problem of trust: How do you distribute encrypted keys? Public-key encryption systems provide each user with both a private key known only to that user and a public key that the user can distribute widely. With this scheme, if Lisa wants to send Jeremiah confidential mail, she enciphers a message using Jeremiah's public key and sends the enciphered message to him as shown in Figure 1.3.[4] Only Jeremiah, with his private key, can decipher this message; without that key, the attacker cannot read or change Lisa 's message. But suppose the attacker is able to fool Lisa into believing that the attacker's public key is Jeremiah's, say by intercepting the unencoded e-mail message that Jeremiah sent giving Lisa the public key and substituting his own. Thus, Lisa would encipher the message using the attacker's public key and send that message to Jeremiah. The attacker intercepts the message, deciphers it, alters it, and re-encrypts it using Jeremiah's real public key. Jeremiah receives the altered message, deciphers it, and the business deal goes sour.

The situation becomes even more complicated with the World Wide Web. Suppose Lisa uses a Web browser to view a Web site in Italy. The

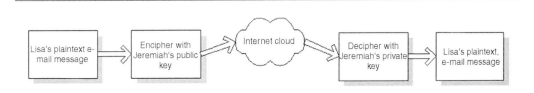

Figure 1.3 In public-key encryption, a user sends a public-key-encrypted message, as shown here, that can be decrypted only with the recipient's private key. Many think such a scheme makes satellite communication secure. But an attacker can defeat it by artfully switching the public key.

Italian Web page, put up by an attacker, has a link on it that says: *Click here to view a graphic image.* When she clicks on the link, an applet that scans her system for personal information (such as a credit card number) and invisibly e-mails it to the attacker, is downloaded along with the image. Here, Lisa trusted the implied promise of the Web page: that only an image would be downloaded. This trust in implied situations (*this program only does what it says it does*) is violated by computer programs containing viruses and Trojan horses. Users of PCs spread viruses by trusting that new programs do only what they are documented to do and have not been altered, so they fail to take necessary precautions.

Auditing

A way of finding the problems discussed in the previous section can be done through auditing. It has five main aims:

- To trace any system or file access to an individual, who may then be held accountable for his or her actions.
- To verify the effectiveness of system protection mechanisms and access controls.
- To record attempts that bypass the system's protection mechanisms.
- To detect users with access privileges inappropriate to the user's role within an organization.
- To deter perpetrators (and reassure system users) by making it known that intrusions are recorded, discovered, and acted upon.[5]

The goals of auditing do not dictate that any particular audit scheme, or model, be followed, even though they are clear in their meaning. Nor do the goals indicate how to perform the auditing. Various ad hoc practices are thus contained in current auditing goals.

Auditing requires that audit events (such as user accesses to protected files and changes in access privileges) be recorded. A log is a collection of audit events. These events are typically arranged in chronological order. They represent the history of the system. Each logged event represents any change in the state of the system that is related to its security.

Audit logs can be voluminous because of the complexity of modern computer systems and the inability to target specific actions. In fact, human analysis of the logs is quite time-consuming because they are often so voluminous. It is therefore quite desirable to have tools that would cull entries of interest from the log. However, development of such automated audit tools for all types of computer systems is hampered by three things: The practice of stating security policies in an ordinary language that does not lend itself to automation (as mentioned initially); a lack of standard formats (such as ASCII or binary); and, semantics (the order in which statements occur) for audit logs.

Some tools are difficult to use, even though they appear to aid in log analysis. As a result, logs are usually inspected manually (often in a cursory manner). This can be done by possibly using some audit browsing tools that employ algorithms which are able to cluster together related satellite transmitted data. However, audit logs all too often are not reviewed at all.

The auditor compares the users' activities to what the security policy says that user may do and reports any policy violations when the log is reviewed. An auditor can also use the log to detect attempts to bypass the protection or attack the system and examine the effectiveness of existing protection mechanisms. Provided the audit log contains sufficient detail, the identities of those behind attempts to violate the policy sometimes can be traced in the history of events.

Tracing the user on satellite connected networked computers may require an audit of logs from several hosts, some quite remote from the system where the intrusion occurred. When prosecution of the perpetrators is warranted, law enforcement agencies may want to use these logs as evidence. This of course can spark jurisdictional and other legal disputes.

In order to keep the infrastructure simple, flexible and robust, the Internet's basic design philosophy is to introduce new resources and capabilities at the end points of the satellite connected network—the client and the server. The disadvantage of this philosophy is that Internet protocol requires only that the network make its best effort to deliver messages. It does not require that messages be delivered at all costs. Nor does it require that records of delivery be kept. As a result, logging on the Internet is merely a function of implementation, not a requirement of the protocols.

Special software running on each node reads and logs satellite transmitted data contained in the packets—in a logging process known as
packet sniffing. Sniffing can use up a lot of processing power and
storage space depending on the amount of traffic on the Net. To
minimize this resource drain, sometimes only the header portion of
packets (which may contain such information as the packet's source,
destination, and the number of packets making up the complete
transmission) is logged and the message data in the packet ignored.
Many extrapolations and assumptions are called into play by deducing
user behavior and the actions caused by the message from the relatively low-level information obtained by sniffing calls.

Whereas there is no standard for every type of system, most World
Wide Web servers do use a standard audit log format, so audit tools
have been developed for a wide range of Web servers. Also, there is
something of a standard for electronic mail. E-mail often has the name
of each computer encountered and some further information that is
carried in the headers of the message as the mail moves over the
Internet. These headers constitute a mini-log of locations and actions
that can be analyzed to diagnose problems or to trace the route of the
message.

Although prevention mechanisms are designed to prohibit violations
of the security policy in the first place, a specter of accountability (the
attacker's fear of being discovered) is raised by detection mechanisms
and thus serves as a deterrent. An audit, then, also may be thought of
as a defense against attacks, albeit a reactive one in which clues to the
identity and actions of the intruder can be detected.

Checking Your System

Fortunately, several tools exist to help administrators check their
systems' security. For Unix systems, three popular tools are Satan,
Tripwire, and Cops. These tools are available free of charge at many
sites on the Internet.

Satan is a World Wide Web–based program that analyzes systems for
several known vulnerabilities exploitable only through network attack.
One such example is the ability of a cracker to make available to any
server files that are supposed to be restricted. It provides a Web

browser interface. Satan also allows scanning of multiple systems simply by clicking on one button. The browser presents a report outlining the vulnerabilities. It also provides tutorials on the causes of each, how to close (or mitigate) the security flaw, and where to get hold of more information from organizations like the Computer Emergency Response Team (CERT). Popularly known as CERT, this is a group within Carnegie Mellon University's Software Engineering Institute in Pittsburgh that issues advisories about computer security.

Another means of verifying security is by checking up on the integrity of the system software — such as log-on programs and libraries. Here, you can check to make sure that the software has not somehow been altered without the administrator's knowledge. Tripwire is an integrity checking program that uses a mathematical function to compute a unique number (*hash*) based on the contents of each file, be it a document or program. Each hash, along with the name of its corresponding file, is then stored for future reference. At random intervals, a system administrator reruns tripwire and compares the results of the new run with the results of the original one. If any of the hashes differ, the corresponding file has been altered and must be scrutinized more closely.

Cops examines the contents of configuration files and directories and decides if either their contents or settings threaten system security. For example, on Jeremiah's Unix system, the contents of a configuration file might state that Lisa need not supply a separate password to use Jeremiah's system. This poses a double security problem at many sites, because anyone who obtains access to Lisa's account also obtains access to Jeremiah's system. Tripwire will not detect this problem. It simply looks for files that changed (and the access control file does not change). However, cops will scan the configuration file, reporting that Lisa does not need a password to log–on to Jeremiah's system, as part of tripwire's analysis of the configuration file's contents.

Detecting Intrusions

Intrusions can be detected either by automated tools that detect certain specific actions or by manual analysis of logs for any suspicious occurrences. Examples are unusual log-in times or unusual system characteristics, such as a very long run time for one supposedly simple program. Automated methods, of course, process lots of satellite transmitted data more quickly and efficiently than humans could. The data come from either logs or from the current state of the system as shown in Figure 1.4.[6]

For a system with a view to uncovering suspicious behavior, human analysis entails looking at all or parts of the logs. As previously mentioned, audit data may be at such a low level that events indicating an intrusion or attack may not be readily detectable as such. Here, detecting attacks may require correlating different sets of audit data, possibly gleaned from multiple logs. Thus, a change in access privileges from the privilege log might be compared with the log-in log's record of the location from which the user who changed the privileges logged-in. The satellite transmitted data may span days or weeks and are often voluminous.

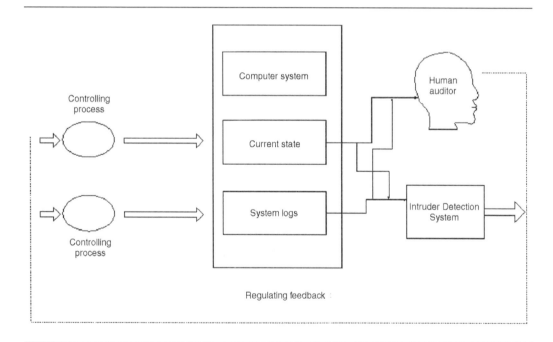

Figure 1.4 An intrusion detection system (IDS) processes information from both the computer system and its logs and reports any problems to a security auditor. The initial information can also be used to determine what other actions should be taken and what further information should be logged.

Another hindrance is that the person conducting the analysis must have special expertise. In order to understand what may have happened and what actually did occur, this expertise must be both in the hardware and software that constitute the system being audited and the particular way in which it is configured.

Modern computers have a capacity to analyze large amounts of satellite transmitted data accurately. They must be told what to look for, provided they are programmed to analyze the right data to detect intrusions correctly. For this purpose, three methods have been established: anomaly detection; misuse detection; and specification-based detection. Among them, there is no one best approach to detecting intrusions. In practice, the particular combination of approaches used is tailored to an organization's specific needs.

Anomaly detection compares the current behavior of a person using a system to the historical behavior of the person authorized to use the system. This technique presumes that deviations from prior behavior (say, different log-in times or the use of different commands) are symptoms of an intrusion by an unauthorized person using a valid account. Similar reasoning suggests that a program altered to violate the security policy (that is, one changed by a virus so it now writes to other executables or to the boot program) will behave differently than the unaltered version of the program.

An intrusion detection system (IDS) based on anomaly detection must first be trained to know the expected behavior of each user — and this could easily be hundreds of users. This normality profile is built using statistical analysis of each user's use of the system and logical rules that define likely behavior for various types of users: programmers; sales managers; support personnel; and so on. Once a normality profile is established, the IDS monitors the system by comparing each user's activity to his or her normality profile. If some activity deviates markedly from the profile, then the IDS flags it as anomalous and, therefore, a possible intrusion.

Admittedly, a legitimate user can be flagged as an intruder (a false positive) because abnormal behavior is not necessarily an attack. For example, a legitimate user may become more proficient in using a program and thus employ commands not previously invoked. False negatives also occur when an intruder's actions closely resemble the normal behavior of the legitimate user whose log-in they have obtained. Finally, establishing the right time period over which to analyze the user's behavior and how often to retrain the IDS system affects its performance.

An anomaly detection system observes the interaction between a program and the operating system. It also builds normality profiles of the short sequences of system calls normally made. Activity outside this is presumed to be part of an intrusion. For example, if an attacker tried to exploit a vulnerability in which unusual input (such as an e-mail message sent to a program rather than a person) caused a mail-receiving program to execute unexpected commands, these commands would be detected as anomalous and a warning given.

Misuse detection does not require user profiling, unlike anomaly detection, in which normal user behavior is taught so that unusual behavior characteristic of an attack can be distinguished. Rather, it requires a priori specification of the behaviors that constitute attacks. The IDS warns the systems administrator if any observed behavior matches a specified attack pattern.

The techniques used to describe the attacks vary. One method is to list events expected to be logged during an attack. A graph-based misuse detection IDS employs a set of rules that describe how to construct graphs based on satellite connected network and host activity. For example, these rules could be a graph of the connections between the systems involved in an attack, the time at which they became involved, and the duration of their involvement. The rules also describe at what point such a graph is considered to represent an attack.

Another is to have an expert write a set of rules describing *felonious* behavior. For example, suppose an attacker gave unusual input to a mail-receiving program to change the way it operated. The expected system calls were *read-input*; *write-file*; but the attacker's input would try to change the set to be *read-input*; *spawn-subprocess*; *overlay-program*. The last two items in the altered set, which tell the system to execute another program, indicate an attack. Were the attacker to try to intrude using that technique, the misuse detection program would detect it.

The misuse detection method can be highly accurate, but, unlike anomaly detection, it cannot detect attacks that fall outside its prepared list of rules describing violations of security. In addition, it depends upon having an expert who is able to specify such rules.

Specification-based detection describes breaches in terms of the system's expected behavior, while anomaly and misuse detection catch security breaches by focusing on the attacker's behavior. Furthermore, there are no false alarms if system behavior has been specified accurately. The first step is to specify formally how the system should behave in all circumstances. Once fully profiled, the system is monitored and all its actions compared against the specification. Any item of system behavior that falls outside what is specified as correct is flagged as a security violation.

One approach to specification-based detection uses a special policy-specification language to describe the security policy in terms of the access privileges assigned to each program in the system. This language indicates under what conditions certain system calls may be made, and it requires knowledge about privileged programs, what system calls they use, and what directories they access. Depending on the particular system for which the policy is being specified and the specification language used, creating specifications of this kind may require expertise, skill, and some time—although some effort might be automated using program analysis. However, if the specifications do not cover all eventualities, false negatives (intrusion alarms) can occur.

Several companies and research groups have developed intrusion detection systems. The Computer Security Laboratory of the University of California, Davis, is designing and developing one such tool, called GrIDS, that will monitor both systems and satellite network traffic—and, look for actions indicating misuse.[7] It also supports analysis of attacks conducted from more than one outside source, even when the attack is spread over a large number of systems.

Other, nonresearch systems are less ambitious, but are currently deployed. The CMDF from Science Applications International Corp. (SAIC), San Diego, CA uses the anomaly approach by building a database of statistical user profiles and looking for deviations from that profile. NetRanger from WheelGroup Corp., San Antonio, TX, and NetStalker from Haystack Labs Inc., Austin, TX, detect attacks by comparing system actions to known exploitations of vulnerabilities.

Damage Assessment and Counterattack

Several responses to security violations are possible, particularly if the attack is detected while it is occurring (typically within a matter of seconds or minutes after an intrusion starts). While a more complex, automated detection system might respond autonomously to any violations of policy, the simplest reaction is to alert other people. The type of response selected depends on the degree of confidence that an attack is actually under way, and upon the nature and severity of the attack.

Gathering information needed to analyze the violation and decide how to respond further, is the first response by a security team to a reported attack. Also, additional auditing (of more user accounts or more system resources) may be turned on. This would be only for those users involved in the violation, or, possibly, if the extent or nature of the violation of policy is not fully understood for the entire system. Moreover, the system can turn defense into offense. This could be done by fooling the attacker by countering his or her activities with misleading or incorrect information. Or, the attacker can even be lured by the security team to a system designed on purpose to monitor intruders.

To determine who is responsible is another common response to a violation. After that, legal action might be taken, or more direct responses (such as blocking further connections from the attacker's site or automatically logging the attacker off) may be appropriate. However, as Internet protocols do not associate users with connections, determining whom to hold accountable can be very difficult. Also, the attack might be laundered through multiple stolen accounts, and might cross multiple administrative domains, as was the case with the attack described in 1989 by Clifford Stohl in his book, *The Cuckoo's Egg.* No formal support infrastructure exists to trace attacks that have been laundered in this way.

The attacked system needs to be analyzed to determine the immediate cause of the system's vulnerability and the extent of the damage, once a violation has been detected. Knowing the vulnerabilities exploited by the attacker can often help to stop ongoing attacks and stop future ones. Knowing the vulnerability's causes helps to determine what to monitor if it cannot be fixed.

Security systems that detect deviations in a user's behavior can indicate only that a user may be an attacker, not what weak points were exploited to violate the security policy. Misuse detection systems catch exploitations of known vulnerabilities. Because the activities that trigger the IDS may not be the root cause of an attack, these detection systems may give only a partial set of those exploited. That is, an attacker may at first use a means to violate the policy that goes undetected. Only subsequent violations, based in part on the initial one, are reported.

Successful assessment depends upon the analysis programs used for the assessment and the integrity of the audit data. Also, a sophisticated attacker may tamper with the audit data or disable or modify the analysis programs to hide the attack. Thus, extra resources are needed to secure those data and programs.

For example, where security is of utmost importance (as in military and financial establishments), audit data may be written to write-only devices. These devices could be write-once, read-many (WORM) optical storage disks. In addition, analysis programs may be put on a dedicated machine that does not have ordinary user accounts or satellite network connections and uses the vendor's distribution of the operating system.

Assessment can be approached using state-based or event-based analysis. The causal relationships in the events recorded in the log are tracked down in event-based analysis. Parent-child processes are a good example. The Unix operating system records each process with an ID that identifies the process that spawned it and the user who started it. Moreover, some versions of Unix record these IDs with the corresponding events in the log.

The processes involved in unauthorized events can be pinned down with the aid of such information. It is often possible to determine the vulnerabilities exploited, and assess the damage caused by the attack, by tracing the parent-child process relationships. Then the user-process associations can be used to identify the user account(s) from which the violation of policy occurred.

To see if the current state of the system is secure in accordance with current requirements, the state-based approach constantly analyzes the system. A state includes the contents of configuration files and the rights of users to access various files.

Recovery

The system can be returned to a secure state (a process referred to as recovery) by using the information obtained through analysis. Recovery may mean a number of things. It may include restarting system services that have been made unavailable; terminating an on-going attack to stop further damage; replacing corrupted files with uncorrupted copies; fixing vulnerabilities to protect the system against future attacks; and, taking appropriate actions (such as notifying affected parties or aborting planned actions).

A common technique used in recovery is rollback, because systems are generally backed up periodically. Here, you would be restoring a system to its state before the attack—using the backup files created before the intrusion occurred. A complete backup of all the files in the system may be done, or, you could have a selective backup in which only copies of recently modified files or critical files are saved. Depending on the level of integrity a site wishes to maintain and the frequency with which files change significantly, different levels of backup may be combined (complete system backup once a week, say, and selective backups once a day).

It may be necessary to use the last complete backup plus any later selective backups in order to reconstruct the preattack state of the system. During rollback, every change made since the last backup may be lost. Because of this, the frequency of the backup is important. Backups may not be needed if the program distribution disks are on hand for unchanging programs.

> **Note:** This rollback technique is useful even if complete damage assessment is not possible.

Another means of returning to a secure state is reconfiguration. This is where the system is modified to bring it to a secure state by fixing all configuration files and, if needed, reinstalling all software. Reconfiguration is appropriate when one cannot roll back to a secure state. This is possible because the system has been in an insecure state since its inception or backups were not done recently.

Once a vulnerability becomes known, many vendors aid recovery by distributing *patches* or fixes for software. Actually, this can be pre-emptive, because system administrators often receive program patches before the vulnerability has been exploited on their system. However, sometimes a weakness cannot be fixed. Perhaps the flaw is one of interaction between the software and another component, requiring modification of the operating system. Or, perhaps no fix is available. Administrators may be forced to disable the offending software or service in such cases. For instance, if an account's password has been compromised, its owner must change the password before it can be used again. Future attacks can be prevented through the compromised account by freezing the account before the password change.

Now that we have taken a look at the potential threat to these orbiting systems from the Internet, it is time to cover how to use encryption to best protect them. The ability to exchange information securely between two users or between a service provider and a user via satellite connection, proves vital to the continued growth and usefulness of satellite communications as we approach the next millennium.

Satellite Encryption Secure Exchange

As previously illustrated, for the sake of public trust in the Internet, an infrastructure must be designed to support the safe use of land-based communication links or ground stations (called gateways, they connect terrestrial networks with satellite systems) as shown in Figure 1.1 and Table 1.1. An *encryption* infrastructure can be effectively designed to solve most of the confidentiality and authentication concerns of satellite transmission with the Internet. However, secure exchange can be either a one-way or a two-way encounter, and the satellite encryption requirements and strategies are quite different for each.

A one-way transaction is typified by e-mail transmissions via to and from satellites over the Internet. Although e-mail messages are frequently answered, each message transmission is a unique, standalone event. A message sender may want assurance that the message can only be read by the intended recipient (confidentiality) and that it came from the alleged sender without being altered en route (authenticity).

Client/server applications, Web exchanges, and many other online applications typify the second class of satellite communications and Internet exchange: two-way transaction. A two-way transaction first involves some sort of a log-on function, in which a user connects to a service, and then an exchange of information between the user and the service. For these two-way transactions, there are again two main security concerns: First, the service wants assurance that the user is not an impostor but actually the person claimed (authenticity). Then, once the service has accepted a user as legitimate and authorized, both the user and the service may wish to ensure that all information exchange between them is safe from eavesdropping (confidentiality).

Although these concerns seem similar, the solutions are quite different. Although many cryptographic tools can provide satellite communications and Internet-security services, the best way to learn how they work is to look at two representative packages: Pretty Good Privacy (PGP) and Kerberos. Pretty Good Privacy (PGP) is a widely used e-mail security package for one-way transactions, while Kerberos is a widely used client/server security package for two-way transactions. Between them, they incorporate the key cryptographic techniques used on the Internet and satellite communications. See Appendix A, "Contributors of Satellite Encryption Software," and Appendix B, "Worldwide Survey of Satellite Encryption Products," for more information on PGP and Kerberos.

Pretty Good Privacy

The PGP package uses the Rivest, Shamir, Adelman (RSA) public-key encryption scheme and the Message Digest 5 (MD5) one-way hash function to form a digital signature, which assures the recipient that an incoming satellite transmission or message is authentic—that it not only comes from the alleged sender but also has not been altered. The sequence for this is as follows:

1. The sender creates a private message.
2. MD5 generates a 128-bit hash code of the message.
3. The hash code is encrypted with RSA using the sender's private key, and the result is attached to the message.
4. The receiver uses RSA with the sender's public key to decrypt and recover the hash code.

5. The receiver generates a new hash code for the message and compares it with the decrypted hash code. If the two match, the message is accepted as authentic.[8]

The combination of MD5 and RSA provides an effective digital-signature scheme. Because of the strength of RSA, the recipient is assured that only the possessor of the matching private key can generate the signature. Because of the strength of MD5, the recipient is also assured that no one else could have generated a new message that matched the hash code and, hence, the signature of the original message. See Sidebar, "Satellite Encryption Cracking," for further information on the private RSA key.

The PGP also solves the confidentiality problem by encrypting messages to be transmitted via satellite or to be stored locally as files. In both cases, PGP uses the confidential International Data Encryption Algorithm (IDEA). This is a relatively new algorithm that is considered to be much stronger than the widely used date encryption standard (DES).

In any conventional satellite encryption system, one must address the problem of key distribution. In PGP, each conventional key is used only once. That is, a new key is generated as a random 128-bit number for each message. This session key is bound to the message and transmitted with it as follows:

1. The sender generates a message and a random 128-bit number to be used as a session key for this message only.
2. The message is encrypted, using IDEA with the session key.
3. The session key is encrypted with RSA using the recipient's public key, and is prepended to the message.
4. The receiver uses RSA with its private key to decrypt and recover the session key.
5. The session key is used to decrypt the message.[9]

Many people underestimate the difficulty of key distribution in a public-key encryption scheme. The common misconception holds that each user simply keeps his or her private key private and publishes the corresponding public key. Unfortunately, life is not this simple.

An impostor can generate a public/private key pair and disseminate the public key as if it belonged to someone else. For example, suppose that

Ellen wishes to send a secure message to Shawn. Meanwhile, Mark has generated a public/private key pair, attached Shawn's name and an e-mail address that Mark can access, and then published this key widely. Ellen picks this key up, uses it to prepare her message for Shawn and then uses the attached e-mail address to send the message. Result: Mark receives and can decrypt the message; Shawn never receives the message, and, even if he had, he could not have read it without the required private key.

The basic tool that permits widespread use of PGP is the public-key certificate. The essential elements of a public-key certificate are the public key itself, a user ID consisting of the key owner's name and e-mail address, and one or more digital signatures for the public key and user ID.

The signer in effect testifies that the user ID associated with the public key is valid. The digital signature is formed using the private key of the signer. Then, anyone in possession of the corresponding public key can verify the validity of the signature. If any change is made to either the public key or the user ID, the signature will no longer compute as valid.

Certificates are used in a number of security applications that require the use of public-key cryptography. In fact, it is the public-key certificate that makes distributed security applications using public keys practical.

The Kerberos Scheme

Kerberos provides a scheme for trusted third-party authentication. Clients and servers trust Kerberos to mediate their mutual authentication. In essence, Kerberos requires users to prove their identity for each service invoked and, optionally, requires servers to prove their identity to clients.

Kerberos uses a protocol that involves clients, application servers, and a Kerberos server. Its complexity reflects the many ways that an opponent can penetrate security. Kerberos counters a variety of threats to the security of a client/server dialogue.

The basic idea is simple. In an unprotected satellite network environ-ment, any client can apply to any server for service, creating an obvious risk of impersonation—an opponent can pretend to be an-other client and obtain unauthorized privileges on server machines. To counter this threat, servers must be able to confirm the identities of clients who request service. Each server can be required to undertake this task for each client/server interaction, but, in an open environ-ment, this places a substantial burden on each server.

One alternative is to use an authentication server (AS) that knows the passwords of all users and stores them in a centralized database. The user can then log onto the AS for identity verification. Once the AS has verified the user's identity, it can pass this information on to an application server, which will then accept service requests from the client.

The trick is how to do all this in a secure way. It simply will not do to have the client send the user's password to the AS over the satellite network: An opponent could observe the password on the network and later reuse it. It also will not do for Kerberos to send a plain message to a server validating a client: An opponent could impersonate the AS and send a false validation.

Kerberos uses the DES algorithm and a set of messages to solve this problem. The AS shares a unique secret key with each server. These keys have been distributed physically or in some other secure manner. This AS can, therefore, send messages to application servers in a secure fashion.

To begin, Ellen logs onto a workstation and requests access to server X via satellite connection. The client sends a satellite transmitted mess-age to the AS that includes Ellen's ID and a request for a ticket-granting ticket (TGT). The AS checks its database to find the password of this user. The AS then responds with a TGT and a one-time session key, both encrypted using the user's password as the key. When this message arrives back at the client, the client prompts the user for his or her password, generates the key, and attempts to decrypt the incoming message. If the correct password has been supplied, the ticket and session key are successfully recovered.

Notice that, while the AS uses a password to verify Ellen's identity, it never crosses the satellite network connection. In addition, the AS has passed information to the client that will later be used to apply to a server for service, and that information is securely encrypted with the user's password.

The ticket constitutes a set of credentials indicating that the AS has accepted both the client and its user. The ticket contains the user ID, the server's ID, a time stamp, a deadline after which the ticket is invalid, and a copy of the same session key sent in the outer message to the client. The entire ticket is encrypted using a secret DES key shared by the AS and the server.

This ticket is good not for a specific application service but for a special ticket granting service (TGS). This ticket can be used by the client to request multiple service-granting tickets. To prevent an impostor from gaining access with this reusable ticket, the time stamp and deadline limit the ticket's life span. Also, the TGT is encrypted with a secret key known only to the AS and the TGS to prevent alteration of the ticket. The ticket is re-encrypted with a key based on the user's password. This ensures that the ticket can be recovered only by the correct user providing the authentication.

Other systems, those available now as well as those being prepared for the future, also use encryption for either one- or two-way satellite communication. The concepts used by PGP and Kerberos serve as a useful guide to how such systems work.

Satellite Encryption Cracking

Recently, a cryptographic consultant found a way around the apparent invulnerability of large-key-length RSA encryption. The consultant proved that an attacker could determine the private RSA key simply by keeping track of how long a computer takes to decipher messages.

The attack the consultant described is easily countered by introducing some random time into the decryption process. However, this and other reports of encryption vulnerabilities (such as Netscape's) have shaken people's confidence that cryptography can keep the bad guys out of confidential messages to and from satellites.

There has always been an arms race in cryptography. On one side are the providers of secure services, who are trying to develop secure encryption algorithms and protocols. On the other side are those trying to find a weak point to attack.

Vulnerabilities in secure applications come in various forms. There may be a weakness in the underlying algorithm. There may be procedural flaws in the secure protocols embedded in an application. There may even be lapses in the security surrounding the generation, satellite communication, and storage of encryption keys. No one can ever be sure.

The attackers have three points in their favor. First, an attacker need only find a single weak spot to penetrate, whereas a defender needs to secure against all forms of attack. Second, knowledge of cryptographic algorithms and protocols has become increasingly widespread, accompanied by increased sophistication on the part of attackers. And finally, the speed of processors continues to rise as the price continues to drop, providing the attacker with powerful computing resources.

Although vendors are strongly motivated to plug leaks as soon as they are discovered, users in general would be wise to take part in forums in which security flaws are discussed so that they can take appropriate countermeasures as soon as a vulnerability is revealed.[10]

New On-line Compressed Satellite Encryption-Based Technology

A new class of on-line compressed satellite encryption-based technologies (already available commercially from third-party vendors) is making it possible to replicate the rich document management and workflow features previously restricted to homogeneous e-mail and groupware environments. These technologies, which use seven levels of encryption that are capable of high-volume data compression, are needed to offset the threat. See Appendix A, "Contributors of Satellite Encryption Software," and Appendix B, "Worldwide Survey of Satellite Encryption Products," for more information on this new class of encryption-based technologies.

These technologies, including software envelopes and electronic authentication services, leverage Internet and intranet infrastructures and facilitate precise management and measurement of document usage and access across heterogeneous systems. Penetration of these technologies will force Information Technology (IT) managers to rethink workflow investments, develop new priorities for tracking, and develop new priorities for tracking and reporting information flows.

Document intensive industries such as financial services have long wrestled with how to compress, streamline, manage, and automate the movement of information around their organizations. In response, systems vendors and integrators invented *work-flow* with the idea that any document intensive business process that could be described could also be automated.

In a similar fashion, Electronic Data Interchange (EDI) has been offered as a way to automate business-to-business supply and ordering functions. Groupware, too, has been marketed as a way to manage business processes, especially messy and unpredictable ones (document creation and editing).

Neither workflow nor EDI, however, has fulfilled its original potential. Workflow systems have thrived in highly structured settings, but these are often inflexible and difficult to link to the outside world. However, although EDI has made it possible to reduce supply management and ordering costs significantly, it too is inflexible: Preplanned bilateral implementations are the best that EDI can offer.

Although groupware has been arguably more successful, it has fallen short of its potential because of huge strategic commitments required to make it effective.

Serendipitous partnering, spontaneous commerce, and adaptability are not strong suits for any of these technologies. Unfortunately, the business world is moving in exactly this direction: toward continuous change, with an increasing premium on high-speed compression and flexibility and a growing requirement to link with external parties.

Until recently, IT managers were faced with two basic choices in automating and managing workflow: commit to a large investment in a unifying solution or accept lowest-common denominator capabilities and/or security levels across systems, and, almost invariably, across organizations.

Now, just as many companies are becoming convinced of the value of workflow, EDI and groupware, a basket of new satellite encryption technologies (based on compressed real-time public key encryption) is poised to alter the cost/feature balance radically, as well as the reach and flexibility of business-process automation. Let us look briefly at one of these key satellite encryption technologies that are beginning to make this possible.

Real-time Compression Router

The goal of this technology is the development of the first *secure gateway router* capable of high-volume satellite transmitted data compression. The technology revolves around a revolutionary online compression software based on a patented scanning process. This technology represents a dramatic departure from traditional off-line compression and far exceeds the performance of hardware-based, online compression, which is the current industry standard. What follows is a detailed look at the software, its application as part of a secure gateway router, and the potential impact on digital wireless communications.

Technology Background

The patented scanning process involves neither a dictionary nor mathematical algorithms. Rather than simply relying on data redundancy, it operates in *real time* by scanning the physical symmetry of each byte of data as it is being transferred to and from a satellite.

The characteristic profile produced by the scanning module is sent to a decision engine, which coordinates the functions of 16 separate relay modules. Each module is designed to address a specific range of characters and programmed to assist the decision engine in making dynamic decisions so as to maximize compression efficiency. Other key performance features include:

- Compression speed of 6.8 megabits per second using 16-bit code run on a Pentium 166. Speed will exceed 10 megabits per second with 32-bit code.
- Bidirectional protocol automatically detects and corrects any corrupt packets as they are being sent.

♦ Compression operates in conjunction with seven levels of encryption with each string of several packets secured by a separate set of codes.

♦ Compression and encryption together require only 20K Random Access Memory (RAM).

> **Note:** In addition to operating in real time, this software compresses with *no packet loss*! Also, because of the patented scanning process, short e-mail messages are compressed as effectively as large files.

Several primary tests have been conducted to measure how well this software performs in a *real-world* satellite networking environment. One test involved a UNIX test over Point-to-Point Protocol (PPP). The primary purpose was to measure compression performance relative to CPU utilization, a major determinant of hardware costs associated with development of a compression router.

Results from repeated tests showed a 486DX-33 with the math co-processor and cache both disabled and settings fixed for *normal* operation capable of compressing a 56K channel in real time with negligible CPU demand. Furthermore, results showed the software effective in compressing a variety of file types. Large text files were consistently compressed in excess of 3 to 1. Tests involving highly compressed gif, jpg and wav files showed consistent compression gains in the area of 10%.

A second test conducted by the software developers involved a Microsoft NT 4.0-Winsock 2 test over Transmission Control Protocol / Internet Protocol (TCP/IP). The primary purpose was to measure real transfer time savings in comparison to V.42bis, a hardware-based, online compression, which is the current international standard for 28.8K modems. Here, results from repeated tests exceeded those over PPP and showed the software to be clearly superior to V.42bis.

The Secure Gateway Router

Designing a *secure gateway router* to maximize the performance of this software will allow real-time compression and encryption of several thousand simultaneous satellite network connections. The extreme speed and efficiency of the software make this possible.

However, both will increase even more as a result of having the software operate as part of the hardware inside a router. Here, it will no longer be necessary to engage RAM. In effect, the processing power inherent to the software program will be continuously active. It is reasonable to expect that by doing this the processing speed would increase well beyond 10 megabits per second using a 32-bit code, to 20 megabits per second or more.

The implication is that a relatively inexpensive PC-based router with a Pentium processor could easily compress and encrypt the entire satellite data transmission flow of a T-1 connection operating at capacity. Depending on how much the processing speed is increased from added efficiency, the same router could have a similar impact on a congested Data Service 3 (DS3).

Obviously, such a device would have value in today's marketplace where there is a growing emphasis on both bandwidth efficiency and security. Key benefits include:

- Maximizes bandwidth efficiency for Local Area Networks (LANs), Wide Area Networks (WANs) and intranets.
- Maximizes security for all satellite transmitted data traveling inside or outside private network.
- Creates infrastructure for an expanded private satellite network or extranet whereby banking or other industries can securely interact with affiliated institutions and customers.
- Creates infrastructure for *Virtual Private Network* whereby the Internet could replace costly dedicated lines as a means of transferring corporate data to and from a satellite.

In addition, the secure gateway router will provide other benefits for increased bandwidth efficiency and security not directly related to compression and encryption. These benefits include:

- Secure packet filtering and multiple-destination firewall.
- Telephone, 100-base-t, Token-Ring and Fiber Distributed Data Interface (FDDI).
- Proxy to shield satellite network behind secure bastion host.
- Internal caching algorithm allowing satellite connected network users to shut off individual memory/disk caches, which aids in reducing bandwidth demand and improving overall workstation performance.

+ Network load balancing to alleviate satellite network connection congestion.
+ Patented IP security for all satellite data transferred outside the corporate network.

Once in place, the secure gateway router paves the way for yet another opportunity.

Note: Each router will be designed to operate across a variety of satellite networks employing extended Border Gateway Protocol (BGP4) routing code. These include: TCP/IP native, frame relay, Asynchronous Transfer Mode (ATM), serial link encapsulation including Cisco High-Level Data Link Control (HDLC), as well as multilink PPP. It will simply be a matter of plugging-in and activating.

Digital Wireless Communications

The fact that the compression process produces tiny packets of data (1500 Maximum Transmission Unit (MTU)) with no packet loss makes this technology uniquely well-suited for wireless and satellite network communications. Furthermore, because both the compression and encryption require just 20K of RAM, the software can easily operate within the limited memory of a Personal Digital Assistant (PDA) or from a Digital Signal Processor (DSP) inside a Personal Communications Service (PCS) phone.

Deploying this technology in conjunction with wireless and satellite networks would dramatically enhance both bandwidth efficiency and satellite communication security. It simply requires a software-enabled phone or modem connected with another enabled device. It does not matter what is in between, just so long as a two-point connection is established. For more information, see the Sidebar that follows here: "Enhanced Bandwidth Efficiency and Security Benefits."

Enhanced Bandwidth Efficiency and Security Benefits

Enhanced bandwidth efficiency and security provide benefits to both users and wireless network providers. Key benefits for users include:

♦ Comfort in knowing you are using the most secure wireless link available.
♦ Reduced connection costs as a result of faster remote satellite data transfers.
♦ More available memory in PDA or laptop.
♦ Protection for data stored in PDA or laptop in case of theft or loss.

Key benefits for satellite network providers include:

♦ Ability to advertise as the most secure wireless link available.
♦ Ability to advertise reduced connection costs without price cutting.
♦ More available bandwidth across satellite network connections.
♦ Makes it possible to encourage businesses and individuals to use wireless communications with greater confidence.

The last point in the preceding sidebar is an important one. Businesses and consumers both have taken to using wireless communication for phone conversations. But, as we move into the future and satellite transmitted data piracy becomes more and more sophisticated and commonplace, it remains to be seen just how willing some will be to rely on wireless as a regular means of transferring the most confidential information.

This applies not only to PCS and satellite, but also to the 39 gigahertz (GHz) broadband microwave link soon to be introduced as the *last mile* to digital wireless communications. Here, providers are counting on hospitals, banks, and other large business operations to use their satellite connected networks for short-distance, high-speed data transmission. For example, 39 GHz is seen as an ideal means for primary care physicians and specialists to take part in interactive video conferences where x-rays, magnetic resonance imaging (MRI) and computerized axial tomography (CAT) scans could be exchanged and discussed.

However, for this scenario to become a reality, providers will have to take every precaution to ensure satellite data transmission security.

They must then also effectively convey to potential users the extent to which they have taken these steps. This means convincing not only hospitals and doctors, banks and bankers, but also patients and customers.

Clearly, when it comes to the most confidential information, the willingness to leave behind the perceived security of wire will depend on the degree to which providers are able to encourage businesses and consumers to use wireless with *greater confidence.* Encryption technology certainly has the potential to have an impact on the marketplace in this regard.

However, there is yet another distinct opportunity presented by a combined approach involving both the software and the secure gateway router. In its design, the router can easily be configured to facilitate a fast and secure connection across a digital wireless network to a corporate LAN or intranet, or a private extranet. It is suggested that a wireless modem working in conjunction with a secure gateway router is the easiest and most secure means of connection with benefits for corporate executives, wireless network providers, and PDA developers. The security provided through this combined approach could be further enhanced by developing custom TCP/IP software, which would be relegated to the internal satellite network.

> **Note:** The attachment of a wireless gateway to an intranet offers a more secure entrance than access through a dial-up connection. Each wireless device that has access to the gateway must be registered on the satellite connected network and enabled by the gateway.

Current versions of the secure gateway router are effective and unique in performance. The approach to design also makes it possible to produce what would be the first secure gateway router capable of high-volume real-time satellite data transmission compression. There would be a need to produce this new compression technology for the following reasons:

♦ For client and server applications it would mean developing specific applications for each operating system, a time-intensive proposition.

- Likely to be considerable opposition to introducing compression and encryption to existing servers.
- Compression router maximizes software performance in satellite connected network environment—would have immediate impact on corporations and Internet service providers faced with congested network connections.
- Compression router maximizes security and bandwidth efficiency by making sure all traffic coming into a private satellite connected network is compressed and encrypted.

The conclusion is that the secure gateway router is the quickest, highest-impact and most cost-effective means of introducing revolutionary real-time compression satellite encryption technology to the marketplace.

Summary

As the need for security on the Internet and satellite transmissions increases, new mechanisms and protocols are being developed and deployed. But a system's security will always be a function of the organization that controls the system. Thus, whether the Internet or satellite communications become more secure depends entirely upon the vendors who sell the systems and the organizations that buy them.

Ultimately, people will decide what, and how much to trust; and so satellite communications security is a nonchallenging, nontechnical, people problem, deriving its strength from the understanding by specifiers, designers, implementers, configurers, and users of what and how far to trust. To begin addressing this challenge, IT managers need to:

- Get to know the key emerging vendors in this field.
- Begin learning about how public key cryptography is being woven into the *soft* infrastructure of the Internet and satellite communications—and by extension, into intranets as well.
- Prepare to respond to business requirements for detailed, real-time measurement and reports of document usage within the organization, as well as the use by outsiders of documents created within the organization.

♦ Spend quality time with business unit managers, educating them on these new technologies and brainstorming applications that make use of them.

From Here

This chapter provided an overview of the current satellite encryption policies; the threat from the Internet; encrypted satellite data transmissions (downlink) and receiving (uplink); and encryption cracking. In addition, the chapter described a new on-line compressed encryption-based technology that uses seven levels of encryption and is capable of high-volume data compression.

The next chapter outlines the problem of growing information vulnerability and the need for technology and policy to mitigate this problem. This chapter will also provide readers with an understanding of the satellite encryption threat, applications, potential solutions, privacy implications of digital signatures, and governmental restrictions on encryption products that put satellite security at risk.

Endnotes

1   Matt Bishop, Steven Cheuing, and Cristopher Wee. Computer Security Laboratory of the Department of Computer Science, University of California, Davis, CA 95616, 1998.

2   Ibid.

3   Ibid.

4   Ibid.

5   Ibid.

6   Ibid.

7   Ibid.

8   William Stallings. "Encryption: Coming and Going," Independent Consultant and Lecturer on Networking Topics, Department of Computer Science, Campus Box 3175, Sitterson Hall, College of Arts and Sciences, The University of North Carolina, Chapel Hill, NC 27599-3175 USA 1995.

9   Ibid.

10   Ibid.

2

Growing Vulnerability in the Information Age

The information age was facilitated by computing and communications technologies (collectively known as information technologies) whose rapid evolution is almost taken for granted today. Computing and communications systems appear in virtually every sector of the economy and increasingly in homes and other locations. These systems focus economic and social activity on information gathering, analyzing, storing, presenting, and disseminating information in text, numerical, audio, image, and video formats as a product itself or as a complement to physical or tangible products.

Today's increasingly sophisticated information technologies cover a wide range of technical progress. This chapter frames a fundamental problem facing the United States today—the need to protect against the growing vulnerability of information via satellite communication uplinks and downlinks to unauthorized access and/or change as the nation makes the transition from an industrial age to an information age. Society's reliance on a changing panoply of information technologies and technology-enabled services, the increasingly global nature of commerce and business, and the ongoing desire to protect traditional freedoms as well as to ensure that government remains capable of fulfilling its responsibilities to the nation all suggest that future needs for information security via satellite encryption will be great. These factors make clear the need for a broadly acceptable national satellite encryption policy that will help to secure vital national interests with regard to the following progressive information technologies.

Workstations and Microprocessors

Microprocessors and workstations are increasingly important to the computing infrastructure of companies and the nation. Further increases in speed and computational power today come from parallel or distributed processing with many microcomputers and processors rather than faster supercomputers.

Electronic Hardware

Special-purpose electronic hardware is becoming easier to develop. Thus, it may make good sense to build specialized hardware optimized for performance, speed, or security with respect to particular tasks. Such specialized hardware will in general be better adapted to these purposes than general-purpose machines applied to the same tasks.

Media

Media for transporting digital information are rapidly becoming faster (fiber optics instead of coaxial cables), more flexible (the spread of wireless communications media), and less expensive (the spread of CD-ROMs as a vehicle for distributing digital information). Thus, it becomes feasible to rely on the electronic transmission of larger and larger volumes of information and on the storage of such volumes on ever-smaller physical objects.

Convergence of Technologies

Today, the primary difference between communications and computing is the distance traversed by data flows in communications. The traversed distance is measured in miles (two people talking to each other), while in computing it is measured in microns (between two subcomponents on a single integrated circuit). A similar convergence affects companies in satellite communications and computing—their boundaries are blurring, their scopes are changing, and their production processes overlap increasingly.

Software

Software increasingly carries the burden of providing functionality in information technology. In general, software is what gives hardware its functional capabilities. Additionally different software running on the same hardware can change the functionality of that hardware entirely. As software is intangible, it can be deployed widely on a very short time scale compared to that of hardware. The Sidebar that follows, "The Role of Software, Computing Devices, and Satellite Communications," contains more discussion of this point.

The Role of Software, Computing Devices, and Satellite Communications

Communications and computing devices can be dedicated to a single purpose or may serve multiple purposes. Dedicated single-purpose devices are usually (though not always) hardware devices whose functionality cannot be easily altered. Examples include unprogrammable pocket calculators, traditional telephones, walkie-talkies, pagers, fax machines, and ordinary telephone answering machines.

A multipurpose device is one whose functionality can be altered by the end user. In some instances, a hardware device may be *reprogrammed* to perform different functions simply by the physical replacement of a single chip by another chip or by the addition of a new circuit board. Open bus architectures and standard hardware interfaces such as the PC Card are intended to facilitate multipurpose functionality.

Despite such interfaces and architectures for hardware, software is the primary means for implementing multipurpose functionality in a hardware device. With software, physical replacement of a hardware component is unnecessary — a new software program is simply loaded and executed. Examples include personal computers (which do word processing or mathematical calculations, depending on what software the user chooses to run); programmable calculators (which solve different problems, depending on the programming given to them); and, even many modern telephones (which can be programmed to execute functions such as speed dialing). In these instances, the software is the medium in which the expectations of the user are embedded.

Today, the lines between hardware and software are blurring. For example, some *hardware* devices are controlled by programs stored in semipermanent read-only memory. *Read-only memory* (ROM) originally referred to memory for storing instructions and data that could never be changed; but, this characteristic made ROM-controlled devices less flexible. Thus, the electronics industry responded with *read-only* memory whose contents take special effort to change (such as exposing the memory chip to a burst of ultraviolet light or sending only a particular signal to a particular pin on the chip). The flexibility and inexpensive nature of today's electronic devices make them ubiquitous. Most homes now have dozens of microprocessors in coffee makers, TVs, refrigerators, and virtually anything that has a control panel.[1]

As these examples suggest, information technologies are ever more affordable and ubiquitous. In all sectors of the economy, they drive demand for information systems. Such demand will continue to be strong and experience significant growth rates. High-bandwidth and/or wireless media are becoming more and more common. Interest in and use of the Internet and similar public networks will continue to experience very rapid growth.

Increasing Interconnections and Interdependence

As the availability and use of computer-based systems grow, so too, does their interconnection. The result is a shared infrastructure of information, computing, and satellite communications resources that facilitates collaboration at a distance, geographic dispersal of operations, and sharing of data. With the benefits of a shared infrastructure also come costs. Changes in the technology base have created more vulnerabilities, as well as the potential to contain them. For example, easier access for users in general implies easier access for unauthorized users.

The design, mode of use, and nature of a shared infrastructure create vulnerabilities for all users. For national institutions such as banking, new risks arise as the result of greater public exposure through such interconnections. For example, a criminal who penetrates one bank interconnected to the world's banking system can steal much larger amounts of money than are stored at that one bank. The Sidebar that follows, "Citicorp's Attempted Electronic Theft," describes a recent electronic bank robbery. Reducing vulnerability to breaches of security will depend on the ability to identify and authenticate people, systems and processes, and to assure with high confidence that information is not improperly manipulated, corrupted, or destroyed.

Citicorp's Attempted Electronic Theft

Electronic money transfers are among the most closely guarded activities in banking. In 1994, an international group of criminals penetrated Citicorp's computerized electronic transfer system and moved about $12 million from legitimate customer accounts to their own accounts in banks around the world. According to Citicorp, this was the first time its computerized cash-management system had been breached. Corporate customers access the system directly to transfer funds for making investments, paying bills, and extending loans, among other purposes. The Citicorp system moves about $700 billion worldwide each day. Authority to access the system is verified with a cryptographic code that only the customer knows.

The case began in June 1994, when Vladimir Levin of St. Petersburg, Russia, allegedly accessed Citicorp computers in New York through the international telephone network, posing as one of Citicorp's customers. He moved some customer funds to a bank account in Finland, where an accomplice withdrew the money in person. In the next few months, Levin moved various Citicorp customers' funds to accomplices' personal or business accounts in banks in St. Petersburg, San Francisco, Tel Aviv, Rotterdam, and Switzerland.

Accomplices had withdrawn a total of about $400,000 by August 1994. By that time, bank officials and their customers were on alert. Citicorp detected subsequent transfers quickly enough to warn the banks into which funds were moved to freeze the destination accounts. Bank officials noted that they could have blocked some of these transfers, but they permitted and covertly monitored them as part of the effort to identify the perpetrators. Other perpetrators were arrested in Tel Aviv and Rotterdam. They revealed that they were working with someone in St. Petersburg. An examination of telephone-company records in St. Petersburg showed that Citicorp computers had been accessed through a telephone line at AO Saturn, a software company. A person arrested after attempting to make a withdrawal from a frozen account in San Francisco subsequently identified Levin, who was an AO Saturn employee. Russia has no extradition treaty with the United States. However, Levin traveled to Britain in March 1995 and was arrested there.

Levin allegedly penetrated Citicorp computers using customers' user identifications and passwords. In each case, Levin electronically impersonated a legitimate customer, such as a bank or an investment capital firm. Some investigators suspect that an accomplice inside Citicorp provided Levin with the necessary information. Otherwise, it is unclear how he could have succeeded in accessing customer accounts. He is believed to have penetrated Citicorp's computers 40 times in all. Citicorp says it has upgraded its system's security to prevent future break-ins.[2]

Although society is entering an era that abounds with new capabilities, many societal practices today remain similar to those of the 1960s and 1970s, when computing was dominated by large, centralized mainframe computers. In the 1980s and 1990s, they have not evolved to reflect the introduction of personal computers, portable computing, and increasingly ubiquitous satellite communications networks. Thus, people continue to relinquish control over substantial amounts of personal information through credit card transactions, proliferating uses of Social Security numbers, and participation in frequent-buyer programs with airlines and stores. Organizations implement trivial or no protection for proprietary data and critical systems, trusting policies to protect portable storage media or relying on simple passwords to protect information.

These practices have endured against a backdrop of relatively modest levels of commercial and individual risk. For example, the liability of a credit-card owner for credit card fraud perpetrated by another party is limited by law to $70. Yet most computer and satellite communications hardware and software systems are subject to a wide range of vulnerabilities, as described in the Sidebar that follows. Moreover, information on how to exploit such vulnerabilities is often easy to obtain. As a result, a large amount of information that people say they would like to protect is in fact available through entirely legal channels (purchasing a credit report on an individual) or in places that can be accessed improperly through technical attacks requiring relatively modest effort.

Vulnerabilities

Information systems and networks can be subject to four generic vulnerabilities:

Eavesdropping or Data Browsing: By surreptitiously obtaining the confidential data of a company or by browsing a sensitive file stored on a computer to which one has obtained improper access, an adversary could be in a position to undercut a company bid, learn company trade secrets (knowledge developed through proprietary company research) that would eliminate a competitive advantage of the company, or obtain the company's client list in order to steal customers. Moreover, stealth is not always necessary for damage to occur—many companies would be damaged if their sensitive data were disclosed, even if they knew that such a disclosure had occurred.

Clandestine Alteration of Data: By altering a company's data clandestinely, an adversary could destroy the confidence of the company's customers in the company, disrupt internal operations of the company, or subject the company to shareholder litigation.

Spoofing: By illicitly posing as a company, an adversary could place false orders for services, make unauthorized commitments to customers, defraud clients, and cause no end of public relations difficulties for the company. Similarly, an adversary might pose as a legitimate customer, and a company (with an interest in being responsive to user preferences to remain anonymous under a variety of circumstances) could then find itself handicapped in seeking proper confirmation of the customer's identity.

Denial of Service: By denying access to electronic services, an adversary could shut down company operations, especially time-critical ones. On a national scale, critical infrastructures controlled by electronic networks (the air traffic control system, the electrical power grid) involving many systems linked to each other are particularly sensitive.[3]

Today, the rising level of familiarity with computer-based systems is combining with an explosion of experimentation with information and satellite communications infrastructure in industry, education, health care, government, and personal settings to motivate new uses of and societal expectations about the evolving infrastructure. A key feature of the new environment is connection or exchange. Organizations are connecting internal private facilities to external public ones. They are using public networks to create virtual private networks, and they are allowing outsiders such as potential and actual customers, suppliers, and business allies to access their systems directly. One vision of a world of electronic commerce and what it means for interconnection is described in the following Sidebar.

Interconnectivity Implications of Electronic Commerce

A number of reports have addressed the potential nature and impact of electronic commerce. Out of such reports, several common elements can be distilled:

- ◆ The interconnection of geographically dispersed units into a *virtual* company.

- The linking of customers, vendors, and suppliers through videoconferencing, electronic data interchange, and electronic networks.
- The creation of temporary or more permanent strategic alliances for business purposes.
- A vastly increased availability of information and information products on line, both free and for a fee, that is useful to individuals and organizations.
- The electronic transaction of retail business, beginning with today's toll-free catalog shopping and extending to electronic network applications that enable customers to:

  —Apply for bank loans.
  —Order tangible merchandise (groceries) for later physical delivery.
  —Order intangible merchandise (music, movies) for electronic delivery.
  —Obtain information and electronic documents (official documents such as driver's licenses and birth certificates).

- The creation of a genuinely worldwide marketplace that matches buyers to sellers largely without intermediaries.
- New business opportunities for small entrepreneurs that could sell low-value products to the large numbers of potential customers that an electronic marketplace might reach.

In general, visions of electronic commerce offer an attempt to leverage the competitive edge that information technologies can provide for commercial enterprises. Originally used exclusively to facilitate internal communications, information technology is now used by corporations to connect directly with their suppliers and business partners.

Note: For example, in manufacturing, collaborative information technologies can help to improve the quality of designs and reduce the cost and time needed to revise designs. Product designers will be able to create a virtual product, make extensive computer simulations of its behavior without supplying all of its details, and show it to the customer for rapid feedback. Networks will enable the entire manufacturing enterprise to be integrated all along the supply chain, from design shops to truck fleets that deliver the finished products.

In the future, corporate networks will extend all the way to customers, enabling improvements in customer service and more direct channels for customer feedback. Furthermore, information technologies will facilitate the formation of ad hoc strategic alliances among diverse enterprises and even among competitors on a short time scale, driven by changes in business conditions that demand prompt action. This entire set of activities is already well under way.

In the delivery of services, the more effective use and transmission of information has had dramatic effects. Today's air transportation system would not exist without rapid and reliable information flows regarding air traffic control, sales, marketing, maintenance, safety, and logistics planning. Retailers and wholesalers depend on the rapid collection and analysis of sales data to plan purchasing and marketing activities, to offer more differentiated services to customers, and to reduce operational costs. The insurance industry depends on rapid and reliable information flows to its sales force and to customize policies and manage risks.[4]

Whereas a traditional national security perspective might call for keeping people out of sensitive stores of information or satellite communications networks, national economic and social activity increasingly involves the exact opposite: inviting people from around the world to come in — with varying degrees of recognition that all who come in may not be benevolent. The Sidebar, that follows describes some of the tensions between security and openness. Such a change in expectations and perspective is unfolding in a context in which controls on system access have typically been deficient, beginning with weak operating system security. The distributed and internetworked satellite communications systems that are emerging raise questions about protecting information regardless of the path traveled (end-to-end security), and as close to the source and destination as possible.

Tense Moments between Openness and Security

Businesses have long been concerned about the tension between openness and security. An environment that is open to everyone is not secure, while an environment that is closed to everyone is highly secure but not useful. A number of trends in business today tend to exacerbate this conflict.

For example, modern competitive strategies emphasize openness to interactions with potential customers and suppliers. Such strategies would demand that a bank present itself as willing to do business with anyone, everywhere, and at any time. However, such strategies also offer potential adversaries a greater chance of success, because increasing ease of access often facilitates the penetration of security measures that may have been taken.

Many businesses today emphasize decentralized management that pushes decision-making authority toward the customer and away from the corporate hierarchy. Yet security often has been (and is) approached from a centralized perspective. For instance, access controls are necessarily hierarchical (and thus centralized) if they are to be maintained uniformly.

Furthermore, many businesses rely increasingly on highly mobile individuals. When key employees were tied to one physical location, it made sense to base security on physical presence (to have a user present a photo ID card to an operator at the central corporate computer center). Today, mobile computing and satellite communications are common, with not even a physical wire to ensure that the person claiming to be an authorized user is accessing a computer from an authorized location or to prevent passive eavesdropping on unencrypted transmissions with a radio scanner.[5]

The international dimensions of business and the growing importance of competitiveness in the global marketplace complicate the picture further. Although *multinationals* have long been a feature of the U.S. economy, the inherently international nature of communications networks and the growing capabilities for distributing and accessing information worldwide are helping many activities and institutions to transcend national boundaries. See the Sidebar that follows here.

Today's International Dimensions of Business and Commerce

United States firms increasingly operate in a global environment, obtaining goods and services from companies worldwide, participating in global virtual corporations, and working as part of international strategic alliances. One key dimension of increasing globalization has been the dismantling of barriers to trade and investment. Over the past 43 years, tariffs among developed countries have been reduced by more than two-thirds. After reductions were phased-in, tariffs in these countries were under 3%, with 46% of current trade free of any customs duties. While tariffs of developing countries are at higher levels, they have recently begun to decline substantially. Tariffs in these countries will average 9% by agreement and will be even lower as a result of unilateral reductions. In response to the reductions in trade barriers, trade has grown rapidly. From 1950 to 1998, U.S. and world trade grew at an average compound rate of 11% annually.

Investment has also grown rapidly in recent years, stimulated by the removal of restrictions and by international rules that provide assurances to investors against discriminatory or arbitrary treatment. United States foreign direct investment also has grown at almost 13% annually during the past 23 years and now totals about a trillion dollars. Foreign direct investment in the United States has risen even faster over the same period at almost 22% annually—and now also totals almost $800 billion.

The expansion of international trade and investment has resulted in a much more integrated and interdependent world economy. For the United States, this has meant a much greater dependence on the outside world. More than a quarter of the U.S. gross domestic product is now accounted for by trade in goods and services and returns on foreign investment. Over 14 million jobs are now directly or indirectly related to our merchandise trade.

Because the U.S. economy is mature, the maintenance of a satisfactory rate of economic growth requires that the United States compete vigorously for international markets, especially in the faster growing regions of the world. Many sectors of our economy are now highly dependent on export markets. This is particularly the case for, but is not limited to high-technology goods, as indicated in Table 2.1.

A second international dimension is the enormous growth in recent years of multinational enterprises. Such firms operate across national boundaries, frequently in multiple countries. According to the 1998 World Investment Report of the United Nations, transnational corporations (TNCs) with varying degrees of integration account for about one-third of the world's private sector productive assets.

The number of TNCs has more than tripled in the last 23 years. At the outset of this decade, about 40,000 U.S. firms had a controlling equity interest in some 280,000 foreign affiliates. This does not include nonequity relationships, such as management contracts, subcontracting, franchising or strategic alliances. There are some 600 TNCs based in the U.S. and almost 18,000 foreign affiliates, of which some 13,000 are nonbank enterprises.

The strategies employed by TNCs vary among firms. They may be based on trade in goods and services alone or, more often, involve more complex patterns of integrated production, outsourcing, and marketing. One measure of the extent of integration by U.S. firms is illustrated by the U.S. Census Bureau, which reported that in 1997, 57% of U.S. imports and 43% of U.S. exports were between related firms. Of U.S. exports to Canada and Mexico, 55% were between related parties. For the European Union and Japan, the share was 38%.

With respect to imports, the shares of related-party transactions were 86.6% for Japan, 58.3% for the European Union, 55.7% for Canada and 80.3% for Mexico. Among those sectors with the highest levels of interparty trade are data processing equipment (including computers); and, parts and telecommunications equipment ranging from 60% to 95%.[6]

Table 2.1 High-Technology Goods

Area of export	Exports as a percentage U.S. output
Electronic computing and parts	55%
Semiconductors and related devices	50%
Magnetic and optical recording media (includes software products)	43%

At the same time, export markets are at least as important as domestic U.S. markets for a growing number of goods and service producers, including producers of information technology products as well as a growing variety of high- and low-technology products. The various aspects of globalization include identifying product and merchandising needs that vary by country; establishing and maintaining employment, customer, supplier, and distribution relationships by country; coordinating activities that may be dispersed among countries but result in products delivered to several countries; and so on. This places new demands on U.S.-based and U.S.-owned information, satellite communication, organizational, and personal resources and systems.

Information Vulnerability

Solutions to cope with the vulnerabilities previously described require both appropriate technology and user behavior and are as varied as the needs of individual users and organizations. Encryption—a technology described more fully in Chapter 3, "Encryption: Roles, Market and Infrastructure" is an important element of many solutions to information vulnerability that can be used in a number of different ways. The national satellite encryption policy—the focus of this chapter—concerns how and to what extent government affects the development, deployment, and use of this important technology. To date, public discussion of the national satellite encryption policy has focused on one particular application of cryptography, namely its use in protecting the confidentiality of information and satellite communications.

Accordingly, consideration of a national satellite encryption policy must take into account two fundamental issues: First of all, if the public information and satellite communications infrastructure continues to evolve with very weak security throughout, reflecting both deployed technology and user behavior, the benefits from encryption for confidentiality will be significantly less than they might otherwise be.

Second, the vulnerabilities implied by weak overall security, affects the ability of specific mechanisms such as satellite encryption to protect not only confidentiality, but also the integrity of information and systems and the availability of systems for use when sought by their users. Simply protecting (encrypting) sensitive information from disclosure can still leave the rest of a system open to attacks that can undermine the satellite encryption (the lack of access controls that could prevent the insertion of malicious software) or destroy the sensitive information.

Satellite encryption thus must be considered in a wider context. It is not a panacea, but it is extremely important to ensuring security and can be used to counter several vulnerabilities.

Both recognition of the need for system and infrastructure security and demand for solutions are growing. Although demand for solutions has yet to become widespread, the trend is away from a marketplace in which the federal government was the only meaningful customer.

Note: The more general statement is that the market historically involved national governments in several countries as the principal customers.

Growing reliance on a shared information and satellite communications infrastructure means that all individuals and organizations should be, and industry analysts believe will become, the dominant customers for better security. That observation is inherent in the concept of infrastructure as something on which people rely.

What may be less obvious is that as visions of ubiquitous access and interconnection are increasingly realized, individual, organizational and governmental needs may become aligned. Such an alignment would mark a major change from the past. Again, the sharing of a common infrastructure is the cause. Everyone, individual or organization, public or private sector is a user. As significantly, all of these parties face a multitude of threats to the security of information as discussed in the Sidebar that follows.

Sources Of Threat

Foreign National Agencies and Intelligence Services: Foreign intelligence operations target key U.S. businesses. For example, two former directors of the French intelligence service have confirmed publicly that the French intelligence service collects economic intelligence information, including classified government information and information related to or associated with specific companies of interest.

Note: Two former directors of the DGSE (the French intelligence service), have publicly stated that one of the DGSE's top priorities was to collect economic intelligence. During a September 1991 NBC news program, Pierre Marion, former DGSE Director, revealed that he had initiated an espionage program against U.S. businesses for the purpose of keeping France internationally competitive. Marion justified these actions on the grounds that the United States and France, although political and military allies, are economic and technological competitors. During an interview in March 1993, then DGSE Director Charles Silberzahn stated that political espionage was no longer a real priority for France, but that France was interested in economic intelligence, a field which is crucial to the world's evolution. Silberzahn advised that the French had some success in economic intelligence but stated that much work is still needed because of the growing global economy. Silberzahn advised during a subsequent interview that theft of classified information, as well as information about large corporations, was a long-term French Government policy. These statements were seemingly corroborated by a DGSE targeting document prepared in late 1989 and leaked anonymously to the U.S. Government and the press in May 1993. It alleged that French intelligence had targeted numerous U.S. Government agencies and corporations to collect economic and industrial information. Industry leaders such as Boeing, General Dynamics, Hughes Aircraft, Lockheed, McDonnell Douglas, and Martin Marietta were all on the list. Heading the U.S. Government listing was the Office of the U.S. Trade Representative.

Foreign intelligence agencies may break into facilities such as the foreign offices of a U.S. company or the hotel suite of a U.S. executive and copy computer files from within that facility (from a laptop computer in a hotel room to a desktop computer connected to a network in an office). Having attained such access, they can also insert malicious code that will enable future information theft.

Note: According to a report from the National Communications System, countries that currently have significant intelligence operations against the U.S. for national security and/or economic pur-

poses include Russia, the People's Republic of China, Cuba, France, Taiwan, South Korea, India, Pakistan, Israel, Syria, Iran, Iraq, and Libya. All of the intelligence organizations listed here have the capability to target telecommunications and information systems for information or clandestine attacks. The potential for exploitation of such systems may be significantly larger.

Disgruntled or Disloyal Employees that Work from the Inside: Such parties may collude with outside agents. Threats involving insiders are particularly pernicious because they are trusted with critical information that is not available to outsiders. Such information is generally necessary to understand the meaning of various data flows that may have been intercepted, even when those data flows are received in the clear.

Network Hackers and Electronic Vandals that are Having Fun or Making Political Statements through the Destruction of Intellectual Property without the Intent of Theft: Information terrorists may threaten to bring down an information network unless certain demands are met. Extortionists may threaten to bring down an information network unless a ransom is paid. Disgruntled customers seeking revenge on a company also fall into this category.

Thieves Attempting to Steal Money or Resources from Businesses: Such individuals may be working for themselves or acting as part of a larger conspiracy (in association with organized crime). The spreading of electronic commerce will increase the opportunities for new and different types of fraud, as illustrated by the large increase in fraud seen as the result of increased electronic filing to the Internal Revenue Service. Even worse, customers traditionally regarded as the first line of defense against fraud (because they check their statements and alert the merchants or banks involved to problems) may become adversaries as they seek to deny a signature on a check or alter the amount of a transaction.

It is difficult to know the prevalence of such threats, because many companies do not discuss for the record specific incidents of information theft. In some cases, they fear stockholder ire and losses in customer confidence over security breaches; in others, they are afraid of inspiring *copycat* attacks or revealing security weaknesses. In still other cases, they simply do not know that they have been the victim of such theft. Finally, only a patchwork of state laws applies to the theft of trade secrets and the like (and not all states have such laws). There is no federal statute that protects trade secrets or that address commercial information theft. Federal authorities probing the theft of commercial information must rely on proving violations of other statutes, such as the wire and mail fraud laws; interstate transport of stolen property; conspiracy; or, computer fraud and abuse laws. As a result, documentation of what would be a federal offense if such a law were present is necessarily spotty. For all of these reasons, what is known on the public record about economic losses from information theft almost certainly understates the true extent of the problem.[7]

Consideration of the nation's massive dependence on the public switched telecommunications network (which is one of many components of the information and communications infrastructure) provides insight into the larger set of challenges posed by a more complex infrastructure. See the Sidebar, "Public Switched Telecommunications Network Vulnerabilities" that follows for more information on this subject.

To illustrate the broad panorama of stakeholder interests in which a national satellite encryption policy is formulated, the next several sections of this chapter examine different aspects of society from the standpoint of needs for information security.

The Economic and Business Perspective

For purposes of this part of the chapter, the relationship of U.S. businesses to the information society has two main elements. One element is that of protecting information important to the success of U.S. businesses in a global marketplace. The second element is ensuring the nation's continuing ability to exploit U.S. strengths in information technology on a worldwide basis.

Public Switched Telecommunications Network Vulnerabilities

The nation's single most critical national-level component of information infrastructure vulnerable to compromise is the public switched telecommunications network (PSTN). The PSTN provides information transport services for geographically dispersed and national assets such as the banking system and financial markets; and, the air traffic control system uses leased lines to connect regional air traffic control centers.

Note: These private networks for banking include Fedwire (operated by the Federal Reserve banks); the Clearinghouse for Interbank Payment Systems (CHIPS; operated by New York Clearinghouse, an association of money center banks); the Society for Worldwide Interbank Financial Telecommunication (SWIFT; an international messaging system that carries instructions for wire transfers between pairs of correspondent banks); and, the Automated Clearing House (ACH) systems for domestic transfers, typically used for routine smaller purchases and payments. In the 1980s, several U.S. banks aggressively developed global networks with packet switches, routers, and so on, to interconnect their local and wide area networks; or, they used third-party service providers to interconnect. In the 1990s, there are signs that U.S. international banks are moving to greater use of carrier-provided or hybrid networks because of the availability of virtual private networks from carriers. Carrier-provided networks are more efficient than networks built on top of dedicated leased lines, because they can allocate demand dynamically among multiple customers.

Even the traditional military is highly dependent on the PSTN. Parties connected to the PSTN are therefore vulnerable to failure of the PSTN itself and to attacks transmitted over the PSTN.

Note: Over 96% of U.S. military and intelligence community voice and data communications are carried over facilities owned by public carriers. Of course, the 96% figure includes some noncritical military communications; however, only 33% of the telecommunications networks that would be used during wartime operate in the classified environment (and are presumably more secure), while the other 67% are based on the use of unclassified facilities of public carriers.

The fundamental characteristic of the PSTN from the standpoint of information vulnerability is that it is a highly interconnected network of heterogeneously controlled and operated computer-based switching stations. Network connectivity implies that an attacker (which might range from a foreign government to a teen-aged hacker) can in principle connect to any network site (including sites of critical importance for the entire network) from any other network site (which may be geographically remote and even outside the U.S.). The sites of critical importance for the PSTN are the switching nodes that channel the vast majority of telecommunications traffic in the United States. Access to these critical nodes, and to other switching facilities, is supposed to be limited to authorized personnel, but in practice these nodes are often vulnerable to penetration. Once in place on a critical node, hostile and unauthorized users are in a position to disrupt the entire network.

The systemic vulnerabilities of the PSTN are the result of many factors. One is the increasing accessibility of network software to third parties other than the common carriers, resulting from the Federal Communications Commission requirement that the PSTN support open, equal access for third-party providers of enhanced services as well as for the common carriers. Such accessibility offers intruders many opportunities to capture user information, monitor traffic, and remotely manipulate the network. A second reason is that service providers are allowing customers more direct access to network elements, in order to offer customer-definable services such as call forwarding. A third reason is that advanced services made possible by Signaling System 7 are dependent on a common, out-of-band signaling system for control of calls through a separate packet-switched data network that adds to network vulnerability. Finally, space-based PSTN components (satellites) have few control centers, are susceptible to electronic attack, and generally do not encrypt their command channels, making the systems vulnerable to hackers copying their commands and disrupting service. These conditions imply that the PSTN is a system that would benefit from better protection of system integrity and availability.

Threats to the PSTN affect all national institutions whose ability to function fully and properly depends on being able to communicate, be it through telephony, data transmission, video, or all of these. Indeed, many data networks operated *privately* by large national corporations or national institutions such as those previously described are private only in the sense that access is supposed to be limited to corporate purposes. In fact, national institutions or corporations generally use all forms of communications, including those physically carried by the PSTN.

Note: *Both shared circuits and private networks are expected to grow dramatically in the next several years.*

However, the physical and computational infrastructure of these networks is in general owned by the telecommunications service provider, and this infrastructure is part of the larger PSTN infrastructure. Thus, like the Internet, the *private* data network of a national corporation is in general not physically independent of the PSTN. Similarly, it is dependence on the PSTN that has led to failures in the air traffic control system and important financial markets.

Back in 1991, the accidental severing of an AT&T fiber-optic cable in Newark, New Jersey, led to the disruption of FAA air traffic control communications in the Boston-Washington corridor and the shutdown of the New York Mercantile Exchange and several commodities exchanges. Also, the severing of a fiber-optic cable led to the shutdown of 4 of the Federal Aviation Administration's 20 major air traffic control centers with *massive operational impact*. Additionally, the failure of a PSTN component in New York caused the loss of connectivity between a major securities house and the Securities Industry Automation Corporation, resulting in an inability to settle the day's trades over the network.

Examples of small-scale activities by the computer *underground* against the PSTN demonstrate capabilities that if coupled to an intent to wage serious information warfare against the United States, pose a serious threat to the U.S. information infrastructure. For instance, back in 1990, several members of the Legion of Doom's Atlanta branch were charged with penetrating and disrupting telecommunications network elements. They were accused of planting *time bomb* programs in network elements in Denver, Atlanta, and New Jersey. These were designed to shut down major switching hubs, but were defused by telephone carriers before causing damage.

Furthermore, back in 1992, members of a group known as MOD (various spell-outs) were indicted on 11 accounts of *programmed attacks*. It is significant that they appear to have worked as a team. Their alleged activities included developing and unleashing *programmed attacks* on telephone company computers and accessing telephone company computers to create new circuits and add services with no billing records.

Reported (but not well documented) is a growing incidence of *programmed attacks*. These have been detected in several networks and rely on customized software targeting specific types of computers or network elements. They are rarely destructive, but rather seek to add or modify services. *The capability illustrated by this* category of attacks has not fully matured. However, if a coordinated attack using these types of tools was directed at the PSTN with a goal of disrupting national security/emergency preparedness (NS/EP) telecommunications, the result could be significant. The same point probably applies to the goal of disrupting other kinds of telecommunications beyond those used for NS/EP. A number of reports and studies have called attention to the vulnerability of components of the national telecommunications infrastructure.[8]

The Protection of Important Business Information

A wide range of U.S. companies operating internationally are threatened by foreign information-collection efforts. The National Counterintelligence Center (NACIC) reports that:

> the U.S. industries that have been the targets in most cases of economic espionage and other foreign collection activities include biotechnology; aerospace; telecommunications; computer hardware/software, advanced transportation and engine technology; advanced materials and coatings; energy research; defense and armaments technology; manufacturing processes; and semiconductors.

Foreign collectors target proprietary business information such as bid, contract, customer and strategy information, as well as corporate financial and trade data. Of all the information vulnerabilities facing U.S. companies internationally (see Sidebar, "Sources Of Threat"), electronic vulnerabilities appear to be the most significant. For example, the NACIC concluded that *specialized technical operations (including computer intrusions, telecommunications targeting and intercept, and private-sector encryption weaknesses) account for the largest portion of economic and industrial information lost by U.S. corporations.*

The NACIC noted that because U.S. corporations are so easily accessed and intercepted, corporate telecommunications (particularly international telecommunications) provide a highly vulnerable and lucrative source for anyone interested in obtaining trade secrets or competitive information. Because of the increased usage of these links for bulk computer data transmission and electronic mail, intelligence collectors find telecommunications intercepts cost-effective. For example, foreign intelligence collectors intercept facsimile transmissions through government-owned telephone companies, and the stakes are large—approximately half of all overseas telecommunications are facsimile transmissions. Innovative *hackers* connected to computers containing competitive information evade the controls and access companies' information. In addition, many American companies have begun using electronic data interchange, a system of transferring corporate bidding, invoice, and pricing data electronically overseas. Many foreign government and corporate intelligence collectors find this information invaluable.

Note: Further intelligence collections by foreign powers are facilitated when a hostile government interested in eavesdropping controls the physical environment in which a U.S. company may be operating. For example, the U.S. company may be in a nation in which the telecommunications system is under the direct control of the government. When a potentially hostile government controls the territory on which a company must operate, many more compromises are possible.

Why is electronic information so vulnerable? The primary reason is that it is computer-readable and thus much more vulnerable to automated search than are intercepted voice or postal mail transmissions. Once the information is collected (through an existing wiretap or a protocol analyzer on an Internet router), it is relatively simple for computers to search streams of electronic information for word combinations of interest (*IBM, research,* and *superconductivity* in the same message). As the cost of computing drops, the cost of performing such searches drops.

Note: As a rough rule of thumb, government researchers have estimated that 10 billion (10^{10}) words can be searched for $1. This estimate is based on an experiment in which they used the Unix utility program *fgrep* to search a 1 million (10^6) character file for a specific string of 10 characters known to be at the end of the file and nowhere else. It took the NeXT workstation on which this experiment was run approximately 1 second to find these last 10 characters. Since there are approximately 10^5 seconds in a day and 10^3 days (about 3 years) in the useful life of the workstation, it can search roughly 10^{13} over its life. Since such a workstation is worth on the order of $1000 today, this works out to 10^{10} words searched for $1. With the use of specialized hardware, this cost could be reduced significantly.

The threat posed by automated search, coupled with the sensitivity of certain communications that are critical for nongovernment users, is at the root of nongovernment demand for security. Other noncomputer-based technology for the clandestine gathering of information is widely available on the retail market. In recent years, concern over the ready availability of such equipment has grown.

Note: Solutions for coping with information-age vulnerabilities may well create new responsibilities for businesses. For example, businesses may have to ensure that the security measures they take are appropriate for the information they are protecting, and/or that the information they are protecting remains available for authorized use. Failure to discharge these responsibilities properly may result in a set of liabilities that these businesses currently do not face.

Exploiting Global Markets

With the increasing globalization of business operations, information technology plays a key role in maintaining the competitive strengths of U.S. business. In particular, U.S. businesses have proven adept at exploiting information and information technologies to create new market niches and expand old ones. This pattern has deep roots. For example, beginning in the 1960s, American Airlines pioneered in computerized reservations systems and extended use of the information captured and stored in such systems — thus, generating an entire new business that is more profitable than air transport services. More recently, creative uses of information technology have advanced U.S. leadership in the production of entertainment products (movies and videos, recorded music, on-line services) for the world.

United States innovation in using information technology reflects in part the economic vitality that makes new technology affordable. It also reflects proximity to the research and production communities that supply key information technology products. These are communities with which a variety of U.S. industries have successfully exchanged talent, communicated their needs as customers, and collaborated in the innovation process. In other words, it is not an accident that innovation in both use and production of information technology has blossomed in the United States.

The business advantages enjoyed by U.S. companies that use information technology are one important reason that the health of U.S. computer, telecommunications and information industries is important to the economy as a whole. A second important reason is the simple fact that the U.S. information technology sector (the set of industries that supply information technology goods and services) is the world's strongest.

Note: A staff study by the U.S. International Trade Commission found that 8 of the world's top 10 applications software vendors, 7 of the world's top 10 systems software vendors, the top 5 systems integration firms, and, 8 of the top 10 custom programming firms are U.S. firms. The top 9 global outsourcing firms have headquarters in the U.S.

The industry has an impressive record of product innovation; key U.S. products are de facto world standards and U.S. marketing and distribution capabilities for software products are unparalleled. In addition, U.S. companies have considerable strengths in the manufacture of specialized semiconductor technologies and other key components. A strong information technology sector makes a significant contribution to the U.S. balance of payments and is responsible for large numbers of high-paying jobs. These strengths establish a firm foundation for continued growth in sales for U.S. information technology products and services as countries worldwide assimilate these technologies into their economies.

Finally, because of its technological leadership, the U.S. should be better positioned to extend that lead, even if the specific benefits that may result are not known in advance. The head start in learning how to use information technology provides a high baseline on which U.S. individuals and organizations can build.

The telecommunications industry believes that information technology is one of a few high-technology areas (others might include aerospace and electronics) that play a special role in the economic health of the nation, and that leadership in this area is one important factor underlying U.S. economic strength in the world today. Furthermore, the industry acknowledges that there is a wide range of judgment among responsible economists on this matter. Some argue that the economy is so diverse that the fate of a single industry or even a small set of industries has a relatively small effect on broader economic trends. Others argue that certain industries are important enough to warrant subsidy or industrial policy to promote their interests. Industry analysts have been discussing this specific issue to a considerable extent and found a middle ground between these two extremes—that information technology is one important industry among others, and that the health and well-being of that industry are important to the nation.

Note: This position is also supported by the U.S. government, which notes that telecommunications and computer hardware/software are among a number of industries that are of strategic interest to the U.S. because they produce classified products for the government and dual use technology used in both the public and private sectors, and are responsible for leading-edge technologies critical to maintaining U.S. economic security.

To the extent that this belief is valid, the economic dimension of national security and perhaps even traditional national security itself may well depend critically on a few key industries that are significant to military capabilities, the industrial base, and the overall economic health of the nation. Policy that acts against the health and global viability of these industries or that damages the ability of the private sector to exploit new markets and identify niches globally thus deserves the most careful scrutiny.

Because it is inevitable that other countries will expand their installed information technology bases and develop their own innovations and entrepreneurial strengths, U.S. leadership is not automatic. Evidence of such development is already available, as these nations build on the falling costs of underlying technologies (microprocessors, aggregate communications bandwidth) and worldwide growth in relevant skills. The past three decades of information technology history provide enough examples of both successful first movers and strategic missteps to suggest that U.S. leadership can be either reinforced or undercut. Leadership is an asset, and it is sensitive to both public policy and private action.

Public and private factors affecting the competitive health of U.S. information technology producers are most tightly coupled in the arena of foreign trade. Of course, many intrafirm and intraindustry factors shape competitive strength, such as good management, adequate financing, good fit between products and consumer preferences, and so on.

United States producers place high priority on ease of access to foreign markets. That access reflects policies imposed by U.S. and foreign governments, including governmental controls on what can be exported to whom. Export controls affect foreign trade in a variety of hardware, software and communications systems.

> **Note:** Export controls are the subject of chronic complaints from industry, to which government officials often respond by pointing to other, industry-centered explanations (deficiencies in product design or merchandising) for observed levels of foreign sales and market shares. Chapter 5, "Export Controls," addresses export controls in the context of cryptography and a national satellite encryption policy.

Privacy: Personal and Individual Interests

The emergence of the information age affects individuals as well as businesses and other organizations. As numerous reports argue, the nation's information infrastructure promises many opportunities for self-education, social exchange, recreation, personal business, cost-effective delivery of social programs, and entrepreneurship. Yet the same technologies that enable such benefits may also convey unwanted side effects. Some of those can be considered automated versions of problems seen in the paper world; others are either larger in scale or different in kind. For individuals, the area relevant to this report is privacy and the protection of personal information. Increasing reliance on electronic commerce and the use of networked communication for all manner of activities suggest that more information about more people will be stored in network-accessible systems and will be communicated more broadly and more often, thus raising questions about the security of that information.

Privacy is generally regarded as an important American value, a right whose assertion has not been limited to those *with something to hide*. Indeed, assertion of the right to privacy as a matter of principle has figured prominently in U.S. political and social history; it is not merely abstract or theoretical.

In the context of an information age, an individual's privacy can be affected on two levels: privacy in the context of personal transactions (with businesses or other institutions and with other individuals); and privacy vis-à-vis governmental units. Both levels are affected by the availability of tools, such as satellite encryption in the context of information and communications systems, that can help to preserve privacy. Today's information security technology, for example, makes it possible to maintain or even raise the cost of collecting information about individuals. It also provides more mechanisms for government to help protect that information. The Clinton Administration has recognized concerns about the need to guard individual privacy, incorporating them into the security and privacy guidelines of its Information Infrastructure Task Force. These guidelines represent an important step in the process of protecting individual privacy.

Privacy in an Information Economy

Today, the prospect for easier and more widespread collection and use of personal data as a byproduct of ordinary activities raises questions about inappropriate activities by industry, nosy individuals, and/or criminal elements in society. Criminals may obtain sensitive financial information to defraud individuals (credit card fraud, for example, amounts to approximately $23 per card per year). Insurance companies may use health data collected on individuals to decide whether to provide or deny health insurance—putting concerns about business profitability in possible conflict with individual and public health needs. On the other hand, much of the personal data in circulation is willingly divulged by individuals for specific purposes; the difficulty is that once shared, such information is available for additional uses. Controlling the further dissemination of personal data is a function both of procedures for how information should be used and of technology (including but not limited to satellite encryption) and procedures for restricting access to those authorized. Given such considerations, individuals in an information age may wish to be able to:

- Keep specific information private.
- Ensure that a party with whom they are transacting business is indeed the party he or she claims to be.
- Prevent the false repudiation of agreed-to transactions.
- Communicate anonymously (carry out the opposite of authenticated communication).
- Ensure the accuracy of data that is relevant to them.

Keep Specific Information Private

Disclosure of information of a personal nature that could be embarrassing if known, whether or not such disclosure is legal, is regarded as an invasion of privacy by many people. For example, a letter to Ann Landers from a reader described his inadvertent eavesdropping on some very sensitive financial transactions being conducted on a cordless telephone. Home banking services using telephone lines or network connections and personal computers will result in the flow on public networks of large amounts of personal information regarding finances. Even the ad copy in some of today's consumer catalogues contains references to information security threats.

Note: A catalogue from Comtrad Industries notes that *burglars use* Code Grabbers to open electric garage doors and break into homes, defining code grabbers as devices that can record and play back the signal produced from your garage door remote control.

Authentication of parties in Business Transactions

Likewise, parties in business transactions may seek to authenticate their own identity with confidence that such authentication will be accepted by other parties, and that anyone lacking such authentication will be denied the ability to impersonate them. For example, a journalist who had reported on the trafficking of illegally copied software on America Online was the victim of hackers who assumed his or her on-line identity, thereby intercepting his or her e-mail messages and otherwise impersonating him or her. Other cases of *stolen identities* have been reported in the press, and while these cases remain relatively isolated, they are still a matter of public concern. Thieves forge signatures and impersonate identities of law-abiding citizens to steal money from bank accounts and to obtain credit cards in the name of those citizens.

Such capabilities are needed to transfer money among mutual funds with a telephone call or to minimize unauthorized use of credit card accounts. For example, security concerns have been raised by the ease of access to 401(k) retirement accounts (for which there is no cap on the liability incurred if a third party with unauthorized access to it transfers funds improperly). In an electronic domain without face-to-face communications or recognizable indicators such as voices and speech patterns (as used today in telephone calls), forgery of identity becomes increasingly easy.

Preventing False Repudiation of Agreed-to Transactions

It is undesirable for a party to a transaction to be able to repudiate (deny) his or her agreement to the terms of the transaction. For example, an individual may agree to pay a certain price for a given product; he or she should not then be able to deny having made that agreement (as he or she might be tempted to do upon finding a lower price elsewhere).

Communicate Anonymously

Individuals may wish to communicate anonymously to criticize the government or a supervisor, report illegal or unethical activity without becoming further involved, or obtain assistance for a problem that carries a social stigma. In other instances, they may simply wish to speak freely without fear of social reprisal or for the entertainment value of assuming a new digital identity in cyberspace.

Ensuring the Accuracy of Relevant Data

Many institutions such as banks, financial institutions, and hospitals keep records on individuals. These individuals often have no personal control of these records, even though the integrity of the data in these records can be of crucial significance. Occasional publicity attests to instances of the inaccuracy of such data (credit records) and to the consequences for individuals. Thus, practical safeguards for privacy such as those just outlined may be more compelling than abstract or principled protection of a right to privacy.

Privacy for Citizens

Public protection of privacy has been less active in the U.S. than in other countries, but the topic is receiving increasing attention. In particular, it has become an issue in the political agenda of people and organizations that have a wide range of concerns about the role and performance of government at all levels. It is an issue that attracts advocates from across the spectrum of political opinion. The politicization of privacy may inhibit the orderly consideration of relevant policy, including satellite encryption policy, because it revolves around the highly emotional issue of trust in government. The trust issue surfaced in the initial criticisms of the Clipper chip initiative proposal in 1993 (see Chapter 13, "Escrowed Satellite Encryption and Related Issues") and continues to color discussion of privacy policy generally and satellite encryption policy specifically.

To many people, freedom of expression and association, protection against undue governmental, commercial or public intrusion into their personal affairs, and fair treatment by various authorities are concerns shaped by memories of highly publicized incidents in which such rights were flouted. Some incidents that are often cited include the surveillance of political dissidents, such as Martin Luther King, Jr., Malcolm X, and the Student Nonviolent Coordinating Committee (SNCC) in the mid- to late-1960s; the activities of the Nixon *plumbers* in the late 1960s, including the harassment and surveillance of sitting and former government officials and journalists and their associates in the name of preventing leaks of sensitive national security information; U.S. intelligence surveillance of the international cable and telephone communications of U.S. citizens from the early 1940s through the early 1970s in support of FBI and other domestic law enforcement agencies; and, the creation of FBI dossiers on opponents of the Vietnam War in the mid-1960s.

It can be argued that such incidents were detectable and correctable precisely because they involved government units that were obligated to be publicly accountable—and indeed, these incidents prompted new policies and procedures as well as greater public vigilance. It is also easy to dismiss them as isolated instances in a social system that for the most part works well. But where these episodes involve government, many of those skeptical about government believe that they demonstrate a capacity of government to violate civil liberties of Americans who are exercising their constitutional rights. This perception is compounded by attempts to justify past incidents as having been required for purposes of national security. Such an approach both limits public scrutiny and vitiates policy-based protection of personal privacy.

> **Note:** At the 4th Conference on Computers, Freedom, and Privacy in Chicago, IL, held in 1994, a government speaker asked the audience if they were more concerned about government abuse and harassment or about criminal activity that might be directed at them. An overwhelming majority of the audience indicated greater concern about the first possibility.

It is hard to determine with any kind of certainty the prevalence of the sentiments described in this part of the chapter. By some measures, over half of the public is skeptical about government in general, but whether that skepticism translates into widespread public concern about government surveillance is unclear.

Industry analysts believe that most people acting as private individuals feel that their electronic communications are secure and do not generally consider it necessary to take special precautions against threats to the confidentiality of those satellite communications. These attitudes reflect the fact that most people, including many who are highly knowledgeable about the risks, do not give much conscious thought to these issues in their day-to-day activities.

At the same time, industry analysts acknowledge the concerns of many law-abiding individuals about government surveillance. The analysts believe that such concerns and the questions they raise about individual rights and government responsibilities must be taken seriously. It would be inappropriate to dismiss such individuals as paranoid or overly suspicious. Moreover, even if only a minority is worried about government surveillance, it is an important consideration, given the nation's history as a democracy, for determining whether and how access to and use of satellite encryption may be considered a citizen's right (see Chapter 26, "Satellite Encryption Technology Policy Options, Analysis and Forecasts").

Protecting communications from government surveillance is a time-honored technique for defending against tyranny. A most poignant example is the U.S. insistence in 1945 that the postwar Japanese Constitution include protection against government surveillance of the communications of Japanese citizens. In the aftermath of the Japanese surrender in World War II, the United States drafted a constitution for Japan. General Douglas MacArthur, who was supervising the drafting of the new Japanese constitution, insisted that the original provision regarding communications secrecy and most other provisions of the original U.S. draft be maintained. The Japanese agreed, this time requesting only minor changes in the U.S. draft, and accepting fully the original U.S. provision on communications secrecy.

The Government's Special Needs

The government encompasses many functions that generate or depend on information, and current efforts to reduce the scope and size of government depend heavily on information technology. In many areas of government, the information and information security needs resemble those of industry. The government also has important responsibilities beyond those of industry, including those related to public safety. For two of the most important and least understood in detail (law enforcement and national security), the need for strong information security has long been recognized.

Domestic law enforcement authorities in our society have two fundamental responsibilities: preventing crime and prosecuting individuals who have committed crimes. Crimes committed and prosecuted are more visible to the public than crimes prevented (see Chapter 4, "Need For Access To Satellite Encrypted Information"). The following areas relevant to law enforcement require high levels of information security:

- ◆ Prevention of information theft from businesses and individuals, consistent with the transformation of economic and social activities previously outlined.

- Tactical law enforcement communications. Law enforcement officials working in the field need secure communications. At present, police scanners available at retail electronics stores can monitor wireless communications channels used by police; criminals eavesdropping on such communications can receive advance warning of police responding to crimes that they may be committing.
- Efficient use by law enforcement officials of the large amounts of information compiled on criminal activity. Getting the most use from such information implies that it be remotely accessible and not be improperly modified (assuming its accuracy and proper context, a requirement that in itself leads to much controversy.
- Reliable authentication of law enforcement officials. Criminals have been known to impersonate law enforcement officials for nefarious purposes, and the information age presents additional opportunities.[9]

In the domain of national security, traditional missions involve protection against military threats originating from other nation-states and directed against the interests of the United States or its friends and allies. These traditional missions require strong protection for vital information.

United States military forces require secure communications. Without cryptography and other information security technologies in the hands of friendly forces, hostile forces can monitor the operational plans of friendly forces to gain an advantage.

Note: The compromise of the BLACK code used by Allied military forces in World War II enabled German forces in Africa in 1942, led by General Erwin Rommel, to determine the British order of battle (quantities, types, and locations of forces), estimate British supply and morale problems, and know the tactical plans of the British. For example, the compromise of one particular message enabled Rommel to thwart a critical British counterattack. In July of that year, the British switched to a new code, thus denying Rommel an important source of strategic intelligence. Rommel was thus surprised at the Battle of Alamein, widely regarded as a turning point in the conflict in the African theater.

Force planners must organize and coordinate flows of supplies, personnel, and equipment. Such logistical coordination involves databases whose integrity and confidentiality as well as remote access must be maintained.

Sensitive diplomatic communications between the United States and its representatives or allies abroad, and/or between critical elements of the U.S. government, must be protected as part of the successful conduct of foreign affairs, even in peacetime. For example, an agreement on Palestinian self-rule was reached in September 1995. According to public reports, the parties involved, Yasser Arafat (leader of the Palestinian Liberation Organization) and Shimon Peres (then Foreign Minister of Israel), depended heavily on the telephone efforts of Dennis Ross (a U.S. negotiator) in mediating the negotiations that led to the agreement. Obviously, in such circumstances, the security of these telephone efforts was critical.

In addition, the traditional missions of national security have expanded in recent years to include protection against terrorists and international criminals, especially drug cartels. Furthermore, recognition has been growing that in an information age, economic security is part of national security.

Note: Terrorist threats generally emanate from nongovernmental groups, though at times they involve the tacit or implicit (but publicly denied) support of sponsoring national governments. Furthermore, the United States is regarded by many parties as a particularly important target for political reasons by virtue of its prominence in world affairs. Thus, terrorists in confrontation with a U.S. ally may wish to make a statement by attacking the United States directly rather than its ally.

More broadly, there is a practical convergence under way among protection of individual liberties, public safety, economic activity, and military security. For example, the nation is beginning to realize that critical elements of the U.S. civilian infrastructure (including the banking system, the air traffic control system, and the electric power grid) must be protected against the threats previously described, as must the civilian information infrastructure that supports the conduct of sensitive government communications. Because the civilian infrastructure provides a significant degree of functionality on which the military and defense sector is dependent, traditional national security interests are at stake as well. As well, concerns have grown about the implications of what has come to be known as information warfare (see the Sidebar, "Information Warfare" that follows). More generally, the need for more secure systems, updated security policies, and effective procedural controls is taking on truly nationwide dimensions.

Information Warfare

Information warfare is a term used in many different ways. Of most utility for this book is the definition of information warfare (IW) as hostile action that targets the information systems and information infrastructure of an opponent (offensive actions that attack an opponent's satellite communications, weapon systems, command and control systems, intelligence systems, and information components of the civil and societal infrastructure such as the power grid and banking system) coupled with simultaneous actions seeking to protect U.S. and allied systems and infrastructure from such attacks. Other looser uses of the term information warfare include the following:

♦ The use of information and tactical intelligence to apply weapon systems more effectively; IW may be used in connection with information-based suppression of enemy air defenses or *smart* weapons using sensor data to minimize the volume of ordnance needed to destroy a target.

♦ The targeting of companies' information systems for IW attacks. As industrial espionage spreads and/or international competitiveness drives multinational corporations into military-like escapades, the underlying notion of information-based probing of and attack on a competitor's information secrets could take on a flavor of intergovernment military or intelligence activities.

♦ The fight against terrorism, organized crime, and even street crime, which might be characterized as IW to the extent that information about these subjects is used to prosecute the battle. This usage is not widespread, although it may develop in the future.

Usage of the term has shifted somewhat as federal agencies, notably the Department of Defense, struggle to fully appreciate this new domain of warfare (or low-intensity conflict) and to create relevant policy and doctrine for it. Conversely, there is some discussion of the vulnerabilities of the U.S. civil information infrastructure to such offense. The range of activities that can take place in information warfare is broad:

- Physical destruction of information-handling facilities to destroy or degrade functionality.
- Denial of use of an opponent's important information systems.
- Degradation of effectiveness (accuracy, speed of response) of an opponent's information systems.
- Insertion of spurious, incorrect or otherwise misleading data into an opponent's information systems (to destroy or modify data, or to subvert software processes via improper data inputs).
- Withdrawal of significant tactical or strategic data from an opponent's information systems.
- Insertion of malicious software into an opponent's system to affect its intended behavior in various ways, and perhaps, to do so at a time controlled by the aggressor.
- Subversion of an opponent's software and/or hardware installation to make it an in-place selfreporting mole for intelligence purposes.

As an operational activity, information warfare is clearly related closely to, but yet distinct from, intelligence functions that are largely analytical; IW is also related to information security, because its techniques are pertinent both to prosecution of offensive IW and to protection for defensive IW.[10]

Let us now go on the offensive with regards to information warfare by taking a closer look at satellite encryption: the threat, applications, and potential solutions afforded by the National Security Agency (NSA). This next section of the chapter also provides additional information concerning the various satellite encryption applications now being used and the potential approaches and methodologies to deal with them.

Satellite Encryption:
The Dawn of a New Tomorrow

The successful conduct of electronic surveillance is crucial to effective law enforcement, the preservation of public safety, and to the maintenance of national security according to the National Security Agency (NSA). The NSA feels that recent advances in satellite communications technology, particularly telecommunications technology, and the increased availability and use of encryption threaten to curtail significantly, and in many instances, preclude effective law enforcement. Efforts have been made to develop, where available, technical solutions to the problems posed by advanced satellite communications technologies and encryption in order to preserve the electronic surveillance technique.

Encryption is, or can be used in a number of applications to secure voice, data communications and stored information. The type of satellite encryption used and the way it is implemented varies depending upon the nature of the application. Encryption applications are available to secure satellite communications transmitted both in analog and digital formats. Digital communications, in particular, support and accommodate the use of encryption. Thus, encryption can be and is employed easily and inexpensively in computer-based applications. To date, its use has been somewhat limited in certain areas such as in voice communications. However, as the transition proceeds from analog to digital telephony, and as consumers migrate from wireline (basic telephone) to wireless (cordless or cellular telephone) communications devices, the use of encryption by telecommunications service providers and end-users can be expected to increase markedly in the near future. Hence, it is expected that encryption will soon be more widely available and used with all communications applications.

Wireless telecommunications devices, such as cordless telephones and cellular telephones however, are vulnerable to unauthorized interception. Consequently, there is a fundamental need to apply some form of enhanced security to wireless telephone devices. As a result, there appears to be a widespread and growing recognition that additional security features, such as encryption, need to be incorporated into these devices. In this vein, NSA's Privacy and Technology Task Force submitted a report back in 1991 to the Subcommittee on Privacy and

Technology. The report recommended that cordless telephones be afforded privacy protection under Title III (cordless telephones currently are not statutorily protected because of the ease with which they can be intercepted). The Task Force noted that is projected that cordless phones will be in 76% of American households by the year 2001. The report also states that a number of task force members indicated that *technical privacy enhancing features for radio-based systems should be more rapidly deployed by manufacturers and service providers.* Currently, AT&T, Motorola, and other service providers and manufacturers are offering encryption for cellular devices or service.

Law enforcement's decryption requirements, particularly real-time intelligibility of communications content, are the same for wireless and wireline voice communications. Hence, according to NSA, a solution to the threat posed by encryption in wireless as well as wireline devices is imperative.

Applications

A variety of PC communications, including e-mail, increasingly are being used not only by businesses but also by individuals. In 1997, approximately 95 million e-Mail users sent nearly 75 billion messages. With increased computer networking and with the recent acceptance of new e-mail standards, electronic messaging via satellite communications will continue to increase dramatically. Existing e-mail standards generally support text transmissions, however, emerging e-mail systems can support voice, facsimile and video capabilities. These electronic communications are fast replacing real-time voice conversations and consequently will increasingly become the subject of electronic surveillance. As these types of communications are more frequently and widely used, the use of encryption to protect the satellite communication content can be expected to increase.

Low-speed data transmissions typically run at speeds less than 64,000 bits of information per second (64Kb/s). The use of encryption on these low-speed applications can be either software- or hardware-based. With respect to certain satellite communications such as facsimile and e-mail, law enforcement typically requires real-time access to these communications— the same way as it does for voice communications.

Software-based encryption on the other hand is more widely used in these low-speed data transmission-related applications for the reasons previously discussed: cost and ease of use. Encryption for functions such as e-mail and individual (nonbulk) file transfers across a local area network (LAN) can be provided, and typically is provided as part of a satellite communications software package. Thus, this encryption is essentially free to mass market software publishers as previously discussed.

Voice/Data and Stored Information Applications

Real-time access to and decryption of stored electronic information secured by hardware-based encryption could be performed utilizing the Clipper technique according to the NSA. Technical solutions, such as they are, will only work if they are incorporated into all encryption products. To ensure that this occurs, legislation mandating the use of government-approved encryption products or adherence to government encryption criteria is required according to the NSA.

Secret High-Speed Data Transmissions

As satellite communications expand and as the requirements to support geographically widespread networks increase, there will be an increased demand for the development of faster speed transmissions to benefit from these high-speed networks. As a result, users will be able to take advantage of these high-speed satellite communications highways to transmit increased amounts of data associated with video, high-volume data retrieval, and other high-speed data services. These types of data services are typically used by large commercial, banking and government institutions. Because of the sensitive banking data and personnel information, there is a need to utilize satellite encryption. By way of example, major interbank data transmissions typically utilize Data Encryption Standard (DES)-based or comparable encryption.

High-speed satellite transmissions today typically run in the range of 50–100 million bits per second (50–100 Mbits/s). At these data rates, hardware-based encryption is the only feasible approach to data security. In this regard, according to NSA (and not supported by this author), the *Clipper* technique offers a suitable solution. In its current configuration, *Clipper* is designed to run at speeds of 50 Mbits/s and if necessary, it can easily be engineered to run at speeds up to 200 Mbits/s.

High-speed satellite transmissions can be viewed from a law enforcement interception standpoint in two ways. According to NSA, if voice communications have been intercepted, and the satellite transmissions are composed of individual data communications that have be multiplexed or bundled, law enforcement has a need for real-time access to and decryption of the specific communications that are the subject of the interception. If, on the other hand, the high-speed of the satellite transmissions were of a bulk file or other voluminous information transfer, it would not be physically possible or even desirable to process or view the product of the interception in real time. In these instances, access to the communications would be practically obtained *after the fact*, under circumstances where the communications are no longer in transit but rather in storage.

Solutions

In brief (really brief), the technical solutions and approaches developed to satisfy law enforcement's satellite decryption requirements with regard to the main encryption applications are as follows according to NSA:

- Voice/Data Applications: Classified as per NSA.
- Real-time access to and decryption of voice/data communications secured by software-based encryption: Classified as per NSA.
- Stored Information Applications: Real-time access to and decryption of stored electronic information secured by hardware-based encryption could be performed utilizing the Clipper technique.
- Technical solutions, such as they are, will only work if they are incorporated into all satellite encryption products. To ensure that this occurs, legislation mandating the use of government-approved satellite encryption products or adherence to government encryption criteria is required.

As you can see, our government does not really offer any clear solutions to the management of satellite encryption for the good of the public sector. Their solutions are based on the monitoring of satellite encryption for widespread electronic surveillance with little if any regard for the public's privacy. But what about satellite encryption on a worldwide scale? Are worldwide governmental restrictions on satellite encryption products putting security at risk? Let us take a look!

Worldwide Satellite Encryption Solutions

Worldwide, there is a political debate regarding the virtue or otherwise of a control of satellite encryption, in particular whether the import, export, and production of cryptographic tools and their use should be restricted. In several countries legal regulations exist, in some others steps are being taken towards such regulations. At present an Organization for Economic Cooperation and Development (OECD) Committee is drafting guidelines on cryptographic policy. There are concerns however. The Council of European Professional Informatics Societies (CEPIS)—with nearly 400,000 professionals in its 40 member societies (the largest European association of professionals working in information technology (IT)), has agreed to the following statement:

> **Note:** The Organization for Economic Co-operation and Development (OECD) is composed of 29 member countries in an organization that, most importantly, provides governments a setting in which to discuss, develop and perfect economic and social policy. Exchanges between OECD governments flow from information and analysis provided by a Secretariat in Paris. Parts of the OECD Secretariat collect data, monitor trends, analyze and forecast economic developments, while others research social changes or evolving patterns in trade, environment, agriculture, technology, taxation and more. The overriding committee is the Council, which has the decision-making power. It is composed of one representative for each member country (as well as a representative of the European Commission). The Secretariat is directed by a Secretary-General, assisted by four deputy Secretaries-General. The Secretary-General also chairs the Council, providing the crucial link between national delegations and the Secretariat.

Should one wish to employ electronic communication as the main vehicle for commercial and personal interaction, then one ought to be assured, and be able to prove, that messages are:

- Not disclosed to unauthorized recipients (confidentiality).
- Not tampered with (integrity).
- Shown to be from the senders stated (authenticity).[11]

It has always been an aim of secure reliable communication to comply with these requirements. The more the information society becomes a reality, the greater the need arises for enterprises, administrations, and private persons to feel an absolute assurance that these requirements are met.

To achieve this, so-called *strong* encryption is available. Several tools based on strong crypto-algorithms are in the public domain and offered on the Internet, others are integrated within commercial products.

A different technique for confidential and even unobservable satellite communication is to use steganography—where secret data are hidden within larger inconspicuous everyday data in such a way that third parties are unable even to detect their existence. Hence there is no way of preventing unobservable secret communication.

To enable surveillance of electronic messaging, many criminal and national security investigators (police and secret services), demand access to keys used for encrypted communication. In order for this to be effective, escrowing (bonding) of these keys is advocated. However, for the reasons previously given, key escrow (depositing copies of the keys with a *trusted third party*, including back-ups) cannot even guarantee effective monitoring. Moreover, key escrow already constitutes a risk for the secrecy of the keys and therefore for the secrecy of the data. This risk is exacerbated in cases of central escrowing.

Furthermore, the burdens of cost and administrative effort as well as the loss of trust in communications could be significant and are prone to deter individuals and organizations, especially small business users, from gaining the benefits of modern information and communications systems.

Effective electronic surveillance of digital networks is difficult and time consuming, and requires extensive resources. In particular, closed groups such as criminal organizations might even use steganographic techniques to avoid any detection short of physical access to the terminals they use. Thus restrictions on satellite encryption may be of very limited help in the fight against organized crime. On the other hand, the essential security of business and private communication may be seriously imperiled and economically hampered should they be subjected to insufficiently secured key escrow. On these grounds, CEPIS recommends the following:

- The use of satellite encryption for identifying data corruption or authenticating people/organizations should be free of restrictions and encouraged by governments.
- All individuals and organizations in the private and public sectors should be able to store and transmit data to others, with confidentiality protection appropriate for their requirements, and should have ready access to the technology to achieve this.
- The opportunity for individuals or organizations in the private and public sectors to benefit from information systems should not be reduced by incommensurable measures considered necessary for the enforcement of law.
- The governments of the world should agree on a policy relating to their access to other people's computerized data, while seeking the best technical advice available in the world on:
    - Whether and which access mechanisms to computerized data are an effective, efficient and adequate way to fight (organized) crime and mount effective prosecution of criminals.
    - How to implement the policy while minimizing the security risks to organizations and individual citizens. Of course, evaluation and implementation of the policy will require regular review as the technology evolves.[12]

Summary

This chapter underscores the need for attention to be paid to the protection of vital U.S. interests and values in an information age characterized by a number of trends—the first being that the world economy is in the midst of a transition from an industrial to an information age in which information products are extensively bought and sold; information assets provide leverage in undertaking business activities; and communications assume ever greater significance in the lives of ordinary citizens. At the same time, national economies are increasingly interlinked across national borders, with the result that international dimensions of public policy are important. Second, trends in information technology suggest an ever-increasing panoply of technologies and technology-enabled services characterized by high degrees of heterogeneity, enormous computing power, and large data storage and transmission capabilities.

Third, given the transition to a global information society and trends in information technology, the future of individuals and businesses alike is likely to be one in which information of all types plays a central role. Electronic commerce in particular is likely to become a fundamental underpinning of the information future. Finally, government has special needs for information security that arise from its role in society, including the protection of classified information and its responsibility for ensuring the integrity of information assets on which the entire nation depends.

Collectively, these trends suggest that future needs for information security will be large. Threats to information security will emerge from a variety of different sources, and they will affect the confidentiality and integrity of data and the reliable authentication of users. These threats do and will affect businesses, government, and private individuals.

From Here

This chapter outlined the problem of growing information vulnerability and the need for technology and policy to mitigate this problem. It also provided readers with an understanding of the satellite encryption threat, applications and potential solutions, and governmental restrictions on encryption products that put satellite security at risk.

Chapter 3 describes how cryptography may help to address all of these problems. It examines possible roles for satellite encryption in reducing information vulnerability and places cryptography into context as one element of an overall approach to ensuring information security. This chapter also covers a simple distributed encrypted infrastructure, the international market for satellite encryption software, legal protection for satellite encrypted services in the internal market, and a European initiative in electronic commerce.

Endnotes

1   John Young Architect, "Cryptography's Role in Securing the Information Society," © 1996 by the National Academy of Sciences. Courtesy of the National Academy Press, 2101 Constitution Avenue NW, Washington D.C. 20418.

2   Ibid.

3   Ibid.

4   Ibid.

5   Ibid.

6   Ibid.

7   Ibid.

8   Ibid.

9   Ibid.

10 Ibid.

11 "Governmental Restrictions on Encryption Products Put Security at Risk," Council of European Professional Informatics Societies (CEPIS), 7 Mansfield Mews, GB London W1M 9FJ, United Kingdom, 1996.

12 Ibid.

3

Encryption: Roles, Market and Infrastructure

Satellite encryption is a technology that can play important roles in addressing certain types of information vulnerability, although it is not able at this time to deal with all threats to communications security. As a technology itself, satellite encryption will in the very near future be embedded into products that will be purchased by a large number of users. Thus, it is important to examine various aspects of the market for satellite encryption. This chapter describes satellite encryption as a technology that is and will be used with products within a larger market context and with reference to the infrastructure needed to support its large-scale use.

Satellite Encryption in Context

Communications system security, and its extension intranet (network) security, are intended to achieve many purposes. Among them are safeguarding physical assets from damage or destruction and ensuring

99

that resources such as computer time, intranet connections, and access to databases are available only to individuals (or to other systems or even software processes) authorized to have them.

Note: The term communications security and shortened versions such as INFOSEC, COMPSEC, and NETSEC are also in use.

Overall communications security is dependent on many factors, including various technical safeguards, trustworthy and capable personnel, high degrees of physical security, competent administrative oversight, and good operational procedures. Of the available technical safeguards, satellite encryption has been one of the least utilized to date.

In general, the many security safeguards in a communications system or intranet not only fulfill their principal task but also act collectively to mutually protect one another. In particular, the protection or operational functionality that can be afforded by the various satellite encryption safeguards treated in this chapter will inevitably require that the hardware or software in question be embedded in a secure environment. To do otherwise is to risk that the encryption might be circumvented, subverted, or misused—hence leading to a weakening or collapse of its intended protection.

As individual stand-alone computer systems have been incorporated into ever larger intranets (local area networks (LANs), wide area networks (WANs), the Internet), the requirements for satellite encryption safeguards have also increased. For example, users of the earliest computer systems were almost always clustered in one place and could be personally recognized as authorized individuals and communications associated with a computer system usually were contained within a single building. Today, users of computer systems can be connected with one another worldwide, through the public switched telecommunications network, a LAN, satellites, microwave towers, and radio transmitters. Operationally, an individual or a software process in one place can request service from a system or a software process in a far distant place. Connectivity among systems is impromptu and occurs on demand. The Internet has demonstrated how to achieve it. Thus, it is now imperative for users and systems to identify themselves to one another with a high degree of certainty and for distant systems to know with certainty what privileges for accessing databases or software processes a remote request brings. Protection that could once be obtained by geographic proximity and personal recognition of users must now be provided electronically and with extremely high levels of certainty.

What Is Satellite Encryption and What Can It Do?

The word *encryption* is derived from Greek words that mean secret writing. Historically, encryption has been used to hide information from access by unauthorized parties, especially during communications when it would be most vulnerable to interception. By preserving the secrecy, or confidentiality of information, encryption has played a very important role over the centuries in military and national affairs.

In the traditional application of encryption for confidentiality, an originator (the first party) creates a message intended for a recipient (the second party), protects (encrypts) it by a cryptographic process, and transmits it as ciphertext. The receiving party decrypts the received ciphertext message to reveal its true content, the plaintext. Anyone else (the third party) who wishes undetected and unauthorized access to the message must penetrate (by cryptanalysis) the protection afforded by the cryptographic process.

In the classical use of encryption to protect satellite communications, it is necessary that both the originator and recipient(s) have common knowledge of the encryption process (the algorithm or cryptographic algorithm) and that both share a secret common element—typically, the key or cryptographic key, which is a piece of information, not a material object. In the encryption process, the algorithm transforms the plaintext into the ciphertext using a particular key; the use of a different key results in a different ciphertext. In the decryption process, the algorithm transforms the ciphertext into the plaintext, using the key that was used to encrypt the original plaintext.

Note: Here, the term *encrypt*, and throughout the chapter, describes the act of using an encryption algorithm with a given key to transform one block of data, usually plaintext, into another block, usually ciphertext.

A scheme in which both communicating parties must have a common key is now called *symmetric encryption* or *secret-key encryption*; it is the kind that has been used for centuries and written about widely. It has the property, usually an operational disadvantage, of requiring a safe method of distributing keys to relevant parties (*key distribution* or *key management*).

It can be awkward to arrange for symmetric and secret keys to be available to all parties with whom one might wish to communicate, especially when the list of parties is large. However, a scheme called *asymmetric encryption* (or, equivalently, *public-key cryptography*), developed in the mid-1970s, helps to mitigate many of these difficulties through the use of different keys for encryption and decryption. Each participant actually has two keys. The public key is published, is freely available to anyone, and is used for encryption; the private key is held in secrecy by the user and is used for decryption. Because the two keys are inverses, knowledge of the public key enables the derivation of the private key in theory. However, in a well-designed public-key system, it is computationally infeasible in any reasonable length of time to derive the private key from knowledge of the public key.

A significant operational difference between symmetric and asymmetric encryption is that with asymmetric encryption, anyone who knows a given person's public key can send secure message to that person. With symmetric encryption, only a selected set of people (those who know the private key) can communicate. While it is not mathematically provable, all known asymmetric encryption systems are slower than their symmetric encryption counterparts. And the more public nature of asymmetric systems lends credence to the belief that this will always be true. Generally, symmetric encryption is used when a large amount of data needs to be encrypted or when the encryption must be done within a given time period. Asymmetric encryption is used for short messages — for example, to protect key distribution for a symmetric encryption system.

Regardless of the particular approach taken, the applications of encryption have gone beyond its historical roots as secret writing. Today, encryption serves as a powerful tool in support of satellite communications security. Satellite encryption can provide many useful capabilities:

- ♦ Confidentiality.
- ♦ Authentication.
- ♦ Integrity check.
- ♦ Digital signature.
- ♦ Digital date/time stamp.
- ♦ Access controls.[1]

Confidentiality

Confidentiality is the characteristic that information is protected from being viewed in transit during communications (satellite uplink or downlink) and/or when stored in a satellite communications system. With cryptographically provided confidentiality, encrypted information can fall into the hands of someone not authorized to view it without being compromised. It is almost entirely the confidentiality aspect of satellite encryption that has posed public policy dilemmas.

Authentication

Authentication is a cryptographically based assurance that an asserted identity is valid for a given person (or communication system). With such assurance, it is difficult for an unauthorized party to impersonate an authorized one.

Integrity Check

Integrity check is cryptographically based assurance that a message or computer file has not been tampered with or altered. With such assurance, it is difficult for an unauthorized party to alter data.

Note: Digital signatures and integrity checks use a condensed form of a message or file—called a digest—which is created by passing the message or file through a one-way hash function. The digest is of fixed length and is independent of the size of the message or file. The hash function is designed to make it highly unlikely that different messages (or files) will yield the same digest, and to make it computationally very difficult to modify a message (or file) but retain the same digest.

Digital Signature

Digital signature is cryptographically based assurance that a message or file was sent or created by a given person. A digital signature cryptographically binds the identity of a person with the contents of the message or file, thus providing nonrepudiation—the inability to deny the authenticity of the message or file. The capability for non-repudiation results from encrypting the digest (or the message or file itself) with the private key of the signer. Anyone can verify the signature of the message or file by decrypting the signature using the public key of the sender. As only the sender should know his or her own private key, assurance is provided that the signature is valid and the sender cannot later repudiate the message. If a person divulges his or her private key to any other party, that party can impersonate the person in all electronic transactions.

Digital Date/Time Stamp

A digital date/time stamp is cryptographically based assurance that a message or file was sent or created at a given date and time. Generally, such assurance is provided by an authoritative organization that appends a date/time stamp and digitally signs the message or file.

These cryptographic capabilities can be used in complementary ways. For example, authentication is basic to controlling access to system or intranet resources. For example, a person may use a password to authenticate his own identity; only when the proper password has been entered will the system allow the user to *log on* and obtain access to files, e-mail, and so on.

Note: An example more familiar to many is that the entry of an appropriate personal identification number (PIN) into an automatic teller machine (ATM) gives the ATM user access to account balances or cash.

Passwords, on the other hand, have many limitations as an access control measure (people tell others their passwords or a password is learned via eavesdropping), and cryptographic authentication techniques can provide much better and more effective mechanisms for limiting system or resource access to authorized parties.

Access controls can be applied at many different points within a system. For example, the use of a dial-in port on a communication system or intranet can require the use of cryptographic access controls to ensure that only the proper parties can use the system or intranet. Many systems and intranets accord privileges or access to resources depending on the specific identity of a user. Thus, a hospital communications system may grant physicians access that allows entering orders for patient treatment, whereas laboratory technicians may not have such access. Authentication mechanisms can also be used to generate an audit trail identifying those who have accessed particular data, thus facilitating a search for those known to have compromised confidential data.

In the event that access controls are successfully bypassed, the use of encryption on data stored and communicated in a system provides an extra layer of protection. Specifically, if an intruder is denied easy access to stored files and satellite communications, he or she may well find it much more difficult to understand the internal workings of the system and thus be less capable of causing damage or reading the contents of encrypted inactive data files that may hold sensitive information. Of course, when an application opens a data file for processing, that data is necessarily unencrypted and is vulnerable to an intruder who might be present at that time.

Authentication and access control can also help to protect the privacy of data stored on a satellite communications system or intranet. For example, a particular database application storing data files in a specific format could allow its users to view those files. If the access control mechanisms are set up in such a way that only certain parties can access that particular database application, access to the database files in question can be limited, and thus the privacy of data stored in those databases protected. On the other hand, an unauthorized user may be able to obtain access to those files through a different, uncontrolled application, or even through the operating system itself. Thus, encryption of those files is necessary to protect them against such *back-door* access.

> **Note:** The measure-countermeasure game can continue indefinitely. In response to file encryption, an intruder can insert into an operating system a Trojan horse program that waits for an authorized user to access the encrypted database. Because the user is authorized, the database will allow the decryption of the relevant file and the intruder can simply *piggyback* on that decryption. Thus, those responsible for satellite communications system security must provide a way to check for Trojan horses, and so the battle goes round.

The various encryption capabilities previously described may be used within a satellite communications system in order to accomplish a set of tasks. For example, a banking system may require confidentiality and integrity assurances on its satellite communications links, authentication assurances for all major processing functions, and integrity and authentication assurances for high-value transactions. On the other hand, merchants may need only digital signatures and date/time stamps when dealing with external customers or cooperating banks when establishing contracts. Furthermore, depending on the type of capability to be provided, the underlying cryptographic algorithms may or may not be different.

Finally, when considering what satellite encryption can do, it is worth making two practical observations. First, the initial deployment of any technology often brings out unanticipated problems, simply because the products and artifacts embodying that technology have not had the benefit of successive cycles of failure and repair. Similarly, human procedures and practices have not been tested against the demands of real-life experience. Satellite encryption is unlikely to be any different, and so it is probable that early large-scale deployments of encryption will exhibit exploitable vulnerabilities.

Second, when up against a determined opponent who is highly motivated to gain unauthorized access to data, the use of encryption in satellite communications may well simply lead that opponent to exploit some other vulnerability in the system or intranet on which the relevant data is communicated or stored, and such an exploitation may well be successful. But the use of encryption can help to raise the cost of gaining improper access to data and may prevent a resource-poor opponent from being successful at all.

How Satellite Encryption Fits into the Big Security Picture

In the context of confidentiality, the essence of satellite communication security is a battle between information protectors and information interceptors. Protectors (who may be motivated by *good* reasons if they are legitimate businesses or *bad* reasons if they are criminals) are those who wish to restrict access to information to a group that they select. Interceptors (who may also be motivated by *bad* reasons if they are unethical business competitors or *good* reasons if they are law enforcement agents investigating serious crimes) are those who wish to obtain access to the information being protected whether or not they have the permission of the information protectors. It is this dilemma that is at the heart of the public policy controversy and is addressed in greater detail in Chapter 4, "Need for Access to Satellite Encrypted Information."

From the perspective of the information interceptor, satellite encryption is only one of the problems to be faced. In general, the complexity of today's satellite communication systems poses many technical barriers. On the other hand, the information interceptor may be able to exploit product features or specialized techniques to gain access.

Now, let us address the technical factors that inhibit access to information. But technical measures are only one class of techniques that can be used to improve satellite communication security. For example, statutory measures can help contribute to satellite communications security. Laws that impose criminal penalties for unauthorized access to computer systems have been used to prosecute intruders. Such laws are intended to deter attacks on satellite communication systems, and to the extent that individuals do not exhibit such behavior, system security is enhanced.

Technical Factors Inhibiting Access to Information

Compared to the task of tapping an analog telephone line, obtaining access to the content of a digital information stream can be quite difficult. With analog *listening* (traditional telephony or radio interception), the technical challenge is obtaining access to the satellite communications channel. When satellite communications are digitized, gaining access to the channel is only the first step: One must then unravel the digital format, a task that can be computationally very complex. Furthermore, the complexity of the digital format tends to increases over time because more advanced information technology generally implies increased functionality and a need for more efficient use of available satellite communications capacity.

Increased complexity is reflected, in particular, in the interpretation of the digital stream that two satellite systems might use to communicate with each other or the format of a file that a system might use to store data. Consider, for example, one particular sequence of actions used to communicate information. The original application in the sending system (uplink) might have started with a plaintext message; then a compressed message (to make it smaller); an encrypted message (to conceal its meaning); and, then error-control bits appended to the compressed encrypted message (to prevent errors from creeping in during transmission).

Note: Error control is a technique used both to detect errors in transmission and sometimes to correct them as well.

In addition, a party attempting to intercept a communication between the sender (uplink) and the receiver (downlink) could be faced with a data stream that would represent the combined output of many different operations that transform the data stream in some way. The interceptor would have to know the error-control scheme and the decompression algorithms as well as the key and the algorithm used to encrypt the message.

When an interceptor moves onto the lines that carry bulk traffic, isolating the bits associated with a particular communication of interest is itself quite difficult. A high-bandwidth line (a long-haul fiber-optic cable) typically carries hundreds or thousands of different satellite communications; any given message may be broken into distinct packets and intermingled with other packets from other contemporaneously operating applications. The traffic on the line may be encrypted in *bulk* by the line provider, thus providing an additional layer of protection against the interceptor. Moreover, since a message traveling from point A to point B may well be broken into packets that traverse different physical paths en route, an interceptor at any given point in between A and B may not even see all of the packets pass by.

Another factor inhibiting access to information is the use of technologies that facilitate anonymous satellite communications. For the most part, intercepted satellite communications are worthless if the identity of the communicating parties is not known. In telephony, call forwarding and pager callbacks from pay telephones have sometimes frustrated the efforts of law enforcement officials conducting wiretaps. In data communications, so-called anonymous remailers can strip out all identifying information from an Internet e-mail message sent from person A to person B in such a way that person B does not know the identity of person A. Some remailers even support return satellite communications from person B to person A without the need for person B to know the identity of person A.

Access is made more difficult because an information protector can switch satellite communications from one medium to another very easily without changing end-user equipment. Some forms of media may be easily accessed by an interceptor (conventional radio), whereas other forms may be much more challenging (fiber-optic cable, spread-spectrum radio). The proliferation of different media that can interoperate smoothly even at the device level will continue to complicate the interceptor's attempts to gain access to satellite communications.

Finally, obtaining access also becomes more difficult as the number of service providers increases (see the Sidebar that follows). In the days when AT&T held a monopoly on voice communications and criminal communications could generally be assumed to be carried on AT&T-operated lines, law enforcement and national security authorities needed only one point of contact with whom to work. As the telecommunications industry becomes increasingly heterogeneous, law enforcement authorities may well be uncertain about what company to approach about implementing a wiretap request.

The Evolution of the Telecommunications Industry

Prior to 1984, the U.S. telecommunications industry was dominated by one primary player—AT&T. An elaborate regulatory structure had evolved in the preceding decades to govern what had become an essential national service on which private citizens, government, and business had come to rely.

By contrast, the watchword in telecommunications a mere decade later has become competition. AT&T is still a major player in the field, but the regional Bell operating companies (RBOCs), separated from AT&T as part of the divestiture decision of 1984, operate entirely independently, providing local services. Indeed, the current mood in Congress toward deregulation is already causing increasingly active competition and confrontation among all of the players involved, including cable TV companies, cellular and mobile telephone companies, the long-distance telecommunications companies (AT&T, MCI, Sprint, and hundreds of others), the RBOCs and other local exchange providers, TV and radio broadcast companies, entertainment companies, and, satellite communications companies. Today, all of these players compete for a share of the telecommunications pie in the same geographic area. Even railroads, gas companies (which own geographic rights of way along which transmission lines can be laid), and power companies (which have wires going to every house) have dreams of profiting from the telecommunications boom. The playing field is even further complicated by the fact of reselling—where institutions often buy telecommunications services from *primary* providers in bulk to serve their own needs and resell the excess to other customers. In short, today's telecommunications industry is highly heterogeneous and widely deployed with multiple public and private service providers, and will become more so in the future.[1]

Factors Facilitating Access to Information: System or Product Design

Unauthorized access to protected information can inadvertently be facilitated by product or system features that are intended to provide legitimate access but instead create unintentional loopholes or weaknesses that can be exploited by an interceptor. Such points of access that may be deliberately incorporated into product or system designs include the following:

- ♦ Maintenance and monitoring ports.
- ♦ Master keys.
- ♦ Mechanisms for key escrow or key backup.
- ♦ Weak encryption defaults.

Maintenance, Monitoring Ports, and Master Keys

A port is a point of connection to a given satellite communications system to which another party (another system, an individual) can connect. For example, many telephone switches and computer systems have dial-in ports that are intended to facilitate monitoring and remote maintenance and repair by off-site technicians. Master keys on the other hand are products that can have a single master key that allows its possessor to decrypt all ciphertext produced by the product.

Mechanisms for Key Escrow or Key Backup

A third party, for example, may store an extra copy of a private key or a master key. Under appropriate circumstances, the third party releases the key to the appropriate individual(s), who is (are) then able to decrypt the ciphertext in question. This subject is discussed at length in Chapter 13, "Escrowed Satellite Encryption and Related Issues."

Weak Encryption Defaults

A product capable of providing very strong encryption may be designed in such a way that users invoke those capabilities only infrequently. For example, encryption on a secure telephone may be designed so that the use of encryption depends on the user pressing a button at the start of a telephone call. The requirement to press a button to invoke encryption is an example of a weak default, because the telephone could be designed so that encryption is invoked automatically when a call is initiated. When weak defaults are designed into systems, many users will forget to press the button.

Despite the good reasons for designing systems and products with these various points of access (facilitating remote access through maintenance ports to eliminate travel costs of system engineers), any such point of access can be exploited by unauthorized individuals as well.

Methods Facilitating Access to Information

Surreptitious access to satellite communications can also be gained by methods such as the following:

- ◆ Interception in the ether.
- ◆ Use of pen registers.
- ◆ Wiretapping.
- ◆ Exploitation of related data.
- ◆ Reverse engineering.
- ◆ Cryptanalysis.
- ◆ Product penetration.
- ◆ Monitoring of electronic emissions.
- ◆ Device penetration.
- ◆ Infrastructure penetration.

Interception in the Ether: Many point-to-point satellite communications make use of a wireless (usually radio) link at some point in the process. As it is impossible to ensure that a radio broadcast reaches only its intended receiver(s), communications carried over wireless links — such as those involving cellular telephones and personal pagers — are vulnerable to interception by unauthorized parties.

Use of Pen Registers: Telephone communications involve both the content of a call and call-setup information, such as numbers called, originating number, time and length of call, and so on. Setup information is often easily accessible, some of it even to end users.

Wiretapping: To obtain the contents of a call carried exclusively by nonwireless means, the information carried on a circuit (actually, a replica of the information) is sent to a monitoring station. A call can be wiretapped when an eavesdropper picks up an extension on the same line, hooks up a pair of alligator clips to the right set of terminals, or obtains the cooperation of telephone company officials in monitoring a given call at a chosen location.

Exploitation of Related Data: A great deal of useful information can be obtained by examining in detail a digital stream that is associated with a given communication. For example, people have developed satellite communications protocol analyzers that examine traffic as it flows by a given point for passwords and other sensitive information.

Reverse Engineering: Decompilation or disassembly of software can yield in–depth understanding of how that software works. One implication is that any algorithm built into software cannot be assumed to be secret for very long, as disassembly of the software will inevitably reveal it to a technically trained individual.

Cryptanalysis: Cryptanalysis is the task of recovering the plaintext corresponding to a given ciphertext without knowledge of the decrypting key. Successful cryptanalysis can be the result of inadequately sized keys. A product with encryption capabilities that implements a strong cryptographic algorithm with an inadequately sized key is vulnerable to a *brute-force* attack. The Sidebar that follows provides more detail.

Note: A brute-force attack against an encryption algorithm is a computer-based test of all possible keys for that algorithm undertaken in an effort to discover the key that actually has been used. Hence, the difficulty and time to complete such attacks increase markedly as the key length grows (specifically, the time doubles for every bit added to the key length).

Fundamentals of Cryptographic Strength

Cryptographic strength depends on two factors: the size of the key, and the mathematical structure of the algorithm itself. For well-designed symmetric cryptographic systems, *brute-force* exhaustive search (trying all possible keys with a given decryption algorithm until the (meaningful) plaintext appears) is the best publicly known cryptanalytic method. For such systems the work factor (the time to cryptanalyze) grows exponentially with key size. Hence, with a sufficiently long key, even an eavesdropper with very extensive computing resources would have to take a very long time (longer than the age of the universe) to test all possible combinations. Adding one binary digit (bit) to the length of a key doubles the length of time required to undertake a brute-force attack while adding only a very small increment to the time it takes to encrypt the plaintext.

How long is a *long* key? To decipher by brute force a message encrypted with a 40-bit key requires $2^{\wedge}40$ (approximately $10^{\wedge}12$) tests. If each test takes $10^{\wedge}\text{-}6$ seconds to conduct, 1 million seconds of testing time on a single computer are required to conduct a brute-force attack, or about 11.5 days. A 56-bit key increases this time by a factor of $2^{\wedge}16$, or 65,536; under the same assumptions, a brute-force attack on a message encrypted with a 56-bit key would take over 2000 years.

From the perspective of the interceptor, two important consider-ations mitigate the bleakness of this conclusion. One is that com-puters can be expected to grow more powerful over time. Speed increases in the underlying silicon technology have exhibited a predictable pattern for the past 50 years—where computational speed doubles every 18 months, equivalent to increasing by a factor of 10 every 5 years. Thus, if a single test takes $10^{\wedge}\text{-}6$ seconds today, in 15 years, it can be expected to take $10^{\wedge}\text{-}9$ seconds. Additional speedup is possible using parallel processing. Some supercomputers use tens of thousands of microprocessors in parallel, and cryp-tanalytic problems are particularly well-suited to parallel process-ing. Even 1000 processors working in parallel, each using the underlying silicon technology of 15 years hence, would be able to decrypt a single 56-bit encrypted message in 18 hours rather than the 2000 years required today.

As for the exploitation of alternatives to brute-force search, all known asymmetric (public-key) cryptographic systems allow shor-tcuts to exhaustive search. Because more information is public in such systems, it is also likely that shortcut attacks will exist for any new systems invented. Shortcut attacks also exist for poorly design-ed symmetric systems. Newly developed shortcut attacks constitute unforeseen breakthroughs, and so by their very nature introduce an unpredictable *wild card* into the effort to set a reasonable key size. Because such attacks are applicable primarily to public-key systems, larger key sizes and larger safety margins are needed for such systems than for symmetric cryptographic systems. For example, factoring a 512-bit number by exhaustive search would take $2^{\wedge}256$ tests (since at least one factor must be less than $2^{\wedge}256$). Known shortcut attacks would allow such numbers to be factored in ap-proximately $2^{\wedge}65$ operations—a number on the order of that re-quired to undertake a brute-force exhaustive search of a message encrypted with a 64-bit symmetric cryptographic system. While symmetric 64-bit systems are considered relatively safe, fear of future breakthroughs in cryptanalyzing public-key systems has led many cryptographers to suggest a minimum key size of 1024 bits for public-key systems, thereby providing in key length a factor-of-two safety margin over the safety afforded by 512-bit keys.

Successful cryptanalysis can also be the result of weak encryption algorithms or poorly designed products. Some encryption algorithms and products have weaknesses that, if known to an attacker, require the testing of only a small fraction of the keys that could in principle be the proper key.

Product Penetration: Like weak encryption, certain design choices such as limits on the maximum size of a password, the lack of a reasonable lower bound on the size of a password, or, use of a random-number generator that is not truly random, may lead to a product that presents a work factor for an attacker that is much smaller than the theoretical strength implied by the algorithm it uses. For example, work factor is defined here to mean a measure of the difficulty of undertaking a brute-force test of all possible keys against a given ciphertext (and known algorithm). A 40-bit work factor means that a brute-force attack must test at most 2^{40} keys to be certain that the corresponding plaintext message is retrieved.

Note: The term *work factor* is also used to mean the ratio of work needed for brute-force cryptanalysis of an encrypted message to the work needed to encrypt that message.

Monitoring of Electronic Emissions: Most satellite communications devices emit electromagnetic radiation that is highly correlated with the information carried or displayed on them. For example, the contents of an unshielded computer display or terminal can in principle be read from a distance (estimates range from tens of meters to hundreds of meters) by equipment specially designed to do so. Coined by a U.S. government program, Transient Electro-Magnetic Pulse Emanation Standard (TEMPEST) is the name of a class of techniques to safeguard against monitoring of emissions.

Device Penetration: A software-controlled device can be penetrated in a number of ways. For example, a virus may infect it, thereby making a clandestine change. A message or a file can be sent to an unwary recipient who activates a hidden program when the message is read or the file is opened. Such a program, once active, can record the keystrokes of the person at the keyboard, scan the mass storage media for sensitive data and transmit it, or, make clandestine alterations to stored data.

Infrastructure Penetration: The infrastructure used to carry satellite communications is often based on software-controlled devices such as routers. Router software can be modified as previously described to copy and forward all (or selected) traffic to an unauthorized interceptor.

The last two techniques (Device Penetration and Infrastructure Penetration) can therefore be categorized as invasive, because they alter the operating environment in order to gather or modify information. In an intranet environment, the most common mechanisms of invasive attacks are called viruses and Trojan horses. A virus gains access to a system, hides within that system, and replicates itself to infect other systems. A Trojan horse exploits a weakness from within a system. Either approach can result in intentional or unintentional denial of services for the host system.

> **Note:** On November 2, 1988, Robert T. Morris, Jr. released a *worm* program that spread itself throughout the Internet over the course of the next day. At his trial, Morris maintained that he had not intended to cause the effects that had resulted, a belief held by many in the Internet community. Morris was convicted on a felony count of unauthorized access.

Modern techniques for combining both techniques to covertly exfiltrate data from a system are becoming increasingly powerful and difficult to detect. Such attacks will gain in popularity as intranets become more highly interconnected.

For example, the popular World Wide Web provides an environment in which an intruder can act to steal data. For example, an industrial spy wishing to obtain data stored on the communications intranet of a large aerospace company or even NASA can set up a Web page containing information of interest to engineers at the aerospace company (e.g., information on foreign aerospace business contracts (NASA with the Russia Federation) in the making), thereby making the page an attractive site for those engineers to visit.

Once an engineer from the company has visited the spy's Web page, a channel is set up by which the Web page could send back a Trojan horse (TH) program for execution on the workstation being used to look at the page. The TH could be passed as part of any executable program (Java and Postscript provide two such vehicles) that otherwise did useful things but on the side collected data resident on that workstation (and any other computers to which it might be connected). Once the data were obtained, they could be sent back to the spy's Web page during the same session, or e-mailed back, or sent during the next session that was used to connect to that Web page.

Furthermore, because contacts with a Web page by design provide the specific address from which the contact is coming, the TH could be sent only to the aerospace company or NASA (and to no one else), thus reducing the likelihood that anyone else would stumble upon it. Furthermore, the Web page contact also provides information about the workstation that is making the contact, thus permitting a customized and specially debugged TH to be sent to that workstation.

The Market for Satellite Encryption

Satellite encryption is a product as well as a technology. Products offering cryptographic capabilities can be divided into two general classes: First, security-specific or stand-alone products that are generally add-on items (often hardware, but sometimes software) and often require that users perform an operationally separate action to invoke the encryption capabilities. Examples include an add-on hardware board that encrypts messages or a program that accepts a plaintext file as input and generates a ciphertext file as output.

The second general class are integrated (often *general-purpose*) products in which cryptographic functions have been incorporated into some software or hardware application package as part of its overall functionality. An integrated product is designed to provide a capability that is useful in its own right, as well as encryption capabilities that a user may or may not use. Examples include a modem with on-board encryption or a word processor with an option for protecting (encrypting) files with passwords.

> **Note:** From a system design perspective, it is reasonable to assert that word processing and database applications do not have an intrinsic requirement for encryption capabilities and that such capabilities could be better provided by the operating system on which these applications operate. But as a practical matter, operating systems often do not provide such capabilities and so vendors have significant incentives to provide encryption capabilities that are useful to customers who want better security.

In addition, an integrated product may provide sockets or hooks to user-supplied modules or components that offer additional cryptographic functionality. An example is a software product that can call upon a user-supplied package that performs certain types of file manipulation such as encryption or file compression. Cryptographic sockets are discussed in Chapter 26, "Satellite Encryption Technology Policy Options, Analysis and Forecasts," as cryptographic applications programming interfaces.

A product with cryptographic capabilities can be designed to provide data confidentiality, data integrity, and user authentication in any combination. A given commercial cryptographic product may implement functionality for any or all of these capabilities. For example, a PC-Card may integrate cryptographic functionality for secure authentication and for encryption onto the same piece of hardware, even though the user may choose to invoke these functions independently. A groupware program for remote collaboration may implement cryptography for confidentiality (by encrypting messages sent between users) and cryptography for data integrity and user authentication (by appending a digital signature to all messages sent between users). Furthermore, this program may be implemented in a way that these features can operate independently (either, both, or neither may be operative at the same time).

Because cryptography is usable only when it is incorporated into a product, whether integrated or security-specific, issues of supply and demand affect the use of cryptography. The remainder of this part of the chapter addresses both demand and supply perspectives on the encryption market.

The Demand Side of the Encryption Market

Chapter 2 discussed vulnerabilities that put the information assets of businesses and individuals at risk. Despite the presence of such risks, however, many organizations do not undertake adequate satellite communications security efforts, whether those efforts involve cryptography or any other tool. This part of the chapter explores some of the reasons for this behavior.

Lack of Security Awareness

Most people who use satellite communications behave as though they regard their communications as confidential. Even though they may know in some sense that their communications are vulnerable to compromise, they fail to take precautions to prevent breaches in communications security. Even criminals aware that they may be the subjects of wiretaps have been overheard by law enforcement officials to say, "This call is probably being wiretapped, but . . . ," after which they go on to discuss incriminating topics (remember what happened to Mafia Don, John Gotti).

Note: A case in point is that the officers charged in the Rodney King beating used their electronic communications system as though it were a private telephone line, even though they had been warned that all traffic over that system was recorded. In 1992, Rodney King was beaten by members of the Los Angeles Police Department. A number of transcripts of police radio conversations describing the incident were introduced as evidence at the trial. Had they been fully cognizant at the moment of the fact that all conversations were being recorded as a matter of department policy, the police officers in question most likely would not have said what they did.

The impetus for thinking seriously about security is usually an event that is widely publicized and significant in impact. It is widely believed that only a few percent of computer break-ins are detected.

An example of responding to publicized problems is the recent demand for encryption of cellular telephone communications. In the past several years, the public has been made aware of a number of instances in which traffic carried over cellular telephones was monitored by unauthorized parties. In addition, cellular telephone companies have suffered enormous financial losses as the result of *cloning*: an illegal practice in which the unencrypted ID numbers of cellular telephones are recorded off the air and placed into cloned units, thereby allowing the owner of the cloned unit to masquerade as the legitimate user. Even though many users today are aware of such practices and have altered their behavior somewhat (by avoiding discussion of sensitive information over cellular telephone lines), more secure systems such as the Groupe Special Mobile/Global System(s) for Mobile (GSM) communication (the European standard for mobile telephones) have gained only a minimal foothold in the U.S. market.

A second area in which people have become more sensitive to the need for satellite communications security is in international commerce. Many international business users are concerned that their international business communications are being monitored, and indeed such concerns motivate a considerable amount of today's demand for secure satellite communications.

It is true that the content of the vast majority of telephone communications in the U.S. (making a dinner date, taking an ordinary business call) and data communications (transferring a file from one computer to another, sending an e-mail message) is simply not valuable enough to attract the interest of most eavesdroppers. Moreover, most communications links for point-to-point communications in the U.S. are hard wired (fiber-optic cable) rather than wireless (microwave). Hardwired links are much more secure than wireless links.

Note: A major U.S. manufacturer reported to a congressional committee that in the late 1980s it was alerted by the U.S. government that its microwave communications were vulnerable. In response, this manufacturer took steps to increase the capacity of its terrestrial communication links, thereby reducing its dependence on microwave communications. A similar situation was faced by IBM in the 1970s.

In some instances, compromises of satellite communications security do not directly damage the interests of the persons involved. For example, an individual whose credit card number is improperly used by another party (who may have stolen his wallet or eavesdropped on a conversation) is protected by a legal cap on his liability.

Other Barriers Influencing Demand for Cryptography

Even when a user is aware that satellite communications security is threatened and wishes to take action to forestall the threat, a number of practical considerations can affect the decision to use cryptographic protection. These considerations include the following:

- Lack of critical mass.
- Uncertainties over government policy.
- Lack of a supporting infrastructure.
- High cost.
- Reduced performance.
- A generally insecure environment.
- Usability.
- Lack of independent certification or evaluation of products.
- Electronic commerce.
- Uncertainties arising from intellectual property issues.
- Lack of interoperability and standards.
- The heterogeneity of the satellite communications infrastructure.

Lack of Critical Mass

A secure telephone is not of much use if only one person has it. Ensuring that satellite communications are secure requires collective action—some critical mass of interoperable devices is necessary in order to stimulate demand for secure communications. To date, such a critical mass has not yet been achieved.

Uncertainties over Government Policy

Policy often has an impact on demand. A number of government policy decisions on satellite encryption have introduced uncertainty, fear, and doubt into the marketplace and have made it difficult for potential users to plan for the future. Seeing the controversy surrounding policy in this area, potential vendors are reluctant to bring to market products that support security. And, potential users are reluctant to consider products for security that may become obsolete in the future in an unstable legal and regulatory environment.

Lack of a Supporting Infrastructure

The mere availability of devices is not necessarily sufficient. For some applications such as secure interpersonal communications, a national or international infrastructure for managing and exchanging keys could be necessary. Without such an infrastructure, encryption may remain a niche feature that is usable only through ad hoc methods replicating some of the functions that an infrastructure would provide and for which demand would thus be limited. Later in the chapter, a description of some infrastructure issues are discussed in greater detail.

High Cost

To date, hardware-based cryptographic security has been relatively expensive, in part, because of the high cost of stand-alone products made in relatively small numbers. A user that initially deploys a system without security features and subsequently wants to add them can be faced with a very high cost barrier, and consequently there is a limited market for security add-on products.

On the other hand, because the cost of implementing cryptographic capabilities in software at the outset is rapidly becoming marginal, it is only a minor part of the overall cost. Therefore, cryptographic capabilities are likely to appear in all manner and types of integrated software products where there might be a need.

Reduced Performance

The implementation of cryptographic functions often consumes computational resources (time, memory). In some cases, excessive consumption of resources makes encryption too slow or forces the user to purchase additional memory. If encrypting the satellite communications link over which a conversation is carried delays that conversation by more than a few tenths of a second, users may well choose not to use the encryption capability.

A Generally Insecure Environment

A given intranet or operating system may be so inherently insecure that the addition of cryptographic capabilities would do little to improve overall security. Moreover, retrofitting security measures atop an inherently insecure system is generally difficult.

Usability

A product's usability is a critical factor in its market acceptability. Products with encryption capabilities that are available for use but are in fact unused do not increase satellite communications security. Such products may be purchased but not used for the encryption they provide because such use is too inconvenient in practice; or, they may not be purchased at all because the capabilities they provide are not aligned well with the needs of their users. In general, the need to undertake even a modest amount of extra work or to tolerate even a modest inconvenience for cryptographic protection that is not directly related to the primary function of the device is likely to discourage the use of such protection. When cryptographic features are well integrated in a way that does not demand case-by-case user intervention (when such capabilities can be invoked transparently to the average user), demand may well increase.

Note: For example, experience with current secure telephones such as the Secure Telephone Unit (STU)-III, suggests that users of such phones may be tempted because of the need to contact many people to use them in a nonsecure mode more often than not.

Lack of Independent Certification or Evaluation of Products

Certification of a product's quality is often sought by potential buyers who lack the technical expertise to evaluate product quality or who are trying to support certain required levels of security (as the result of bank regulations). Many potential users are also unable to detect failures in the operation of such products. With one exception discussed in Chapter 14, "Other Dimensions of National Satellite Encryption Policy," independent certification for products with integrated encryption capabilities is not available, leading to market uncertainty about such products.

Note: Even users who do buy security products may still be unsatisfied with them. For example, in two consecutive surveys in 1993 and 1994, a group of users reported spending more and being less satisfied with the security products they were buying.

Electronic Commerce

An environment in which secure communications were an essential requirement would do much to increase the demand for cryptographic security. However, the demand for secure satellite communications is currently nascent.

Note: AT&T plans to take a nontechnological approach to solving some of the security problems associated with retail Internet commerce. AT&T has announced that it will insure its credit-card customers against unauthorized charges, as long as those customers were using AT&T's service to connect to the Internet. This action was taken on the theory that the real issue for consumers is the fear of unauthorized charges, rather than fears that confidential data per se would be compromised.

Uncertainties Arising from Intellectual Property Issues

Many of the algorithms that are useful in encryption (especially public-key cryptography) are protected by patents. Some vendors are confused by the fear, uncertainty, and doubt caused by existing legal arguments among patent holders. Moreover, even when a patent on a particular algorithm is undisputed, many users may resist its use because they do not wish to pay the royalties.

Lack of Interoperability and Standards

For cryptographic devices to be useful, they must be interoperable. In some instances, the implementation of cryptography can affect the compatibility of systems that may have interoperated even though they did not conform strictly to interoperability standards. In other instances, the specific cryptographic algorithm used is yet another function that must be standardized in order for two products to interoperate. Nevertheless, an algorithm is only one piece of a cryptographic device, thus two devices that implement the same cryptographic algorithm may still not interoperate. Only when two devices conform fully to a single interoperability standard (a standard that would specify how keys are to be exchanged, the formatting of the various data streams, the algorithms to be used for encryption and decryption, and so on) can they be expected to interoperate seamlessly.

For instance, consider the Data Encryption Standard (DES) as an example. The DES is a symmetric encryption algorithm (first published in 1975 by the U.S. government) that specifies a unique and well-defined transformation when given a specific 56-bit key and a block of text. However, the various details of operation within which DES is implemented can lead to incompatibilities with other systems that include DES, with stand-alone devices incorporating DES, and even with software-implemented DES.

Specifically, how the information is prepared prior to being encrypted (how it is blocked into chunks) and after the encryption (how the encrypted data is modulated on the communications line), will affect the interoperability of satellite communications devices that may both use DES. In addition, key management may not be identical for DES-based devices developed independently. Many DES-based systems for file encryption generally require a user-generated password to generate the appropriate 56-bit DES key. However, as the DES standard does not specify how this aspect of key management is to be performed, the same password used on two independently developed DES-based systems may not result in the same 56-bit key. For these and similar reasons, independently developed DES-based systems cannot necessarily be expected to interoperate.

Also, an approach gaining favor among product developers is protocol negotiation, which calls for two devices or products to mutually negotiate the protocol that they will use to exchange information. For example, the calling device may query the receiving device to determine the right protocol to use. Such an approach frees a device from having to conform to a single standard and also facilitates the upgrading of standards in a backward-compatible manner.

Note: Transmitting a digital bit stream requires that the hardware carrying that stream be able to interpret it. Interpretation means that regardless of the content of the satellite communications (voice, pictures), the hardware must know what part of the bit stream represents information useful to the ultimate receiver and what part represents information useful to the carrier. A satellite communications protocol is an agreed-upon convention about how to interpret any given bit stream and includes the specification of any encryption algorithm that may be used as part of that protocol.

Heterogeneity of the Satellite Communications Infrastructure

Satellite communications are ubiquitous, but they are implemented through a patchwork of systems and technologies and communications protocols rather than according to a single integrated design. In some instances, they do not conform completely to the standards that would enable full interoperability. In other instances, interoperability is achieved by intermediate conversion from one data format to another. The result can be transmission of encrypted data across interfaces with achieving connectivity among disparate systems. Under these circumstances, users may be faced with a choice of using unencrypted satellite communications or not being able to communicate with a particular other party at all.

Note: In an analogous example, two Internet users may find it very difficult to use e-mail to transport a binary file between them because the e-mail systems on either end may well implement standards for handling binary files differently, even though they may conform to all relevant standards for carrying ASCII text.

Supply Side of the Encryption Market

The supply of products with encryption capabilities is inherently related to the demand for them. Cryptographic products result from decisions made by potential vendors and users as well as standards determined by industry and/or government. Use depends on availability as well as other important factors such as user motivation, relevant learning curves, and other nontechnical issues. As a general rule, the availability of products to users depends on decisions made by vendors to build or not to build them. Further, all of the considerations faced by vendors of all type of products are relevant to products with encryption capabilities. In addition to user demand, vendors need to consider the following issues before deciding to develop and market a product with encryption capabilities:

- ♦ Accessibility of the basic knowledge underlying encryption.
- ♦ The skill to implement basic knowledge of encryption.
- ♦ The skill to integrate the encryption into a usable product.
- ♦ The cost of developing, maintaining, and upgrading an economically viable product with encryption capabilities.
- ♦ The suitability of hardware versus software.
- ♦ Nonmarket considerations and export controls.

Accessing the Basic Knowledge Underlying Encryption

Given that various books, technical articles, and government standards on the subject of encryption have been published widely over the past 20 years, the basic knowledge needed to design and implement cryptographic systems that can frustrate the best attempts of anyone (including government intelligence agencies) to penetrate them is available to government and nongovernment agencies and parties both here and abroad. For example, because a complete description of DES is available worldwide, it is relatively easy for anyone to develop and implement an encryption system that involves multiple uses of DES to achieve much stronger security than that provided by DES alone.

Implementing the Basic Knowledge of Encryption

A product with encryption capabilities involves much more than a cryptographic algorithm. An algorithm must be implemented in a system, and many design decisions affect the quality of a product even if its algorithm is mathematically sound. Indeed, efforts by multiple parties to develop products with encryption capabilities based on the same algorithm could result in a variety of manufactured products with varying levels of quality and resistance to attack.

For example, although cryptographic protocols are not part and parcel of a cryptographic algorithm per se, these protocols specify how critical aspects of a product will operate. Thus, weaknesses in cryptographic protocols—such as a key generation protocol specifying how to generate and exchange a specific encryption key for a given message to be passed between two parties, or a key distribution protocol specifying how keys are to be distributed to users of a given product, can compromise the confidentiality that a real product actually provides, even though the cryptographic algorithm and its implementation are flawless.

> **Note:** An incident that demonstrates the importance of the nonalgorithm aspects of a product is the failure of the key-generation process for the Netscape Navigator Web browser that was discovered in 1995. A faulty random number generation used in the generation of keys would enable an intruder exploiting this flaw to limit a brute-force search to a much smaller number of keys than would generally be required by the 40-bit key length used in this product.

Integrating Encryption into a Usable Product

Even a product that implements a strong cryptographic algorithm in a competent manner is not valuable if the product is unusable in other ways. For integrated products with encryption capabilities, the non-cryptographic functions of the product are central because the primary purpose of an integrated product is to provide some useful capability to the user (word processing, database management, communications) that does not involve encryption per se. If encryption interferes with this primary functionality, it detracts from the product's value.

In this area, U.S. software vendors and system integrators have distinct strengths, even though engineering talent and cryptographic expertise are not limited to the United States. For example, foreign vendors do not market integrated products with encryption capabilities that are sold as mass-market software, whereas many such U.S. products are available.

> **Note:** For example, the Department of Commerce and the NSA found no general-purpose software products with encryption capability from non-U.S. manufacturers.

Costs Of Developing, Maintaining, and Upgrading Encryption

The technical aspects of good encryption are increasingly well understood. As a result, the incremental cost of designing a software product so that it can provide cryptographic functionality to end users is relatively small. As cost barriers to the inclusion of cryptographic functionality are reduced dramatically, the long-term likelihood increases that most products that process digital information will include some kinds of cryptographic functionality.

Suitability of Hardware Versus Software

The suitability of hardware versus software as a medium in which to implement a product with encryption capabilities is very important here. The duplication and distribution costs for software are very low compared to those for hardware, and yet, trade secrets embedded in proprietary hardware are easier to keep than those included in software. Moreover, software cryptographic functions are more easily disabled.

Nonmarket Considerations and Export Controls

Vendors may withhold or alter their products at government request. For example, a well-documented instance is the fact that AT&T voluntarily deferred the introduction of its 3600 Secure Telephone Unit (STU) at the behest of government (see Chapter 14, "Other Dimensions Of National Satellite Encryption Policy" on government influence). Export controls also affect decisions to make products available even for domestic use, as described in Chapter 5, "Export Controls."

Infrastructure for Widespread Use of Satellite Encryption

The widespread use of satellite encryption requires a support infrastructure that can service organizational or individual user needs with regard to cryptographic keys. Let us look at the key management infrastructure first.

Key Management Infrastructure

In general, to enable use of satellite encryption across an enterprise, there must be a mechanism that:

♦ Periodically supplies all participating locations with keys (typically designated for use during a given calendar or time period—the cryptoperiod) for either stored materials or satellite communications.
♦ Permits any given location to generate keys for itself as needed (to protect stored files).
♦ Can securely generate and transmit keys among communicating parties (for data transmissions, telephone conversations).[3]

In the most general case, any given location will have to perform all three functions. With symmetric systems, the movement of keys from place to place obviously must be done securely and with a level of protection adequate to counter the threats of concern to the using parties. Whatever the distribution system, it clearly must protect the keys with appropriate safeguards and must be prepared to identify and authenticate the source. The overall task of securely assuring availability of keys for symmetric applications is often called key management.

If all secure satellite communications take place within the same corporation or among locations under a common line of authority, key management is an internal or possibly a joint obligation. For parties that communicate occasionally or across organizational boundaries, mutual arrangements must be formulated for managing keys. One possibility might be a separate trusted entity whose line of business could be to supply keys of specified length and format, on demand and for a fee.

With asymmetric systems, the private keys are usually self-generated, but they may also be generated from a central source, such as a corporate security office. In all cases, however, the handling of private keys is the same for symmetric and asymmetric systems. They must be guarded with the highest levels of security. Although public keys need not be kept secret, their integrity and association with a given user are extremely important and should also be supported with extremely robust measures.

The costs of a key management infrastructure for national use are not known at this time. One benchmark figure is that the cost of the Defense Department infrastructure needed to generate and distribute keys for approximately 430,000 STU-III telephone users is somewhere in the range of $20 million–$24 million per year.

Certificate Infrastructures

The association between key information (such as the name of a person and the related public key) and an individual or organization is an extremely important aspect of a cryptographic system. That is, it is undesirable for one person to be able to impersonate another. To guard against impersonation, two general types of solutions have emerged: an organization-centric approach consisting of certificate authorities and a user-centric approach consisting of a web of trust.

A certificate authority serves to validate information that is associated with a known individual or organization. Certificate authorities can exist within a single organization, across multiple related organizations, or across society in general. Any number of certificate authorities can coexist. They may or may not have agreements for crosscertification—whereby if one authority certifies a given person, then another authority will accept that certification within its own structure. Certificate authority hierarchies are defined in the Internet RFCs 1421-1424, the X.509 standard, and other emerging commercial standards, such as that proposed by MasterCard/Visa. A number of private certificate authorities, such as VeriSign, have also begun operation to service secure massmarket software products, such as the Netscape Navigator Web browser.

Among personal acquaintances, validation of public keys can be passed along from person to person or organization to organization, thus creating a web of trust in which the entire ensemble is considered to be trusted based on many individual instances of trust. Such a chain of trust can be established between immediate parties, or from one party to a second to establish the credentials of a third. This approach has been made popular by the Pretty Good Privacy (PGP) encryption software product. All users maintain their own *key-ring*, which holds the public keys of everyone with whom they want to communicate.

Importantly, it should be noted that both the certificate authority approach and the web of trust approach replicate the pattern of trust that already exists among participating parties in societal and business activities. In a sense, the certificate infrastructure for satellite encryption simply formalizes and makes explicit that to which society and its institutions are already accustomed.

At some point, banks, corporations, and other organizations already generally trusted by society will start to issue certificates. At that time, individuals especially, may begin to feel more comfortable about the cryptographic undergirding of society's electronic infrastructure — at which point the webs of trust can be expected to evolve according to individual choices and market forces. However, it should be noted that different certificates will be used for different functions, and it is unlikely that a single universal certificate infrastructure will satisfy all societal and business needs. For example, because an infrastructure designed to support electronic commerce and banking may do no more than identify valid purchasers, it may not be useful for providing interpersonal communication or corporate access control.

Certificate authorities already exist within some businesses, especially those that have moved vigorously into an electronic way of life. Generally, there is no sense of a need for a legal framework to establish relationships among organizations, each of which operates its own certificate function. Arrangements exist for them to cross-certify one another. In general, the individual(s) authorizing the arrangement will be a senior officer of the corporation, and the decision will be based on the existence of other legal agreements already in place — notably, contracts that define the relationships and obligation among organizations.

For the general business world, where any individual or organization wishes to conduct a transaction with any other individual or organization (such as the sale of a house), a formal certificate infrastructure has yet to be created. There is not even one to support just a digital signature application within government. Hence, it remains to be seen how, in the general case, individuals and organizations will make the transition to an electronic society.

Certificate authorities currently operate within the framework of contractual law. That is, if some problem arises as the result of improper actions on the part of the certification authority, its subscribers would have to pursue a civil complaint. As certificate authorities grow in size and service a greater part of society, it will probably be necessary to regulate their actions under law, much like those of any major societal institutions. It is interesting to observe that the legal and operational environment that will have to exist for certificate organizations involves the same set of issues that are pertinent to escrow organizations (as discussed in Chapter 13, "Escrowed Satellite Encryption and Related Issues").

Summary

Satellite encryption provides important capabilities that can help deal with the vulnerabilities of electronic information. Encryption can help to assure the integrity of data, authenticate the identity of specific parties, prevent individuals from plausibly denying that they have signed something, and preserve the confidentiality of information that may have improperly come into the possession of unauthorized parties. At the same time, encryption is not a silver bullet, and many technical and human factors other than encryption can improve or detract from satellite communications security. In order to preserve satellite communications security, attention must be given to all of these factors. Moreover, people can use encryption only to the extent that it is incorporated into real products and systems; unimplemented cryptographic algorithms cannot contribute to satellite communications security. Many factors other than raw mathematical knowledge contribute to the supply of and demand for products with cryptographic functionality. Most importantly, the following aspects influence the demand for cryptographic functions in products:

- Critical mass in the marketplace.
- Government policy.
- Supporting infrastructure.
- Cost.
- Performance.
- Overall security environment.
- Usability.
- Quality certification and evaluation.
- Interoperability standards.[4]

Finally, any large-scale use of satellite encryption, with or without key escrow (discussed in Chapter 13, "Escrowed Satellite Encryption and Related Issues"), depends on the existence of a substantial supporting infrastructure—the deployment of which raises a different set of problems and issues.

From Here

This chapter described how cryptography may help to address information vulnerability issues. It examined possible roles for satellite encryption in reducing information vulnerability and placed cryptography into context as one element of an overall approach to ensure satellite communications security. This chapter also covered a simple distributed encrypted infrastructure, the international market for satellite encryption software, legal protection for satellite encrypted services in the internal market, and a European initiative in electronic commerce.

Chapter 4 discusses nongovernment needs for access to encrypted satellite information and related public policy issues, specifically those related to information gathering for law enforcement and national security purposes. The chapter also discusses the right to satellite encryption by the financial community, international law firms, international management consulting firms and CEOs and Corporate Senior Managers.

Endnotes

1 © 1996 by the National Academy of Sciences. Courtesy of the National Academy Press, 2101 Constitution Avenue NW, Washington D.C. 20418.

2 Ibid.

3 "A Study of the International Market for Computer Software with Encryption," Prepared by the U.S. Department of Commerce and the National Security Agency for the Interagency Working Group on Encryption and Telecommunications Policy, Washington, DC, 1996, pp. 1–5.

4 "Legal Protection for Encrypted Services in the Internal Market," European Union (EU) Commission, Research Institute for Symbolic Computation (RISC-Linz), Johannes Kepler University in Linz, Austria, 1996, pp. 1–44.

5 © 1996 by the National Academy of Sciences. Courtesy of the National Academy Press, 2101 Constitution Avenue NW, Washington D.C. 20418.

4

Need for Access to Satellite Encrypted Information

Information protected for confidentiality (satellite encrypted information) is stored or communicated for later use by certain parties with the authorization of the original protector. However, it may happen for various legitimate and lawfully authorized reasons that other parties may need to recover this information as well. This chapter discusses the need for access to encrypted information under exceptional circumstances for legitimate and lawfully authorized purposes from the perspectives of businesses, individuals, law enforcement, and national security. Businesses and individuals may want access to encrypted data or communications for their own purposes, and thus may cooperate in using products to facilitate such access, while law enforcement and national security authorities may want access to the encrypted data or communications of criminals and parties hostile to the United States.

Terminology

It is useful to conceptualize satellite communications and data storage using the language of transactions. For example, one individual may telephone another; the participants in the transaction are usually referred to as the calling party and the called party. Or, a person makes a purchase; the participants are called the buyer and seller. Or, a sender mails something to the recipient. Adopting this construct, consider communications in which the first party (Party A) sends a message (uplink) and the second party (Party B) receives it. *Party* does not necessarily imply a person; a *party* can be a computer system, a satellite communication system, or a software process. In the case of data storage, Party A stores the data, while Party B retrieves it.

> **Note:** Party A and Party B can be the same party (as is the case when an individual stores a file for his or her own later use).

Under some circumstances, a third party may be authorized for access to data stored or being communicated. For example, law enforcement authorities may be granted legal authorization to obtain redundant access to a telephone conversation or a stored data file or record without the knowledge of Parties A or B. The employer of Party A may have the legal right to read all data files for which Party A is responsible or to monitor all communications in which Party A participates. Party A might inadvertently lose access to a data file and wish to recover that access.

In cases when the data involved is unencrypted, the procedures needed to obtain access can be as simple as identifying the relevant file name or as complex as seeking a court order for legal authorization. But when the data involved is encrypted, the procedures needed to obtain access will require the possession of certain critical pieces of information, such as the relevant cryptographic keys.

Third-party access has many twists and turns. When it is necessary for clarity of exposition or meaning, this chapter uses the phrase *exceptional access* to stress that the situation is not one that was included within the intended bounds of the original transaction, but is an unusual subsequent event. Exceptional access refers to situations in which an authorized party needs and can obtain the plaintext of encrypted data (for storage or satellite communications). The word *exceptional* is used in contrast to the word *routine* and connotes something unusual about the circumstances under which access is required. Exceptional access can be divided into three generic categories:

- Government exceptional access.
- Employer (or corporate) exceptional access.
- End-user exceptional access.

Exceptional Government Access

Exceptional government access refers to the case in which government has a need for access to information under specific circumstances authorized by law. For example, a person might store data files that law enforcement authorities need to prosecute or investigate a crime. Alternatively, two people may be communicating with each other in the planning or commission of a serious crime. Government exceptional access thus refers to the government's need to obtain the relevant information under circumstances authorized by law, and requires a court order (for access to voice or data communications) or a subpoena or search warrant (for access to stored records). Government exceptional access is discussed in greater detail later in the chapter.

Exceptional Employer Access

Employer (or corporate) exceptional access refers to the case in which an employer (the corporate employer) has the legal right to access to information encrypted by an employee. If an employee who has encrypted a file is indisposed on a certain day, for example, the company may need exceptional access to the contents of the file. Alternatively, an employee may engage in communications whose content the company may have a legitimate need to know (the employee may be leaking proprietary information). Employer exceptional access would then refer to the company's requirement to obtain the key necessary to obtain the contents of the file or communications, and may require the intervention of another institutional entity. Employer or corporate exceptional access is covered further later in the chapter.

Exceptional End-User Access

End-user exceptional access refers to the case in which the parties primarily intended to have access to plaintext have lost the means to obtain such access. For example, a single user may have stored a file for later retrieval, but encrypted it to ensure that no other party would have access to it while it was in storage. However, the user might also lose or forget the key used to encrypt that file. End-user exceptional access refers to such a user's requirement to obtain the proper key, and may require that the individual who has lost a key prove his or her identify to a party holding the backup key and verify his or her authorization to obtain a duplicate copy of his or her key. End-user exceptional access is also discussed later in the chapter.

The need for exceptional access when the information stored or communicated is encrypted, has led to an examination of a concept generically known as escrowed satellite encryption (the subject of Chapter 13, "Escrowed Satellite Encryption and Related Issues"). Escrowed satellite encryption, loosely speaking, uses agents other than the parties participating in the communication or data storage to hold copies of or otherwise have access to relevant cryptographic keys *in escrow* so that needs for end-user, corporate and government exceptional access can be met. These agents are called escrow agents.

Law Enforcement: Investigation and Prosecution

Obtaining information (both evidence and intelligence) has always been a central element in the conduct of law enforcement investigations and prosecutions. Accordingly, criminals have always wished to protect the information relevant to their activities from law enforcement authorities.

Value of Access to Information for Law Enforcement

Many criminals keep records related to their activities. Such records can be critical to the investigation and prosecution of criminal activity. For example, criminals engaged in white-collar crimes such as fraud often leave paper trails that detail fraudulent activities. Drug dealers often keep accounting records of clients, drop-offs, supplies, and income. Reconstruction of these paper trails is often a critical element in building a case against these individuals. The search-and-seizure authority of law enforcement to obtain paper records is used in a large fraction of criminal cases. Law enforcement officials believe that wiretapping is a crucial source for information that could not be obtained in any other way or obtained only at high risk (see the Sidebar that follows).

Examples of the Utility of Wiretapping

The El Rukn Gang in Chicago, acting as a surrogate for the Libyan government and in support of terrorism, planned to shoot down a commercial airliner within the United States using a stolen military weapon. This act of terrorism was prevented through the use of telephone wiretaps.

The 1988 *Ill Wind* public corruption and Defense Department fraud investigation relied heavily on court-ordered telephone wiretaps. To date, this investigation has resulted in the conviction of 76 individuals and more than a quarter of a billion dollars in fines, restitutions, and recoveries.

Numerous drug trafficking and money laundering investigations, such as the *Polar Cap* and *Pizza Connection* cases, utilized extensive telephone wiretaps in the successful prosecution of large-scale national and international drug trafficking organizations. *Polar Cap* resulted in the arrest of 44 subjects and the recovery of $60 million in assets. Additionally, in a 1992 Miami raid, which directly resulted from wiretaps, agents confiscated 26,000 pounds of cocaine and arrested 33 subjects.

The investigation of convicted spy Aldrich Ames relied heavily on wiretaps ordered under the Foreign Intelligence Surveillance Act (FISA) authority.

In a 1990 *Sexual Exploitation of Children* investigation, the FBI relied heavily on wiretaps to prevent violent individuals from abducting, torturing, and murdering a child in order to make a *Snuff Murder* film.[1]

For instance, the FBI has testified that without law enforcement's ability to effectively execute court orders for electronic surveillance, the country would be unable to protect itself against foreign threats, terrorism, espionage, violent crime, drug trafficking, kidnapping, and other crimes. According to the FBI, they may be unable to intercept a terrorist before he or she sets off a devastating bomb, thwart a foreign spy before he or she can steal secrets that endanger the entire country, and arrest drug traffickers smuggling in huge amounts of drugs that will cause widespread violence and death. Court-approved electronic surveillance is of immense value, and often is the only way to prevent or solve the most serious crimes facing today's society. Alleged criminals often discuss their past criminal activity and plans for future criminal activity with other parties. Obtaining *inside information* on such activities is often a central element of building a case against alleged perpetrators. A defendant who describes in his or her own words how he or she committed a crime or the extent to which he or she was involved in it gives prosecutors a powerful weapon that juries tend to perceive as fair.

Other methods of obtaining *inside information* have significant risks associated with them. For example, informants are often used to provide inside information. However, the credibility of informants is often challenged in court, either because the informants have shady records themselves or because they may have made a deal with prosecutors by agreeing to serve as informants in return for more lenient treatment. By contrast, challenges to evidence obtained through wiretaps are far more frequently based on their admissibility in court rather than their intrinsic credibility. Informants may also be more difficult to find when a criminal group is small in size.

Surreptitiously planted listening devices are also used to obtain inside information. However, they generally obtain only one side of a conversation (use of a speaker-phone presents an exception). Further, because listening devices require the use of an agent to plant them, installation of such devices is both highly intrusive (arguably more so than wiretapping) for the subject of the device and risky for the planting agent. Requests for the use of such devices are subject to the same judicial oversight and review as wiretaps.

This discussion is not intended to suggest that wiretaps are a perfect source of information and always useful to law enforcement. An important difficulty in using wiretaps is that context is often difficult for listeners to establish when they are monitoring a telephone conversation that assumes shared knowledge between the communicators.

> **Note:** Indeed, in some instances, wiretap evidence has been used to exculpate defendants. There are many difficulties in ascribing meaning to particular utterances that may be captured on tape recordings of conversations.

Because of the legal framework regulating wiretaps, and the fact that communications are by definition transient whereas records endure, wiretapping is used in far fewer criminal cases than is seizure of records. Although the potential problems of denying law enforcement access to communications have been the focus of most of the public debate, encryption of data files in a way that denies law enforcement authorities access to data files relevant to criminal activity arguably presents a much larger threat to their capabilities.

Legal Framework Governing Surveillance

An evolving legal framework governs the authority of government authorities to undertake surveillance of satellite communications that take place within the United States or that involve U.S. persons. Surveillance within the U.S. is authorized only for certain legislatively specified purposes: the enforcement of certain criminal statutes and the collection of foreign intelligence.

Domestic Satellite Communications Surveillance for Domestic Law Enforcement Purposes

Satellite communications surveillance can involve surveillance for traffic analysis and/or content. These separate activities are governed by different laws and regulations. Traffic analysis, a technique that establishes patterns of connections and satellite communications, is performed with the aid of pen registers that record the numbers dialed from a target telephone and trap-and-trace devices that identify the numbers of telephones from which calls are placed to the target telephone. Orders for the use of these devices may be requested by any federal attorney and granted by any federal district judge or magistrate, and are granted on a more or less pro forma basis.

Surveillance of satellite communications for content for purposes of domestic law enforcement is governed by Title 18, United States Code, Sections 2510-2521 concerning *wire and satellite communications interceptions and interception of all communications*, generally known as Title III. These sections of the U.S. code govern the use of listening devices (usually known as *bugs*); wiretaps of communications involving human speech (called *oral communications* in Title III) carried over a wire or wire-like cable, including optical fiber; and, other forms of satellite transmitted communication, including various forms of data, text, and video that may be communicated between or among people as well as computers or satellite communications devices. Under Title III, only certain federal crimes may be investigated (murder, kidnapping, child molestation, racketeering, narcotics offenses) through the interception of oral communications. In addition, 38 states have passed laws that are similar to Title III, but they include such additional restrictions as allowing only a fixed number of interceptions per year (Connecticut) or only for drug-related crimes (California). State wiretaps account for the majority of wiretaps in the United States.

Surveillance of oral communications governed under Title III generally requires a court order (a warrant) granted at the discretion of a judge. Because electronic surveillance of oral communications is both inherently intrusive and clandestine, the standards for granting a warrant for such surveillance are more stringent than those required by the Fourth Amendment. These additional requirements are specified in Title III. They are enforced by criminal and civil penalties applicable to law enforcement officials or private citizens. They are also enforced by a statutory exclusionary rule that states *that violations of the central features of requirements may lead to suppression of evidence in a later trial, even if such evidence meets the relevant Fourth Amendment test.*

Note: Emergency intercepts may be performed without a warrant in certain circumstances, such as physical danger to a person or conspiracy against the national security. There has been *virtually no use* of the emergency provision, and its constitutionality has not been tested in court.

Because of the resources required, the administrative requirements for the application procedure, and the legal requirement that investigators exhaust other means of obtaining information, wiretaps are not often used. Approximately 2000 orders (both federal and state) are authorized yearly (a small number compared to the number of felonies investigated, even if such felonies are limited to those specified in Title III as eligible for investigation with wiretaps). About 2500 conversations are intercepted per order, and the total number of conversations intercepted is a very small fraction of the annual telephone traffic in the United States.

> **Note:** Some analysts critical of the U.S. government position on wiretaps have suggested that the actual distribution of crimes investigated under Title III intercept or surveillance orders may be skewed as a result of somewhat inconsistent government claims as to which potential crimes are more important. For example, no cases involving arson, explosives, or weapons were investigated using Title III wiretaps in 1988. The majority of Title III orders have involved drug and gambling crimes.

Surveillance of nonvoice communications, including fax and satellite communications, is also governed by Title III. The standard for obtaining an intercept order for electronic communications is less stringent than that for intercepting voice communications. For example, any federal felony may be investigated through electronic interception. In addition, the statutory exclusionary rule of Title III for oral and wire communications does not apply to satellite communications.

> **Note:** When there is no reasonable expectation of privacy, law enforcement officials are not required to undertake any special procedure to monitor such communications. For example, a law enforcement official participating in an on line *chat* group is not required to identify himself as such, nor must he or she obtain any special permission at all to monitor the traffic in question. However, as a matter of policy, the FBI does not systematically monitor electronic forums such as Internet relay chats.

Despite the legal framework previously outlined, it is nevertheless possible that unauthorized or unlawful surveillance, whether undertaken by rogue law enforcement officials or overzealous private investigators, also occurs. Concerns over such activity are often expressed by critics of the current administration policy, and they focus on two scenarios:

First of all, with current telephone technology, it is sometimes technically possible for individuals (private investigators, criminals, rogue law enforcement personnel) to undertake wiretaps on their own initiative (e.g., by placing alligator clips on the proper terminals in the telephone box of an apartment building). Such wiretaps would subject the personnel involved to Title III criminal penalties, but detection of such wiretaps might well be difficult. On the other hand, it is highly unlikely that such a person could obtain the cooperation of major telephone service providers without a valid warrant or court order, and so these wiretaps would have to be conducted relatively close to the target's telephone, and not in a telephone switching office.

Second, information obtained through a wiretap in violation of Title III can be suppressed in court, but such evidence may still be useful in the course of an investigation. Specifically, such evidence may cue investigators regarding specific areas that would be particularly fruitful to investigate. And if the illegal wiretap is never discovered, a wiretap that provides no court-admissible evidence may still prove pivotal to an investigation. Even if it is discovered, different judges enforce the *fruit of the poisons tree* doctrine with different amounts of rigor.

Note: Such concerns are raised by reports of police misconduct as described in Chapter 2.

The extent to which these and similar scenarios actually occur is hard to determine. Information provided by the FBI to the Congressional committee indicates a total of 298 incidents of various types (including indictment/complaints and convictions/pretrial diversions) involving charges of illegal electronic surveillance (whether subsequently confirmed or not) over the past six fiscal years (1993 through 1998).

For instance, the U.S. congressional committee on privacy and technology recognizes the existence of controversy over the question of whether such reports should be taken at face value. For example, critics of the U.S. government who believe that law enforcement authorities are capable of systematically abusing wiretap authority argue that law enforcement authorities would not be expected to report figures that reflected such abuse. Alternatively, it is also possible that cases of improper wiretaps are in fact more numerous than reported and have simply not come to the attention of the relevant authorities.

The congressional committee discussed such matters and concluded that it had no reason to believe that the information it received on this subject from law enforcement authorities was in any way misleading.

Domestic Satellite Communications Surveillance for Foreign Intelligence Purposes

The statute governing interception of satellite communications for purposes of protecting national security is known as the Foreign Intelligence Surveillance Act (FISA), which has been codified as Sections 1801 to 1811 in Title 18 of the U.S. Code. Passed in 1978, FISA was an attempt to balance Fourth Amendment rights against the constitutional responsibility of the executive branch to maintain national security. The FISA is relevant only to satellite communications occurring at least partly within the United States (wholly, in the case of radio and satellite communications), although listening stations used by investigating officers may be located elsewhere. And FISA surveillance may be performed only against foreign powers or their agents. Interception of communications, when the communications occur entirely outside the United States (whether or not the participants include U.S. persons), is not governed by FISA, Title III, or any other statute. However, when a U.S. person is outside the United States, Executive Order 12333 governs any satellite communications intercepts targeted against such individuals. The basic framework of FISA is similar to that of Title III, with certain important differences, among which are the following:

First, the purpose of FISA surveillance is to obtain foreign intelligence information, defined in terms of U.S. national security, including defense against attack, sabotage, terrorism, and clandestine intelligence activities, among others. The targeted satellite communications need not relate to any crime or be relevant as evidence in court proceedings.

Second of all, in most instances, a FISA surveillance application requires a warrant based on probable cause that foreign intelligence information will be collected. Surveillance of a U.S. person (defined as a U.S. citizen, U.S. corporation or association, or legal resident alien) also requires probable cause showing that the person is acting as a foreign agent. Political and other activities protected by the First Amendment may not serve as the basis for treating a U.S. person as a foreign agent.

Note: Surveillance may take place without a court order for up to 1 year if the U.S. Attorney General certifies that there is very little likelihood of intercepting satellite communications involving U.S. persons and that the effort will target facilities used exclusively by foreign powers. Under limited circumstances, emergency surveillance may be performed before a warrant is obtained.

Third, targets of FISA surveillance might never be notified that satellite communications have been intercepted. Since 1979, there have been an average of over 600 FISA orders per year. In 1992, 595 were issued. Other information about FISA intercepts is classified.

The Nature of Surveillance Needs of Law Enforcement

In cooperation with the National Technical Investigators Association, the FBI has articulated a set of requirements for its electronic surveillance needs (see Sidebar, "Law Enforcement Requirements for the Surveillance of Satellite Communications" that follows). Of course, access to surveillance that does not meet all of these requirements is not necessarily useless. For example, surveillance that does not meet the transparency requirement may still be quite useful in certain cases (if the subjects rationalize the lack of transparency as *static on the line*). The basic point is that these requirements constitute a set of continuous metrical structures by which the quality of a surveillance capability can be assessed, rather than a list that defines what is or is not useful surveillance. Of these requirements, the real-time requirement is perhaps the most demanding. The FBI has noted that some satellite encryption products put at risk efforts by federal, state and local law enforcement agencies to obtain the contents of intercepted communications by precluding real-time decryption. Real-time decryption is often essential so that law enforcement can rapidly respond to criminal activity and, in many instances, prevent serious and life-threatening criminal acts.

Note: The transparency requirement is part of the Law Enforcement Requirements for the Surveillance of Electronic Communications. It is the transparent access to the communications that is undetectable to all parties to the communications (except to the monitoring parties) and implementation of safeguards to restrict access to intercept information.

An illustrative example is an instance in which the FBI was wiretapping police officers who were allegedly guarding a drug shipment. During that time, the FBI overheard a conversation between the police chief and several other police officials that the FBI believes indicated a plot to murder a certain individual who had previously filed a police brutality complaint against the chief. However, the FBI was unable to decode the police chief's *street slang and police jargon* in time to prevent the murder.

For instance, the statement of FBI Special Agent in Charge, Special Operations Division, New York Field Division, Federal Bureau of Investigation on *Security Issues in Computers and Communications*, before the Subcommittee on Technology, Environment, and Aviation of the Committee on Science, Space, and Technology, U.S. House of Representatives, May 3, 1994:

Law Enforcement Requirements for the Surveillance of Satellite Communications

- ◆ Prompt and expeditious access both to the contents of the electronic communications and *setup* information necessary to identify the calling and called parties.
- ◆ Real-time, full-time monitoring capability for intercepts. Such capability is particularly important in an operational context, in which conversations among either criminal conspirators (regarding a decision to take some terrorist action) or criminals and innocent third parties (regarding a purchase order for explosives from a legitimate dealer) may have immediate significance.
- ◆ Delivery of intercepted satellite communications to specified monitoring facilities.
- ◆ Transparent access to the communications (access that is undetectable to all parties to the communications except to the monitoring parties) and implementation of safeguards to restrict access to intercept information.
- ◆ Verification that the intercepted satellite communications are associated with the intercept subject.
- ◆ Capabilities for some number of simultaneous intercepts to be determined through a cooperative industry/law enforcement effort.

♦ Reliability of the services supporting the intercept at the same (or higher) level as that of the reliability of the satellite communication services provided to the intercept subject.

♦ A quality of service for the intercept that complies with the performance standards of the service providers.[2]

Real-time surveillance is generally less important for crimes that are prosecuted or investigated than for crimes that are prevented because of the time scales involved. Prosecutions and investigations take place on the time scales of days or more, whereas prevention may take place on the time scale of hours. In some instances, the longer time scale is relevant: Because Title III warrants can be issued only when *probable cause* exists that a crime has been committed, the actual criminal act is committed before the warrant is issued, and thus prevention is no longer an issue. In other instances, information obtained under a valid Title III warrant issued to investigate a specific criminal act can be used to prevent a subsequent criminal act, in which case the shorter time scale may be relevant. The situation is similar under FISA, in which warrants need not necessarily be obtained in connection with any criminal activity. A good example is terrorism cases, in which it is quite possible that real-time surveillance could provide actionable information useful in thwarting an imminent terrorist act.

Impact of Satellite Encryption and New Media on Law Enforcement

Satellite encryption can affect information collection by law enforcement officials in a number of ways. However, for perspective, it is important to keep in mind a broader context—namely that advanced information technologies (of which satellite encryption is only one element) have potential impacts across many different dimensions of law enforcement. The Sidebar that follows provides some discussion of this point.

How Noncryptography Applications of Information Technology Could Benefit Law Enforcement

As acknowledged elsewhere in this book, ubiquitous satellite encryption would create certain difficulties for law enforcement. Nevertheless, it is important to place into context the overall impact on law enforcement of the digital information technologies that enable satellite encryption and other capabilities that are not the primary subject of this book.

Chapter 3 suggested how satellite encryption capabilities can be a positive force for more effective law enforcement (secure police satellite communications). But information technology is increasingly ubiquitous and could appear in a variety of other applications less obvious than satellite encryption. For example:

Video technology has become increasingly inexpensive. Thus, it is easy to imagine police cruisers with video cameras that are activated upon request when police are responding to an emergency call. Monitoring those cameras at police headquarters would provide a method for obtaining timely information regarding the need of the responding officers for backup. Equipping individual police officers with even smaller video cameras attached to their uniforms and recording such transmissions would provide objective evidence to corroborate (or refute) an officer's description of what he or she saw at a crime scene.

The number of users of cellular telephones and wide-area wireless communications services will grow rapidly. As such technologies enable private citizens to act as responsible eyes and ears that observe and report emergencies in progress, law enforcement officials will be able to respond more quickly.

Electronically mediated sting operations help to protect the covers of law enforcement officials. For example, the Cybersnare sting operation resulted in the arrest of six individuals who allegedly stole cellular telephone numbers en masse from major companies, resulting in millions of dollars of industry losses. Cybersnare was based on an underground bulletin board that appealed to cellular telephone and credit card thieves. Messages were posted offering for sale cellular telephone *cloning* equipment, stolen cellular telephone numbers, and included contact telephone numbers that were traced to the individuals in question.

The locations of automobiles over a metropolitan area could be tracked automatically, either passively or actively. An active technique might rely on a coded beacon that would localize the position of the automobile on which it was mounted. A passive technique might rely on automatic scanning for license plates that were mounted on the roofs of cars. As an investigative technique, the ability to track the location of a particular automobile over a period of time could be particularly important.

Even today, information technology enables law enforcement officials to conduct instant background checks for handgun purchases and arrest records when a person is stopped for a traffic violation. Retail merchants guard against fraud by using information technology to check driving records when cars are rented and make credit checks for big purchases. The Department of the Treasury uses sophisticated information technology to detect suspicious patterns that might indicate large-scale money laundering by organized crime.

All such possibilities involve important social as well as technical issues. For example, the first two examples featured here seem relatively benign, while the last two raise serious entrapment and privacy issues. Even the *instant background checks* of gun buyers have generated controversy. The mention of these applications (potential and actual) is not meant as endorsement, recommendation, or even suggestion. They do, however, place into better context the potential of information technology in some overall sense to improve the capabilities of law enforcement while at the same time illustrate that concerns about excessive government power are not limited to the issue of cryptography.[3]

Encrypted Satellite Communications

As far as the U.S. congressional committee of privacy and technology has been able to determine, criminal use of digitally encrypted voice satellite communications has not presented a significant problem to law enforcement to date. On rare occasions, law enforcement officials conducting a wiretap have encountered *unknown signals* that could be encrypted satellite traffic or simply a data stream that was unrecognizable to the intercept equipment. For example, a high-speed fax transmission might be transported on a particular circuit. A monitoring agent might be unable to distinguish between the signal of the fax and an encrypted voice signal with the equipment available to him.

Note: In this regard, it is important to distinguish between *voice scramblers* and encrypted voice satellite communications. Voice scramblers are a relatively old and widely available technology for concealing the contents of a voice communication; they transform the analog waveform of a voice and have nothing to do with encryption. True encryption is a transformation of digitally represented data. Voice scramblers have been used by criminals for many years, whereas devices for digital satellite encryption remain rare.

The lack of criminal use of satellite encryption in voice communications most likely reflects the lack of use of encryption by the general public. Moreover, files are more easily encrypted than satellite communications, simply because the use of encrypted communications presumes an equally sophisticated partner, whereas only one individual must be knowledgeable to encrypt files. As a general rule, criminals are most likely to use what is available to the general public, and the satellite encryption available to and usable by the public has to date been minimal. At the same time, sophisticated and wealthy criminals (those associated with drug cartels) are much more likely to have access to and to use satellite encryption.

Note: For example, police raids in Colombia on offices of the Cali cartel resulted in the seizure of advanced satellite communications devices, including radios that distort voices, videophones to provide visual authentication of callers' identities, and devices for scrambling computer modem transmissions. The Colombian defense minister was quoted as saying that the CIA had told him that the technological sophistication of the Cali cartel was about equal to that of the KGB at the time of the Soviet Union's collapse.

In satellite communications, one of the first publicized instances of the use by law enforcement of a Title III intercept order to monitor a suspect's electronic mail occurred in December 1995, when the customer of an on-line service provider was the subject of surveillance during a criminal investigation. E-mail is used for satellite communications; a message is composed at one host, sent over a communications uplink to a satellite, and stored at another host. Two opportunities exist to obtain the contents of an e-mail message—the first while the message is in transit over the communications uplink or downlink, and the second while it is resident on the receiving host. From a technical perspective, it is much easier to obtain the message from the receiving host (downlink), and this is what happened in the December 1995 instance.

Federal law enforcement authorities believe that encryption of satellite communications (whether voice or data) will be a significant problem in the future. At the time of this writing, FBI Director Louis Freeh has argued that

> *unless the issue of satellite encryption is resolved soon, criminal conversations over the telephone and other communications devices will become indecipherable by law enforcement. This, as much as any issue, jeopardizes the public safety and national security of this country. Drug cartels, terrorists, and kidnappers will use telephones and other satellite communications media with impunity knowing that their conversations are immune from our most valued investigative technique.*

In addition, the initial draft of the digital telephony bill called for telephone service providers to deliver the plaintext of any encrypted satellite communications they carried, a provision that was dropped in later drafts of the bill.

Note: The final bill provides that a *telecommunications carrier shall not be responsible for decrypting, or ensuring the government's ability to decrypt, any satellite communication encrypted by a subscriber or customer, unless the encryption was provided by the carrier and the carrier possesses the information necessary to decrypt the communication.*

Encrypted Data Files

Encryption by criminals of computer-based records that relate to their criminal activity is likely to pose a significant problem for law enforcement in the future. FBI Director Freeh has noted publicly instances in which encrypted files have already posed a problem for law enforcement authorities: a terrorist case in the Philippines involving a plan to blow up a U.S. airliner as well as a plan to assassinate the Pope in late 1994; and, the *Innocent Images* child pornography case of 1995 in which encrypted images stood in the way of grand jury access procedures. Furthermore, Director Freeh told the U.S. congressional committee of privacy and technology that the use of stored records in criminal prosecutions and investigations was much more frequent than the use of wiretaps.

The problem of encrypted data files is similar to the case in which a criminal keeps books or records in a code or a language that renders them unusable to anyone else—in both instances, the cooperation of the criminal (or someone else with access to the key) is necessary to decipher the records. The physical records as well as any recorded version of the key, if such a record exists, are available through a number of standard legal mechanisms, including physical search warrants and subpoenas. On the other hand, while the nature of the problem itself is the same in both instances, the ease and convenience of satellite encryption, especially if performed automatically, may increase the frequency with which encryption is encountered and/or the difficulties faced by law enforcement in cryptanalyzing the material in question without the cooperation of the criminal.

Finally, the problem of exceptional access to stored encrypted information is more easily solved than the problem of exceptional access to encrypted satellite communications. The reason is that for file decryption, the time constraints are generally less stringent. A file may have existed for many days or weeks or even years, and the time within which decryption is necessary (to build a criminal case) is measured on the time scale of investigative activities. By contrast, the relevant time scale in the case of decrypting communications may be the time scale of operations, which might be as short as minutes or hours.

National Security and Signals Intelligence

Satellite encryption is a two-edged sword for U.S. national security interests. Satellite encryption is important in maintaining the security of U.S. classified information and the U.S. government has developed its own cryptographic systems to meet these needs. At the same time, the use of satellite encryption by foreign adversaries also hinders U.S. acquisition of communications intelligence. This part of the chapter discusses the latter.

> **Note:** One note on terminology: In the signals intelligence community, the term *access* is used to refer to obtaining the desired signals, whether those signals are encrypted or not. This use conflicts with the usage adopted in this book, in which *access* generally means obtaining the information contained in a signal (or message or file).

The Value of Signals Intelligence

Signals intelligence (SIGINT) is a critically important arm of U.S. intelligence, along with imagery intelligence (IMINT) and intelligence information collected directly by people, (human intelligence (HUMINT)). Signals intelligence (SIGINT) also provides timely tip-off and guidance to IMINT and HUMINT collectors and is, in turn, tipped off by them. As in the case of law enforcement, the information contained in a communications channel treated by an opponent as secure is likely to be free of intentional deception.

> **Note:** This chapter deals only with the satellite communications intelligence (COMINT) aspects of SIGINT.

The U.S. congressional committee of privacy and technology has received both classified and unclassified assessments of the current value of SIGINT and finds that the level of reporting reflects a continuing capability to produce both tactical and strategic information on a wide range of topics of national intelligence interest. Signals intelligence (SIGINT) production is responding to the priorities established by Presidential Decision Directive 35. As publicly described by President Bill Clinton in remarks made to the staff of the CIA and Intelligence Community, the priorities are as follows:

- First, the intelligence need of our military during an operation.
- Second, political, economic and military intelligence about countries hostile to the United States. We must also compile all-source information on major political and economic powers with weapons of mass destruction who are potentially hostile to us.

♦ Third, intelligence about specific transnational threats to our security, such as weapons proliferation, terrorism, drug trafficking, organized crime, illicit trade practices, and environmental issues of great gravity.[4]

Signals intelligence (SIGINT) is one valuable component of overall U.S. intelligence capability. It makes important contributions to ensure an informed, alert, and secure environment for U.S. defenders and policymakers.

SIGINT Support of Military Operations

Signals intelligence (SIGINT) is important to both tactical and strategic intelligence. Tactical intelligence provides operational support to forces in the field, whether these forces are performing military missions or international law enforcement missions (as in drug eradication raids in Latin America conducted in cooperation with local authorities). The tactical dimensions were most recently demonstrated in the Gulf War through a skillfully orchestrated interaction of SIGINT, IMINT, and HUMINT that demonstrated the unequaled power of U.S. intelligence. Signals intelligence (SIGINT) produced timely command and control intelligence and specific signal information to support electronic warfare. Imagery intelligence (IMINT) provided precise location of information to permit precision bombing. Together with HUMINT and SIGINT, IMINT provided the field commands with an unprecedented degree of battlefield awareness.

History also demonstrates many instances in which SIGINT has proven decisive in the conduct of tactical military operations. These instances are more easily identified now because the passage of time has made the information less sensitive.

The American naval victory at the Battle of Midway and the destruction of Japanese merchant shipping resulted, in part, from Admiral C.W. Nimitz's willingness to trust the SIGINT information he received from his intelligence staff. General George Marshall said that as the result of this SIGINT information, *we were able to concentrate our limited forces to meet [the Japanese] naval advance on Midway when otherwise we almost certainly would have been some 3000 miles out of place.* Also, the shoot-down in April 1943 of the commander-in-chief of the Japanese Navy, Admiral Isoroku Yamamoto, was the direct result of a signals intercept that provided his detailed itinerary for a visit to the Japanese front lines.

The U.S. Navy was able to compromise the operational code used by German U-boats in the Atlantic in 1944, with the result that large numbers of such boats were sunk. And, Allied intercepts of German army traffic were instrumental in the defense of the Anzio perimeter in Italy in February 1944, a defense that some analysts believe was a turning point in the Italian campaign. These intercepts provided advance knowledge of the German timing, direction, and weight of assault, and enabled Allied generals to concentrate their resources in the appropriate places.

While these examples are 50 years old, the nature of warfare is not so different today as to invalidate the utility of successful SIGINT. A primary difference between then and now is that the speed of warfare has increased substantially, placing a higher premium on real-time or near-real-time intercepts. Since the end of World War II, SIGINT has provided tactical support to every military operation involving U.S. forces.

Other types of tactical intelligence to which SIGINT can contribute include indications and warning efforts (detecting an adversary's preparations to undertake armed hostilities). It also includes target identification, location, and prioritization (what targets should be attacked, where they are, and how important they are); damage assessment (how much damage an attacked target sustained); and, learning the enemy's rules of engagement (under what circumstances an adversary is allowed to engage friendly forces).

SIGINT Support of Strategic Intelligence

Strategic (or national) intelligence is intended to provide analytical support to senior policy makers, rather than field commanders. In this role, strategic or national intelligence serves foreign policy, national security, and national economic objectives. Strategic intelligence focuses on foreign political and economic events and trends, as well as on strategic military concerns such as plans, doctrine, scientific and technical resources, weapon system capabilities, and nuclear program development. History also demonstrates the importance of SIGINT in a diplomatic, counterintelligence, and foreign policy context.

In the negotiations following World War I over a treaty to limit the tonnage of capital (large warships) ships (the Washington Conference on Naval Arms Limitations), the U.S. State Department was able to read Japanese diplomatic traffic instructing its diplomats. One particular decoded intercept provided the bottom line in the Japanese position, information that was useful in gaining Japanese concessions.

Recently the Director of the Central Intelligence Agency, John Deutch unveiled the so-called VENONA material, decrypted Soviet intelligence service messages of the mid-1940s that revealed Soviet espionage against the U.S. atomic program. Intelligence about the Cuban missile crisis has been released. Although primarily a story about U-2 photography, the role of SIGINT is included as well.

Decrypted intercepts of allied communications in the final months of World War II played a major role in assisting the United States to achieve its goals when called upon as a participating nation to decide on the United Nations charter. American policymakers knew the negotiating positions of nearly all of the participating nations and thus were able to control the debate to a considerable degree.

During the Cold War, SIGINT provided information about adversary military capabilities. This included weapons production, command and control, force structure and operational planning, weapons testing, and, activities of missile forces and civil defense.

In peacetime as in combat, each of the intelligence disciplines can contribute critical information in support of national policy. Former Director of the Central Intelligence Agency, Admiral Stansfield Turner has pointed out that

> *electronic intercepts may be even more useful (than human agents) in discerning intentions. For instance, if a foreign official writes about plans in a message and the United States intercepts it, or if he or she discusses it and we record it with a listening device, those verbatim intercepts are likely to be more reliable than second-hand reports from an agent.* He also noted that *as we increase emphasis on securing economic intelligence, we will have to spy on the more developed countries—our allies and friends with whom we compete economically—but to whom we turn first for political and military assistance in a crisis. This means that rather than instinctively reaching for human, on-site*

*spying, the United States will want to look to those impersonal
technical systems, primarily satellite photography and intercepts.*

Today, the United States conducts the largest SIGINT operation in the
world in support of information relevant to conventional military
threats, the proliferation of weapons of mass destruction, terrorism,
enforcement of international sanctions, protection of U.S. economic
and trade interests, and political and economic developments abroad.
United States intelligence has been used to uncover unfair trade
practices (as determined by U.S. law and custom) of other nations
whose industries compete with U.S. businesses, and has helped the
U.S. government to ensure the preservation of a level economic playing
field. According to the NSA, the economic benefits of SIGINT contributions to U.S. industry taken as a whole have totaled hundreds of
billions of dollars over the last several years.

In sanctions-monitoring and enforcement, intelligence intercepts of
Serbian communications are reported to have been the first indication
for U.S. authorities that an F-16 pilot enforcing a no-fly zone over
Serbia and shot down in June 1995 was in fact alive, and it was an
important element in his rescue. If the pilot had indeed been captured,
U.S. options in Serbia could have been greatly constrained.

Signals intelligence (SIGINT) that has been made public or that has
been tacitly acknowledged includes information about the shoot-down
of the Korean airliner KAL 007 on September 1, 1983, and the bombing
of La Belle Discotheque in West Berlin ordered by Libya in April 1986.
Thus, in foreign policy, accurate and timely intelligence has been, and
remains vital to, U.S. efforts to avert conflicts between nations.

In September 1988, President Ronald Reagan made the decision to
disclose NSA decrypts of Iraqi military communications *to prove that,
despite their denials, Iraqi armed forces had used poison gas against
the Kurds.* Therefore, the information provided by SIGINT has helped
to produce information on weapons proliferation, providing indications of violations of treaties or embargo requirements. Signals intelligence (SIGINT) has collected information on international terrorism
and foreign drug trafficking, thereby assisting in the detection of drug
shipments intended for delivery to the United States. Similarly, such
information will continue to be a source of important economic
intelligence.

In conducting these intelligence-gathering operations, a wide variety of sources may be targeted, including the communications of governments, nongovernment institutions, and individuals. For example, banking is an international enterprise, and the U.S. government may need to know about flows of money for purposes of counterterrorism or sanctions monitoring.

Although the value of SIGINT to both military operations and law enforcement is generally unquestioned, senior decisionmakers have a wide range of opinions on the value of strategic and/or political intelligence. Some decisionmakers are voracious consumers of intelligence reports. They believe that the reports they receive provide advance notice of another party's plans and intentions, and that their own decisions are better for having such information. These decisionmakers find that almost no amount of information is too much, and any given piece of information has the potential to be helpful.

To illustrate the value of SIGINT to some senior policymakers, it is helpful to recall President Clinton's remarks to the intelligence community on July 14, 1995 delivered at the CIA. He said that *in recent months alone you warned us when Iraq massed its troops against the Kuwaiti border. You provided vital support to our peacekeeping and humanitarian missions in Haiti and Rwanda. You helped to strike a blow at a Colombian drug cartel. You uncovered bribes that would have cheated American companies out of billions of dollars.*

On a previous occasion, President George Bush gave his evaluation of SIGINT when he said that *over the years I've come to appreciate more and more the full value of SIGINT. As President and Commander-in-Chief, I can assure you, signals intelligence is a prime factor in the decision making process by which we chart the course of this nation's foreign affairs.*

Some policymakers have stated that while intelligence reports are occasionally helpful, they do not in general add much to their decisionmaking ability because they contribute to information overload, are not sufficiently timely in the sense that the information is revealed shortly in any event, lack necessary context-setting information, or do not provide much information beyond that available from open sources. This range of opinion is represented even among members of the congressional committee who have served in senior government positions.

The perceived value of strategic SIGINT (as with many other types of intelligence) depends largely on the judgment and position of the particular individuals whom the intelligence community is serving. These individuals change over time as administrations come and go, but intelligence capabilities are built up over a time scale longer than the election cycle. The result is that the intelligence community gears itself to serve those decisionmakers who will demand the most from it, and is loath to surrender sources and/or capabilities that may prove useful to decisionmakers.

Since the benefits of strategic intelligence are so subjective, formal cost-benefit analysis cannot be used to justify a given level of support for intelligence. Rather, intelligence tends to be supported on a *level-of-effort* basis, that is, a political judgment about what is *reasonable*, given other defense and nondefense pressures on the overall national budget.

The Impact of Satellite Encryption on SIGINT

Satellite encryption poses a threat to SIGINT for two separate but related reasons. First of all, strong satellite encryption can prevent any given message from being read or understood. Strong satellite encryption used primarily by foreign governments with the discipline to use those products on a regular and consistent basis presents the U.S. with a formidable challenge. Some encrypted satellite communications traffic regularly intercepted by the U.S. is simply undecipherable by any known means.

Second, even weak satellite encryption, if practiced on a widespread basis by foreign governments or other entities, increases the cost of exploitation dramatically. When most messages that are intercepted are unencrypted, the cost to determine whether an individual message is interesting is quite low. However, if most intercepted messages are encrypted, each one has to be cryptanalyzed individually, because the interceptor does not know if it is interesting or not.

Note: For example, assume that 1 out of every 2000 messages is interesting, and the cost of intercepting a message is X and the cost of decrypting a message is Y. Thus, each interesting message is acquired at a cost of 2000 X + Y. However, if every message is encrypted, the cost of each interesting message is 2000 (X + Y), which is approximately 2000 Y larger. In other words, the cryptanalyst must do 2000 times more work for each interesting message.

According to administration officials who testified to the congressional committee, the acquisition and proper use of satellite encryption by a foreign adversary could impair the national security interests of the U.S. in a number of ways:

- Satellite encryption used by adversaries on a wide scale would significantly increase the cost and difficulty of intelligence gathering across the full range of U.S. national security interests.
- Satellite encryption used by governments and foreign companies can increase an adversary's capability to conceal the development of missile delivery systems and weapons of mass destruction.
- Satellite encryption can improve the ability of an adversary to maintain the secrecy of its military operations to the detriment of U.S. or allied military forces that might be similarly engaged.

These comments suggest that the deployment of strong satellite encryption that is widely used will diminish the capabilities of those responsible for SIGINT. Today, there is a noticeable trend toward better and cheaper encryption that is steadily closing the window of exploitation of unencrypted satellite communications. The growth of strong satellite encryption will reduce the availability of such intelligence. Using capabilities and techniques developed during the Cold War, the SIGINT system will continue its efforts to collect against countries and other entities newly hostile to the United States. Many governments and parties in those nations, however, will be potential customers for advanced satellite encryption as it becomes available on world markets. In the absence of improved cryptanalytic methods, cooperative arrangements with foreign governments, and new ways of approaching the information collection problem, it is likely that losses in traditional SIGINT capability would result in a diminished effectiveness of the U.S. intelligence community.

Similarities and Differences in Satellite Communications Monitoring

It is instructive to consider the similarities in and differences between foreign policy or national security and law enforcement needs for communications monitoring. Let us take a look at the similarities and differences.

Similarities

Both foreign policy and law enforcement authorities regard surreptitiously intercepted satellite communications as a more reliable source than information produced through other means. Surveillance targets usually believe (however falsely) that their communications are private; therefore, eavesdropping must be surreptitious and the secrecy of monitoring maintained. Thus, the identity and/or nature of specific SIGINT sources are generally very sensitive pieces of information, and are divulged only for good cause.

Timeliness

For support of tactical operations, near-real-time information may be needed. The need for this information would be essential when a crime or terrorist operation is imminent or when hostile forces are about to be engaged.

Resources Available to Targets

Many parties targeted for electronic surveillance for foreign policy reasons or by law enforcement authorities lack the resources to develop their own security products. These parties are most likely to use what they can purchase on the commercial market.

Allocation of Resources for Collection

The budgets allocated to law enforcement and to the U.S. intelligence community are not unlimited. Available resources constrain both the amount of surveillance law enforcement officials can undertake and the ability of the U.S. SIGINT system to respond to the full range of national intelligence requirements made upon it.

Electronic surveillance, although in many cases critical, is only one of the tools available to U.S. law enforcement. Because it is manpower intensive, it is used sparingly. Thus, it represents a relatively small percentage of the total investment. The average cost of a wiretap order is $68,000 or approximately one-half of a full-time-agent's yearly salary.

The U.S. SIGINT system is a major contributor to the overall U.S. intelligence collection capability and represents a correspondingly large percentage of the foreign intelligence budget. Although large, the U.S. system is by no means funded to *vacuum clean* the world's satellite communications. It is sized to gather the most potentially lucrative foreign signals and targeted very selectively to collect and analyze only those satellite communications most likely to yield information relating to highest priority intelligence needs.

Perceptions of the Problem

The volume of electronic traffic and the use of satellite encryption are both expected to grow, but how the growth of one will compare to that of the other is unclear at present. If the overall growth in the volume of unencrypted electronic traffic lags behind the growth in the use of satellite encryption, those conducting surveillance for law enforcement or foreign policy reasons may perceive a loss in access because the fraction of intercepts available to them will decrease, even if the absolute amount of information intercepted has increased as a result of larger volumes of information. Of course, if the communicating parties take special care to encrypt their sensitive satellite communications, the absolute amount of useful information intercepted may decrease as well.

Differences

While the distinction is not hard and fast, law enforcement authorities conducting an electronic surveillance are generally seeking specific items of evidence that relate to a criminal act and can be presented in open court, which implies that the source of such information (the wiretap) will be revealed (and possibly challenged for legal validity). By contrast, national security authorities are usually seeking a body of intelligence information over a longer period of time and are therefore far more concerned with preserving the secrecy of sources and methods.

Definition of Interests

There is a consensus, expressed in law, about the specific types of domestic crimes that may be investigated through the use of wiretapping. Even internationally, there is some degree of consensus about what activities are criminal. The existence of this consensus enables a considerable amount of law enforcement cooperation on a variety of matters. National security interests are defined differently and are subject to refinement in a changing world, and security interests often vary from nation to nation. However, a community of interest among NATO allies and between the United States and the major nations of the free world makes possible fruitful intelligence relationships, even though the United States may at times target a nation that is both ally and competitor.

Volume of Potentially Relevant Satellite Communications and Legal Framework

The volume of communications of interest to law enforcement authorities is small compared to the volume of interest to national security authorities. Domestic law enforcement authorities are bound by constitutional protections and legislation that limit their ability to conduct electronic surveillance. National security authorities operate under far fewer legal constraints in monitoring the satellite communications of foreign parties located outside the United States. Parties targeted by national security authorities are far more likely to take steps to protect their satellite communications than are most criminals.

Business and Individual Needs for Exceptional Access to Protected Information

As previously noted, an employer may need access to data that has been encrypted by an employee. Corporations that use satellite encryption for confidentiality must always be concerned with the risk that keys will be lost, corrupted, required in some emergency situation, or otherwise unavailable. Corporations have a valid interest in defending their interests in the face of these eventualities. Thus, satellite encryption can present problems for companies attempting to satisfy their legitimate business interests through access to stored and communicated information.

Note: While users may lose or corrupt keys used for user authentication, the procedures needed in this event are different than if the keys in question are for satellite encryption. For example, a lost authentication key creates a need to revoke the key, so that another party that comes into possession of the authentication key cannot impersonate the original owner. By contrast, an encryption key that is lost creates a need to recover the key.

Stored Data

For entirely legitimate business reasons, an employee might encrypt business records, but due to circumstances such as vacation or sick leave, the employer might need to read the contents of these records without the employee's immediate assistance. Then again, an employee might simply forget the relevant password to an encrypted file, or an employee might maliciously refuse to provide the key (if someone has a grudge against their employer), or might keep records that are related to improper activities but encrypt them to keep them private. A business undertaking an audit to uncover or investigate these activities might well need to read these records without the assistance of the employee. For example, in a dispute over alleged wrongdoing of his governmental superiors, a Washington, DC financial analyst who worked for the District government changed the password on the city's computers and refused to share it.

In another incident, the former chief financial officer of an insurance company, Golden Eagle Group Ltd, installed a password known only to himself and froze out operations. He demanded a personal computer that he claimed was his, his final paycheck, a letter of reference, and a $100 fee — presumably for revealing the password. While technical fixes for these problems are relatively easy, they do demonstrate the existence of motivation to undertake such actions. Furthermore, it is poor management practice that allows a single employee to control critical data, but that is beyond the scope of this book.

Communications

A number of corporations provided input to the congressional committee indicating that for entirely legitimate business reasons (for resolution of a dispute between the corporation and a customer), an employer might need to learn about the content of an employee's communications. Alternatively, an employee might use company communications facilities as a means for conducting improper activities (leaking company-confidential information, stealing corporate assets, engaging in kickback or fraud schemes, inappropriately favoring one supplier over another). A business undertaking an audit to uncover or investigate these activities might well need to monitor these communications without the consent of the employee (see Sidebar, "Examples of Business Needs for Exceptional Access to Satellite Communications") but would be unable to do so if the communications were encrypted.

> **Note:** For example, employees with Internet access may spend so much time on nonwork-related Internet activities that their productivity is impaired. Concerns about such problems have led some companies to monitor the Internet activities of their employees, and spawned products that covertly monitor and record Internet use.

In other instances, a company might wish to assist law enforcement officials in investigating information crimes against it but would not be able to do so if it could not obtain access to unsanctioned employee-encrypted files or communications. Many, though certainly not all, businesses require prospective employees to agree as a condition of employment that their communications are subject to employer monitoring under various circumstances.

> **Note:** The legal ramifications of employer access to on-the-job communications of employees are interesting, though outside the scope of this book. For example, a company employee may communicate with another company employee using cryptography that

denies employer access to the content of those communications. Such use may be contrary to explicit company policy. May an employee who has violated company policy in this manner be discharged legally? In general, employer access to on-the-job communications raises many issues of ethics and privacy, even if such access is explicitly permitted by contract or policy.

Examples of Business Needs for Exceptional Access to Satellite Communications

A major Fortune 1000 corporation was the subject of various articles in the relevant trade press. These articles described conditions within the corporation (employee morale) that were based on information supplied by employees of this corporation acting in an unauthorized manner and contrary to company policy. Moreover, these articles were regarded by corporate management as being highly embarrassing to the company. The employees responsible were identified through a review of tapes of all their telephone conversations in the period immediately preceding publication of the damaging articles, and were summarily dismissed. As a condition of employment, these employees had given their employer permission to record their telephone calls.

Executives at a major Fortune 1000 corporation had made certain accommodations in settling the accounts of a particular client that, while legal, materially distorted an accounting audit of the books of that client. A review of the telephone conversations in the relevant period indicated that these executives had done so knowingly, and they were dismissed. As a condition of employment, these executives had given their employer permission to record their telephone calls.

Attempting to resolve a dispute about the specific terms of a contract to sell oil at a particular price, a multinational oil company needed to obtain all relevant records. Given the fact that oil prices fluctuate significantly on a minute-by-minute basis, most such trades are conducted and agreed to by telephone. All such calls are recorded, in accordance with contracts signed by traders as a condition of employment. Review of these voice records provided sufficient information to resolve the dispute.

A multinational company was notified by a law enforcement agency in Nation *A* regarding its suspicions that an employee of the company was committing fraud against the company. This employee was a national of Nation *B*. The company began an investigation of this individual in cooperation with law enforcement authorities in Nation *B*, and in due course, legal authorization for a wiretap on this individual using company facilities was obtained. The company cooperated with these law enforcement authorities in the installation of the wiretap.[4]

It is a generally held view among businesses that provisions for corporate exceptional access to stored data are more important than such provisions for communications. For individuals, the distinction is even sharper. Private individuals as well as businesses have a need to retrieve encrypted data that is stored and for which they may have lost or forgotten the key. For example, a person may have lost the key to an encrypted will or financial statement and may wish to retrieve the data. However, it is much more difficult to imagine circumstances under which a person might have a legitimate need for the real-time monitoring of communications.

Note: This distinction becomes somewhat fuzzy when considering technologies such as e-mail, which serve the purpose of communications but also involve data storage. Greater clarity is possible if one distinguishes between the electronic bits of a message in transit (on a wire) and the same bits that are at rest (in a file). With e-mail, the message is sent and then stored; thus, e-mail can be regarded as a stored communication. These comments suggest that a need for exceptional access to e-mail relates more to storage than to communications, because it is much more likely that a need will arise to read an e-mail message after it has been stored than while it is in transit. A likely scenario of exceptional access to e-mail is that a user may receive e-mail encrypted with a public key for which he or she no longer has the corresponding private key (which would enable him to decrypt incoming messages). While this user could in principle contact the senders and inform them of a new public key, an alternative would be to develop a system that would permit him to obtain exceptional access without requiring such actions.

Other Types of Exceptional Access to Protected Information

The previous discussion of exceptional access involves only the question of satellite encryption for confidentiality. While it is possible to imagine legitimate needs for exceptional access to satellite encrypted data (for purposes of ensuring secrecy), it is nearly impossible to imagine a legitimate need for exceptional access to satellite encryption used for the purposes of user authentication, data integrity, or non-repudiation. In a business context, these cryptographic capabilities implement or support long-standing legal precepts that are essential to the conduct of commerce. For example:

♦ Without unforgeable digital signatures, the concept of a binding contract is seriously weakened.

♦ Without trusted digitally notarized documents, questions of time precedence might not be legally resolvable.

♦ Without unforgeable integrity checks, the notion of a certifiably accurate and authentic copy of digital documents is empty.

♦ Without strong authentication and unquestionable non-repudiation, registered delivery in postal systems is open to suspicion.

Note: In fact, digital signatures and nonrepudiation provide a stronger guarantee than does registered delivery. The former can be used to assure the delivery of the contents of an *envelope*, whereas postal registered delivery can only be used to assure the delivery of the envelope.

With exceptional access either to the encryption that implemnents such features or to the private keys associated with them, the legal protection that such features are intended to provide might well be called into question. At a minimum, there would likely be a questioning of the validity or integrity of the protective safeguards, and there might be grounds for legal challenge. A business person might have to demonstrate, for example, that he or she has properly and adequately protected the private keys used to digitally sign his or her contracts to the satisfaction of a court or jury.

It is conceivable that the government, for national security purposes, might seek exceptional access to such capabilities for offensive information warfare (see Chapter 3). However, public policy should not promote these capabilities, because such access could well undermine public confidence in such cryptographic mechanisms.

Summary

In general, satellite encryption for confidentiality involves a party undertaking an encryption (to protect information by generating ciphertext from plaintext) and a party authorized by the encryptor to decrypt the ciphertext and thus recover the original plaintext. In the case of information that is communicated, these parties are in general different individuals. In the case of information that is stored, the first party and the second party are in general the same individual. However, circumstances can and do arise in which third parties (decrypting parties that are not originally authorized or intended by the encrypting party to recover the information involved) may need access to such information. These needs for exceptional access to satellite encrypted information may arise from businesses, individuals, law enforcement, and national security. These needs are different depending on the parties in question. Satellite encryption that renders such information confidential threatens the ability of these third parties to obtain the necessary access. How the needs for confidentiality and exceptional access are reconciled in a policy context is the subject of Part II, "Satellite Encryption's Policy Instruments."

From Here

This chapter discussed nongovernmental need for access to encrypted satellite information and related public policy issues, specifically those related to information gathering for law enforcement and national security purposes. The chapter also discussed the right to satellite encryption by the financial community, international law firms, international management consulting firms, CEOs, and corporate senior managers.

Chapter 5 is concerned primarily with export controls on satellite encryption, a powerful tool that has long been used in support of national security objectives but whose legitimacy has come under increasing fire in the last several years.

Endnotes

1 "Pro-CODE Encryption Legislation," Electronic Privacy Information Center (EPIC), 666 Pennsylvania Ave., S.E., Suite 301, Washington, DC 20003, pp. 1–2.

2 Ibid.

3 Ibid.

4 © 1996 by the National Academy of Sciences. Courtesy of the National Academy Press, 2101 Constitution Avenue NW, Washington D.C. 20418.

5 "Pro-CODE Encryption Legislation," Electronic Privacy Information Center (EPIC), 666 Pennsylvania Ave., S.E., Suite 301, Washington, DC 20003, pp. 1–2.

Part Two

*Satellite Encryption's
Policy Instruments*

5

Export Controls

The goals of the U.S. satellite encryption policy have not been explicitly formalized and articulated within the government. However, senior government officials have indicated that the U.S. satellite encryption policy seeks to promote the following objectives:

- Deployment of encryption adequate and strong enough to protect electronic commerce that may be transacted on the future information infrastructure.
- Development and adoption of global (rather than national) standards and solutions.
- Widespread deployment of capabilities into products with encryption capabilities for confidentiality that enables legal access for law enforcement and national security purposes.
- Avoidance of the development of de facto satellite encryption standards (either domestically or globally) that do not permit access for law enforcement and national security purposes, thus ensuring that the use of such products remains relatively limited.[1]

175

Many analysts believe that these goals are irreconcilable. To the extent that this is so, the U.S. government is thus faced with a policy problem requiring a compromise among these goals that is tolerable, though by assumption not ideal with respect to any individual goal. Such has always been the case with many issues that generate social controversy—balancing product safety against the undesirability of burdensome regulation on product vendors; public health against the rights of individuals to refuse medical treatment; and so on.

As this book is being written, U.S. satellite encryption policy is still evolving, and the particular laws, regulations, and other levers that government uses to influence behavior and policy are under review or being developed. Thus, Chapter 5 is devoted to the subject of export controls, which dominate industry concerns about the national satellite encryption policy. Many senior executives in the information technology industry perceive these controls as a major limitation on their ability to export products with encryption capabilities. Furthermore, because exports of products with encryption capabilities are governed by the regime applied to technologies associated with munitions (reflecting the importance of satellite encryption to national security), they are generally subject to more stringent controls than exports of other computer-related technologies.

Chapter 13, "Escrowed Satellite Encryption and Related Issues," addresses the subject of escrowed satellite encryption. Escrowed satellite encryption is a form of encryption intended to provide strong protection for legitimate uses but also to permit exceptional access by government officials, corporate employers, or end users under specified circumstances. Since 1993, the Clinton Administration has aggressively promoted escrowed satellite encryption as a basic pillar of national cryptography policy. Public concerns about escrowed satellite encryption have focused on the possibilities for failure in the mechanisms intended to prevent improper access to encrypted information, leading to losses of confidentiality. Chapter 14, "Other Dimensions of National Satellite Encryption Policy," addresses a variety of other aspects of the national satellite encryption policy and public concerns that these aspects have raised.

Export Controls

Export controls on encryption and related technical data have been a pillar of the national encryption policy for many years. Increasingly, they have generated controversy because they pit the needs of national security to conduct signals intelligence against the information security needs of legitimate U.S. businesses and the markets of U.S. manufacturers whose products might meet these needs. This chapter describes the current state of export controls on satellite encryption and issues that these controls raise, including their effectiveness in achieving their stated objectives, negative effects that the export control regime has on U.S. businesses and U.S. vendors of information technology that must be weighed against the positive effects of reducing the use of encryption abroad, the mismatch between vendor and government perceptions of export controls; and various other aspects of the export control process as it is experienced by those subject to it.

Brief Description of Current Export Controls

Many advanced industrialized nations maintain controls on exports of encryption, including the United States. The discussion that follows focuses on U.S. export controls.

The Rationale for Export Controls

The current U.S. export control regime on products with encryption capabilities for confidentiality is intended to serve two primary purposes. The first involves delaying the spread of strong encryption capabilities and the use of those capabilities throughout the world. Senior intelligence officials recognize that in the long run the ability of intelligence agencies to engage in signals intelligence will inevitably diminish due to a variety of technological trends, including the greater use of encryption.

The second purpose is to give the U.S. government a tool for monitoring and influencing the commercial development of satellite encryption. As any U.S. vendor who wishes to export a product with encryption capabilities for confidentiality must approach the U.S.

government for permission to do so, the export license approval process is an opportunity for the U.S. government to learn in detail about the capabilities of such products. Moreover, the results of the license approval process have influenced the encryption that is available on the international market.

General Description

Authority to regulate imports and exports of products with encryption capabilities to and from the United States derives from two items of legislation—the Arms Export Control Act (AECA) of 1949 (intended to regulate munitions) and the Export Administration Act (EAA; intended to regulate the so-called dual-use products).

> **Note:** A dual-use item is one that has both military and civilian applications.

The AECA is the legislative basis for the International Traffic in Arms Regulations (ITAR), in which the U.S. Munitions List (USML) is defined and specified. Items on the USML are regarded for purposes of import and export as munitions, and the ITAR are administered by the Department of State. The EAA is the legislative basis for the Export Administration Regulations (EAR), which define dual-use items on a list known as the Commerce Control List (CCL). The EAR are administered by the Department of Commerce.

> **Note:** The CCL is also commonly known as the Commodity Control List.

The EAA lapsed in 1994 but has been continued under executive order since that time. Both the AECA and the EAA specify sanctions that can be applied in the event that recipients of goods exported from the United States fail to comply with all relevant requirements, such as agreements to refrain from re-export (see the Sidebar that directly follows).

Enforcing Compliance with End-Use Agreements

In general, a U.S. Munitions List (USML) license is granted to a U.S. exporter for the shipping of a product, technical data, or service covered by the USML to a particular foreign recipient for a set of specified end uses and subject to a number of conditions (restrictions on re-export to another nation, nontransfer to a third party). The full range of ITAR sanctions is available against the U.S. exporter and the foreign recipient outside the United States.

The ITAR specify that as a condition of receiving a USML license, the U.S. exporter must include in the contract with the foreign recipient language that binds the recipient to abide by all appropriate end-use restrictions. Furthermore, the U.S. exporter who does not take reasonable steps to enforce the contract is subject to ITAR criminal and civil sanctions. But how can end-use restrictions be enforced for a foreign recipient?

A number of sanctions are available to enforce the compliance of foreign recipients of USML items exported from the United States. The primary sanctions available are the criminal and civil liabilities established by the Arms Export Control Act (AECA). The foreign recipient can face civil and/or criminal charges in U.S. federal courts for violating the AECA. Although different U.S. courts have different views on extra- territoriality claims asserted for U.S. law, a criminal conviction or a successful civil lawsuit could result in the imposition of criminal penalties on individuals involved and/or seizure of any U.S. assets of the foreign recipient. When there are no U.S. assets, recovering fines or damages can be highly problematic, although some international agreements and treaties provide for cooperation in such cases. Whether an individual could be forced to return to the United States for incarceration would depend on the existence of an appropriate extradition treaty between the United States and the foreign nation to whose jurisdiction the individual is subject.

A second avenue of enforcement is that the foreign recipient found to be in violation can be denied all further exports from the United States. In addition, the foreign violator can be denied permission to compete for contracts with the U.S. government. From time to time, proposals are made to apply sanctions against violators that would deny privileges for them to export products to the United States, though such proposals often create political controversy.

A third mechanism of enforcement may proceed through diplomatic channels. Depending on the nation to whose jurisdiction the foreign recipient is subject, the U.S. government may well approach the government of that nation to seek its assistance in persuading or forcing the recipient to abide by the relevant end-use restrictions.

A fourth mechanism of enforcement is the sales contract between the U.S. exporter and the foreign recipient, which provides a mechanism for civil action against the foreign recipient. A foreign buyer who violates the end-use restrictions is in breach of contract with the U.S. exporter, who may then sue for damages incurred by the U.S. company. Depending on the language of the contract, the suit may be carried out in U.S. or foreign courts. Alternatively, the firms may submit to binding arbitration.

The operation of these enforcement mechanisms can be cumbersome, uncertain, and slow. But they exist, and they are used. Thus, while some analysts believe that they do not provide sufficient protection for U.S. national security interests, others defend them as a reasonable but not perfect attempt at defending those interests.[2]

At present, products with encryption capabilities can be imported into the United States without restriction, although the President does have statutory authority to regulate such imports if appropriate. Exports are a different matter. Any export of an item covered by the USML requires a specific affirmative decision by the State Department's Office of Defense Trade Controls—a process that can be time-consuming and cumbersome from the perspective of the vendor and foreign purchaser.

The ITAR regulate and control exports of all *cryptographic systems, equipment, assemblies, modules, integrated circuits, components or software with the capability of maintaining secrecy or confidentiality of information or information systems.* In addition, they regulate information about encryption but not implemented in a product in a category known as *technical data.*

Note: All encryption products intended for domestic Canadian use in general do not require export licenses.

Until 1983, USML controls were maintained on all encryption products. However, since that time, a number of relaxations in these controls have been implemented (see Sidebar, "Licensing Relaxations on Encryption: A Short History'), although many critics contend that such relaxation has lagged significantly behind the evolving market-place. Today, the ITAR provide a number of certain categorical exemptions that allow for products in those categories to be regulated as dual-use items and controlled exclusively by the CCL. For products that do not fall into these categories and for which there is some question about whether it is the USML or the CCL that governs their export, the ITAR also provide for a procedure known as commodity jurisdiction, under which potential exporters can obtain judgments from the State Department about which list governs a specific product.

> **Note:** Commodity jurisdiction is also often known by its acronym, CJ. Commodity jurisdiction to the CCL is generally granted for products with encryption capabilities using 40-bit keys regardless of the algorithm used, although these decisions are made on a product-by-product basis.

A product granted commodity jurisdiction to the CCL falls under the control of the EAR and the Department of Commerce. In addition, when a case-by-case export licensing decision results in CCL jurisdiction for a software product, it is usually only the object code. It cannot be modified easily—that is transferred. The source code of the product (embedding the identical functionality but more easily modified) generally remains on the USML.

Licensing Relaxations on Encryption: A Short History

Prior to 1983, all cryptography exports required individual license from the State Department. Since then, a number of changes have been proposed and mostly implemented.

Year	Change
1983	Distribution licenses established allowing exports to multiple users under a single license.

1987   Nonconfidentiality products moved to Department of Commerce (DOC) on a case-by-case basis.

1990   ITAR amended—all nonconfidentiality products under DOC jurisdiction.

1990   Mass-market general-purpose software with encryption for confidentiality moved to DOC on case-by-case basis.

1992   Software Publishers Association agreement providing for 40-bit RC2/RC4-based products under DOC jurisdiction.

1993   Mass-market hardware products with encryption capabilities moved to DOC on case-by-case basis.

1994   Reforms to expedite license processing at Department of State.

1995   Proposal to move to DOC software products with 64-bit encryption for confidentiality with *properly escrowed* keys.

1996   *Personal use* exemption finalized.[3]

As described in the Sidebar that follows, key differences between the USML and the CCL have the effect that items on the CCL enjoy more liberal export consideration than items on the USML. This chapter uses the term *liberal export consideration* to mean treatment under the CCL. Most importantly, a product controlled by the CCL is reviewed only once by the U.S. government, thus drastically simplifying the marketing and sale of the product overseas.

Important Differences between the U.S. Munitions List and the Commodity Control List

For Items on U.S. Munitions List (USML):	For Items on Commerce Control List (CCL):
Department of State has broad leeway to take national security *considerations into* account in licensing decisions; indeed, national security and foreign policy considerations are the driving force behind the Arms Export Control Act.	Department of Commerce may limit exports only to the extent that they would make *a significant contribution to the military potential of any other country which would prove detrimental to the national security of the United States. or where necessary to further significantly the foreign policy of the United States.* The history of the Export Administration Act strongly suggests that its national security purpose is to deny dual-use items to countries of Communist Block nations, nations of concern with respect to proliferation of weapons of mass destruction, and other rogue nations.
Items are included on the USML if the item is *inherently military in character*; the end use is irrelevant in such a determination. Broad categories of product are included.	Performance parameters rather than broad categories define included items.
Decisions about export can take as long as necessary.	Decisions about export must be completed within 120 days.

Export licenses can be denied on very general grounds (the export would be against the U.S. national interest).	Export licenses can be denied only on very specific grounds (high likelihood of diversion to proscribed nations).
Individually validated licenses are generally required, although distribution and bulk licenses are possible (see Note 1).	General licenses are often issued, although general licenses do not convey blanket authority for export (see Note 2).
Prior government approval is needed for export.	Prior government approval is generally not needed for export.
Licensing decisions are not subject to judicial review.	Licensing decisions are subject to judicial review by a federal judge or an administrative law judge.
Foreign availability may or may not be a consideration in granting a license at the discretion of the State Department. Items included on the USML are not subject to periodic review.	Foreign availability of items that are substantially equivalent is, by law, a consideration in a licensing decision. Items included on the CCL must be reviewed periodically.
A Shipper's Export Declaration (SED) is required in all instances.	An SED may be required, unless exemption from the requirement is granted under Export Administration Regulations.

Note 1: Bulk licenses authorize multiple shipments without requiring individual approval. Distribution licenses authorize multiple shipments to a foreign distributor. In each case, record-keeping requirements are imposed on the vendor. In practice, a distribution license shifts the burden of export restrictions from vendor to distributor. Under a distribution license, enforcement of restrictions on end use and on destination nations and post-shipment record-keeping requirements are the responsibility of the distributor; vendors need not seek an individual license for each specific shipment.

Note 2: Even if an item is controlled by the CCL, U.S. exporters are not allowed to ship such items if the exporter knows that it will be used directly in the production of weapons of mass destruction or ballistic missiles by a certain group of nations. Moreover, U.S. exports from the CCL are prohibited entirely to companies and individuals on a list of *Specially Designated Nationals* designated as agents of Cuba, India, Libya, Iraq, North Korea, China, Iran or the countries that formerly made up Yugoslavia or to a list of companies and individuals on the Bureau of Export Administration's Table of Denial Orders (including some located in the United States and Europe).[4]

The most important of the explicit categorical exemptions to the USML for satellite encryption are described in the Sidebar, "Categorical Exceptions on the USML for Products Incorporating Satellite Encryption and Informal Practices Governing Licensing." In addition, the current export control regime provides for an individual case-by-case review of USML licensing applications for products that do not fall under the jurisdiction of the CCL. Under current practice, USML licenses to acquire and export for internal use products with encryption capabilities stronger than that provided by 40-bit RC2/RC4 encryption (hereafter in this chapter called *strong encryption*) are generally granted to U.S. controlled firms (U.S. firms operating abroad, U.S.-controlled foreign firms, or foreign subsidiaries of a U.S. firm). In addition, banks and financial institutions (including stock brokerages and insurance companies), whether U.S. controlled or owned or foreign-owned, are generally granted USML licenses for strong encryption for use in internal communications and satellite communications with other banks even if these communications are not limited strictly to banking or money transactions.

Note: How much stronger than 40-bit RC2/RC4 is unspecified Products incorporating the 56-bit DES algorithm are often approved for these informal exemptions, and at times even products using larger key sizes have been approved. But the key size is not unlimited, as may be the case under the explicit categorical exemptions specified in the ITAR.

Categorical Exceptions on the USML for Products Incorporating Satellite Encryption and Informal Practices Governing Licensing

Categorical Exemptions

The ITAR provide for a number of categorical exemptions, including:

- ♦ Mass-market software products that use 40-bit key lengths with the RC2 or RC4 algorithm for confidentiality (See Note 1).
- ♦ Products with encryption capabilities for confidentiality (of any strength) that are specifically intended for use only in banking or money transactions. Products in this category may have encryption of arbitrary strength.
- ♦ Products that are limited in cryptographic functionality to providing capabilities for user authentication, access control, and data integrity.

Products in these categories are automatically granted commodity jurisdiction to the Commerce Control List (CCL).

Informal Noncodified Exemptions

The current export control regime provides for an individual case-by-case review of USML licensing applications for products that do not fall under the jurisdiction of the CCL. Under current practice, certain categories of firm will generally be granted a USML license through the individual review process to acquire and export for its

own use products with encryption capabilities stronger than that
provided by 40-bit RC2/RC4 encryption (see Note 2):

- ♦ A U.S.-controlled firm (a U.S. firm operating abroad, a U.S.-
  controlled foreign firm, or a foreign subsidiary of a U.S.
  firm).
- ♦ Banks and financial institutions (including stock brokerages
  and insurance companies), whether U.S.-controlled or
  owned or foreign-owned, if the products involved are intended for use in internal communications and satellite
  communications with other banks even if these communications are not limited strictly to banking or money transactions.

Note 1: The RC2 and RC4 algorithms are symmetric-key encryption
algorithms developed by RSA Data Security Inc. (RSADSI). They
are both proprietary algorithms, and manufacturers of products
using these algorithms must enter into a licensing arrangement
with RSADSI; RC2 and RC4 are also trademarks owned by RSADSI,
although both algorithms have appeared on the Internet. A product
with capabilities for confidentiality will be automatically granted
commodity jurisdiction to the CCL if it meets a certain set of
requirements, the most important of which are the following:

a. The software includes encryption for data confidentiality and
   uses the RC4 and/or RC2 algorithms with a key space of 40
   bits.
b. If both RC4 and RC2 are used in the same software, their
   functionality must be separate; that is, no data can be
   operated on by both routines.
c. The software must not allow the alteration of the data encryption mechanism and its associated key spaces by the user
   or by any other program.
d. The key exchange used in the data encryption must be based
   on either a public-key algorithm with a key space less than
   or equal to a 512-bit modulus and/or a symmetrical algorithm with a key space less than or equal to 64 bits.
e. The software must not allow the alteration of the key management mechanism and its associated key space by the user
   or any other program.

To ensure that the software has properly implemented the approved encryption algorithm(s), the State Department requires that the product pass a *vector test*, in which the vendor receives test data (the vector) and a random key from the State Department, encrypts the vector with the product using the key provided, and returns the result to the State Department. If the product-computed result is identical to the known correct answer, the product automatically qualifies for jurisdiction under the CCL.

Note 2: How much stronger than 40-bit RC2/RC4 is unspecified? Products incorporating the 56-bit DES algorithm are often approved for these informal exemptions, and at times even products using larger key sizes have been approved. But the key size is not unlimited, as may be the case under the explicit categorical exemptions specified in the ITAR.[5]

In September 1994, the Administration promulgated regulations that provided for U.S. vendors to distribute approved products with satellite encryption capabilities for confidentiality directly from the United States to foreign customers without using a foreign distributor and without prior State Department approval for each export. It also announced plans to finalize a *personal use exemption* to allow license-free temporary exports of products with encryption capabilities when intended for personal use. A final rule on the personal use exemption was announced in early 1996 and is discussed later in the chapter. Last, it announced a number of actions intended to streamline the export control process to provide more rapid turnaround for certain *preapproved* products.

Note: Prior to this rule, almost every encryption export required an individual license. Only those exports covered by a distribution arrangement could be shipped without an individual license. This distribution arrangement required a U.S. vendor of products with cryptographic capabilities to export to a foreign distributor who could then resell them to multiple end users. The distribution arrangement had to be approved by the State Department and included some specific language. Under the new rule, a U.S. vendor without a foreign distributor can essentially act as his own distributor, and avoid having to obtain a separate license for each sale. Exporters are required to submit a proposed arrangement identifying, among other things, specific items to be shipped, proposed end users and end use, and countries to which the items are destined. Upon approval of the arrangement, exporters are permitted to ship the specified products directly to end users in the approved countries based on a single license.

In August 1995, the Administration announced a proposal to liberalize export controls on software products with encryption capabilities for confidentiality that use algorithms with a key space of 64 or fewer bits, provided that the key(s) required to decrypt messages and files are *properly escrowed.* Such products would be transferred to the CCL. However, since an understanding of this proposal requires some background in escrowed encryption, discussion of it is deferred to Chapter 13, "Escrowed Satellite Encryption and Related Issues."

Discussion of Current Licensing Practices: The Categorical Exemptions

The categorical exemptions described in the Sidebar, "Categorical Exceptions on the USML for Products Incorporating Satellite Encryption and Informal Practices Governing Licensing" raise a number of issues:

In the case of the 40-bit limitation, the committee was unable to find a specific analytical basis for this figure. Most likely, it was the result of a set of compromises that were politically driven by all of the parties involved. However, whatever the basis for this key size, recent successful demonstrations of the ability to undertake brute-force cryptanalysis on messages encrypted with a 40-bit key (see the Sidebar, "Successful Challenges to 40-Bit Encryption") have led to a widespread perception that such key sizes are inadequate for meaningful information security.

Note: It is worth noting a common argument among many non-government observers that any level of encryption that qualifies for export (that qualifies for control by the CCL, or that is granted an export license under the USML) must be easily defeatable by NSA, or else NSA would not allow it to leave the country. The subtext of this argument is that such a level of encryption is per force inadequate. Of course, taken to its logical conclusion, this argument renders impossible any agreement between national security authorities and vendors and users regarding acceptable levels of encryption for export.

Successful Challenges to 40-Bit Encryption

In the summer of 1995, a message encoded with the 40-bit RC4 algorithm was successfully decrypted without prior knowledge of the key by Damien Doligez of the INRIA organization in France. The message in question was a record of an actual submission of form data that was sent to Netscape's electronic shop order form in *secure* mode (including a fictitious name and address). The challenge was posed to break the encryption and recover the name and address information entered in the form and sent securely to Netscape. Breaking the encryption was accomplished by a brute-force search on a network of about 120 workstations and a few parallel computers at INRIA, Ecole Polytechnique, and ENS. The key was found after scanning a little more than half the key space in 8 days, and the message was successfully decrypted. Doligez noted that many people have access to the amount of computing power that he used, and concluded that the exportable Secure Sockets Layer (SSL) protocol is not strong enough to resist the attempts of amateurs to decrypt a *secure* message.

In January 1996, an MIT undergraduate student used a single $83,000 graphics computer to perform the same task in 8 days. Testing keys at an average rate of more than 830,000 keys per second, the program running on this computer would take 15 days to test every key.[6]

In the case of products intended for use only in banking or money transactions, the exemption results from the recognition by national security authorities that the integrity of the world's financial system is worth protecting with high levels of encryption security. Given the primacy of the U.S. banking community in international financial markets, such a conclusion makes eminent sense. Furthermore, at the time this exemption was promulgated, the financial community was the primary customer for products with encryption capabilities.

This rationale for protecting banking and money transactions naturally calls attention to the possibilities inherent in a world of electronic commerce, in which routine satellite communications will be increasingly likely to include information related to financial transactions. Banks (and retail shops, manufacturers, suppliers, end customers, and so on) will engage in such communications across national borders. In a future world of electronic commerce, connections among nonfinancial institutions may become as important as the banking networks are today. At least one vendor has been granted authority to use strong encryption in software intended for export that would support international electronic commerce (though under the terms of the license, strong encryption applies only to a small portion of the transaction message).

> **Note:** For example, the Kerberos operating system is designed to provide strong cryptographic authentication of users (and hence strong access control for system resources). Typically, Kerberos is distributed in the United States in source code through the Internet to increase its usability on a wide range of platforms, to accommodate diverse user needs, and to increase maintainability. Source code distribution is a common practice on the Internet. However, since Kerberos uses the DES algorithm as the encryption engine to support its authentication features, the source code for Kerberos is controlled under the USML and is not available through the Internet to foreign end users. It is thus fair to say that Kerberos is less used by foreign users than it might be if there were no export controls on products with encryption capabilities, even though the primary purpose of Kerberos is authentication. Kerberos is also designed with operating system calls that support confidentiality. These calls are stripped out of the exportable version of Kerberos, which is only available in object form in any event.

In the case of products useful only for user authentication, access control, and data integrity, the exemption resulted from a judgment that the benefits of more easily available technology for these purposes outweigh whatever costs there might be to such availability. Thus, in principle, these nonconfidentiality products from U.S. vendors should be available overseas without significant restriction.

In practice, however, this is not entirely the case. Export restrictions on confidentiality have some *spillover* effects that reduce somewhat the availability of products that are intended primarily for authentication.

Note: Another example is a company that had eliminated all encryption capabilities from a certain product because of its perceptions of the export control hurdles to be overcome. The capabilities eliminated included those for authentication. While it can be argued that the company was simply ignorant of the exemptions in the ITAR for products providing authentication capabilities, the fact remains that much of the vendor community is either not familiar with the exemptions or does not believe that they represent true *fast-track* or *automatic* exceptions.

Another spillover effect arises from a desire among vendors and users to build and use products that integrate multiple encryption capabilities (for confidentiality and for authentication/integrity) with general-purpose functionality. In many instances, it is possible for encryption for authentication/integrity and encryption for confidentiality to draw on the same algorithm. Export control regulations may require that a vendor weaken or even eliminate the encryption capabilities of a product that also provides authentication/integrity capabilities, with all of the consequent costs for users and vendors as described later in the chapter. Such spillover effects suggest that government actions that discourage capabilities for confidentiality may also have some negative impact on the development and use of products with authentication/integrity capabilities even if there is no direct prohibition or restriction on export of products with capabilities only for the latter.

Informal Noncodified Practices

As previously described, it is current practice to grant USML licenses for exports of strong encryption to firms in a number of categories described in the Sidebar, "Categorical Exceptions on the USML for Products Incorporating Satellite Encryption and Informal Practices Governing Licensing." However, the fact that this practice is not explicitly codified contributes to a sense of uncertainty among vendors and users about the process and in practice leads to unnecessary delays in license processing.

In addition, there is uncertainty about whether or not a given foreign company is *controlled* by a U.S. firm. Specifically, vendors often do not know (and cannot find out in advance) whether a proposed sale to a particular foreign company falls under the protection of this unstated exemption. As a practical rule, the U.S. government has a specific set of guidelines that are used to make this determination. However, these rules require considerable interpretation and thus do not provide clear guidance for U.S. vendors.

Note: Under Defense Department guidelines for determining foreign ownership, control or influence (FOCI), a U.S. company is considered under FOCI *whenever a foreign interest has the power, direct or indirect, whether or not exercised, and whether or not exercisable through the ownership of the U.S. company's securities, by contractual arrangements or other means, to direct or decide matters affecting the management or operations of that company in a manner which may result in unauthorized access to classified information or may affect adversely the performance of classified contracts.* A FOCI determination for a given company is made on the basis of a number of factors, including whether a foreign person occupies a controlling or dominant minority position. This includes the identification of immediate, intermediate and ultimate parent organizations. According to ITAR Regulation 122.2, *ownership* means that more than 50% of the outstanding voting securities of the firm are owned by one or more foreign persons. *Control* means that one or more foreign persons have the authority or ability to establish or direct the general policies or day-to-day operations of the firm. Control is presumed to exist where foreign persons own 25% or more of the outstanding voting securities if no U.S. persons control an equal or larger percentage. The standards for control specified in 22 CFR 60.2(c) also provide guidance in determining whether control in fact exists. Defense Department Form 4415, August 1990, requires answers to 11 questions in order for the Defense Department to make a FOCI determination for any given company.

A third issue that arises with current practice is that the lines between *foreign* and *U.S.* companies are blurring in an era of transnational corporations, ad hoc strategic alliances, and close cooperation between suppliers and customers of all types. For example, U.S. companies often team with foreign companies in global or international ventures. It would be desirable for U.S. products with encryption capabilities to be used by both partners to conduct business related to such alliances without requiring a specific export licensing decision.

Note: In one instance reported to the committee, a major multinational company with customer support offices in China experienced a break-in in which Chinese nationals apparently copied paper documents and computer files. File encryption would have mitigated the impact associated with this *bag job.* Then-current export restrictions hampered deployment of encryption to this site because the site was owned by a foreign (Chinese) company rather than a U.S.-controlled company and therefore not easily covered under then-current practice.

In some instances, USML licenses have granted U.S. companies the authority to use strong encryption rather freely (in the case of a U.S. company with worldwide suppliers). But these licenses are still the result of a lengthy case-by-case review whose outcome is uncertain. Finally, the State Department and NSA explicitly assert control over products without any encryption capability at all but developed with *sockets,* or more formally, encryption applications programming interfaces into which a user can insert his or her own cryptography. Such products are regarded as having an inherent encryption capability (although such capability is latent rather than manifest), and as such are controlled by the USML, even though the text of the ITAR does not mention these items explicitly. In general, vendors and users understand this to be the practice and do not challenge it, but they dislike the fact that it is not explicit.

Note: Specifically, the ITAR place on the USML *encryption devices, software, and components specifically designed or modified therefore, including: cryptographic (including key management) systems, equipment, assemblies, modules, integrated circuits, components or software with the capability of maintaining secrecy or confidentiality of information or information systems.* Note that these categories do not explicitly mention systems without encryption but with the capability of accepting *plug-in* cryptography.

Effectiveness of Export Controls
on Satellite Encryption

One of the most contentious points in the debate over export controls on satellite encryption concerns their effectiveness in delaying the spread of strong cryptographic capabilities and the use of those capabilities throughout the world. Supporters of the current export control regime believe that these controls have been effective. They point to the fact that encryption is not yet in widespread commercial use abroad and that a significant fraction of the satellite communications traffic intercepted globally is unencrypted. Further, they argue that U.S. products with encryption capabilities dominate the international market to an extent that impeding the distribution of U.S. products necessarily affects worldwide usage. Critics of current policy assert that export controls have not been effective in limiting the availability of satellite encryption abroad.

Furthermore, critics of U.S. export controls argue that sources other than U.S. commercial vendors (specifically foreign vendors, the in-house expertise of foreign users, Internet software downloads, and pirated U.S. software) are capable of providing very good encryption that is usable by motivated foreign users. In assessing the arguments of both supporters and critics of the current export control regime, it is important to keep in mind that the ultimate goal of export controls on encryption is to keep strong cryptography out of the hands of potential targets of signals intelligence. Set against this goal, Congress believes that the arguments of both supporters and critics have merit but require qualification.

The supporters of the current export regime are right in asserting that U.S. export controls have had a nontrivial impact in retarding the use of satellite encryption worldwide. This argument is based on three linked factors.

First of all, U.S. export controls on satellite encryption have clearly limited the sale of United States products with encryption capabilities in foreign markets. Indeed, it is this fact that drives the primary objection of U.S. information technology vendors to the current export control regime on satellite encryption.

Second, very few foreign vendors offer integrated products with encryption capabilities. U.S. information technology products enjoy a very high reputation for quality and usability and United States information technology vendors, especially those in the mass-market software arena, have marketing and distribution skills that are as yet unparalleled by their foreign counterparts. As a result, foreign vendors have yet to fill the void left by an absence of U.S. products.

Note: The Department of Commerce and the National Security Agency found no general-purpose software products with encryption capability from non-U.S. manufacturers.

Third, U.S. information technology products account for a large fraction of global sales. For example, a recent U.S. International Trade Commission staff report points out that over half of all world sales in information technology come from the United States. Actions that impede the flow of U.S. products to foreign consumers are bound to have significant effects on the rate at which those products are purchased and used.

On the other hand, it is also true that some foreign targets of interest to the U.S. government today use encryption that is for all practical purposes unbreakable. Major powers tend to use *home-grown* encryption that they procure on the same basis that the United States procures cryptography for its own use; and, export controls on U.S. products clearly cannot prevent these powers from using such encryption.

Furthermore, the fact that satellite encryption is not being widely used abroad does not necessarily imply that export controls are effective (or will be in the near future) in restraining the use of cryptography by those who desire the protection it can provide. The fact is that satellite encryption is not used widely either in the United States or abroad; and, so it is unclear whether it is the lack of information security consciousness described in Chapter 3 or the U.S. export control regime for cryptography that is responsible for such nonuse. Most probably, it is some combination of these two factors.

The critics of the current export regime are right in asserting that foreign suppliers of encryption are many and varied, software products with encryption capabilities are quite available through the Internet (probably hundreds of thousands of individuals have the technical skill needed to download such products), and encryption does pose special difficulties for national authorities wishing to control such technology (see Sidebar, "Difficulties in Controlling Satellite Encryption"). Yet, most products with encryption capabilities available on the Internet are not integrated products. Using security-specific products is generally less convenient than using integrated products (as described in Chapter 3); and, because such products are used less often, their existence and availability pose less of a threat to the collection of signals intelligence.

Difficulties in Controlling Satellite Encryption

Hardware products with encryption capabilities can be controlled on approximately the same basis as traditional munitions. But software products with encryption capabilities are a different matter. A floppy disk containing programs involving encryption is visually indistinguishable from one containing any other type of program or data files. Furthermore, software products with encryption capabilities can be transported electronically, with little respect for physical barriers or national boundaries, over telephone lines, satellites, and the Internet with considerable ease. Encryption algorithms, also controlled by the International Traffic in Arms Regulations as *technical data,* represent pure knowledge that can be transported over national borders inside the heads of people or via letter.

As is true for all other software products, those with encryption capabilities are infinitely reproducible at low cost and with perfect fidelity; hence, a controlled item can be replicated at a large number of points. This fact explains how vast amounts of software piracy can occur both domestically and abroad. In principle, one software product with encryp-tion capabilities taken abroad can serve as the seed for an unlimited number of reproductions that can find their way to hostile parties. Finally, it can be argued that the rogue nations that pose the most important targets for U.S. signals intelligence collection are also the least likely to refrain from pirating U.S. software.[7]

Furthermore, Internet products are, as a general rule, minimally supported and do not have the backing of reputable and established vendors. Whether major vendors will continue to avoid the Internet as a distribution medium remains to be seen. Even today, a number of important products, including Adobe's Acrobat Reader, Microsoft's Word Viewer and Internet Assistant, and the Netscape Navigator are distributed through the Internet. Some vendors make products freely available in limited functionality versions as an incentive for users to obtain full-featured versions. Others make software products freely available to all takers in order to stimulate demand for other products from that vendor for which customers pay.

Users who download software from the Internet may or may not know exactly what code the product contains and may not have the capability to test it to ensure that it functions as described. Corporate customers, the primary driver for large-scale deployment of products, are unlikely to rely on products that are not sold and supported by reputable vendors; and, it is products with a large installed base (those created by major software vendors) that would be more likely to have the high-quality satellite encryption that poses a threat to signals intelligence. The Sidebar, "Key Differences between Commercial Products And Freeware" describes the primary differences between commercial products and *freeware* available on the Internet.

Note: Indeed, the lack of quality control for Internet-available software provides an opportunity for those objecting to the proliferation of good products with encryption capability to flood the market with their own products anonymously or pseudoanonymously. Such products may include features that grant clandestine access with little effort.

Key Differences between Commercial Products and Freeware

	Products from Major Commercial Vendors	*Freeware* Products
Stake of reputation of product offer	Higher	Lower
Scale of operation	Larger	Smaller
Cost of distribution	Higher	Lower
Support for products	Greater	Lesser
Role of profit-making motive	Higher	Lower
Ability to integrate cryptography into useful and sophisticated general-purpose software	Greater	Lesser
Vulnerability to regulatory and legal constraints	Higher	Lower
Likelihood of market *staying power*	Higher	Lower *r*
Likelihood of wide distribution and use	Higher	Lower
Financial liability for poor product performance	Higher	Lower
Cost of entry into markets	Higher	Lower

Note: All of the characterizations listed are tendencies rather than absolutes, and are relative (determined by comparing products from major commercial vendors to freeware).[8]

A brief NSA survey of product literature describing foreign stand-alone security-specific products with encryption capabilities (see the Sidebar, "A Partial Survey of Foreign Encryption Products on the NSA Survey") also indicated many implementations that were unsound from a security standpoint — even taking for granted the mathematical strength of the algorithms involved and the proper implementation of the indicated algorithms. The NSA analysis of foreign stand-alone products for encryption was based on material it had collected through its survey. This material was limited to product brochures and manuals, which NSA believes puts the best possible face on a product's quality. Thus, NSA's identification of security defects in these products is plausibly regarded as a minimum estimate of their weaknesses — more extensive testing (involving disassembly) would be likely to reveal additional weaknesses, as implementation defects would not be written up in a product brochure. Moreover, the availability of a product brochure does not ensure the availability of the corresponding product. The NSA has brochures for all of the 800-plus products identified in its survey, but due to limited resources, it has been able to obtain physical versions (a disk, a circuit board) of fewer than 10% of the products described in those brochures.

The NSA has no reason to believe that the stand-alone security-specific products with encryption capabilities made by U.S. vendors are on average better at providing security. An *amateur* review of encryption for confidentiality built into several popular U.S. mass-market software programs noted that the encryption facilities did not provide particularly good protection. The person who reviewed these programs was not skilled in encryption but was competent in his understanding of programming and how the Macintosh manages files. By using a few commonly available programming tools (a file compare program, a *debugger* that allows the user to trace the flow of how a program executes, and a *disassembler* that turns object code into source code that can be examined), the reviewer was able to access in less than two hours the *protected* files generated by four out of eight programs.

Note: One well-publicized encryption security flaw found in the Netscape Corporation's Navigator Web browser is discussed in Chapter 3. Because of a second flaw, Netscape Navigator could also enable a sophisticated user to damage information stored on the host computer to which Navigator is connected.

Nevertheless, large established software vendors in the United States do have reputations for providing relatively high quality in their products for features unrelated to security. Without an acceptable product certification service, most users have no reliable way of determining the quality of any given product for themselves.

Note: In addition, a product with a large installed base is subject to a greater degree of critical examination than a product with a small installed base, and hence flaws in the former are more likely to be noticed and fixed. Large installed bases are more characteristic for products produced by established vendors than of freeware or shareware producers.

A Partial Survey of Foreign Encryption Products on the NSA Survey

A British product manual notes that *a key can be any word, phrase, or number from 1 to 78 characters in length, though for security purposes keys shorter than six characters are not recommended.* Only alphanumeric characters are used in the key, and alpha characters do not distinguish between upper and lower case. While the longer pass phrases can produce keys with the full 56 bits of uncertainty (changing *can* to *do* would require more extensive tests), passwords of even 6 characters are woefully inadequate. It is dangerous to allow users to enter such keys, much less the single-character keys allowed by this product.

One British product is a DES implementation that recommends cipher block chaining, but uses electronic codebook (ECB) mode as the default. The use of ECB as the default is dangerous because ECB is less secure than cipher block chaining.

A Danish product uses DES with an 8-character key, but limits each character to alphanumeric and punctuation symbols. Hence the key is less than a full 56 bits long. With this restriction, many users are likely to use only upper or lower case alpha characters, resulting in a key less than 40 bits long.

A foreign product uses the FEAL algorithm as well as a proprietary algorithm. Aside from the question of algorithm strength, the key is 1 to 8 characters long and does not distinguish between upper and lower case. The result is a ridiculously short key, a problem that is compounded by the recommendation in the manual to use a 6- to 8-letter artificial word as the key (it suggests that for the name Bill, *billbum* might be used as the key).

A product from New Zealand uses DES plus a public-key system similar to RSA, but based on Lucas functions. The public-key portion limits the key size to 1024 bits, but does not seem to have a lower bound, a potentially dangerous situation. The DES key can be 1 to 24 characters in length. If the key is 1 to 8 characters, then single DES is used, otherwise triple DES is used. The lack of a lower bound on key length is dangerous.

An Israeli product uses DES or QUICK, a proprietary algorithm. The minimum key length is user selectable between 0 and 8 characters. Allowing such small lower bounds on key length is dangerous. The product also has a *super-password* supplied by the vendor, another potentially dangerous situation. This product is available both in hardware and in software.

A German hardware product has user-settable S-boxes, and the key can be entered either as 8 characters or 16 hexadecimal characters to yield a true 64-bit key (which will be reduced by the algorithm to 56 bits). The use of 16 hexadecimal character keys will result in higher security, but if the key can also be entered as 8 alphanumeric characters, many users are likely to do so, thus severely reducing the security level. User-selectable S-boxes can have advantages (if they are unknown to a cryptanalyst) and disadvantages (if they are poorly chosen and either are known to or can be guessed by a cryptanalyst). On balance, the danger is arguably greater than the advantage.

One British product recommends one master key per organization so that files can be shared across personal computers. This practice is very dangerous.

To summarize, the defects in these products are related to poor key management practices, because they either employ or allow poor key management that would enable a determined and knowledgeable adversary to penetrate the security they offer with relative ease. As previously discussed, U.S. products are not necessarily more secure.[9]

As a general rule, a potential user of encryption faces the choice of buying commercially available products with encryption capabilities on the open market (perhaps custom-made, perhaps produced for a mass market) or developing and deploying those products independently. The arguments previously discussed suggest that global dissemination of knowledge about satellite encryption makes independent development an option, but the problems of implementing knowledge as a usable and secure product drive many potential users to seek products available from reputable vendors. In general, the greater the resources available to potential users and the larger the stakes involved, the more likely they are to attempt to develop their own encryption resources. Thus, large corporations and First World governments are, in general, more likely than small corporations and Third World governments to develop their own encryption implementations.

Finally, the text of the ITAR seems to allow a number of entirely legal actions that could have results that the current export control regime is intended to prevent (see the Sidebar that directly follows). For example, RSA Data Security Inc. has announced a partnership with the Chinese government to fund an effort by Chinese government scientists to develop new encryption software. This software may be able to provide a higher degree of confidentiality than software that qualifies today for liberal export consideration under the CCL.

Circumventions of the ITAR

Current export controls on encryption can apparently be circumvented in a number of entirely legal and/or hard-to-detect ways. For example, a U.S. company can develop a product without encryption capabilities and then sell the source code of the product to a friendly foreign company that incorporates additional source code for encryption into the product for resale from that foreign country (assuming that that country has no (or weaker) export controls on satellite encryption).

In addition, a U.S. company possessing products with encryption capabilities can be bought by a foreign company. In general, no attempt is made to recover those products. Finally, a U.S. company can work with legally independent counterparts abroad that can incorporate cryptographic knowledge available worldwide into products.[10]

The Impact of Export Controls on U.S. Information Technology Vendors

United States export controls have a number of interrelated effects on the economic health of U.S. vendors and on the level of cryptographic protection available to U.S. firms operating domestically. The impact of foreign import controls on U.S. vendors is discussed in Chapter 14, "Other Dimensions of National Satellite Encryption Policy."

De Facto Restrictions on the Domestic Availability of Satellite Encryption

Current law and policy place no formal restrictions whatever on products with encryption capabilities that may be sold or used in the United States. In principle, the domestic market can already obtain any type of satellite encryption it wants. For stand-alone security-specific products, this principle is true in practice as well. However, the largest markets are not for stand-alone security-specific products, but rather for integrated products with encryption capabilities.

For integrated products with encryption capabilities, export controls do have an effect on domestic availability. For example, the Netscape Communications Corporation distributes a version of Netscape Navigator over the Internet and sells a version as shrink-wrapped software. Because the Internet version can be downloaded from abroad, its encryption capabilities are limited to those that will allow for liberal export consideration; the shrink-wrapped version is under no such limitation and in fact is capable of much higher levels of encryption. Because it is so much more convenient to obtain, the Internet version of Netscape Navigator is much more widely deployed in the United States than is the shrink-wrapped version, with all of the consequences for information security that its weaker encryption capability implies.

Note: The shrink-wrapped version of Netscape Navigator sold within the United States and Canada supports several different levels of encryption, including 40-bit RC4, 128-bit RC4, 56-bit DES, and triple-DES. The default for a domestic client communicating with a domestic server is 128-bit RC4.

When Microsoft Corporation received permission to ship Windows NT Version 4.0, it shipped a product that incorporated a cryptographic applications programming interface approved by the U.S. government for commodity jurisdiction to the CCL. Furthermore, this product was being shipped worldwide with a cryptographic module that provided encryption capabilities using 40-bit RC4. While domestic users may replace the default module with one providing stronger encryption capabilities, many will not, and the result is a weaker encryption capability for those users.

For example, another major U.S. software vendor distributes its major product in modular form in such a way that the end user can assemble a system configuration in accordance with local needs. However, since the full range of USML export controls on encryption is applied to modular products into which cryptographic modules may be inserted, this vendor has not been able to find a sensible business approach to distributing the product in such a way that it would qualify for liberal export consideration. The result has been that the encryption capabilities provided to domestic users of this product are much less than they would otherwise be in the absence of export controls.

What factors underlie the choices made by vendors that result in the outcomes just previously described? At one level, the previously described examples are simply the result of market decisions and preferences. At a sufficiently high level of domestic market demand, U.S. vendors would find it profitable and appropriate to develop products for the domestic market alone. Similarly, given a sufficiently large business opportunity in a foreign country (or countries) that called for a product significantly different from that used by domestic users, vendors would be willing to develop a customized version of a product that would meet export control requirements. Furthermore, many other manufacturers of exportable products must cope with a myriad of different requirements for export to different nations (differing national standards for power, safety, and electromagnetic interference), as well as differing languages in which to write error messages or user manuals. From this perspective, export controls are simply one more cost of doing business outside the United States.

On the other hand, the fact that export controls are an additional cost of doing business outside the United States is not an advantage for U.S. companies planning to export products. A vendor incurs less expense and lower effort for a single version of a product produced for both domestic and foreign markets than it does when multiple versions are involved. While the actual cost of developing two different versions of a product with different key lengths and different algorithms is relatively small, a much larger part of the expense associated with multiple versions relates to marketing, manufacture, support, and maintenance of multiple product versions after the initial sale has been made.

> **Note:** Development and support concerns are even more significant when a given product is intended for cross-platform use (for use in different computing environments such as Windows, Mac OS, Unix, and so on), as is the case for many high-end software products (such as database retrieval systems). When a product is intended for use on 50 different platforms, and you multiply by a factor of two, the effort required on the part of the vendor entails much more of an effort than if the product were intended for use on only one platform.

Since a vendor may be unable to export a given product with encryption capabilities to foreign markets, domestic market opportunities must be such that it warrants a domestic-only version. Given that about half of all sales of U.S. information technology vendors are made to foreign customers, the loss of foreign markets can be quite damaging to a U.S. vendor. When they are not, vendors have every incentive to develop products with encryption capabilities that would easily qualify for liberal export consideration. As a result, the domestic availability of products with strong encryption capability is diminished.

While a sufficiently high level of domestic market demand would make it profitable for U.S. vendors to develop products for the domestic market alone, the *sufficient* qualifier is a strong one indeed. Also, users are affected by an export control regime that forces foreign and domestic parties in communication with each other to use encryption systems based on different algorithms and/or key lengths. In particular, an adversary attempting to steal information will seek out the weakest point. If that weakest point is abroad because of the weak encryption allowed for liberal export, then that is where the attack will be. In businesses with worldwide satellite communication connections, it is critical that security measures be taken abroad, even if key information repositories and centers of activity are located in the continental United States. Put differently, the use of weak encryption abroad means that sensitive information communicated by U.S. businesses to foreign parties faces a greater risk of compromise abroad because stronger cryptography integrated into U.S. information technology is not easily available abroad.

Finally, the export licensing process can have a significant impact on how a product is developed. For example, until recently, products developed to permit the user to substitute easily his or her own encryption module were subject to the USML and the ITAR.

> **Note:** The use of object-oriented software technology can in general facilitate the use of applications programming interfaces that provide *hooks* to modules of the user's choosing. A number of vendors have developed or are developing general purpose applications programming interfaces that will allow the insertion of a module to do almost anything. Since these programming interfaces are not specialized for encryption, but instead enable many useful functions (file compression, backups), it is very difficult to argue the basis on which applications incorporating these interfaces should be denied export licenses simply because they *could* be used to support encryption. A further discussion of recent developments involving satellite encryption modules and cryptographic applications programming interfaces is contained in Chapter 26, "Satellite Encryption Technology Policy Options, Analysis and Forecasts."

For example, one vendor pointed out that its systems were designed to be assembled *out of the box* by end users in a modular fashion, depending on their needs and computing environment. This vendor believed that such systems would be unlikely to obtain liberal export consideration, because of the likelihood that a foreign user would be able to replace an *export-approved* encryption module with a cryptography module that would not pass export review. Under these circumstances, the sensible thing from the export control perspective would be to deny exportability for the modularized product even if its capabilities did fall within the *safe harbor* provisions for products with encryption capabilities.

The considerations just previously discussed led the National Research Council's (NRC) National Cryptographic Study Committee to conclude that U.S. export controls have had a negative impact on the cryptographic strength of many integrated products with encryption capabilities available in the United States. Export controls tend to drive major vendors to a *least common denominator* cryptographic solution that will pass export review as well as sell in the United States. The committee also believes that export controls have had some impact on the availability of cryptographic authentication capabilities around the world. Export controls distort the global market for satellite encryption, and the product decisions of vendors that might be made in one way in the absence of export controls may well be made another way in their presence.

> **Note:** A similar conclusion was reached by the FBI, whose testimony to the committee noted that *the use of export controls may well have slowed the speed, proliferation, and volume of encryption products sold in the U.S.*

Some of the reasons for this vendor choice are explored next.

Regulatory Uncertainty Related to Export Controls

A critical factor that differentiates the costs of complying with export controls from other costs of doing business abroad is the unpredictability of the export control licensing process. Other dimensions of uncertainty for vendors not related to export controls are discussed in Chapter 14, "Other Dimensions of National Satellite Encryption Policy."

Nevertheless, a company must face the possibility that despite its best efforts, a USML export license or a commodity jurisdiction to the CCL will not be granted for a product. Uncertainties about the decisions that will emerge from the export control regime force vendors into very conservative planning scenarios. In estimating benefits and costs, corporate planners must take into account the additional costs that could be incurred in developing two largely independent versions of the same product or limit the size of the potential market to U.S. purchasers. When such planning requirements are imposed, the number of product offerings possible is necessarily reduced.

The United States Munitions List licensing is particularly unpredictable, because the reasons that a license is denied in any given instance are not necessarily made available to the applicant. In some cases, the rationale for specific licensing decisions is based on considerations that are highly classified and by law cannot be made available to an uncleared applicant. Because such rationales cannot be discussed openly, an atmosphere of considerable uncertainty pervades the development process for vendors seeking to develop products for overseas markets. Furthermore, there is no independent adjudicating forum to which a negative licensing decision can be appealed.

As USML licensing is undertaken on a case-by-case basis, it requires the exercise of judgment on the part of the regulatory authorities. A judgment-based approach has the disadvantage that it requires a considerable degree of trust between the regulated and the regulator.

Note: In contrast to a judgment-based approach, a clarity-based approach would start from the premise that regulations and laws should be as clear as possible, so that a party that may be affected knows with a high degree of certainty what is and is not permitted or proscribed. The downside of a clarity-based approach is that affected parties tend to go *right up to the line* of what is prohibited and may seek ways to *design around* any stated limitations. Furthermore, a clarity-based approach would require the specification, in advance, of all acts that are prohibited, even when it may not be possible to define in advance all acts that would be undesirable.

To the extent that an individual regulated party believes that the regulator is acting in the best interests of the entire regulated community, it is natural that it would be more willing to accept the legitimacy of the process that led to a given result. However, in instances in which those that are regulated do not trust the regulator, judgments of the regulator are much more likely to be seen as arbitrary and capricious.

Note: For example, critics of the uncertainty engendered by the export regime point out that uncertainty is helpful to policy makers who wish to retain flexibility to modify policy without the work or publicity required for a formal regulatory change.

This situation currently characterizes the relationship between encryption vendors/users and national security authorities responsible for implementing the U.S. export control regime for cryptography. In input received by NSA, virtually all industry representatives, from large to small companies, testified about the unpredictability of the process. From the vendor point of view, the resulting uncertainty inhibits product development and allows negative decisions on export to be rendered by unknown forces and/or government agencies with neither explanation nor a reasonable possibility of appeal.

The need to stay far away from the vague boundaries of what might or might not be acceptable is clearly an inhibitor of technological progress and development. Vendor concerns are exacerbated in those instances in which export control authorities are unwilling to provide a specific reason for the denial of an export license or any assurance that a similarly but not identically configured product with encryption capabilities would pass export review. Even worse from the vendor perspective, product parameters are not the only determinant of whether a licensing decision will be favorable except in a very limited and narrow range of cryptographic functionality.

The uncertainty just previously described is not limited to new and inexperienced vendors encountering the U.S. export control regime for the first time. Large and sophisticated institutions with international connections have also encountered difficulties with the current export control regime. For example, a representative from a major U.S. bank with many international branches reported that export controls affect internally developed bank software with encryption capabilities. A U.S. citizen who works on bank software with encryption capabilities in England may *taint* that software so that it falls under U.S. export control guidelines. Thus, despite the fact that the current export control regime treats banks and other financial institutions relatively liberally, major banks have still struggled under the limitations of the export control regime.

The situation is worse for smaller companies. While large companies have experience and legal staffs that help them to cope with the export control regime, small companies do not. New work on information technology often begins in garage-shop operations, and the export control regime can be particularly daunting to a firm with neither the legal expertise nor the contacts to facilitate compliance of a product with all of the appropriate regulations. These companies in particular are the ones most likely to decide in the end to avoid entirely the inclusion of cryptographic features due to concern of running afoul of the export control rules. The following three examples illustrate how the unpredictability of the export control licensing process has affected U.S. vendors and their products.

Modularity

As previously noted, cryptographic applications programming interfaces that are directly and easily accessible to the user are in general subject to USML licensing. However, even *closed* interfaces that are not easily accessible to the user are sometimes perceived to pose a risk for the vendor. One major product vendor reported to NSA that it was reluctant to use modular development for fear that even an internal module interface could keep a product from passing export control review. Any software product that uses modular techniques to separate the basic product functionality from the encryption has a well-defined interface between the two. Even when the software product is converted to object code, that interface is still present (though it is hidden from the casual user). However, the interface cannot in general be hidden from a person with strong technical skills, and such a person would be able to find it and tamper with it in such a way that a different encryption module could be used. A number of similar considerations apply for hardware products, in which the encryption capabilities might be provided by a *plug-in* chip.

> **Note:** Of course, such considerations obviously apply to software products with encryption capabilities that are designed to be shipped in source code. Not only can the encryption module be easily identified and replaced, but it can also be pulled out and adapted to other purposes. This point was raised earlier in the chapter.

The alternative to the use of modular techniques in the development of integrated products would complicate the *swap-in/swap-out* of encryption capabilities. This consists of lines of code (if software) and wires (if hardware) that implemented cryptographic capabilities would be highly interwoven with lines of code and wires that implemented the primary capabilities of the product. On the other hand, this approach would be tantamount to the development of two largely distinct products with little overlap in the work that was required to produce them.

The NSA has spoken publicly about its willingness to discuss with vendors from the early stages of product design features and capabilities of proposed products with encryption capabilities for confidentiality so that the export license approval process can be facilitated; and, its willingness to abide by nondisclosure agreements to reassure vendors that their intellectual property rights will be protected.

> **Note:** For example, NSA representatives have made comments to this effect at every RSA Data Security Conference they have attended.

Nonetheless, the receipt of an export control license useful for business purposes is not guaranteed by such cooperation. For example, while decisions about commodity jurisdiction often provide CCL jurisdiction for object code and USML jurisdiction for source code (and thus need not inhibit modular product development if the product is to be distributed in object form only), the fact remains that such decisions are part of a case-by-case review whose outcome is uncertain. Different vendors are willing to tolerate different levels of risk in this regard, depending on the magnitude of the investments involved.

As a general rule, NSA does not appear willing to make agreements in advance that will assure licenses for a product that has not yet been instantiated or produced. Such a position is not unreasonable given NSA's stance toward products with encryption capabilities in general, and the fact that the true capabilities of a product may depend strongly on how it is actually implemented in hardware or software. Thus, vendors have no indemnification against the risk that a product might not be approved.

> **Note:** Although other industries also have to deal with the uncertainties of regulatory approval regarding products and services, the export control process is particularly opaque, because clear decisions and rationales for those decisions are often not forthcoming (and indeed are often classified and/or unrelated to the product per se).

The Definition of Export

There is uncertainty about what specific act constitutes the *export* of software products with encryption capabilities. It is reasonably clear that the act of mailing to a foreign country a disk with a product with encryption capabilities on it constitutes an export of that product. But if that product is uploaded to an Internet site located in the United States and is later downloaded by a user located in another country, is the act of export the upload or the download? What precautions must be taken by the uploader to remain on the legal side of the ITAR?

The NSA has been unable to find any formal document that indicates answers to these questions. However, a March 1994 letter from the State Department Office of Defense Trade Controls appears to indicate that a party could permit the posting of cryptographic software on an Internet host located in the United States if

> *(a) the host system is configured so that only people originating from nodes in the United States and Canada can access the cryptographic software, or (b) if the software is placed in a file or directory whose name changes every few minutes, and the name of the file or directory is displayed in a publicly known and readable file containing an explicit notice that the software is for U.S. and Canadian use only.*

Of course, such a letter does not provide formal guidance to parties other than the intended addressee (indeed, under the ITAR, advisory opinions provided to a specific party with a given set of circumstances are not binding on the State Department even with respect to that party), and so the issue remains murky.

The Speed of the Licensing Process

Uncertainty is also generated by a lengthy licensing process without time lines that allow vendors to make realistic schedules. The Sidebar, "Problems Arising from a Lengthy Export Licensing Process" describes some of the problems reported to NSA. To summarize, the perceptions of many vendors about the excessive length of time it takes to obtain a license reflects the time required for discussions with NSA about a product before an application is formally submitted. The prospect of facing the export control process deters some vendors entirely from creating certain products at all. By contrast, NSA starts the clock only when it receives a formal application; and, in fact the usual time between receipt of a formal application and rendering of a decision is relatively short (a few weeks). The reason that such a fast turnaround is possible is that by the time the application is received, enough is known about the product involved that processing is routine because there is no need for negotiation about how the product must be changed for a license to be approved.

Problems Arising from a Lengthy Export Licensing Process

Some foreign customers know it will take a long time to obtain a positive licensing decision, and as a consequence do not bother to approach U.S. vendors at all. Products to market are delayed; even when export licenses are eventually granted, they are often granted too late to be useful, because the area of information technology is so fast-moving.

Rapid decisions are not rendered. In one instance reported to NSA, a U.S. information technology company wanted permission to use its own software (with strong encryption capabilities) to communicate with its foreign offices. Such cases are in theory expedited because of a presumptive approval in these circumstances. This vendor's government contacts agreed that *such an application would be no problem* and that an approval would be a rapid *rubber-stamp* one; but, in fact, this vendor is still awaiting a license after more than a year.

System integrators intending to ship complete systems rather than individual products face particular difficulties in obtaining a speedy turnaround. Why? Because, the task for national security authorities involves an assessment of the entire system into which a given product (or products) with encryption capabilities will be integrated, rather than an assessment of just the products with encryption capabilities alone.

Even vendors that manufacture encryption software not intended for export are required to register with the State Department Office of Defense Trade Controls. This is primarily *to provide the U.S. government with necessary information on who is involved in certain manufacturing and exporting activities.*

> **Note:** See International Traffic in Arms Regulations, Section 122.1 (c).[11]

In response to some of these concerns, the U.S. government has undertaken a number of reforms of the export control regime (described earlier in the chapter) to reduce the hassle and red tape involved in obtaining export licenses. These reforms are important.

> **Note:** For example, according to NSA, the detailing of an NSA representative to work with the State Department Office of Defense-Trade Controls has resulted in a considerable reduction in the time needed to process a license.

Nevertheless, the pace at which new information technology products develop and the increasing complexity of those products will complicate product review efforts in the future. Given relatively fixed staffing, these factors will tend to increase the length of time needed to conduct product reviews at a time when vendors are feeling pressures to develop and market products more rapidly.

One particular reform effort that deserves discussion is the *personal use* exemption. For many years, Americans traveling abroad were required under the ITAR to obtain *temporary export licenses* for products with encryption capabilities carried overseas for their personal use. The complexity of the procedure for obtaining such a license was a considerable burden for U.S. business people traveling abroad; and, these individuals were subject to significant criminal penalties for an act that was widely recognized to be harmless and well within the intent of the export control regime.

In February 1994, the Administration committed itself to promulgating regulations to support a personal-use exemption from the licensing requirement. Two years later, on February 16, 1996, the *Federal Register* contained a notice from the Department of State, Bureau of Political Military Affairs, announcing the final rule of an amendment to the International Traffic in Arms Regulation (ITAR) allowing U.S. persons to temporarily export cryptographic products for personal use without the need for an export license.

> **Note:** According to the regulation, the product must not be intended for copying, demonstration, marketing, sale, re-export, or transfer of ownership or control. It must remain in the possession of the exporting person, which includes being locked in a hotel room or safe. While in transit, it must be with the person's accompanying baggage. Exports to certain countries are prohibited — currently Cuba, Iran, Iraq, China, India, Libya, North Korea, Sudan, and Syria. The exporter must maintain records of each temporary export for 5 years.

Some critics of government policy have objected to the particular formulation of the record-keeping requirement. All parties involved — including senior Administration officials — have agreed that 2 years was far too long a period for promulgation of so simple a rule.

The Size of The Affected Market for Satellite Encryption

Since export controls on products with satellite encryption capabilities constrain certain aspects of sales abroad, considerable public attention has focused on the size of the market that may have been affected by export controls. Vendors in particular raise the issue of market share with considerable force:

First of all, *the only effect of the export controls is to cause economic harm to US software companies that are losing market share in the global encryption market to companies from the many countries that do not have export controls.* Second, *the government's current policy on encryption is anti-competitive. The government's encryption export policy jeopardizes the future of the software industry — one of the fastest growing and most successful industries.*

The size of the market for products with encryption capabilities cuts across many dimensions of cryptography policy, but because it is raised most often in the context of the export control debate, it is addressed in this part of the chapter. Therefore, plausible arguments can be made that the market ranges from no more than the value of the security-specific products sold annually (several hundred million dollars per year-a low-end estimate) to the total value of all hardware and software products that might include encryption capabilities (many tens of billions of dollars—a high-end estimate). The NSA was unable to determine the size of the information technology market directly affected by export controls on encryption to within a factor of more than 100—a range of uncertainty that renders any estimate of the market quite difficult to use as the basis for a public policy decision.

Of course, it is a matter of speculation what fraction of the information technology market (on the order of $282 billion in 1997; see next) might usefully possess encryption capabilities. Good arguments can be made to suggest that this fraction is very small or very large. A number of information technology trade organizations have also made estimates. The Software Publishers Association cited a survey by the National Computer Security Association that quoted a figure of $234 million in aggregate known losses in 1997 because of export controls. In 1997, the Business Software Alliance estimated that *approximately $10–13 billion in U.S. company revenues are currently at risk because of the inability of those companies to be able to sell worldwide generally available software with encryption capabilities employing DES or other comparable strength algorithms.* The Computer Systems Policy Project (CSPP) estimated that in 2005, the potential annual revenue exposure for U.S. information technology vendors would range from $8 billion to $11 billion on sales of cryptographic products, including both hardware and software. CSPP also estimated $80 billion to $110 billion in potential revenue exposure on sales of associated computer systems.

The $282 billion figure is taken from the Department of Commerce, and includes computers and peripherals ($91.6 billion), packaged software ($46.9 billion), information services ($20.0 billion), data processing and network services ($67.9 billion), and, systems integration/custom programming services ($55.6 billion). Note that this figure does not include some other industry sectors that could, in principle, be affected by regulations regarding secure communications. In 1997, U.S. companies provided telecommunications services valued at $15.2 billion to foreign nations and shipped $25.6 billion (1996 dollars) in telephone equipment worldwide.

Nevertheless, although it is not large enough to be decisive in the policy debate, the floor of such estimates—a few hundred million dollars per year—is not a trivial sum. Furthermore, all trends point to growth in this number, growth that may well be very large and nonlinear in the near future. To the extent that both of these observations are valid, it is only a matter of a relatively short time before even the floor of any estimate will be quite significant in economic terms. The next part of the chapter describes some of the factors that confound the narrowing of the large range of uncertainty in any estimate of the size of the market affected by export controls.

Defining a Lost Sale

A number of vendors have pointed to specific instances of lost sales as a measure of the harm done to the vendors as the result of export controls on cryptography. National security officials believe that these figures are considerably overstated.

> **Note:** For example, in a presentation to the NRC National Cryptography Policy Committee on July 19, 1995, the Software Publishers' Association documented several specific instances in which a U.S. company had lost a sale of a product involving encryption to a foreign firm. These instances included a company that lost one-third of its total revenues because export controls on DES-based encryption prevented sales to a foreign firm; a company that could not sell products with encryption capability to a European company because that company resold products to clients other than financial institutions; a U.S. company whose European division estimated at 50% the loss of its business among European financial institutions, defense industries, telecommunications companies, and government agencies because of inadequate key sizes; and, a U.S. company that lost the sale of a DES based system to a foreign company with a U.S. subsidiary.

Administration officials and congressional staff have expressed considerable frustration in pinning down a reliable estimate of lost sales. It is important to begin with the understanding that the concept of a *lost sale* is intrinsically soft. Trying to define the term *lost sales* raises a number of questions.

For instance, what events count as a sale lost because of export restrictions? Several possibilities illustrate the complications.

First of all, a U.S. vendor is invited along with foreign vendors to bid on a foreign project that involves encryption, but declines because the bid requirements are explicit. And the U.S. vendor knows that the necessary export licenses will not be forthcoming on a time scale compatible with the project.

Again, a U.S. vendor is invited along with foreign vendors to bid on a foreign project that involves encryption. In order to expedite export licensing, the U.S. vendor offers a bid that involves 40-bit encryption (thus ignoring the bid requirements), and the bid is rejected.

Still again, a U.S. vendor is invited along with foreign vendors to bid on a foreign project that involves encryption. A foreign vendor emerges as the winner. The sale is certainly a lost sale, but as customers often make decisions with a number of reasons in mind and may not inform losing vendors of their reasons, it is difficult to determine the relationship of export controls to the lost sale.

In the fourth illustration, no U.S. vendor is invited to bid on a foreign project that involves encryption. In such an instance, the potential foreign customer may have avoided U.S. vendors, recognizing that the encryption would subject the sale to U.S. export control scrutiny — possibly compromising sensitive information or delaying contract negotiations inordinately. On the other hand, the potential customer may have avoided U.S. vendors for other reasons (because the price of the U.S. product was too high).

What part of a product's value is represented by the cryptographic functionality that limits a product's sales when export controls apply? As noted in Chapter 3, stand-alone products with encryption capabilities are qualitatively different from general-purpose products integrated with encryption capabilities. A security-specific stand-alone product provides no other functionality, and so the value of the cryptography is the entire cost of the product. But such sales account for a very small fraction of information technology sales. Most sales of information technology products with encryption capabilities are integrated products. Many word processing and spreadsheet programs may have encryption capabilities, but users do not purchase such programs for those capabilities — they purchase them to enhance their ability to work with text and numbers. Integrated products intended for use in networked environments (*groupware*) may well have encryption capability, but such products are purchased primarily to serve collaboration needs rather than encryption functions. In these instances, it is the cost of the entire integrated product (which may not be exportable if encryption is a necessary but secondary feature) that counts as the value lost.

How does a vendor discover a *lost sale*? In some cases, a specific rejection counts as evidence. But in general there is no systematic way to collect reliable data on the number or value of lost sales.

An often-unnoticed dimension of *lost sales* does not involve product sales at all, but rather services whose delivery may depend on cryptographic protection. For example, a number of U.S. on-line service providers (America Online, Compuserve, Prodigy) are intending to offer or expand access abroad. The same is true for U.S. providers of telecommunications services. To the extent that maintaining the security of foreign interactions with these service providers depends on the use of strong encryption, the ability of these companies to provide these services may be compromised by export restrictions and thus sales of service potentially reduced.

Latent versus Actual Demand

In considering the size of the market for encryption, it is important to distinguish between *actual* demand and *latent* demand. Actual demand reflects what users spend on products with encryption capabilities. While the value of *the market for encryption* is relatively well defined in the case of stand-alone security-specific products (it is simply the value of all of the sales of such products), it is not well defined when integrated products with encryption capabilities are involved. The reason is that for such products, there is no demand for encryption per se. Rather, users have a need for products that do useful things. Encryption is a feature added by designers to protect users from outside threats to their work, but as a purely defensive capability, encryption does not so much add functional value for the user as protection against reductions in the value that the user sees in the product. Lotus Notes, for example, would not be a viable product in the communications software market without its encryption capabilities, but users buy it for the group collaboration capabilities that it provides rather than for the encryption per se.

Latent demand (inherent demand that users do not realize or wish to acknowledge but that surfaces when a product satisfying this demand appears on the market) is even harder to measure or assess. Recent examples include Internet usage and faxes. In these instances, the underlying technology has been available for many years, but only recently have large numbers of people been able to apply these technologies for useful purposes. Lower prices and increasing ease of use, prompted in part by greater demand, have stimulated even more demand. To the extent that there is a latent demand for encryption, the inclusion of cryptographic features into integrated products might well stimulate a demand for encryption that grows out of knowledge and practice, and out of learning by doing.

Determining the extent of latent demand is complicated greatly by the fact that latent demand can be converted into actual demand on a relatively short time scale. Indeed, such growth curves — very slow growth in use for a while and then a sudden explosion of demand — characterize many critical mass phenomena. In other words, some information technologies (networks, faxes, telephones) are valuable only if some critical mass of people use them. Once that critical mass is reached, other people begin to use those technologies, and demand takes off. Linear extrapolations 5 or 10 years into the future based on 5 or 10 years in the past miss this very nonlinear effect.

Of course, it is difficult to predict a surge in demand before it actually occurs. In the case of encryption, market analysts have been predicting significantly higher demand for many years. Today, growth rates are high, but demand for information security products including encryption is not yet ubiquitous.

Two important considerations bearing directly on demand are increasing system complexity and the need for interoperability. Users must be able to count on a high degree of interoperability in the systems and software they purchase if they are to operate smoothly across national boundaries (as described in Chapter 2). Users understand that it is more difficult to make different products interoperate, even if they are provided by the same vendor, than to use a single product. For example, the complexity of a product generally rises as a function of the number of products with which it must interoperate, because a new product must interoperate with already deployed products. Increased complexity almost always increases vulnerabilities in the system or network that connects those products. In addition, more complex products tend to be more difficult to use and require greater technical skill to maintain and manage. Thus, purchasers tend to shy away from such products. This reluctance, in turn, dampens demand, even if the underlying need is still present.

From the supply side, vendors feel considerable pressure from users to develop interoperable products. But, greater technical skills are needed by vendors to ensure interoperability among different product versions than to design a single product that will be used universally—just as they are for users involved in operation and maintenance of these products. Requirements for higher degrees of technical skill translate into smaller talent pools from which vendors can draw and thus fewer products available that can meet purchasers' needs for interoperability.

Problems relating to interoperability and system complexity, as well as the size of the installed base, have contributed to the slow pace of demand to date for products with encryption capabilities. Nevertheless, NSA believes it is only a matter of time until a surge occurs—at the same time acknowledging the similarity between this prediction and other previous predictions regarding demand. This belief is based on projections regarding the growth of networked applications and the trends discussed in Chapter 2:

- ♦ Increasing demand for all kinds of information technology.
- ♦ Increasing geographic dispersion of businesses across international boundaries.
- ♦ Increasing diversity of parties wishing/needing to communicate with each other.
- ♦ Increasing diversity in information technology applications and uses in all activities of a business.

Note: For example, a survey by industry analysts indicated that the installed base of users for work-group applications (involving communications among physically separated users) is expected to grow at a rate of about 85% annually between 1998 and 2003. It is true that a considerable amount of remote collaboration is done via e-mail without cryptographic protection, but work-group applications provide much higher degrees of functionality for collaboration because they are specifically designed for that purpose. As these applications become more sophisticated (as they begin to process large assemblies of entire documents rather than the short messages for which e-mail is best suited), the demand for higher degrees of protection is likely to increase.

Furthermore, NSA believes that computer users the world over have approximately the same computing needs as domestic users. Therefore, domestic trends in computing (including demand for more information security) will be reflected abroad, though perhaps later (probably years later but not decades later).

Market Development

A third issue in assessing the size of the market for encryption is the extent to which judgments should be made on the basis of today's market conditions (which are known with a higher certainty) rather than markets that may be at risk tomorrow (which are known with a much lower degree of certainty). Therefore, the market for certain types of software tends to develop in a characteristic manner. In particular, the long-term success of infrastructure software (software that supports fundamental business operations such as operating systems or groupware) depends strongly on the product's market timing. Once such software is integrated into the infrastructure of the installing organization, demands for backward-compatibility make it difficult for the organization to install any alternative.

> **Note:** Many products require backward-compatibility for market-place acceptance. Demands for backward-compatibility even affect products intended for operation in a stand-alone environment—an institution with 2 million spreadsheet files is unlikely to be willing to switch to a product that is incompatible with that existing database unless the product provides reasonable translation facilities for migrating to the new product. Network components are even harder to change, because stations on a network must interoperate. For example, most corporate networks have servers deployed with workstations that communicate with those servers. Any change to the software for the servers must not render it impossible for those workstations to work smoothly with the upgrade.

In other words, an existing software infrastructure inhibits technological change even if better software might be available. It is for this reason that in some software markets, major advantages accrue to the first provider of a reasonable product.

These pressures complicate life for government policy makers who would naturally prefer a more deliberate approach to policy making, because it is only during a small window of time that their decisions are relevant—the sooner they act, the better. The longer they wait, the higher will be the percentage of companies that have already made their technology choices; and, these companies will face large change-over costs if policy decisions entail incompatible alternatives to their currently deployed infrastructure. If the initial choices of companies involve putting non-U.S. software in place, U.S. vendors fear that they will have lost huge future market opportunities.

> **Note:** The deployment of Lotus Notes provides a good example. Lotus marketing data suggests fairly consistently that once Notes achieves a penetration of about 200 users in a given company, an explosion of demand follows, and growth occurs until Notes is deployed company-wide.

Inhibiting Vendor Responses to User Needs

In today's marketing environment, volume sales (licensing) to large corporate or government customers, rather than purchases by individuals, tend to drive sales of business software products. Because corporate customers have large leverage in the marketplace (because one purchasing decision can result in thousands of product sales to a single corporation), major software vendors are much more responsive to the needs of corporate users. Of particular relevance to the export control debate are three perceptions of corporate users:

> **Note:** The Department of Commerce noted that *civil use of software-based encryption will significantly increase in the next five years, with corporate customers dominating this new marketplace.*

Corporate users do not see that different levels of encryption strength (as indicated, for example, by the key length of foreign and domestic versions of a product) provide differential advantages. Put differently, the market reality is that users perceive domestic-strength versions as the standard and liberally exportable versions of encryption as weak, rather than seeing liberally exportable versions of encryption as the standard and domestic-strength versions as stronger.

Corporate users weigh all features of a product in deciding whether or not to buy it. Thus, the absence of a feature such as strong encryption that is desired but not easily available because of U.S. export controls, counts as a distinct disadvantage for a U.S. product. Although other features may help to compensate for this deficiency, the deficiency may pose enough of a barrier to a product's acceptance abroad that sales are significantly reduced.

Corporate users see cryptographic strength as an important parameter in their assessments of the information security that products offer. It is true that encryption is only one dimension of information security; that export controls do not affect certain approaches to increasing overall information security; and, that vendors often do not address these other approaches. But encryption is a visible aspect of the information security problem, and vendors feel an obligation to respond to market perceptions even if these perceptions may not be fully justified by an underlying technical reality. Moreover, many of the information security measures that do not involve export controls are more difficult and costly than encryption to implement, and so it is natural for vendors to focus their concerns on export controls on encryption. Thus, U.S. vendors who are unable to respond in a satisfactory manner to these perceptions have a natural disadvantage in competing against vendors that are able to respond.

The Impact of Export Controls on U.S. Economic and National Security Interests

By affecting U.S. industries abroad that might use encryption to protect their information interests and U.S. vendors of a critical technology (namely, information technology), export controls have a number of potentially negative effects on national security that policymakers must weigh against the positive effects of reducing the use of encryption by hostile parties.

Direct Economic Harm to U.S. Businesses

While acknowledging economic benefits to U.S. businesses from signals intelligence (as described in Chapter 4), NSA notes that protection of the information interests of U.S. industries is also a dimension of national security, especially when the threats emanate from foreign sources.

If the potential value of proprietary information is factored into the debate over export controls, it dominates all other figures of merit. A figure of $560 billion to $840 billion was placed by the Computer Systems Policy Project on the value of future revenue opportunities as the result of electronic distribution and commerce and future opportunities to reengineer business processes by 2005. Opponents of export controls on encryption argue that if electronic channels and information systems are perceived to be vulnerable, businesses may well be discouraged from exploiting these opportunities, thereby placing enormous potential revenues at risk.

On the other hand, it is essentially impossible to ascertain with any degree of confidence what fraction of proprietary information would be at risk in any practical sense if businesses did move to exploit these opportunities. Current estimates of industrial and economic espionage provide little guidance.

For example, in today's world in which a country's power and stature are often measured by its economic/industrial capability, foreign government ministries—such as those dealing with finance and trade— and major industrial sectors are increasingly looked upon to play a more prominent role in their respective country's collection efforts. An economic competitor steals a U.S. company's proprietary business information or government trade strategies, and foreign companies and commercially oriented government ministries are the main beneficiaries of U.S. economic information. The aggregate losses that can mount as a result of such efforts can reach billions of dollars per year, constituting a serious national security concern.

Furthermore, there is no formal mechanism for determining the full qualitative and quantitative scope and impact of the loss of this targeted information. Industry victims have reported the loss of hundreds of millions of dollars, jobs, and market share. Thus, even this report, backed by all of the counterintelligence efforts of the U.S. government, is unable to render a definitive estimate to within an order of magnitude. Of course, it may well be that these estimates of loss are low, because companies are reluctant to publicize occurrences of foreign economic and industrial espionage as such publicity can adversely affect stock values, customers' confidence, and ultimately competitiveness and market share. Or, also because clandestine theft of information may not be detected. In addition, because all business trends point to greater volumes of electronically stored and communicated information in the future, it is clear that the potential for information compromises will grow—the value of information that could be compromised through electronic channels is only going to increase.

Damage to U.S. Leadership in Information Technology

The strength of the U.S. information technology industry has been taken as a given for the past few decades. But as knowledge and capital essential to the creation of a strong information technology industry become more available around the world, such strength can no longer be taken for granted.

In other words, it is impossible to predict with certainty whether export controls will stimulate the growth of significant foreign competition for U.S. information technology vendors. But the historical evidence suggests some reason for concern. For example, a 1991 report by the National Research Council (NRC), found that *unilateral embargoes on exports (of technologies for commercial aircraft and jet engines) to numerous countries not only make sales impossible but actually encourage foreign competitors to develop relationships with the airlines of the embargoed countries. By the time the U.S. controls are lifted, those foreign competitors may have established a competitive advantage.* The same report also found that for computer technology, marginal supplier disadvantages can lead to significant losses in market position, and it is just such marginal disadvantages that can be introduced by export controls. An earlier study pointed out that the emergence of strong foreign competition in a number of high-technology areas appeared in close temporal proximity to the enforcement of strong export controls in these areas for U.S. vendors. While the correlation does not prove that export controls necessarily influenced or stimulated the growth of foreign competition, the history suggests that they may have had some causal relationship. In the financial arena (not subject to export controls), U.S. financial controls associated with the Trading-with-the-Enemy Act may have led to the rise of the Eurodollar market — a set of foreign financial institutions, markets, and instruments that eroded the monopoly held on dollar-denominated instruments and dollar-dominated institutions by U.S. firms.

The likelihood of foreign competition being stimulated for encryption may be larger than suggested by some of these examples because, at least in the software domain, product development and distribution are less capital-intensive than in traditional manufacturing industries. Lower capital intensity would mean that competitors would be more likely to emerge.

Finally, while it is true that some foreign nations also impose export controls on encryption, those controls tend to be less stringent than those of the United States. In particular, it is more difficult to export encryption from the United States to the United Kingdom than the reverse; and, the U.S. market is an important market for foreign vendors. Further, it takes only one nation with weak or nonexistent controls to spawn a competitor in an industry such as software.

If and when foreign products become widely deployed and well integrated into the computing and communications infrastructure of foreign nations, even better versions of U.S. products will be unable to achieve significant market penetration. One example of such a phenomenon may be the growing interest in the United States in personal communications systems based on GSM — the European standard for digital cellular voice communications. Further, as the example of Microsoft vis-a-vis IBM in the 1980s demonstrated, industry dominance once lost is quite difficult to recover in rapidly changing fields.

The development of foreign competitors in the information technology industry could have a number of disadvantageous consequences from the standpoint of U.S. national security interests.

First of all, foreign vendors, by assumption, will be more responsive to their own national governments than to the U.S. government. To the extent that foreign governments pursue objectives involving encryption that are different from those of the United States, U.S. interests may be adversely affected. Specifically, foreign vendors could be influenced by their governments to offer for sale to U.S. firms products with weak or poorly implemented encryption. If these vendors were to gain significant market share, the information security of U.S. firms could be adversely affected. Furthermore, the United States is likely to have less influence and control over shipments of products with encryption capabilities between foreign nations than it has over similar U.S. products that might be shipped abroad. Indeed, many foreign nations are perfectly willing to ship products (missile parts, nuclear reactor technology) to certain nations in contravention to U.S. or even their own interests. In the long run, the United States may have even less control over the products with satellite encryption capabilities that wind up on the market than it would have if it promulgated a more moderate export control regime.

Detailed information about the workings of foreign products with satellite encryption capabilities is much less likely to be available to the U.S. government than comparable information about similar U.S. products that are exported. Indeed, as part of the export control administration process, U.S. products with satellite encryption capabilities intended for export are examined thoroughly by the U.S. government. As a result, large amounts of information about U.S. products with encryption capabilities are available to it.

Note: For example, U.S. vendors are more likely than foreign vendors to reveal source code of a program to the U.S. government (for purposes of obtaining export licenses). While it is true that the object code of a software product can be decompiled, decompiled object code is always much more difficult to understand than the original source code that corresponds to it.

Export controls on satellite encryption are not the only factor influencing the future position of U.S. information technology vendors in the world market. Yet, the NRC National Cryptography Policy Committee believes that these controls do pose a risk to their future position that cannot be ignored, and that relaxation of controls will help to ensure that U.S. vendors are able to compete with foreign vendors on a more equal footing.

The Mismatch between the Perceptions of Government/National Security and Those of Vendors

The NSA has observed what can only be called a disconnect between the perceptions of the national security authorities that administer the export control regulations on satellite encryption and the vendors that are affected by it. This disconnect was apparent in a number of areas.

First of all, the national security authorities asserted that export controls did not injure the interests of U.S. vendors in the foreign sales of products with encryption capabilities. United States vendors asserted that export controls had a significant negative effect on their foreign sales.

Second, national security authorities asserted that nearly all export license applications for a product with encryption capabilities are approved. Vendors told the NRC National Cryptography Policy Committee that they refrained from submitting products for approval because they had been told on the basis of preliminary discussions that their products would not be approved for export.

Third, national security authorities presented data showing that the turnaround time for license decisions had been dramatically shortened (to a matter of days or a few weeks at most). Vendors noted that these data took into account only the time from the date of formal submission of an application to the date of decision, and did not take into account the much greater length of time required to negotiate product changes that would be necessary to receive approval. See "Regulatory Uncertainty Related to Export Controls," earlier in the chapter for additional information on this subject.

Fourth, national security authorities asserted that they wished to promote good information security for U.S. companies, pointing out the current practice described earlier in the chapter that presumes the granting of USML licenses for stronger encryption to U.S.-controlled companies, banking and financial institutions. Vendors pointed to actions taken by these authorities to weaken the cryptographic security available for use abroad, even in business ventures in which U.S. firms had substantial interests. Potential users often told the committee that even under presumptive approval, licenses were not forthcoming, and that for practical purposes, these noncodified categories were not useful.

Fifth, national security authorities asserted that they took into account foreign competition and the supply of products with satellite encryption capabilities when making decisions on export licenses for U.S products with encryption capabilities. Vendors repeatedly pointed to a substantial supply of foreign products with encryption capabilities.

Finally, national security authorities asserted that they wished to maintain the worldwide strength and position of the U.S. information technology industry. Vendors argued that when they are prevented from exploiting their strengths—such as being the first to develop integrated products with strong encryption capabilities—their advantages are in fact being eroded.

The NRC committee believes that to some extent, these differences can be explained as the result of rhetoric by parties intending to score points in a political debate. But the differences are not merely superficial; they reflect significantly different institutional perspectives. For example, when national security authorities *take into account foreign supplies of cryptography,* they focus naturally on what is available at the time the decision is being made. On the other hand, vendors are naturally concerned about incorporating features that will give their products a competitive edge, even if no exactly comparable foreign products with encryption are available at the moment. Thus, different parties focus on different areas of concern—national security authorities on the capabilities available today, and vendors on the capabilities that might well be available tomorrow.

Perceptions by NSA of vendors and users of encryption may well be clouded by an unwillingness to speak publicly about the full extent of vendor and user unhappiness with the current state of affairs. National security authorities asserted that their working relationships with vendors of products with encryption capabilities are relatively harmonious. Vendors contended that since they are effectively at the mercy of the export control regulators, they have considerable incentive to suppress any public expression of dissatisfaction with the current process. A lack (or small degree) of vendor outcry against the encryption export control regime cannot be taken as vendor support for it. More specifically, the committee received input from a number of private firms on the explicit condition of confidentiality.

For example, companies with interests in encryption affected by export control were reluctant to express fully their dissatisfaction with the current rules governing export of products with encryption capabilities or how these rules were actually implemented in practice. They were concerned that any explicit connection between critical comments and their company might result in unfavorable treatment of a future application for an export license for one of their products.

Companies that had significant dealings with the Department of Defense were reluctant to express fully their unhappiness with policy that strongly promoted classified encryption algorithms and government-controlled key-escrow schemes. These companies were concerned that expressing their unhappiness fully might result in unfavorable treatment in competing for future DOD business.

Many companies have expressed dissatisfaction publicly, although a very small number of firms did express to the NRC committee their relative comfort with the way in which the current export control regime is managed. The NRC committee did not conduct a systematic survey of all firms affected by export regulations, and it is impossible to infer the position of a company that has not provided input on the matter.

Note: The Department of Commerce study is the most systematic attempt to date to solicit vendors' input on how they have been affected by export controls, and the solicitation received a much smaller response than expected.

Export of Technical Data

The rules regarding *technical data* are particularly difficult to understand. An encryption algorithm (if described in a manner that is not machine-executable) is counted as technical data, whereas the same algorithm if described in machine-readable form (source or object code) counts as a product. Legally, the ITAR regulate products with encryption capabilities differently than technical data related to cryptography, although the differences are relatively small in nature. For example, technical data related to cryptography enjoys an explicit exemption when distributed to U.S.-controlled foreign companies, whereas products with encryption capabilities are in principle subject to a case-by-case review in such instances (although in practice, licenses for products with encryption capabilities under such circumstances are routinely granted).

Private citizens and academic institutions and vendors are often unclear about the legality of actions such as:

♦ Discussing encryption with a foreign citizen in the room.
♦ Giving away software with encryption capabilities over the Internet (see a further discussion of this subject later in the chapter).
♦ Shipping products with encryption capabilities to a foreign company within the United States that is controlled but not owned by a U.S. company.
♦ Selling a U.S. company that makes products with strong encryption capabilities to a foreign company.
♦ Selling products with encryption capabilities to foreign citizens on U.S. soil.
♦ Teaching a course on encryption that involves foreign graduate students.
♦ Allowing foreign citizens residing in the United States to work on the source code of a product that uses embedded encryption.

Note: For example, one vendor argues that because foreign citizens hired by U.S. companies bring noncontrolled knowledge back to their home countries anyway, the export control regulations on technical data make little sense as a technique for limiting the spread of knowledge. In addition, other vendors note that in practice the export control regulations on technical data have a much more severe impact on the employees that they may hire than on academia, which is protected at least to some extent by presumptions of academic freedom.

The Sidebar that follows provides excerpts from the only document known to the NRC committee that describes the U.S. government explanation of the regulations on technical data related to encryption. In practice, these and other similar issues regarding technical data do not generally pose problems because these laws are for the most part difficult to enforce and in fact are not generally enforced. Nevertheless, the vagueness and broad nature of the regulations may well put people in jeopardy unknowingly.

On the Export of Technical Data Related to Satellite Encryption

Cryptologic technical data refers only to such information as is designed or intended to be used, or which reasonably could be expected to be given direct application, in the design, production, manufacture, repair, overhaul, processing, engineering, development, operation, maintenance or reconstruction of items in such categories. This interpretation includes, in addition to engineering and design data, information designed or reasonably expected to be used to make such equipment more effective, such as encoding or enciphering techniques and systems, and communications or signal security techniques and guidelines, as well as other encryption and cryptanalytic methods and procedures. It does not include general mathematical, engineering or statistical information. It also does not include basic theoretical research data. It does, however, include algorithms and other procedures purporting to have advanced cryptologic application.

The public is reminded that professional and academic presentations and informal discussions, as well as demonstrations of equipment, constituting disclosure of cryptologic technical data to foreign nationals are prohibited without prior approval. Approval is not required for publication of data within the United States.

The interpretation set forth should exclude from the licensing provisions of the ITAR's most basic scientific data and other theoretical research information, except for information intended or reasonably expected to have a direct cryptologic application. Because of concerns expressed by vendors that licensing procedures for proposed disclosures of cryptologic technical data contained in professional and academic papers and oral presentations could cause burdensome delays in exchanges with foreign scientists, the

Office Of Munitions Control of the Department of State will expedite
consideration as to the application of ITAR to such disclosures. If
requested, the Office Of Munitions Control will on an expedited
basis provide an opinion as to whether any proposed disclosure, for
other than commercial purposes, of information relevant to encryption, would require licensing under the ITAR.[13]

In any event, a recently filed suit seeks to bar the government from
restricting publication of cryptographic documents and software
through the use of the export control laws. The plaintiff in the suit is
Dan Bernstein, a graduate student in mathematics at the University of
California at Berkeley. Bernstein developed an encryption algorithm
that he wished to publish and to implement in a computer program
intended for distribution, and he wanted to discuss the algorithm and
program at open, public meetings. Under the current export control
laws, any individual or company exporting unlicensed encryption
software may be in violation of the export control laws that forbid the
unlicensed export of defense articles; and, any individual discussing
the mathematics of encryption algorithms may be in violation of the
export control laws that forbid the unlicensed export of *technical data.*
The lawsuit argues that the export control scheme as applied to
encryption software is an *impermissible prior restraint on speech, in
violation of the First Amendment* and that the current export control
laws are vague and overbroad in denying people the right to speak
about and publish information about encryption freely. A decision by
the Northern District Court of California on April 15, 1996, by Judge
Marilyn Patel, denied the government's motion to dismiss this suit,
and found that for the purposes of First Amendment analysis, source
code should be treated as speech. The outcome of this suit is unknown
at the time of this writing.

The constitutionality of export controls on technical data has not been
determined by the United States Supreme Court. A ruling by the
United States Ninth Circuit Court of Appeals held that the ITAR, when
construed as "*prohibiting only the exportation of technical data significantly and directly related to specific articles on the Munitions List, do
not interfere with constitutionally protected speech, are not overbroad
and the licensing provisions of the Act are not an unconstitutional
prior restraint on speech.* (See 579 F.2d 516, U.S. vs Edler, United
States Court of Appeals, Ninth Circuit, July 31, 1978.)

Another suit filed by Philip Karn directly challenging the constitutionality of the ITAR was dismissed by the U.S. District Court for the District of Columbia on March 22, 1996. The issue at hand was the fact that Karn had been denied CCL jurisdiction for a set of floppy diskettes containing source code for cryptographic confidentiality.

Some scholars argue to the contrary that export controls on technical data may indeed present First Amendment problems, especially if these controls are construed in such a way that they inhibit academic discussions of encryption with foreign nationals or prevent academic conferences on encryption held in the United States from inviting foreign nationals.

Foreign Policy Considerations

A common perception within the vendor community is that the National Security Agency is the sole *power behind the scenes* for enforcing the export control regime for encryption. While NSA is indeed responsible for making judgments about the national security impact of exporting products with encryption capabilities, it is by no means the only player in the export license application process.

The Department of State plays a role in the export control process that is quite important. For example, makers of foreign policy in the U.S. government use economic sanctions as a tool for expressing U.S. concern and displeasure with the actions of other nations. Such sanctions most often involve trade embargoes of various types. Violations of human rights by a particular nation, for example, represent a common issue that can trigger a move for sanctions. Such sanctions are sometimes based on presidential determinations (that the human rights record of country X is not acceptable to the United States) undertaken in accordance with law. In other cases, sanctions against specific nations are determined directly by congressional legislation. In still other cases, sanctions are based entirely on the discretionary authority of the President.

The imposition of sanctions is often the result of congressional action that drastically limits the discretionary authority of the State Department. In such a context, U.S. munitions or articles of war destined for particular offending nations (or to the companies in such nations) are the most politically sensitive; and in practice, the items on the USML are the ones most likely to be denied to the offending nations. In all such cases, the State Department must determine whether a particular item on the USML should or should not qualify for a USML license. A specific example of such an action given to the NRC committee in testimony involved the export of encryption by a U.S. bank for use in a branch located in the People's Republic of China. Because of China's human rights record, the Department of State delayed the export, and the contract was lost to a Swiss firm. The sale of cryptographic tools that are intended to protect the interests of a U.S. company operating in a foreign nation was subject to a foreign policy stance that regarded such a sale as equivalent to supplying munitions to that nation.

Thus, even when NSA has been willing to grant an export license for a given cryptography product, the State Department has sometimes denied a license because encryption is on the USML. In such cases, NSA takes the blame for a negative decision, even when it had nothing to do with it.

Critics of the present export control regime have made the argument that encryption, as an item on the USML that is truly dual-use, should not necessarily be included in such sanctions. Such an argument has some intellectual merit, but under current regulations it is impossible to separate encryption from the other items on the USML.

Technology-Policy Mismatches

Two cases are often cited in the encryption community as examples of the mismatch between the current export control regime and the current state of encryption technology (See the Sidebar that follows.) Moreover, they are often used as evidence that the government is harassing innocent law-abiding citizens.

Two Export Control Cases

The Zimmermann PGP Case:

Philip Zimmermann is the author of a software program known as PGP (for Pretty Good Privacy). PGP is a program that is used to encrypt mail messages end-to-end based on public-key cryptography. Most importantly, PGP includes a system for key management that enables two users who have never interacted to communicate securely based on a set of trusted intermediaries that certify the validity of a given public key. Across the Internet, PGP is one of the most widely used systems for secure e-mail communication.

Zimmermann developed PGP as a *freeware* program to be distributed via diskette. Another party subsequently posted PGP to a USENET newsgroup (a commercial version licensed from but not supplied by Zimmermann has since emerged). In 1993, Zimmermann was determined to be the target of a criminal investigation probing possible violations of the export control laws. Zimmermann was careful to state that PGP was not to be used or downloaded outside the United States, but of course international connections to the Internet made for easy access to copies of PGP located within the United States. In January 1996, the U.S. Department of Justice closed its investigation of Zimmermann without filing charges against him.

The Bruce Schneier-*Applied Cryptography* Case:

Bruce Schneier wrote a book called *Applied Cryptography* (John Wiley and Sons, 1994) that was well received in the encryption community. It was also regarded as useful in a practical sense because it contained printed on its pages source code that could be entered into a computer and compiled into a working encryption program. In addition, when distributed within the United States, the book contained a floppy disk that contained source code identical to the code found in the book. However, when another party (Philip Karn) requested a ruling on the exportability of the book, he (Karn) received permission to export the book but not the disk. This decision has been greeted with considerable derision in the academic encryption community, with comments such as: *They think that terrorists can't type?*—thus, expressing the general dismay of the community.

Note: A USENET newsgroup is in effect a mailing list to which individuals around the world may subscribe. Posting is thus an act of transmission to all list members.[14]

Taken by themselves and viewed from the outside, both of the cases outlined in the preceding Sidebar, suggest an approach to national security with evident weaknesses. In the first instance, accepting the premise that programs for encryption cannot appear on the Internet because a foreigner might download them seems to challenge directly the use of the Internet as a forum for exchanging information freely even within the United States. Under such logic (claim the critics), international telephone calls would also have to be shut down because a U.S. person might discuss encryption with a foreign national on the telephone. In the second instance, the information contained in the book (exportable) is identical to that on the disk (not exportable). Since it is the information about encryption that is technically at issue (the export control regulations make no mention of the medium in which that information is represented), it is hard to see why one would be exportable and the other not.

On the other hand, taking the basic assumptions of the national security perspective as a given, the decisions have a certain logic that is not only the logic of selective prosecution or enforcement. In the case of Zimmermann, the real national security issue is not the program itself, but rather the fact that a significant PGP user base may be developing. Two copies of a good encryption program distributed abroad pose no plausible threat to national security. But 20 million copies might well pose a threat. However, the export control regulations as written do not mention potential or actual size of the user base, and so the only remaining leverage is the broad language that brings encryption under the export control laws.

In the case of Schneier, the real national security issue relates to the nature of any scheme intended to deny capabilities to an adversary. Typing the book's source code into the computer is an additional step that an adversary must take to implement an encryption program and a step at which an adversary could make additional errors. No approach to denial can depend on a single *silver bullet*. Instead, denial rests on the erection of multiple barriers, all of which taken together are expected to result in at least a partial denial of a certain capability. Moreover, if one begins from the premise that export controls on software encryption represent appropriate national policy, it is clear that allowing the export of the source code to Schneier's book would set a precedent that would make it very difficult to deny permission for the export of other similar software products with encryption capabilities. Finally, the decision is consistent with a history of commodity jurisdiction decisions that generally maintains USML controls on the source code of a product whose object code implementation of confidentiality has been granted commodity jurisdiction to the CCL.

These comments are not intended to excoriate or defend the national security analysis of these cases. But the controversy over these cases does suggest quite strongly that the traditional national security paradigm of export controls on encryption (one that is biased toward denial rather than approval) is stretched greatly by current technology. Put differently, when the export control regime is pushed to an extreme, it appears to be manifestly ridiculous.

Summary

Current export controls on products with satellite encryption capabilities are a compromise between the needs of national security to conduct signals intelligence and the needs of U.S. and foreign businesses operating abroad to protect information and the needs of U.S. information technology vendors to remain competitive in markets involving products with encryption capabilities that might meet these needs. These controls have helped to delay the spread of strong encryption capabilities and use of those capabilities throughout the world, to impede the development of standards for cryptography that would facilitate such a spread, and to give the U.S. government a tool for monitoring and influencing the commercial development of satellite encryption. Export controls have clearly been effective in limiting the foreign availability of products with strong encryption capabilities made by U.S. manufacturers, although enforcement of export controls on certain products with encryption capabilities appears to have created many public relations difficulties for the U.S. government; and circumventions of the current regulations appear possible. The dollar cost of limiting the availability of encryption abroad is hard to estimate with any kind of confidence, since even the definition of what counts as a cost is quite fuzzy. At the same time, a floor of a few hundred million dollars per year for the market affected by export controls on encryption seems plausible, and all indications are that this figure will only grow in the future.

A second consideration is the possibility that export controls on products with satellite encryption capabilities may well have a negative impact on U.S. national security interests by stimulating the growth of important foreign competitors over which the U.S. government has less influence and possibly by damaging U.S. competitive advantages in the use and development of information technology. In addition, the export control regime is clouded by uncertainty from the vendor standpoint, and there is a profound mismatch between the perceptions of government/national security and those of vendors on the impact of the export control regime. Moreover, even when a given product with encryption capabilities may be acceptable for export on national security grounds, nonnational security considerations may play a role in licensing decisions.

Partly in response to expressed concerns about export controls, the export regime has been gradually loosened since 1983. This relaxation raises the obvious question of how much farther and in what directions such loosening could go without significant damage to national security interests. This subject is addressed in Chapter 26, "Satellite Encryption Technology Policy Options, Analysis and Forecasts."

From Here

This chapter was concerned primarily with export controls on satellite encryption, a powerful tool that has long been used in support of national security objectives but whose legitimacy has come under increasing fire in the last several years.

Chapter 6 addresses the following, sales of individual images from satellites owned and operated by U.S. firms, sales of real-time bit streams from such satellites, satellites owned by third parties (who could, unless otherwise limited, sell to whomever they chose) but operated by U.S. firms, satellites owned and operated by third parties that rely on support from U.S. firms, and satellites owned and operated autonomously by third parties.

Endnotes

1 "Rules and Regulations," Federal Register, Vol. 61, No. 33, Friday, February 16, 1996, Office of the Federal Register, National Archives and Records Administration, United States Government Printing Office, Washington, DC, pp. 1-3.
2 James Hanson and Roger Dean. "European Companies Threatened by U.S. Export Controls on Encryption Technology," © European Electronic Messaging Association, Alexander House, High Street, Inkberrow, Worcestershire, WR7 4DT, United Kingdom, 1998, pp. 1–2.
3 National Security Agency, Fort Meade, Maryland, 11997.
4 Ibid.
5 Ibid.
6 Ibid.
7 Ibid.
8 Ibid.
9 Ibid.
10 Ibid.
11 Ibid.
12 © 1996 by the National Academy of Sciences. Courtesy of the National Academy Press, 2101 Constitution Avenue, NW, Washington D.C. 20418.
13 Ibid.
14 Ibid.

6

Controlling the Use of Encryption on Surveillance Satellites

How far will U.S. producers be allowed to go in offering encrypted imagery from surveillance satellites with 1-meter resolution? From the defense perspective, it would be best if such capabilities were unavailable outside the U.S. The next best outcomes, in descending order of preference are:

- ◆ Sales of individual encrypted images from satellites owned and operated by U.S. firms.
- ◆ Sales of real-time encrypted bit streams—increased timeliness means greater military applicability—from such satellites.
- ◆ Surveillance satellites owned by third parties (who could, unless otherwise limited, sell to whomever they chose), but operated by U.S. firms.

♦ Surveillance satellites owned and operated by third parties that rely on support from U.S. firms.
♦ Surveillance satellites owned and operated autonomously by third parties.[1]

The Commerce Department has licensed four U.S. groups to sell surveillance-encrypted bitstreams. The first, Worldview, offers 3-meter resolution. The other three groups — led by Lockheed, Litton, and Ball, respectively — offer 1-meter resolution. The licenses permit the U.S. government to prevent the capture and transmission of encrypted data in emergencies. Although financing remains an issue for each group, Lockheed's consortium is set to receive the majority funding from Arabian sources.

U.S. policy on surveillance satellites is not made in a strategic vacuum. French officials recognize that 1-meter encrypted imagery could not only imperil French forces, but also threaten the commercial viability of their Spot satellite, which they were allowed to offer only over the initial objections of the French Defense Ministry. Although French government sources have reiterated their opposition to the sale of encrypted imagery finer than the five meters that Spot 5 could offer, Matra, which builds France's surveillance satellites, has publicly considered selling 1-meter imagery on the commercial market in competition with U.S. providers. Russia's reaction is more difficult to assess. Russia dislikes proliferation, but reportedly the same country whose inquiry prompted Lockheed to consider selling 1-meter surveillance satellites has decided to pursue a better deal (0.8-meter resolution) with Russian sources.

How long and at what cost could U.S. policy inhibit the flow of encrypted surveillance information? Many sensors ostensibly built for environmental monitoring have military uses. Surveillance satellites are becoming more accurate, easier to control, cheaper, and more available. Would foreign access to U.S. encrypted image streams inhibit other countries from launching their own satellites? It appears probable that military-relevant surveillance satellite-encrypted imagery will become increasingly available. U.S. forces will have to learn to operate in an environment where they do not have a monopoly on such capabilities.

Now that surveillance satellites are becoming more accurate, easier to control, cheaper, and more available to everyone, the recent launching of the world's first commercial spy satellite would make it possible for just about anybody with a credit card to enjoy an eye in the sky. At least that is what everyone thought!

What Happened to EarlyBird?

The apparent loss of a commercial eye-in-the-sky surveillance satellite is a setback for the public's *right to know* according to aerospace and media experts. However, the general news media (among the most likely consumers of detailed space-based photos) have largely ignored the failure.

The 710-pound EarlyBird 1 surveillance satellite (shown in Figure 6.1), launched December 24, 1998 from eastern Siberia, was expected to deliver the sharpest-ever civilian photographs of Earth's surface.[2] Objects as small as 9 feet wide (cars, trucks, and small buildings), could be picked out from encrypted imagery relayed (downlinked) back to Earth from its 300-mile orbit.

Figure 6.1 EarlyBird 1 was supposed to snap very high-resolution pictures from space for commercial customers. But its owners have not been able to communicate with the spacecraft.

Currently, only military spy satellites are capable of capturing such detailed pictures. The sharpest commercially available imagery captures features no smaller than 32 feet. Such pictures are currently sold by Spot Images of Toulouse, France, as well as by companies in Russia and the United States.

EarthWatch Inc. of Longmont, Colorado, which owns EarlyBird, planned to sell images to private customers through the Internet around the world, allowing journalists (or any approved customer with a credit card) to scrutinize military or environmental sites virtually anywhere on the planet, including politically sensitive areas such as the Persian Gulf. EarthWatch also hoped to end the monopoly of the world's most advanced military and intelligence services in gathering high-resolution pictures from space.

A Variety of Photo Flavors from Space

Prices of black-and-white pictures were to range from $2.65 to $7.15 per square kilometer (0.4 square miles), depending on whether they come from the company's archives or are collected at a specified time and date. Color imagery was also available, although it was less sharp and more expensive.

The pictures were to be used in a wide range of applications, including town planning, map making, disaster relief, mining and giving the media and the public the chance to scrutinize environmental and military crises. With 9-foot capabilities, cars could be distinguished from trucks, for example. In imagery taken at 32-foot resolution, neither cars nor trucks could be identified.

EarthWatch, Spot and firms in India, Israel, Russia and China planned to launch next-generation surveillance satellites in coming years capable of distinguishing ground objects with a diameter just below 36 inches. For example, with such *submeter* resolution, you would be able to see a blanket on the beach but not the person on it. Earth-Watch, despite this failure, still plans to launch such a system called QuickBird in 1999.

The Clinton administration opened the door for U.S. companies to enter the so-called high-resolution *remote sensing* field in 1994, bowing to industry arguments that foreign rivals would otherwise have a free hand. However, the federal government retains the right to switch off the commercial sensors in times of war or international crises. In addition, it bans U.S.-licensed surveillance satellite operators from selling encrypted images to the governments of Cuba, China, India, Libya, North Korea, Iran and Iraq or any of their suspected agents.

The launch of EarlyBird 1 provided a vivid reminder that the Cold War is over—so everyone thought! It was the first commercial launch from the Svobodny Cosmodrome, Russia's newest commercial launch site. The surveillance satellite was fired into space atop a former intercontinental ballistic missile known as Start-1, named for the arms control treaty that made the missile obsolete.

However, what went wrong with EarlyBird 1? Was it sabotage?

Silence is Costly

Within days of reaching polar orbit 300 miles above the Earth, the surveillance satellite's radio suddenly stopped. Frustrated ground controllers sent out a call over the Internet to amateur satellite trackers around the world requesting help in restoring contact. For nearly three weeks the Web buzzed with messages about listening attempts, but no signals have been reported so far.

EarthWatch officials suspect an electronic failure aboard the surveillance satellite itself, but have not ruled out sabotage. It is possible that harsh conditions at the Siberian launch site or the rapid acceleration of the Start-1 ICBM-style rocket that it was launched from may have inadvertently damaged the first-of-its-kind payload. There is some indication that the spacecraft may have been tampered with during its shipment from Colorado to Russia's Svobodny Cosmodrome. As of this writing, an investigation is ongoing.

Although recovery of the spacecraft has still not been ruled out, as of this writing, EarthWatch still hopes to re-establish communications and resume the mission. However the odds against success grow dimmer with each passing day of silence. Nevertheless, similar high-resolution spy satellites will be launched by other American, French, Israeli, Indian and Russian aerospace firms during the coming decade.

Despite U.S. government bans against licensed surveillance satellite operators selling encrypted images to the governments of Cuba, China, India, Libya, North Korea, Iran and Iraq or any of their suspected agents, U.S. companies are lobbying hard to sell sensitive satellite encryption devices to China.

U.S. Companies Bank on China's Encryption Demand

For months, Senate investigators have searched for quid pro quos — major policy changes favorable to China that came as a result of illegal campaign contributions. As it happens, one major policy change is about to take effect. The U.S. is on the verge of allowing the sale of sensitive encryption and surveillance satellite communications technology to China, a development far more beneficial to China than anything yet mentioned in the political-corruption hearings. The prospect of U.S. firms giving China anything sensitive like encryption and sophisticated satellite communication technology, however, has intelligence experts and a good many members of Congress apoplectic. China's continued recklessness in assisting countries such as Iran, Pakistan, India and Algeria makes it the discount store for weapons of mass destruction and a threat to national security.

The proposed sensitive high technology transfers to China are now being fully investigated by Congress for a simple reason: The change happened not at the behest of Chinese spies, but of the White House and large American corporations. Faced with a declining American demand, U.S. high tech surveillance satellite communication companies have been begging for approval to sell the Chinese satellite communication technology worth as much as $80 billion. But a 1985 federal law prohibits them from selling the Chinese sensitive technology until the president *certifies* that China has indeed stopped selling weapons technology to other nations as it has in the past. Government and industry sources say President Clinton has departed from the policy of past presidents and has given China this certification. This is why the White House and other aerospace corporations are now under investigation by Congress amid charges of threats to national security.

Nevertheless, the policy change occured after a year of intense lobbying from American companies. Westinghouse, Loral, General Electric, Bechtel, ABB Combustion Engineering Inc., and others have pressed the White House, Commerce and Energy departments, and members of Congress to approve the exports. For them, a China deal may be a lifesaver. If they do not have an industry in China, they do not think American industry can survive. The numbers are huge. The Chinese still represent anywhere from 50 to 75 percent of the world's new nuclear and high tech satellite communications equipment market.

In the Clinton administration, the nuclear and high tech surveillance satellite communications industry has found an eager partner. Since first taking office, White House officials have argued that legitimate trade was being blocked by unreasonable cold-war-era fears over the export of *dual-use technology*—products that can be put to either commercial or military use. To that end, the administration dramatically revised trade rules, beginning with its 1993 decision to ease limits on the export of high-speed super computers. That change came after pressure from electronics and computer-industry CEOs, several of whom publicly supported Clinton in 1992. They had complained that high-speed computers already were being sold overseas by other countries. Soon, other dual-use restrictions were also lifted.

The result has been a lucrative trade with China, much of it in high-tech goods whose export was unthinkable just a few years ago. In 1996, for instance, the Commerce Department approved the shipment of more than 200 specialized oscilloscopes—instruments that can test advanced electronic circuits but can also diagnose nuclear test results—worth $6.9 million. In 1995, U.S. companies shipped China almost $2 billion in dual-use *digital computers, assemblies, and related surveillance satellite communications equipment*, according to a Commerce Department report. A brisk exchange in rocket, satellite, and avionics technology has opened up for U.S. companies such as Hughes and Boeing. Other exports now include computer-controlled machine tools, laser technology, underwater acoustic equipment, computer-chip manufacturing machines, and code-encryption components. Since 1992, dual-use exports to China have grown to $14 billion.

Summary

The administration sees little danger in selling sensitive encryption and surveillance satellite communications technology to China. It does not feel that there is an urgent need to control the use of encryption on surveillance satellites.

China—already a nuclear state—is adopting its own export controls. The administration and the supporting U.S. companies think the way to deal with China is through engagement, engaging them in better nonproliferation practices. China, though, has a history of misrepresenting the use of U.S. technology it buys. For instance, a Sun Microsystems supercomputer wound up, illegally, in a Chinese military facility (China recently agreed to return the computer). Similarly, machine tools sold by McDonnell Douglas to the Chinese firm CATIC in 1994 ended up not at an airplane plant but at a facility that makes Silkworm cruise missiles.

China also has a record of reselling technology to other nations: It sold chemical-weapons precursors to Iran and missiles and ring magnets used to process uranium to Pakistan. Such trade has led the CIA to conclude that China has become the *most significant supplier* of nuclear and chemical weapons and high-speed surveillance satellite communications technology to foreign countries.

From Here

This chapter addressed the following: Sales of individual images from surveillance satellites owned and operated by U.S. firms; sales of real-time bit streams from such satellites; surveillance satellites owned by third parties (who could, unless otherwise limited, sell to whomever they chose), but operated by U.S. firms; surveillance satellites owned and operated by third parties that rely on support from U.S. firms; and, surveillance satellites owned and operated autonomously by third parties.

Chapter 7 discusses why many commercial users (who constitute the bulk of the GPS system's beneficiaries) have expressed nervousness about relying on a DOD-controlled system. This chapter will also explore the DOD task force recommendation that management (as opposed to operational control) of the GPS be shifted to a joint DOD/DOT executive board that would resolve issues that could not be made by normal interagency coordination. Additionally, Chapter 7 will also discuss why a joint DOD/DOT executive board cannot support close operations such as guiding aircraft into runways, governing port traffic, or intelligent vehicle/highway system applications, which require DGPS.

Endnotes

1   "Key U.S. Security Policy Issues," *institute for National Strategic Studies, Fort Lesley J. McNair, Washington, DC 20319-6000, 1995, pp. 1-2.*

2   *EarthWatch Incorporated, 1900 Pike Road, Longmont, CO 80501, USA, 1998.*

7

Managing Global Navigation Encryption Systems

As commercial users come to constitute the bulk of the Global Positioning System's (GPS) beneficiaries, many have expressed nervousness about relying on a DOD-controlled global navigation encryption system. Might such worries inhibit beneficial investments that take advantage of GPS in areas such as transportation safety? In December 1993, a joint DOD/Department of Transportation (DOT) task force dealt with two key issues: who would manage and fund the system; and how the needs of civilian users would be met.

The task force recommended that management (as opposed to operational control) of the GPS be shifted to a joint DOD/DOT executive board that would resolve issues that could not be made by normal interagency coordination. DOD agreed to pay for the normal cost of operating and maintaining the system (roughly $500 million a year, mostly for replacing satellites ending their useful life). DOT would pay for augmentations to the sytem to support civil navigation needs (such as differential GPS beacons).

The accuracy issue was thornier. Although 100-meter resolution suffices for some purposes, such as preventing airliners from drifting into hostile airspace, it cannot support close operations such as guiding aircraft into runways, governing port traffic, or intelligent vehicle/highway system applications, which require a Differential Global Positioning System (DGPS). The outcome permitted the Coast Guard to install local correction services that will cover coastal and inland waterways, while the Federal Aviation Administration (FAA) will instal similar services for airport use. The issue of global correction services remained unresolved.

In early 1994, the FAA announced that it would replace most current microwave landing systems with DGPS. Will other nations follow? If so, they would have to overcome misgivings about dependence on a system run by the U.S. government, which retains the right to degrade even civilian encrypted signals in an emergency. If not, however, they face the considerable cost of maintaining their own systems, and international aircraft would have to maintain two systems: one for domestic use, the other for use overseas.

In common with other areas in this chapter, trends point to the eventual diffusion of the technologies in question. As other navigation systems emerge, GPS will lose its exclusive place, and U.S. policy will have decreasing relevance over navigation systems investments. Controls can slow down proliferation, but they cannot reverse deep-seated trends.

Nevertheless, collaboration by the U.S. government with the oceanographic community has identified real-time oceanographic encrypted data collected and transmission through the launching of large communication satellites into orbit as a desirable capability for the scientific, industrial, educational, and international government communities. The lack of the satellite communications technology base and the prohibitive cost of deploying oceanographic instrumentation and sensors, developing encrypted data collection capabilities and especially real time, satellite encrypted data transmission capabilities, currently limits the user community.

With the technological advances in antennas, telemetric buoy platforms, and fiber-optic risers, the capability (now technically feasible) of marrying agile Soviet (former Soviet Union) rocket design with the best oil platform technology, may provide an altogether new means of launching big communication satellites into orbit. Let us take a look at how all of this is being done.

Launching Communication Satellites at Sea

An international team of marine and rocket engineers is combining advanced oil platform technology with a Cold War rocket system into a novel means of launching communication satellites at sea. The objective is to heave telecommunications satellites from the equator, the best launch latitude on the planet, into geostationary orbit, the most marketable territory in space.

The job is being tackled by Sea Launch Co., formed in the spring of 1995 and registered in the Cayman Islands. The work of preparing launch vehicles for three-stage Russian Zenit (zenith) rockets is being done at its facilities in Long Beach, California (see Figure 7.1).[1] Then Sea Launch will sail to the equator south of Hawaii to launch them, sending their payloads into an orbit 48,000 km (30,000 miles) above the Earth before returning to Los Angeles to reload. With this fresh approach, the company expects to extend satellite life, cut launch costs, and sidestep the environmental and political problems of building, maintaining, and policing a new launch site on dry land.

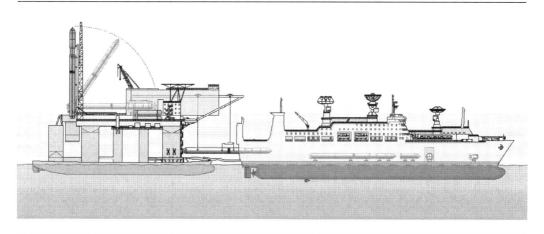

Figure 7.1 A Zenit 3SL rocket is transferred from the 200-meter-long assembly and command ship (right) to the ocean-going Odyssey launch platform (left) while in port at Long Beach, CA. The ship and the platform will be detached before sailing for the launch site in mid-ocean. Not until a few hours before launch will the rocket emerge from the hangar for the firing.

Putting large satellites into geostationary orbit is a big business. Until now, it has depended on launch sites owned by the governments of the U.S., France, or Russia. As the world's first strictly private venture for launching such satellites, Sea Launch is a bold new undertaking. It unites not only space and oceanic technologies, but also former Cold War adversaries. The launch platform and control ship are being built by a Norwegian firm (Kvaerner a.s.); the rocket and launch systems are being provided by Russian (RSC-Energia) and Ukrainian (NPO-Yuzhnoye) firms; and the overall program management of the project is being done by a leadng aerospace manufacture in the U.S. (Boeing) with funding being provided by DOD and DOT.

From a technical viewpoint, the basic concept is to lift off right from the equator, where the Earth's rotational speed is greatest and gives a rocket a greater running start as it heads for geostationary orbit (see the Sidebar that follows). But finding accessible launch sites on land, in friendly and reliable countries, has been a tall order. In fact, all existing launch sites (with the exception of the European Space Agency's facility at Kourou, French Guiana) are situated less than optimally at some distance from the equator. Launching from sea finesses that problem and also, to an extent, frees companies from protectionist pressures that otherwise might constrain their use of encryption technology from competitor countries.

The Early Satellite Gets the Orbital Velocity

The direction of the Earth's rotation gives a satellite an eastward running start toward reaching orbital velocity, which is about 28,000 kilometers (17,500 miles) per hour for an orbit about 200 km (125 miles) high. This boost increases with the distance from the Earth's axis, from zero at the north and south poles to 1676 km/h (1048 miles/h), 6 percent of minimum orbital velocity, at the equator. The extra thrust pays off not so much in conserving launch propellants, which are cheap, but in allowing the designer to pack more electronics into a satellite. Or the satellite can simply carry enough control propellant to extend its life (and revenue production) from six years to nine.

By far the most popular destination for spacecraft is geostationary orbit, the only *fixed* place in space. A satellite in a circular orbit

35,898 km (22,436 miles) high and in an orbital plane with the Earth's equator takes 24 hours to orbit the earth—while staying in one spot over the equator. Thus an antenna on the ground always sees the satellite at the same angle, and the satellite always sees the same half of the globe, making geostationary orbit the best location for encrypted communications, weather, and missile early warning satellites.

Thus, the ideal launch site for geostationary orbit is the equator. Not only does it provide the greatest eastward boost, but it eliminates the need for a costly maneuver to change the satellite's initial orbital plane, which matches the latitude of its launch site, to match the equator. Without changing its orbital plane, the satellite would drift north and south every 24 hours in a figure-eight pattern.

From a mechanical standpoint, any point on the equator will do. But rockets drop stages during launch and even fall out of the sky, thus it is best to have an ocean or desert to the east of the launch site. And because launch facilities are large investments of capital and sensitive technology, the site should be inside a nation's own territory. For example, the first launch sites, established to test ballistic missiles years before rockets could reach geostationary orbit, were distant from the equator and within safe borders for military and political security— the U.S. at Cape Canaveral, Fla., and the USSR and China, far island. In the 1950s, the U.S. government considered several vacant Pacific islands it owns near the equator, but rejected those sites by the high costs of an overseas base.

No launch site has ever been located precisely on the equator. Locations now range from Kourou, French Guiana, at 5° north (for a 1669-km/h (1043-miles/h)) running start), where the Euro-French Ariane is launched, to Plesetsk, Russia, at 63° north 768 km/h (480 miles/h)), where Russia launches military satellites into polar orbits. Cape Canaveral, the world's most famous spaceport, is at 28.5° north (1472 km/h (920 miles/h)) as shown in Figure 7.2.[2]

Sea Launch represents the first effort to build a large launch system that can be taken to the ideal location rather than literally being tied to just one spot. Orbital Sciences Corp., Vienna, VA, utilizes a smaller, airborne launch vehicle, Pegasus, but its payload is limited to a few hundred kilograms. Although most of the launches will take place near the equator, some may start at higher latitudes to reduce the eastward motion for earth-observing satellites that are headed into polar orbits less than 800 km (500 miles) high.[3]

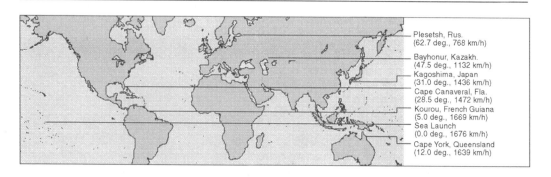

Plesetsh, Rus.
(62.7 deg., 768 km/h)

Bayhonur, Kazakh.
(47.5 deg., 1132 km/h)

Kagoshima, Japan
(31.0 deg., 1436 km/h)

Cape Canaveral, Fla.
(28.5 deg., 1472 km/h)

Kourou, French Guiana
(5.0 deg., 1669 km/h)

Sea Launch
(0.0 deg., 1676 km/h)

Cape York, Queensland
(12.0 deg., 1639 km/h)

Figure 7.2 This map shows the locations of several major launch sites worldwide, including the Sea Launch platform at the equator, and the relative *running start* a rocket would get by the Earth's rotation at that latitude. Cape York is in the proposal stage.

The first attempt to move the hefty Zenit rockets for a launch closer to the equator was made in the mid-1980s by a group of Australian investors, led by United Technologies Corp., Hartford, Conn.. They tried to develop an international spaceport at Cape York, Queensland, located 12° south of the equator. Lacking funds and political support, though, the project soon faded away.

Sea Launch grew out of what were almost casual conversations in 1993 to explore business ventures between Boeing Commercial Space Co., Seattle, WA, and two companies from the former USSR—RSC-Energia, in Moscow, and NPO-Yuzhnoye, in Dneprpetrovsk, Ukraine, which had emerged from the design bureaus that once developed ballistic missiles for the USSR. Conceptual studies for the venture started in the spring of 1994, and on April 3, 1995, joined by Kvaerner a.s., Oslo, Europe's largest shipbuilding firm, they formed Sea Launch with U.S. $600 million in funding and went looking for customers. Kvaerner has since moved its headquarters from Oslo to London.

The company quickly signed its first contract, with Hughes Space and Communications, El Segundo, CA, for the launch of 10 of Hughes's new HS-702 Galaxy communications satellites, starting in 1999. Space Systems/Loral, Palo Alto, CA, also signed up for five launches. The fee has not been disclosed, but it is expected to be competitive with $56 million to $130 million being charged by the U.S. and European launchers.

The task ahead is formidable. The schedule is ambitious, the approach is new, the joining of technologies and cultures from four nations is a challenge, and the three-stage Zenit 3SL has yet to be proven (although the two-stage Zenit 2 has flown successfully). Key to the venture's triumph is its dramatic reduction of infrastructure, as well as its exploitation of the unique capabilities of the Zenit rocket.

Point of Launch

Launching rockets at sea was demonstrated as early as 1949 when the U.S. Navy fired a V-2 rocket from the flight deck of an aircraft carrier. Between 1967 and 1988, the U.S. and Italy sent several satellites aloft atop Scout rockets that rose from the San Marco platform (an oil-drilling rig converted into a launch pad) in Fuji Bay, off the east coast of Kenya, just 3° north of the equator. But the Scout was a small rocket and used solid propellants, and thus was easily readied for launch.

Little more was done with rockets at sea till now. Maritime vessels lacked the size and stability needed to launch large rockets. At the same time, the rockets used commercially were owned by governments or corporations that were in effect bound to use launch pads run by the U.S. or Europe, which dominate the market.

That situation started to change in the late 1980s, when Japan and other nations pioneered the development of massive, self-propelled platforms to drill for petroleum at sea. Also, in 1991, the Soviet Union collapsed, ending the Cold War and easing political pressures that had kept the USSR's low-cost launchers out of Western markets. Finally, with global communications needs (both telephony and television services) expanded, there was a requirement to loft more communications satellites using encrypted signals than the U.S. and Europe could handle.

To launch the Zenit at sea, Kvaerner explored several options, including using supertankers. The company settled on a large catamaran-like platform that offered the required stability. It decided to convert *Odyssey* (see Figure 7.3), one of its existing 28,000-metric-ton, self-propelled, semisubmersible offshore production platforms.[4] Popularly called oil rigs, they are used as often to drill for natural gas.

Figure 7.3 The *Odyssey* launch platform nears completion at Kvaerner a.s.'s Rosenborg Shipyard in Stavanger, Norway. The machinery house of the crane (left foreground) stands approximately where the engines and base of the Zenit 3SL rocket will be positioned on the launch pad. The structure atop the platform (rear) is the hangar where the assembled rocket will be stored during the cruise to the launch site.

Odyssey was built in Japan in the 1980s, when large platforms were being developed. The rig is not the largest production platform in the world, but comes close to it.

Because the platform's pontoons are 66 meters apart and mostly submerged, the support pillars offer little surface area for waves to hit. This setup allows the motion at rocket launch to be only a few degrees off vertical, even in stormy seas with waves as high as 6 meters from crest to trough.

In modifying *Odyssey* for Sea Launch, a launch pad, fuel and oxidizer storage tanks, supply lines, and launch electronics will be added, and the platform's aft will be lengthened at Kvaerner's Rosenborg Shipyard in Stavanger, Norway, to acommodate the launch support structure. The original platform stands on two partially submersible pontoons, from each of which rise two massive pillars and two intermediate pillars. The changes will lengthen the pontoons by 14 meters (45.5 ft) to 133 meters (432.3 ft), and a third large pillar will be added to the end of each pontoon to hold the launch pad itself. These alterations will place the exhaust nozzles of the Zenit rocket about 17 meters (55.3 ft) above the waves. A thrust bucket will divert the exhaust to keep the rockets from lifting the platform between ignition and release.

After Kvaerner finishes renovating the structure, the rig will sail (under its own power) to a shipyard at Vyborg, on Russia's Baltic Sea coast, for installation of some 4000 tons of Russian space support equipment. Included will be the rocket propellants, liquid oxygen and refined kerosene (also called RP-1) nitrogen and helium gas systems to pressurize the tanks, heating, ventilation, and air conditioning systems for the instrumentation bays; machinery to move and erect the rocket; and the launch pad itself.

The last preparatory step will be to mount a Zenit rocket on the platform, load it with propellants, and proceed with the countdown — but without firing. Delivery of the completed platform is set for 1999.

Preparing the Foundations for Launch

To prepare Zenit 3SL for launch at sea, and to control the launch, Kvaerner is building a new 200-meter-long (650 ft) assembly and command ship as shown in Figure 7.4. With antennas atop its superstructure, it will resemble those space tracking ships used by the U.S. and the USSR in the 1950s and 1960s. Although the ship will carry up to three rockets at a time, one launch every other month is the current plan until the launch rate builds.

Figure 7.4 A shipyard worker inspects the hull bottom for the Sea Launch assembly and command ship, a new vessel being built at Kvaerner's Govan Shipyard in Glasgow, Scotland. The design is adapted from designs for RoRo (roll-on and roll-off) ships. In this case, the rockets and their transporters will roll on and off.

The assembly and command ship is a new design based on the firm's experience in *RoRo* cargo ships designed for trucks that roll on and off them. In this case, large transporters carrying Zenit rockets will roll onto the ship, and cranes will carry the rocket, when assembled, out to the launch platform.

The ship is being built at Kvaerner's Govan Shipyard in Glasgow, Scotland, and is to be delivered in 1999 to Russia (the exact yard has not yet been selected). There, it will be outfitted with rocket-handling equipment as well as launch command and control systems. Among them will be two sets of tracking antennas using pointing and control systems similar to those on the Soviet electronic intelligence ships that once haunted the waters off Cape Canaveral, FL. The Russian control and tracking systems will be copies of those systems in use at present at Zenit launch facilities in Baykonur and Plesetsk.

At sea, the ship will have a crew of about 80 people and be able to accommodate 250, including the 30 crew members of the launch platform when they evacuate it before the rocket lifts off. The first sea launches will be certain to draw a large crowd of corporate and government officials, plus reporters wanting to cover the launch, and thus an extra ship might have to be chartered to meet demand.

Power Requirements

One of the challenges in designing the Sea Launch system is accommodating the differing power requirements of East and West. Both control ship and launch platform use 60-Hz (hertz — unit of frequency, cycles per second), 440-V (voltage) electricity, the standard for Western Europe. The Russian equipment, though, uses 50 Hz, 380 V.

Odyssey has a complex power plant comprising eight diesel-powered generators producing 20,000 kW between them. A 6-kV switch plant is provided to accommodate the Russian equipment and also to handle the ship's electric motors. Frequency converters are another item. Finally, a large number of uninterruptible power supplies are installed to ensure a smooth electric current free of the noise that can cause spurious signals in computers or simply wear out components.

The electric power systems on both the assembly and command ship are similar, although the ship has separate engines dedicated to propulsion and electric power. A 6-kV (voltage range) generator on the propeller shaft had been intended for use by some Russian equipment that has since been deleted; thus now it will power steerable thrusters that will hold the ship's position precisely during launch operations and help keep antennas pointed at the platform. Four 50-Hz, 380-V generators, plus switchboards and a great many uninterruptible power supplies, will supply electric power for Russian launch control equipment.

In addition, a special document has been drawn up to set standards for controlling electromagnetic interference, radiation hazards, and grounding. It is being overseen by Det Norsk Veritas, Noway's ship certification agency.

Zenith or Bust

The center of attention at Sea Launch is the Zenit booster, developed by RSC-Energia and NPO-Yuzhnoye when they were part of the former USSR's Ministry of Heavy Industry. Zenit's original purpose was to replace quickly military satellites that might be destroyed by U.S. antisatellite interceptors in time of war. Knocking out an enemy's encrypted communications, navigation, and spy satellites was a standard ingredient in the nuclear war fighting scenarios of the Cold War.

For that reason, Zenit was designed for horizontal assembly and test within a few hours, followed by a quick launch—all with minimal human involvement and without crews having to go to the launch pad, which itself might have been contaminated by radiation. The U.S. never developed these techniques, preferring vertical assembly of its encrypted communication spy satellites, which were designed to have longer lives and a better chance of surviving a war in space.

Zenit was derived from the four large liquid-propellant boosters used by the Energia super-booster that has flown only twice (first to launch the Soviet Buran space shuttle and then in a failed attempt to launch a large space station that would have been manned later). NPO-Yuzhnoye builds the Zenit first and second stages, and Energia builds the Block DM Impuls upper stage, which was developed as the fourth stage of the Proton booster.

Two- and 3-stage versions of Zenit are available. The third stage is required on the Zenit 3SL to insert satellites into geostationary transfer orbit. In size and capability, Zenit 3SL is comparable to the U.S. Atlas IIAS and Delta III, and Europe's Ariane 4 and 5.

The first Zenit 2 orbital flight was made in April, 1985. Of at least 25 launches of Zenit 2, four have failed (a 1990 accident destroyed one of two launch pads at Baykonur). No failures have been recorded since flights resumed in November 1992. Zenit's makers and owners are marketing the vehicle, and business is promising—they have signed

Loral's Globalstar communications satellites for launches (12 at a time on the same rocket) starting in 1999. As yet, no Zenit 3 model has flown, nor has the Zenit 2 been used for other than Russian government payloads.

Geostationary Transfer Orbit

The usual target for Zenit 3SL, as with most other commercial launchers, is geostationary transfer orbit, an ellipse that arcs from a few hundred kilometers above one side of the Earth to 35,898 km (22,436 miles) above the other. Once a satellite is in the desired transfer orbit, its own thrusters finish the job of inserting it into a circular 35,898-km orbit, where, moving in pace with the planet, it hovers over one point on the equator. Generally, the launch service has met its obligations when the satellite is in geostationary transfer orbit.

From near the equator, the 3SL will be able to place 5900 kg in that orbit. By comparison, the U.S. Atlas IIAS can place 3697 kg, and the Ariane 5 will be able to place 6800 kg.

The Zenit 3SL will stand 61.4 meters (199.6 ft) tall and 3.9 meters (12.7 ft) in diameter. Mass at lift-off will be 466,000 kg. All three stages burn liquid oxygen and kerosene, which leave only carbon dioxide and water vapor exhaust. The first-stage engine is a uniquely Russian design, comprising four combustion chambers and exhaust nozzles fed by a common liquid-oxygen and RP-1 pump. The Russians took this design approach in the 1950s when they had difficulty in developing large thrust chambers for rocket engines.

Boeing will build no part of the Zenit rocket proper, although it has the ability to do so. The company has extensive space launch experience. It built the first stage of the Saturn V moon rocket and the U.S. land-based intercontinental ballistic missiles that were once aimed at Baykonur, among other Soviet facilities. It now builds the advanced inertial upper stage for the U.S. Space Shuttle and the Titan rocket, and it is competing for an advanced expendable launch vehicle contract from the U.S. Air Force. It recently acquired the space and defense segments of Rockwell International Corp., builder of the Space Shuttle. These units are located in the Los Angeles area. For Zenit, Boeing will build only the graphite-composite nose cone and payload enclosure and the rocket-to-satellite interfaces at its Seattle-area facilities.

Prelaunch Activities

Part of the Sea Launch plan is to do as much work as possible on board the assembly and command ship and thus reduce the cost of maintaining ashore such facilities as rocket storage and processing hangars, as well as personnel. Sea Launch has already started converting former U.S. Navy facilities at Long Beach for shore operations.

Zenit boosters will be shipped to Long Beach by freighter or by the assembly and command ship, which can shuttle back to Russia to pick up three boosters at a time and install them on its RoRo deck. The communications satellite will be readied by the customer's engineers in a satellite processing and control facility ashore. Before leaving the shore facility, the satellite will be enclosed in the nose cone, which will also serve as an environment container to protect it from the sea environment.

Once it is enclosed, the satellite is rolled out and installed on the assembly and command ship, at which time it is mated with the three-stage rocket shown in Figure 7.1. Although eventually transfers from the ship to the launch platform will be possible at sea, the first transfers will be done with the assembly and command ship docked to the launch platform in Long Beach. The rocket will be rolled aft onto an exposed aft deck so that cranes on the launch platform can hoist it into the hangar.

The voyage to the equator will take 9 to 11 days, depending on the launch point, followed by five to six days for checking out the equipment at the site. A crew of 30 or so will live and work on the platform until about seven hours before launch. Then they will be evacuated by helicopter to the assembly and command ship, standing by about 5 km (3 miles) away.

At this point, all operations will take place automatically, just as they do in regular Zenit launches from Baykonur. The rocket will roll out of the hangar and be erected on the launch pad, fluid and electrical connections will be made, and the propellant tanks will be loaded.

No personnel are on board the platform. That is one of the advantages of the Zenit rocket in that it allows Sea Launch personnel to use it in the manner in which they intended. In contrast to Zenit's few hours of preparation, all Western launch vehicles require large teams to prepare the vehicle over a period of weeks.

Near the end of the countdown, the erector and the cable mast will be retracted, the engines ignited, and the rocket launched. The entire process will be controlled remotely by encrypted radio frequency links from the control center aboard the assembly and command ship. Tracking antennas on the ship will follow the vehicle until it is out of sight. Then a Russian-encrypted data relay satellite system (similar to the U.S. Tracking and Data Relay Satellites) will take over, acquiring encrypted data from Zenit and relaying it to a control station in Long Beach and to a backup station in Russia.

If an on-pad shutdown occurs or a storm blows up, the rocket's propellant tanks can be drained (the platform has enough capacity for three launch attempts) and the vehicle can be returned to the hangar. Even returning to port with liquid oxygen in the storage tanks is no problem. It is not much different from what the port has been using for U.S. Navy operations for years. This has all been ironed out with the port authorities, local shipyard authorities, and the local fire department, all of whom have studied this operation very carefully.

Cultural Differences

Indeed, pulling together so many different players (from Long Beach firemen to Russian engineers) has been Boeing's principal role in the Sea Launch venture. Just working with the Russians and Ukrainians, period, has been very interesting for Boeing. First of all, probably the biggest job for Boeing is establishing the trust necessary between two Eastern powers and two Western powers that have been at odds for so many years. Referring to cultural differences among the various parties, one could consider these as problems, but it is mainly finding out they all have different ways of doing things, and then finding common ground.

A principal concern of any nation in many international ventures is technology transfer. Each government wants to ensure that it preserves the competitive edge it has over other nations while taking full advantage of the others' capabilities. Such concerns have helped keep Western and Soviet space programs at arm's length (and one nation's encrypted comunications satellites off the other's launch vehicles) and helped kill the Cape York launch venture. Now, Russia and Ukraine are concerned that sharing their technology with the Sea Launch partners might help the U.S. launch industry, one of their top competitors.

From the beginning, Boeing, which has experience with international suppliers for its 777 jetliner and as prime contractor for the International Space Station, has sought to reassure its Sea Launch partners. They claim that they are not attempting to take over any of the technologies of Russia and Ukraine. Boeing is relying strictly on their Russian and Ukrainian partners for the processing and manufacturing of their respective rocket stages. The integration of those rocket stages is all done by the Russians who also provided the test equipment and checkout equipment. Existing equipment is being used at Baykonur, in Khazakhstan, to launch Zenit rockets. The real challenge is working peaceably to establish the interface between the payload and the launch vehicle.

One area where U.S. and former Soviet rocket programs have differed is documentation. U.S. engineers often complain that they are awash in forms, while their Soviet counterparts take a *just do it* approach that U.S. managers perceive as more risky. At prelaunch readiness reviews, the Russian engineers usually do what U.S. engineers would consider to be a case of working without a net: they just stand up and describe conditions without using viewgraphs.

Nevertheless, the Russians and Ukrainians appear to be satisfied. Their methods may be different, but the results are good. Energia has worked extensively with Hughes on interfacing the payload and the Zenit's third stage. And, the Energia engineers appear to be very comfortable with what they understand is needed by the U.S. engineers.

Launch Site Operations

Handling operations at the launch site will also be a different challenge. All other launch sites are owned by national governments, which their military and law enforcement agencies to clear the area of traffic or of people who simply want to watch but get too close for their own safety. Launches at Cape Canaveral, for example, have been delayed as the U.S. Coast Guard worked to detain fishing boats that had wandered into the danger area.

Sea Launch, though, will operate under the flag of either Liberia or Panama, nations that provide 90 percent of the world's commercial ships with *flags of convenience*—meaning less stringent regulations. But that also means that the U.S. government will be restricted in the protection it might afford when the Sea Launch platform is in international waters.

And what about that favorite scenario of action films, the hijacked warship? Sea Launch personnel have declined to discuss security details. But, from all indications, they are in contact with the appropriate federal officials and intend to ensure an appropriate level of security for their operations.

Even though Sea Launch will operate on the high seas, from a legal point of view, the launch will originate from U.S. territory, and therefore must be licensed through the Office of Commercial Space Transportation at the U.S. Department of Transportation (DOT), Washington, DC. In turn, the department will inquire about the procedures to guarantee the safety of people (both Sea Launch personnel and observers at the launch site) and property.

Unfortunately, Sea Launch personnel do not have answers to all those questions about security. They have discussed this with DOT (they have to apply for the proper licenses and clearances to even perform such a mission) and that is one of the first questions that always comes up. They know they have a lot of work to do with local authorities as well as international authorities.

Insurance is not expected to be a problem. The insurance industry is interested in covering Sea Launch after the first flight, for which Sea Launch will have to provide its own insurance.

Quiet Launch

Except for people on ships who may be observing the launch, no one should be in the way at the launch site, as it will be just a point in mid-ocean. Although the site will be in the subtropics, the U.S. Coast Guard and Coast and Geodetic Survey data indicate that the weather will be acceptable 95 percent of the time. Sea Launch officials declined to disclose the maximum winds under which they can launch.

There is always the unexpected, of course. One of the advantages of this system is automation. If there is something that comes up unexpectedly, Sea Launch personnel can either download (put the rocket back in the hangar) and ride out any kind of storm. Or, if it is an extreme storm, they can find a lagoon somewhere on a nearby island, or go into a semisubmersible state that stabilizes the platform, as they do in the North Sea, and ride it out as shown in Figure 7.5.[5]

As for salt corrosion, it is worth noting that this has always been a challenge to launch pads at Cape Canaveral, FL, Wallops Island, VA, and Lompoc, CA, all of which are less than a kilometer from the surf. Baykonur can be described as a *million-year-old beach* where pipelines have to be put above ground or else they rust away. At each of these facilities, engineers have long since learned how to design rockets and launch pads to survive sea-like conditions.

At present, Sea Launch engineers are designing and making modifications to existing hardware that is used in the ground environment for purposes of environmental protection. The launch platform itself was designed to operate in the North Sea, where it was originally used.

Corrosion is not a problem, except in compartments that actually contain salt water. Ordinary computers have been used extensively without problem on large ships; and, the air systems for the electronics suites will supply filtered air. Furthermore, the launch platform has an environmentally controlled hangar, so that the rocket is protected at all times—up to a few hours before launch. The nose cone also serves as a shroud that encapsulates the payload in an environment comparable to that in a hangar to protect it from temperature extremes in the Pacific that might affect the satellite. However, the rocket was designed taking such variations into consideration.

Summary

Once all the engineering issues have been resolved, the Sea Launch enterprise must grapple with factors that helped bring it into existence: market forces. Although only Hughes and Loral have signed up for launches so far, other clients are being pursued. At present, Arianespace SA, based in Paris, France, has orders for about 70 percent of the world's expected $20 billion in launches within the next few years.

Figure 7.5 To support the Zenit 3SL launch pad, the *Odyssey* platform had to be cut apart and expanded aft (toward the reader in this view). The massive pontoons on which the platform is mounted can be flooded so they partially submerge and improve stability.

Sea Launch must also deal with launch quotas imposed by the U.S. government (DOD and DOT) to protect the dwindling U.S. launch industry, although Sea Launch officials believe that ultimately Sea Launch will increase employment and revenues for the U.S. aerospace industry. So far, the Clinton administration, DOD and DOT have cleared Sea Launch for 15 launches. All will have to involve commercial or non-U.S. satellites, since U.S. government policy prohibits launching U.S. government satellites on non-U.S. vehicles unles this involves some cooperative arrangement.

With Chase Manhattan Bank, the World Bank and the U.S. government lined up to provide debt financing, the financial community is taking Sea Launch seriously, as is the competition.

From Here

This chapter discussed why many commercial users (who constitute the bulk of the GPS system's beneficiaries), have expressed nervousness about relying on a DOD-controlled system. This chapter also explored the DOD task force recommendation that management (as opposed to operational control) of the GPS be shifted to a joint DOD/DOT executive board that would resolve issues not resolvable by normal interagency coordination. Additionally, the chapter also discussed why a joint DOD/DOT executive board cannot support close operations such as guiding aircraft into runways, governing port traffic, or intelligent vehicle/highway system applications, which require DGPS. Finally, the chapter presented an ongoing satellite communications launching control ship with a launch platform that is being built to lift international commercial satellites into geostationary orbits.

Chapter 8 examines the following: Issues concerning the fall-out from possible widespread deployment of PKE technologies; how encryption helps to preserve vital national secrets, limit attacks on the nation's information infrastructure, and eliminates security and authentication obstacles to electronic commerce; and export controls.

Endnotes

1 Dave Dooling. "Launcher Without a Country," *IEEE SPECTRUM* (October 1996), The Boeing Company, MC 6H-PL, Seattle, WA 98124-2207, IEEE Spectrum, The Institute of Electrical and Electronic Engineers Incorporated. All rights reserved. © 1996 by IEEE. 3 Park Avenue, New York, NY, 10016-5997, Volume 33, Number 10, October 1996, pp. 18–19.

2 Ibid., p. 20.

3 Ibid.

4 Ibid., p. 21.

5 Ibid., p. 24.

8

Collecting Encrypted Signals of Intelligence

Encryption helps to preserve vital national secrets, limit attacks on the nation's information infrastructure, and eliminate security and authentication obstacles to electronic commerce. In the wrong hands, however, it can be used to plan or cover up domestic crimes or overseas military operations. The Federal Bureau of Investigation and the National Security Agency seek to preserve their ability to intercept and decode domestic and international encrypted satellite communications via GPS (see Figure 8.1), and thus would like to inhibit the use of public key encryption (PKE) to generate unbreakable codes—stopping it altogether may be technically impossible and may raise constitutional issues.[1] Yet neither wishes to impede government use of such technologies. Issues in this chapter also concern the fall-out from the possible widespread deployment of PKE technologies with regards to satellite communications.

Figure 8.1 Navistar Global Positioning System (GPS) satellite.

The NSA and FBI's solution is embodied in silicon as the Clipper chip, an encryption device attached to telephones. Each chip carries a law enforcement access field. This information is divided into two parts, each of which is escrowed by a separate federal agency. Upon court order, the pair could be recombined, thus enabling government decryption. If established as a standard, the Clipper chip would let government users encrypt and decrypt at reasonable cost, without worrying about interoperability. Moreover, if the Clipper chip sells well, its cost will drop, making it the most cost-effective route to encryption in the private market as well—albeit one with a legal back door.

The Clipper proposals has been controversial. Escrow arrangements have failed to win complete public confidence. Fears of *Big Brother* dominate the media debate on this issue. DES, the previous encryption standard, was extended for five years, but its fixed key-length means that it can eventually be broken, and thus is not a reliable guarantor of private encryption in the long run.

Will the Clipper policy work? Government sales may permit sought-after economies of scale for the Clipper chip, but the ferocity of private opposition dims the prospects of the Clipper chip coming into wide-spread commercial use in the satellite communications market.

Foreign governments may likewise reject an encryption technology for which the U.S. government retains a key. Even so, any data-encrypted communications technology needs standards to ensure interoperabil-ity. Two users with different encryption methods cannot talk to each other. With the adoption of the Clipper chip, government users have a standard but commercial users do not, thus potentially slowing the development of universally acceptable, interoperable encryption methods. Whether such a delay is good or bad depends on whether the virtues of commercial encryption for satellite communications out-weigh the possibility of the diversion of such capabilities into the wrong hands.

Export Controls

Similar controversy surrounds export controls on encryption software. Legally exportable software with 40-bit keys is 16,000 times less capable than that sold domestically where the legal limit is 56 bits. Vendors argue that such controls hurt the competitiveness of U.S. software (or forces the quality of domestic software down to the legal limit on exported software), a protest that has stirred congressional calls for relaxing export restrictions.

Few doubt that the U.S., in general, and the NSA, in particular, retain a marked edge in information security technologies, particularly in codebreaking. However, the rest of the world is catching up fast and, particularly through the Internet, has been able to access codemaking technologies (such as Pretty Good Privacy (PGP)) as good as if not better than what is available here. All but the most recent versions of PGP violate U.S. but not overseas patents. Forcing the global electronic network to take detours around U.S. export prohibitions may not necessarily be in the larger interests of the U.S.

What Is PGP?

PGP is a very powerful encryption program. Using *public key* encryption people can communicate securely with others without having to agree on a secret key first. It can also be used for authentication of messages. The public key algorithm used is RSA (Rivest-Shamir-Adelman), which is considered impossible to break in a reasonable time, if properly implemented.

Chaining remailers is of little help without encryption. E-mail mesages are normally sent in the clear. Anyone can read the entire message and see who is sending what. If PGP is used on messages to the remailers, that is no longer possible. Each remailer will only know where the message came from and where it is going, but not the identity of anyone else in the chain or the content of the actual message.

More on PGP

How secure is PGP? The big unknown in any encryption scheme based on RSA is whether or not there is an efficient way to factor huge numbers, or if there is some backdoor algorithm that can break the code without solving the factoring problem. Even if no such algorithm exists, it is still believed that RSA is the weakest link in the PGP chain.

Breaking PGP

Can PGP be broken by trying all of the possible keys? This is one of the first questions that people ask when they are first introduced to cryptography. They do not understand the size of the problem. For the International Data Encryption Algorithm (IDEA) encryption scheme, a 128 bit key is required. Any one of the 2^{128} possible combinations would be legal as a key, and only that one key would successfully decrypt all message blocks. Suppose that you had developed a special-purpose chip that could try a billion keys per second. This is far beyond anything that could really be developed today. Suppose also that you could afford to throw a billion such chips at the problem at the same time. It would still require over 10 trillion years to try all of the possible 128 bit keys.[2] That is something like a thousand times the age of the known universe! While the speed of computers continues to increase and their cost decreases at a very rapid pace, it will probably never get to the point that IDEA could be broken by a brute force attack.

The only type of attack that might succeed is one that tries to solve the problem from a mathematical standpoint by analyzing the transformations that take place between plain text blocks, and their cipher text equivalents. IDEA is still a fairly new algorithm, and work still needs to be done on it as it relates to complexity theory. However, so far, it appears that there is no algorithm much better suited to solving an IDEA cipher than a brute force attack, which has already been shown to be unworkable. The nonlinear transformation that takes place in IDEA puts it in a class of extremely difficult to solve mathematical problems.

Securing Conventional Cryptography

Assuming that you are using a good strong random pass phrase, conventional cryptography is actually much stronger than the normal mode of encryption because you have removed RSA that is believed to be the weakest link in the chain. Of course, in this mode, you will need to exchange secret keys ahead of time with each of the recipients using some other secure method of communication, such as an in-person meeting or trusted courier.

This option is especially useful if you want to back up sensitive files, or want to take an encrypted file to another system where you will decrypt it. In this case, you do not have to take your secret key with you, and it will also be useful when you lose your secret key. Also, you can even pick a different pass phrase for each file you encrypt, so that an attacker who manages to get one file decrypted cannot then decrypt all the other files.

Cracking RSA and PGP

If the NSA were able to crack RSA, you would probably never hear about it from them. Now that RSA is becoming more and more popular, it would be a very closely guarded secret. The best defense against this is the fact that the algorithm for RSA is known worldwide. There are many competent mathematicians and cryptographers outside the NSA, and there is much research being done in the field at present. If any of them were to discover a hole in RSA, we would hear about it from them. It would be hard to hide such a discovery. For this reason, when you read messages on USENET saying that *someone told them* that the NSA is able to break PGP, take it with a grain of salt and ask for some documentation on exactly the source of the information.

Nevertheless, RSA has been cracked publicly, specifically, two RSA-encrypted messages have been cracked publicly.

First, there is the RSA-129 key. The inventors of RSA published a message encrypted with a 129-digits (430 bits) RSA public key, and offered $100 to the first person who could decrypt the message. In 1994, an international team coordinated to Paul Leyland, Derek Atkins, Arjen Lenstra, and Michael Graff successfully factored this public key and recovered the plaintext. The message read:

THE MAGIC WORDS ARE SQUEAMISH OSSIFRAGE[3]

They headed a huge volunteer effort in which work was distributed via E-mail, fax, and regular mail to workers on the Internet who process their portion and sent the results back. About 1600 machines took part, with computing power ranging from a fax machine to Cray supercomputers. They used the best-known factoring algorithm of the time. Better methods have been discovered since then, but the results are still instructive in the amount of work required to crack an RSA-encrypted message. The coordinators have estimated that the project took about eight months of real time and used approximately 5000 MIPS-years of computing time.

What does all this have to do with PGP? The RSA-129 key is approximately equal in security to a 426-bit PGP key. This has been shown to be easily crackable by this project. PGP used to recommend 384-bit keys as *casual grade* security. Recent versions offer 512 bits as a recommended minimum security level.

Note: This effort cracked only a single RSA key. Nothing was discovered during the course of the experiment to cause any other keys to become less secure then they had been.

A year later, the first real PGP key was cracked. It was the infamous Blacknet key, a 384-bits key for the anonymous entity known as *Blacknet.* A team consisting of Alec Muffett, Paul Leyland, Arjen Lenstra and Jim Gillogly managed to use enough computation power (approximately 1300 MIPS) to factor the key in three months. It was then used to decrypt a publicly available message encrypted with that key.

The most important thing in this attack is that it was done in almost complete secrecy. Unlike with the RSA-129 attack, there was no publicity of the crack until it was complete. Most of the computers only worked on it in spare time, and the total power is well within reach of large- perhaps even a medium-sized organization.

So what is the best way to crack PGP? Currently, the best attack possible on PGP itself is a dictionary attack on the pass phrase. This is an attack where a program picks words out of a dictionary and strings them together in different ways in an attempt to guess one's pass phrase.

This is why picking a strong pass phrase is so important. Many of these cracker programs are very sophisticated and can take advantage of language idioms, popular phrases, and rules of grammar in building their guesses. Single-word *phrases,* proper names (especially famous ones), or famous quotes are almost always crackable by a program with any *smarts* in it at all.

There is a program available that can *crack* conventionally encrypted files by guessing the pass phrase. It does not do any cryptanalysis, so if you pick a strong pass phrase your files will still be safe.

There are also other methods to obtain the contents of an encrypted message, such as bribery, snooping of electronic emanation from the computers processing the message (often called a TEMPEST attack), blackmail, or *rubber-hose cryptography*—beating you on the head with a rubber hose until you give the pass phrase.

Forgetting, Choosing and Remembering Your Pass Phrase

What if you forget your pass phrase? In a word: don't. If you forget your pass phrase, there is absolutely no way to recover any encrypted files. If you're concerned about forgetting your pass phrase, you could make a copy of your secret keyring, then change the pass phrase to something else, and then store the secret keyring with the changed pass phrase in a safe location.

Note: Can your messages be read if your secret key ring is stolen? No, not unless they have also stolen your secret pass phrase, or if your pass phrase is susceptible to a brute force attack. Neither part is useful without the other. You should, however, revoke that key and generate a fresh key pair using a different pass phrase. Before revoking your old key, you might want to add another user ID that states what your new key ID is so that others can know of your new address.

So, why is the term *pass phrase* used instead of *password*? This is because most people when asked to choose a password, select some simple common word. This can be cracked by a program that uses a dictionary to try out passwords on a system. Because most people really do not want to select a truly random password, where the letters and digits are mixed in a nonsense pattern, the term phrase is used to urge people to at least use several unrelated words in sequence as the pass phrase.

In any event, how does one go about choosing a pass phrase? All of the security that is available in PGP can be made absolutely useless if you do not choose a good pass phrase do encrypt your secret key ring. Too many people use their birthday, their telephone number, the name of a loved one, or some easy to guess common word. While there are a number of suggestions for generating good pass phrases, the ultimate in security is obtained when the characters of the pass phrase are chosen completely at random. It may be a little harder to remember, but the added security is worth it. As an absolute minimum pass phrase, it is suggested that a random combination of at least eight letters and digits be used, with 12 being a better choice. With a 12-character pass phrase made up of the lower case letters a-z plus the digits 0-9, you have about 62 bits of key, which is 6 bits better than the 56-bit DES keys.[4] If you wish, you can mix upper and lower case letters in your pass phrase to cut down the number of characters that are required to achieve the same level of security.

As previously mentioned, a pass phrase that is composed of ordinary words without punctuation or special characters is susceptible to a dictionary attack. Transposing characters or misspelling words makes your pass phrase less vulnerable, but a professional dictionary attack will compensate for this sort of thing.

So, how does one remember his or her pass phrase? This can be quite a problem especially if you have about a dozen different pass phrases that are required in your everyday life. Writing them down someplace so that you can remember them would defeat the whole purpose of pass phrases in the first place. There is really no good way around this. Either remember it, or write it down someplace and risk having it compromised.

It may be a good idea to periodically try out all the pass phrases, or to iterate them in your mind. Repeating them often enough will help keep them from being completely blanked out when the time comes that you need them. If you use long pass phrases, it may be possible to write down the initial portion without risking compromising it, so that you can read the *hint* and remember the rest of the pass phrase.

Finally, can you be forced to reveal your pass phrase in any legal proceedings? The following information applies only to citizens of the U.S. in U.S. courts. The laws in other countries may vary.

There have been several threads on Internet concerning the question of whether or not the Fifth Amendment right, regarding self-incrimination, can be applied to the subject of being forced to reveal your pass phrase. There apparently has not been much case history to set precedents in this area. So if you find yourself in this situation, you should be prepared for a long and costly legal fight on the matter. Do you have the time and money for such a fight? Also, remember that judges have great freedom in the use of *Contempt of Court* citations and could sentence you to jail until you decide to reveal the pass phrase. If only you just had a poor memory!

Verifying PGP

How does one verify that his or her copy of PGP has not been tampered with? If you do not presently own any copy of PGP, use great care on where you obtain your first copy. You should try and get two or more copies from different sources that you feel you can trust. Compare the copies to see if they are absolutely identical. This will not eliminate the possiblity of having a bad copy, but it will greatly reduce the chances.

If you already own a trusted version of PGP, it is easy to check the validity of any future version. Newer binary versions of MIT PGP are distributed in popular archive formats; the archive file you receive will contain only another archive file, a file with the same name as the archive file with the extension .ASC, and a *setup.doc* file. The .ASC file is a stand-alone signature file for the inner archive file that was created by the developer in charge of that particular PGP distribution. Because nobody except the developer has access to his/her secret key, no one can tamper with the archive file without it being detected. Of course, the inner archive file contains the newer PGP distribution.

Note: If you upgrade to MIT PGP from an older copy (2.3a or before), you may have problems verifying the signature.

To check the signature, you must use your old version of PGP to check the archive file containing the new version. If your old vesion of PGP is in a directory called C:\PGP and your new archive file and signature is in C:\NEW (and you have retrieved MIT PGP 2.6.2 or higher), you may execute the following command:[5]

c:\pgp\pgp c:\new\pgp262i.asc c:\new\pgp262i.zip

If you retrieve the source distribution of MIT PGP, you will find two additional files in your distribution: an archive file for the RSAREF library and a signature file for RSAREF. You can verify the RSAREF library in the same way as you verify the main PGP source archive.

Non-MIT versions typcally include a signature file for the PGP.EXE program file only. This file will usually be called PGPSIG.ASC. You can check the integrity of the program itself this way of running your older version of PGP on the new version's signature file and program file.

Phil Zimmermann (creator of PGP) himself signed all versions of PGP up to 2.3a. Since then, the primary developers for each of the different versions of PGP have signed their distributions.

What about verifying the signature on your new copy of MIT PGP with your old PGP 2.3a or higher? If you cannot verify it, the reason might be that the signatures generated by MIT PGP are no longer readable with PGP 2.3a or higher.

First of all, you may not verify the signature and follow other methods for ensuring you are not getting a bad copy. This is not as secure, though. If you are not careful, you could get passed a bad copy of PGP.

If you are intent on checking the signature, you may do an intermediate upgrade to MIT PGP 2.6 or higher. This older version was signed before the *time bomb* took effect, so its signature is readable by the older versions of PGP. Once you have validated the signature on the intermediate version, you can then use that version to check the current version.

As another alternative, you may upgrade to PGP 2.6.2i or 2.6ui or higher, checking their signatures with 2.3a or higher, and use them to check the signature on the newer version.

People living in the U.S. who do this may be violating the RSA patent. Then again, they may have been violating it anyway by using 2.3a or higher, so you are not in much worse shape.

PGP Trap Doors and Back Doors

How do you know that there is no trap door in PGP? The fact that the entire source code for the free versions of PGP is available makes it just about impossible for there to be some hidden trap door. The source code has been examined by countless individuals and no such trap door has been found. To make sure that your executable file actually represents the given source code, all you need to do is to re-compile the entire program.

With regards to back doors, rumors abound that NSA put a back door in MIT PGP, and that they only allowed it to be legal with the back door. First of all, the NSA had nothing to do with PGP becoming *legal*. The legality problems solved by MIT PGP had to do with the alleged patent on the RSA algorithm used in PGP.

Second, all the freeware versions of PGP are released with full source code to both PGP and to the RSAREF library they use (just as was the case with every other freeware version before them). Thus, it is subject to the same peer review mentioned earlier. If there were an intentional hole, it would probably be spotted. After all, if you are really paranoid and *trust no one*, you can read the code yourself and look for holes! *The truth is out there* you know!

Finally, is there a back door in the international version of PGP? No. The international version of PGP is based on an illegally exported version of PGP, and uses an RSA encryption/decryption library

(MPILIB) which may violate a patent that is only valid in the U.S. There are no intentional back doors of any kind in the international version, nor is the encryption strength reduced in any way.

Putting PGP on Multiuser Systems and Operating Systems

Can you put PGP on a multiuser system such as a network or a mainframe? Yes. PGP will compile for several high-end operating systems such as Unix and VMS. Other versions may easily be used on machines connected to a network.

You should be very careful, however. Your pass phrase may be passed over the network in the clear where it could be intercepted by network monitoring equipment, or the operator on a multiuser machine may install *keyboard sniffers* to record your pass phrase as you type it in. Also, while it is being used by PGP on the host system, it could be caught by some Trojan Horse program. Furthermore, even though your secret key ring is encrypted, it would not be good practice to leave it lying around for anyone else to access.

So why distribute PGP with directions for making it on Unix and VMS machines at all? The simple answer is that not all Unix and VMS machines are network servers or *mainframes.* If you use your machine only from the console (or if you use some network encryption package such as Kerberos), you are the only user, you take reasonable system security measures to prevent unauthorized access, and, you are aware of the risks just discussed, you can use PGP securely on one of these systems.

You can still use PGP on multiuser systems or networks without a secret key for checking signatures and encrypting. As long as you do not process a private key or type a pass phrase on the multiuser system, then PGP can be used securely.

Of course, it all comes down to how important you consider your secret key. If it is only used to sign posts to Usenet, and not for important private correspondence, you do not have to be as paranoid about guarding it. If you trust your system administrators, they you can protect yourself against malicious users by making the directory in which the keyrings are only accessible by you.

So, can PGP be used under a *swapping* operator system such as Windows or OS/2? Yes. PGP for DOS runs well in most *DOS windows* for these systems, and PGP can be built natively for many of them as well.

The problem with using PGP on a system that swaps is that the system will often swap PGP out to disk while it is processing your pass phrase. If this happens at the right time, your pass phrase could end up in cleartext in your swap file. How easy it is to swap *at the right time* depends on the operating system; Windows reportedly swaps the pass phrase to disk quite regularly, though it is also one of the most inefficient systems. PGP does make every attempt to not keep the pass phrase in memory by *wiping* memory used to hold the pass phrase before freeing it, but this solution is not perfect.

If you have reason to be concerned about this, you might consider getting a swapfile wiping utility to securely erase any trace of the pass phrase once you are done with the system. Several such utilities exist for Windows and Linux at least.

A Hybrid Mix

Why not use a hybrid mix of IDEA, Message Digest 5 (MD5), & RSA rather than RSA alone? Two reasons: First, the IDEA encryption algorithm used in PGP is actually much stronger than RSA given the same key length. Even with a 1024-bit RSA key, it is believed that IDEA encryption is still longer, and because a chain is no stronger than its weakest link, it is believed that RSA is actually the weakest part of the RSA-IDEA approach. Second, RSA encryption is much slower than IDEA. The only purpose of RSA in most public key schemes is for the transfer of session keys to be used in the conventional secret key algorithm, and to encode signatures.

Summary

Are all of the previous security procedures a little paranoid? That all depends on how much your privacy means to you! Even apart from the government, there are many people out there who would just love to read your private mail. And many of these individuals would be willing to go to great lengths to compromise your mail. Look at the amount of work that has been put into some of the virus programs that have found their way into various computer systems. Even when it does not involve money, some people are obsessed with breaking into systems.

In addition, do not forget that private keys are useful for more than decrypting. Someone with your private key can also sign items that could later prove to be difficult to deny. Keeping your private key secure can prevent, at the least, a bit of embarassment, and at most, charges of fraud or breach of contract.

Moreover, many of the previously mentioned procedures in this chapter are also effective against some common indirect attacks. As an example, the digital signature also serves as an effective integrity check of the file signed. Thus, checking the signature on new copies of PGP ensures that your computer will not get a virus through PGP (unless, of course, the PGP version developer contracts a virus and infects PGP before signing).

From Here

This chapter examined the following: Issues concerning the fall-out from possible wisespread deployment of PKE technologies (PGP); how encryption helps to preserve vital national secrets, limit attacks on the nation's information infrastructure, and eliminates security and authentication obstacles to electronic commerce; and export controls.

Chapter 9 addresses the following: How the history of commercial space launch is a story of continually increasing competition; how the U.S. has taken advantage of the fact that it supplies over two-thirds of all commercial payloads—essentially communications satellites—to persuade China and Russia to restrict their launch rates and limit their price discounts; how the U.S. has taken steps to inhibit Russia's sale of rocket equipment to India; how the idea of using export controls over satellite payloads to stabilize markets for launchers may be counterproductive in the long run by casting doubt on the reliability of U.S. suppliers of satellites; and, if technology comes to favor small satellites for some missions (as may be the case for land-mobile communications and surveillance), then small, quick-turnaround launch vehicles may be preferred over large ones.

Endnotes

1 "Key U.S. Security Policy Issues," National Defense University and the Institute For National Strategic Studies, Fort Lesley J. McNair, Washington, DC 20319-6000, 1995, p. 3.

2 Arnoud P. Engelfreit. "Security: Encryption: PGP," PO Box 448, NL-5660 AK Geldrop, The Netherlands, 1996, p. 2.

3 Ibid., p. 3.

4 Ibid., p. 4.

5 Ibid., p. 5.

9

Encrypting U.S. Commercial Space Launch Capabilities

Commercial space launch history is a story of continually increasing competition. Although the first technological rival to the U.S., the Soviet Union, was not initially active as a commercial vendor, the U.S. monopoly on commercial launch services was breached by France in the early 1980s, China in the late 1980s, Russia after the Cold War ended, and, within a few years, perhaps also by Japan, India and Pakistan. Only a third of the global commercial launch market is now held by U.S. firms.

The U.S. has, however, taken advantage of the fact that it supplies over two-thirds of all commercial payloads (essentially encrypted communications satellites) to persuade China and Russia to restrict their launch rates and limit their price discounts. China agreed to sell launch services only at market prices, and Russia's Proton-based launches go for 93.6 percent of market price. The success of such agreements is open to question. Reports have surfaced of unofficial 40 percent discounts on Chinese launches. In August 1993, the launch picture was further complicated when the U.S. slapped a two-year moratorium on the export of U.S. payloads to Chinese launch sites in retaliation for Chinese sales of prohibited military material to Pakistan (although two satellites were released to export in early 1994), which has resulted in that country now having nuclear weapons launch capabilities. Attempts to renegotiate further restrictions may be politically complicated by the large number of joint ventures being put together between U.S. and foreign companies, such as the recent pact among Boeing, Russia, and Ukraine to use Zenit rockets as discussed in Chapter 7.

The U.S. had also taken steps to inhibit Russia's sale of rocket equipment to India (for whatever good that did). As a result of U.S. pressure, Russia agreed to sell only finished cryogenic engines to India's space program rather than the technology itself. Without Russian help, India was considered unlikely to produce such engines indigenously until after the year 2000. However, recent developments now have left India with the capability to launch a nuclear attack on its neighbor Pakistan as well as any country within a 7500 mile range.

Although the management of U.S. space facilities could be altered to better serve U.S. launchers, how strenuously should U.S. trade policy serve to bolster the market share of domestic firms? The consistent reliability of the French Ariane, the low wages earned by Russian and Chinese rocket workers, the subsidized determination of Japan to become a commercial launch player, the prospect of India and Pakistan's entry into the development of launchers, and, the large stocks of military rockets facing conversion, all dim the prospects for U.S. producers of launchers. The idea of using export controls over satellite payloads to stabilize markets for launchers may be counterproductive in the long run by casting doubt on the reliability of U.S. suppliers of satellites. True, U.S. satellite makers have kept European competitors locked in a small niche, and serious competition from Japan has yet to emerge.

Nevertheless, the situation is subject to change; for example, at least two studies predict that Japan may emerge as a competitive supplier of satellites by the year 2005.

Can the U.S. launch industry be saved by radical technological changes in rockets or satellites? If technology comes to favor small satellites for some missions (as may be the case for land-mobile encrypted communications and surveillance), then small, quick-turnaround launch vehicles may be preferred over large ones. United States' companies such as Orbital Sciences Corporation, CTA, and Lockheed have a comparative advantage in the technology of small vehicles, despite competition from Russia's SS-25 and Israel's Shavit. Alternatively, the development of a completely reusable launch vehicle could, according to its proponents, reduce the cost per unit mass of lofting a payload into orbit by as much as a factor of ten. This, too, is likely to favor U.S. suppliers.

But, what about the security of transmissions to and from (uplink and downlink) these satellites after they are in orbit? A high level of concern by NASA, NSA and DOD has recently surrounded the requirement for protecting the command/control uplinks and downlinks of spacecraft with encryption.

Encrypted Communications Security (COMSEC) for Space Systems

Over the past few years, the NASA Security Management Office has received numerous inquiries regarding the applicability of national and NASA policy on the application of encrypted Communications Security (COMSEC) to NASA space systems. As previously mentioned, a high level of concern has surrounded the requirement for protecting the command/control uplinks of spacecraft with encryption. In the course of answering these questions and facilitating support from the National Security Agency (NSA) in conducting risk assessments for NASA programs, NASA has either not included encrypted COMSEC as part of the Program Commitment Agreement (PCA) or has employed non-NSA approved methods. These inconsistencies have resulted in the unnecessary and costly expenditure of funds and tacit acceptance of unacceptable risk. This practice is in stark contrast to the protective measures implemented in the Department of Defense (DOD) and the commercial space community. See the following Sidebar for further specific information on this issue.

Clarification of NASA Policy on Application of Communications Security to Space Systems

There have been numerous inquiries regarding COMSEC for NASA spacecraft, and the applicability of National and NASA encrypted COMSEC policies to U.S. Civil (non-DoD) space programs. Because of this heightened level of interest, and some inconsistencies associated with decisions concerning the application of encrypted COMSEC to command uplinks (which NASA believes have resulted in the unnecessary expenditure of funds and the acceptance of unnecessary risk), the NASA Security Management Office felt it necessary to clarify the existing policies to eliminate any confusion created by misinterpretation of these policies.

The National Policy on Application of Communication Security to U.S. Civil and Commercial Space Systems, NTISSP No. 1 dated June 17, 1985, contains the applicable and current policy on protecting the command and control uplink for U.S. Government owned spacecraft. This policy has not been rescinded or superseded, nor has it been revised since its original issue in 1985. Moreover, NASA adopted, within the NASA Security Handbook (NHB 1620.3C), the total provisions contained in NTISSP No. 1.

Paragraph 1 states, "Government classified and Government or Government contractor national security related information transmitted over satellite circuits shall be protected by approved techniques from exploitation by unauthorized intercept." This aspect of the policy has led to misconceptions. Although there is no question that telemetry carrying classified data must be protected by encryption, we often hear comments like "NASA is a non-DoD agency, and our missions are totally unclassified, therefore, encryption is not required." Although it is true that most of NASA's mission data is unclassified, there is a great deal of Government sensitive unclassified information handled daily at NASA. NASA's Automated Information Security Handbook, NHB 2410.9, requires justification for not encrypting the most sensitive of this information. The bottom line is that program decision makers must determine the risk balanced against the cost of encrypting sensitive unclassified data with the knowledge that interception of this data is increasingly easy and common when transmitted from one terrestrial location to another via a communications satellite.

Paragraph 2 states, "Government or Government contractor use of U.S. civil (Government-owned but non-DoD) and commercial satellites launched five years from the date of this policy shall be limited to space systems using accepted techniques necessary to protect the command/control uplink." Paragraph 2 embodies another commonly misunderstood aspect of this policy. It discusses the most critical aspect of encrypted COMSEC needs for all NASA space flights, the requirement to protect the command uplink. Threats to the command link range from intentional attack by nation states, terrorist groups, foreign space consortiums, and hackers or space communications enthusiasts, to those posed by operations in the increasingly crowded frequency spectrum where unintentional interference can have dire consequences. Thousands of hacker penetration attempts occur daily on the Internet. With the degree of sophistication in these attacks increasing, coupled with the communications equipment NASA's Security Management Office believes it could only be a matter of time before an unprotected NASA satellite is damaged through the introduction of an unauthorized command. It is for the same reasons that commercial enterprises are routinely applying encrypted COMSEC to their command uplinks to prevent *outages* as a result of unintentional interference that can result in the significant loss of revenue.

The term *accepted techniques* referred to in both of the preceding paragraphs also needs to be explained. Approved techniques as they pertain to space encrypted COMSEC equate to National Security Agency (NSA) endorsed encryption and authentication systems. The NSA, as the national manager for encrypted COMSEC, has thirty years of experience in developing reliable, lightweight, space qualified COMSEC systems which are now unclassified. There is a perception in some circles that, since the command link is unclassified, encryption systems endorsed by the National Institute of Standards and Technology (NIST) are to be used. This is wrong. The NIST has no plans now or in the future to get into the space COMSEC business. It is important to note that this misconception has resulted in the unnecessary expenditure of significant development funds for a nonspace related encryption system, when proven NSA space rated devices were already available, at a fraction of the cost.

Paragraph 3 states, "The need for and means to protect the command/control uplink associated with civil satellite systems, intended exclusively for unclassified missions, will be determined by the organization responsible for the satellite system in coordination with the National Security Agency." The requirement to apply encrypted COMSEC is not automatic in all cases, but it does seek to engage the responsible programs with the NSA for the purpose of discussing threat and associated risk when making these decisions.

Paragraph 6 states, "The Director, National Security Agency, in coordination with other departments or agencies as appropriate, shall assess space systems telecommunications and command/control uplink functions to determine their vulnerability to unauthorized use and provide approved protection techniques and guidance." Additionally, in complying with this policy, Chapter 46 of the NASA Security Handbook (NHB 1620.3C) seeks to facilitate this process by requiring responsible organizations within NASA to also coordinate with the NASA Security Management Office on space COMSEC issues.[1]

Although much of the traditional military threat that the U.S. has faced over the years has changed significantly, the U.S. and NASA continue to face multifaceted threats, particularly in the Information Technology Security (ITS) arena. The increasingly hostile and sophisticated ITS threatened environment includes daily break-ins and attempted penetrations to NASA automated information systems and telecommunications networks. Although it is difficult to predict threat intentions, threat capabilities are well documented and are growing. Threats concerning unauthorized commands and *spoofing* range from adversarial nations or rogue states and terrorist groups that consistently seek higher-visibility targets to potential space communication enthusiasts or radio frequency hackers who may strive to advance their own agendas. These challenges take on added urgency with NASA's planed use of the Internet to allow scientists access to payloads on future space missions. Because NASA has an inherent responsibility in this regard, the following Agency policy clarification is being issued:

♦ It is NASA policy to encrypt telecommunications involving national secure information processed by NASA resources and security spacecraft and command/control uplink with NSA-approved and endorsed techniques.

♦ It is NASA's policy to secure the command/control uplinks of space systems that are intended for unclassified missions but may be used to augment national security operations in the event of a national emergency. Exceptions to this policy may only be granted by the National Security Telecommunications and Information Systems Security Committee (NSTISSC).[2]

The need for and means to protect the command/control uplink associated with satellite systems intended exclusively for unclassified missions will be determined by the organization responsible for the satellite system in coordination with NSA. Exceptions will be considered on a case-by-case basis, only where application of this policy to a particular system would result in an impact on cost of schedule that is judged excessive to the protection afforded and only after a risk assessment consistent with the provisions of NASA handbook 2410.9 has been conducted and documented. Waiver approval shall reside with NASA's Chief Information Officer after consultation with the NASA Security Management Office.

The ITS, in general, and COMSEC requirements, in particular, will be addressed for all space programs as part of the program initialization and will be included as part of the PCA. This policy clarification applies to all NASA's space programs and their contractors who design NASA spacecraft.

In the past, the implementation of encrypted COMSEC to space systems has been somewhat time-consuming, costly, and inefficient in terms of size, weight, and power. The present reality surrounding the integration of COMSEC into space systems has improved dramatically over the last 10 years such that impacts to program development, especially new starts and block changes, are greatly minimized. These improvements make the benefits of system security and the requisite adherence to NASA policy all the more reasonable and valuable. The following Sidebar is a description of available NSA-endorsed COMSEC hardware.

Available Hardware Endorsed by National Security Agency Communications Security

The COMSEC hardware, endorsed by the National Security Agency (NSA), for the protection of satellite command/control uplinks is available as Contractor-Furnished Equipment (CFE) from several vendors. This allows off-the-shelf procurement of ground and flight hardware for integration into candidate systems. The endorsed products are available in small, light, and efficient package formats. For embedded flight applications, single VLSI chips (25MW, hi-rel, rad hard) accomplish the decryption/authentication function (2 chips, 2 PROMS comprising a complete system). And, for ground applications, small rackmount boxes accomplish satellite command/encryption/authentication (19 × 7 × 18 inches, 35 pounds, 75 watts).

Embeddable flight VLSI chips costs are in the $3k to 5k range and ground boxes in the $50k range, depending on quantities. Both embeddable flight and ground boxes are handled as unclassified COMSEC-controlled material that can be handled with much more reduced constraints, as compared with previous classified material. Keying material is also handled as unclassified crypto which eliminates the need for user clearances.[3]

From Here

The chapter addressed the following: How the history of commercial space launch is a story of continually increasing competition; how the

U.S. has taken advantage of the fact that it supplies over two-thirds of all commercial payloads—essentially communications satellites—to persuade China and Russia to restrict their launch rates and limit their price discounts; how the U.S. has taken steps to inhibit Russia's sale of rocket equipment to India; how the idea of using export controls over satellite payloads to stabilize markets for launchers may be counterproductive in the long run by casting doubt on the reliablity of U.S. suppliers of satellites; and, if technology comes to favor small satellites for some missions (as may be the case for land-mobile communications and surveillance), then small, quick-turnaround launch vehicles may be preferred over large ones.

Chapter 10 covers how Direct Broadcast Satellites (DBS) raise potential national security issues because they challenge state sovereignty over communications by bypassing national communications monopolies. This chapter will also cover the military uses of commercial satellites.

Endnotes

1   National Aeronautics and Space Administration, Office of the Administrator, Washington, DC 20546-0001, 1996.
2   Ibid.
3   Ibid.

Coping with the National Security Dimensions of Civilian Encrypted Communications Satellites

As is the case with global cellular telephony, Direct Broadcast Satellite (DBS) raises potential national security issues because it challenges state sovereignty over encrypted communications by bypassing national communications monopolies. Affected nations may respond aggressively and perhaps pursue policies hostile to the world's lines of encrypted communications.

The advent of DBS, notably AsiaSat (with one footprint over China and the other over South/Southeast Asia), portends the proliferation of media whose content cannot easily be regulated by national authorities. How have states in fact reacted? Some, including Persian Gulf states, China, and Singapore, have banned private satellite dishes outright, forcing broadcasts to be channeled through a local cable provider, and thus subjecting them to government control. In China, particularly in the south, the ban is routinely flouted. Malaysia and Thailand, on the other hand, seek to launch their own DBS systems, hoping that few consumers will opt to pay extra for access to unrestricted systems if they can get most of what they want from sanctioned ones.

The DBS broadcasters are, of course, not immune to pressure from national governments. StarTV (which uses AsiaSat), for instance, dropped BBC service from its northern footprint over China. Although the owners deny it, pressure from the Chinese government may have been a factor contributing to this decision. Even Canada has pressed for more Canadian content in encrypted transmissions from Hughes's new DBS by threatening to restrict access to cable for stations that transmit via DBS. Intelsat's plans for DBS service to Latin America were deferred in part due to concerns over the impact of such service on the cultures of targeted countries. If national governments or political groups see DBS as spreading unwanted encrypted social, cultural, or political messages to their population, hostility toward the West may be increased.

How long can governments control access to DBS? Tomorrow's antennas can increasingly be blended into walls and other backgrounds, thus frustrating bans on their possession. Electronic focusing can frustrate terrestrial jamming. Video compression, which multiplies the number of channels that any satellite can host, enhances the economics of narrowcasting. A billion-dollar investment can yield well over a hundred digital stations, which, in turn, could be profitably leased for perhaps $3 million a year. At that price, any of several aggrieved national or political groups (Kurds, radical Shiites, Sikhs, Burmese mountain tribes) could afford to broadcast propaganda 24 hours a day to wide swaths of territory.

Commercial Satellite Services Spectrum Allocation

If spectrum is not allocated in an orderly fashion, encrypted signals would interfere with each other, and radio-based services would be impossible. For geosynchronous satellite systems, orbital space must be allocated at the global level, and methods have already been worked out to resolve disputes. An international organization, the International Telecommunication Union (ITU), assigns satellites to orbital slots that are separated by one-and-a-half degrees, thus allowing 240 distinct geostationary slots for satellites of a given type, which are accessed by directional antennas. For the most part, controversies are avoided although a few rogue cases persist. For instance, Tonga, a small Pacific island, has tried to reserve far more orbital slots than it needs, and is enjoying a thriving business in the resale market. More seriously, China has maneuvered a satellite into a space where it could interfere with some Japanese transmissions. A similar dispute between Tonga and Indonesia was resolved in late 1993.

Global cellular communications satellites, on the other hand, move across the Earth, with the result that the total spectrum itself must be allocated. Despite the existence of an international conference (the World Administrative Radio Conference (WARC)) that assigns broad spectral bands for various uses, the allocation of spectrum to competing satellites has been left for countries to negotiate. This presents few problems if all vendors aspiring to provide global cellular services come from the same country, as is now the case for the two most likely vendors: Motorola's Iridium and Loral's Globalstar, which are both of U.S. origin. However, once another country (such as France or Russia) sponsors a system, controversy can arise.

In response, aspiring vendors have been making deals with various countries for permission to broadcast within their borders. But when countries are close together, it is impossible to service one without overlapping encrypted signals into another. The European Union, which does not have an aspiring global cellular vendor of its own, has nevertheless been hinting that it objects to systems such as Iridium. If other nations follow, equipment usable in the United States could be prohibited overseas, and the prospects for any global cellular system would be substantially reduced. European objections may reflect a zero-sum mentality concerning economic opportunities, rather than spectrum allocation per se: with each U.S. system in operation, the prospects for the future European systems are reduced.

Military Use of Commercial Satellites

With declining defense budgets, and the lack of commercially available transponders, the logic of putting more U.S. military traffic on commercial satellites is becoming compelling (see Figure 10.1).[1] United States

Figure 10.1 Titan IV launch.

commanders already concede that they cannot run operations in remote areas in the absence of commercial satellite transmission capability. The U.S. Navy is mulling the use of commercial DBS for broadcasting noncritical information to ships (see Sidebar, "Air Force Command's Use of DBS" for further information). Also NATO has reportedly decided to get out of the business of owning satellites (it has two at the moment). However, commercial satellites are not inherently secure, even though source-level encryption can reduce the threat of hostile interception. More important, commercial satellites have little protection against jamming and destruction. This has kept DOD's Milstar program going despite its high cost.

Air Force Command's Use of DBS

The Air Force Command, Control, Communications, and Computer Agency technology insertion branch was formed to inject new and emerging technologies into C4 systems. The branch achieved a major technical breakthrough with its Direct Broadcast Satellite demonstration. DBS is a method, which broadcasts a variety of Department of Defense encrypted applications to sites across the U.S.

This demonstration will fundamentally change the way critical information is disseminated to the war fighters in times of crisis. Other communicators throughout DOD and the Central Intelligence Agency have echoed similar sentiments and are anxious to see operational implementation.

The overall idea was approved by Lt. Gen. Carl O'Berry, the former SC, who envisioned digital satellite broadcast technology as a pillar of his Infosphere concept. The term DBS refers to a new simplex digital broadcast satellite capability of transmitting high data rates theater-wide to literally thousands of dispersed war fighters, using antennas as small as 18 inches in diameter. At present, the term Global Broadcast Service is used within DOD.

Air Force Command (AFC)4A conducted a series of demonstrations of digital satellite broadcast technology. The AFC4A team began work on this initiative even before DirecTV commercial broadcast entertainment video was introduced to the consumer market in June 1994.

In the AFC4A demonstration, the set-top receivers used at the five receive sites were the same type as the one million plus units sold. The AFC4A team has participated in all major DOD digital satellite encryption broadcast demonstrations. DOD capabilities in using this technology has increased at a tremendous rate. For instance, the Navy-sponsored *Radiant Storm* demonstration of 1994 used a commercial uplink facility and transferred encrypted data at 1.544 million bits per second.

The next demonstration was conducted during exercise ROVING SANDS in April 1995. For this demonstration, the number of sites within the theater increased to about 25, a commercial mobile Ku-band uplink was used, and live video was transmitted. AFC4A contributed a DBS suite of transmit equipment, satellite antennas and receivers, and engineers to assist the Space Warfare Center at their field sites.

In June 1995, the AFC4A team and the Operational Support Office demonstrated the system used during ROVING SANDS at the Armed Forces Communications-Electronics Association-sponsored TechNet '95 and during a demonstration at the National Information Display Laboratory, Princeton, NJ. Finally, in July 1995, the AFC4A team achieved a major breakthrough by merging Asynchronous Transfer Mode with the DBS technology. ATM is a high-speed switching technology being embraced by the U.S. Government, commercial vendors, and industry, which will provide speeds into the billion bit per second range to both local and wide area networks. Incorporation of ATM sets this demonstration apart from previous DO demonstrations.

To pull off a demonstration of this magnitude, many other AFC4A offices, as well as organizations outside AFC4A, played crucial roles in ensuring the demonstration's success. One major activity involved networking organizations who were going to transmit data to the AFC4A for subsequent satellite broadcast. These sites included the Global Weather Central from Offutt AFB, NE, Headquarters Air Intelligence Agency from Kelly AFB, TX, and 8th Air Force from Barksdale AFB, LA, via the C4 validation office, an AFC4A unit at Barksdale. This network was designed to allow data to be transmitted from remote sites to the uplink station at AFC4A to then be broadcast via satellite to war fighters throughout a theater.

The sites for the DBS were Barksdale, Kelly, Langley AFB, Va., the Pentagon, and at the AFC4A Interoperability Facility and Headquarters Air Weather Service at Scott. To make the demonstration more realistic, a simulated receive site encampment was created with support from the 3rd Combat Communications Group from Tinker AFB, OK. This showed that the equipment could work in a field environment as opposed to strictly a lab environment.

A major technical contributor to the demonstration was the NIDL, co-located with the Sarnoff Research Center in Princeton, NJ. Of all the technical assistance provided by the NIDL, the key was producing an ATM/DBS interface which allowed the incorporation of ATM into the demonstration. This equipment was the key piece necessary for the high impact promised by providing the means necessary to achieve data rates of 15 million bits per second. AFC4A connected and configured the interfaces as AFC4A assumed end-to-end responsibility for the project.

This demonstration was the first ever to achieve a combined throughput of 23 million bits per second to include encrypted video, voice, and data over a satellite transponder while incorporating ATM. A wide variety of applications were demonstrated, including those listed in the 23 Mbps circuit breakdown shown in the following:

♦ **15 Mbps ATM/wideband data interface unit channel**—Air Tasking Order data, imagery, Tactical Forecast System weather data, C2IPS software updates and databases, engineering drawings from Defense Logistics Agency, and ATM video/audio provided by MITRE from the Rome Laboratory from Griffiss AFB, NY.

♦ **3-6 Mbps video channel**—switchable among video tape, live video, CNN, and a PC screen.

♦ **1 Mbps data channel**—distributed database with collaborative planning (provided by Hughes Information Technology Corporation) which was enabled by a low-rate query channel.

♦ **1 Mbps data channel**—weather data from Global Weather Central and Intel products.

♦ Other smaller channels which handled such items as the program guide (similar to a TV guide listing) and audio.

During Operation Desert Storm, it was not uncommon for Air Tasking Orders to take a number of hours to get to the intended war fighter. This demonstration proved that the same ATOs can be transferred in less than five seconds. As another example, a full CD-ROM containing 623 million bytes of Combat Weather System data can be transmitted in about six minutes over the 15-Mbps ATM channel. In contrast, using a 9600 bit per second channel, the same data would take over 144 hours to transmit. During the demonstration (five eight-hour days), approximately three trillion bits of information was broadcast over the satellite.

One major addition for the Global Broadcast Service portions is the incorporation of cryptographic equipment. The Direct Broadcast Satellite demonstration proves that DOD can partner with the commercial world and provide cost-effective, interoperable solutions for the war fighter.[2]

Summary

To summarize, a more serious long-term problem may arise should it become necessary to target satellites used by hostile forces. With knowledge that U.S. military traffic is being carried by commercial satellites, the best way for a hostile force to preserve its own transmission capacity would be to lease a transponder on the same craft. With encryption being ubiquitous, it may be difficult to ascertain that any one transponder was being used for hostile purposes.

From Here

This chapter covered how Direct Broadcast Satellites (DBS) raise potential national security issues because they challenge state sovereignty over communications by bypassing national communications monopolies. The chapter also covered the military uses of commercial satellites.

Chapter 11 covers the growing tendency for U.S. forces to engage in multinational coalitions, coupled with the transition to computer-based information systems. Chapter 11 further suggests that the ability to transmit information among dissimilar systems will be a growing concern in such missions. Additionally, the chapter reviews tactical intelligence coordination that has always been a problem in coalition organizations. Finally, the chapter discusses the Internet privacy coalition.

Endnotes

1   "Key U.S. Security Policy Issues," National Defense University and the Institute for National Strategic Studies, Fort Lesley J. McNair, Washington, DC, 20319-6000, 1995, p. 5.

2   D. C. Brashares and Stan Grant, Capt, "AFC4A demonstrates Direct Broadcast Satellite," Air Force C4 Agency, Scott AFB, Il, 1995.

11

Improving Encrypted Communications in Multinational Coalitions

The growing tendency for U.S. forces to engage in multinational coalitions, coupled with the transition to computer-based information systems, suggests that the ability to transmit encrypted information among dissimilar systems will be a growing concern in such missions. This leads to a series of issues at the tactical, security, and global level.

Tactical intelligence coordination has always been a problem in multinational coalition organizations. The primary barrier remains translating languages. Today's coalitions have developed very structured ways of handling official messages, such as the use of liaison officers. The transition from human language to systems translations will make some tasks easier and others more difficult. On the one hand, English has become the lingua franca of the computer world, permitting at least informal encrypted communications among systems operators. Progress on automatic language translation is also proceeding apace, although only the major languages are being studied in this capacity.

On the other hand, computers are more finicky than people. To get two systems to interact requires common protocols at all levels from radio electronic, to bit-sequencing, packetizing, and encoding, to message formatting, to database definitions. For example, the ability of new sensors to see the battlefield in novel ways may be outstripping the ability to communicate these visions in encrypted standardized formats.

Establishing system-to-system links can force encrypted satellite communications systems with good security to share secrets with less secure systems (see Figure 11.1).[1] It also reveals what command-and-control systems and their associated sensors are capable of, which, in turn, suggests the strategic imperatives and assumptions that went into their design. Such information may be shared comfortably with established allies, but some members of today's ad hoc coalitions may be tomorrow's opponents. Hence, the case-by-case need to weigh operational efficiency against security arises.

If there is going to be a Global Information Infrastructure (GII), the various national encrypted satellite communications structures must

Figure 11.1 Satellite communications dish in Somalia.

find ways of passing information back and forth. Although international organizations are busy developing standards, too often the world bifurcates between the U.S. way of doing things and the international way of doing things. Does this mean that the U.S. is out of step with the world, or possibly that (given that U.S. companies sell twice as much software as the rest of the world combined) the world is out of step with the U.S.? The Clinton administration has put harmonization of the world's encrypted telecommunications infrastructure high on its agenda, as evidenced by the recent conference of G7 nations on this topic. Whether other nations view such efforts as the high road of promoting global satellite communications or the low road of pushing U.S. exports remains to be seen.

In the meantime, the DOD has put together a U.S. defense strategy to deal with improving encrypted communications in multinational coalitions as well as other U.S. defense interests. Let us take a look at this defense strategy, and see how it will thwart future threats to the interests of the U.S., its allies, and its friends.

U.S. Defense Strategy

Since the founding of the republic, the U.S. government has always sought to secure for the American people a set of basic objectives:

♦ The protection of their lives and personal safety, both at home and abroad.
♦ The maintenance of the nation's sovereignty, political freedoms, and independence, with its values, institutions, and territory intact.
♦ Their material well-being and prosperity.[2]

On the eve of the 21st century, the international environment is more complex and integrated than at any other time in history. The number and diversity of nations, organizations, and other actors vying for influence continue to grow. At the same time, the global economy is increasingly interdependent. Not only does this offer the U.S. the promise of greater prosperity, it also ties the security and well-being of Americans to events beyond their borders more than ever before. Today, incidents formerly considered peripheral to American security (the spread of ethnic and religious conflict, the breakdown of law and order, or the disruption of trade in faraway regions) can pose real threats to the U.S. Likewise, new opportunities have arisen for the U.S. (in concert with other like-minded nations) to advance its long-term interests and promote stability in critical regions via multinational coalitions.

In order to shape the international security environment in ways that protect and advance U.S. interests, the U.S. must remain engaged and exert leadership abroad. United States leadership can deter aggression, foster the peaceful resolution of dangerous conflicts, help stabilize foreign markets, encourage democracy, inspire others to create a safer world, and resolve global problems. Without active U.S. leadership and engagement abroad, threats to U.S. security will worsen and opportunities will narrow. If the U.S. chooses not to lead in the post-Cold War world, it will become less able to secure the basic objectives outlined in the foregoing.

Threats to the interests of the U.S., its allies, and its friends can come from a variety of sources. Prominent among these are:

- Attempts by regional powers hostile to U.S. interests to gain hegemony in their regions through aggression or intimidation.
- Internal conflicts among ethnic, national, religious, or tribal groups that threaten innocent lives, force mass migration, and undermine stability and international orders.
- Threats by potential adversaries to acquire or use nuclear, chemical, or biological weapons and their means of delivery.
- Threats to democracy and reform in the former Soviet Union, Central and Eastern Europe, and elsewhere.
- Subversion and lawlessness that undermine friendly governments.
- Terrorism.
- Threats to U.S. prosperity and economic growth.
- Global environmental degradation.
- The illegal drug trade.
- International crime.[3]

Many of these threats are global in scale and cannot be adequately addressed unilaterally, either by the U.S. or any other single nation state. Thus, the U.S. will need to secure the cooperation of a number of groups, nations, and international organizations to protect Americans from such threats.

The National Security Strategy

The Administration's National Security Strategy acknowledges both the inescapable reality of interdependence and the serious threats to U.S. interests posed by actions beyond its borders. To protect and advance U.S. interests, the American government must be able to shape the international environment—influencing the policies and actions of others. This mandates that the U.S. remain engaged abroad, particularly in regions where its most important interests are at stake. At the same time, it is essential that U.S. allies and friends share responsibility via a multinational force for regional and global security. The U.S. and its allies must work together to help build a more peaceful and prosperous world. This means, among other things, taking pragmatic steps to enlarge the world's community of free-market democracies. Therefore, the three principal components of the U.S. strategy of engagement and enlargement are:

- ♦ Enhancing security. The U.S. must maintain a strong defense capability and promote cooperative security measures.
- ♦ Promoting prosperity. The U.S. will work with other countries to create a more open and equitable international trading system and spur global economic growth.
- ♦ Promoting democracy. The U.S. will work to protect, consolidate, and enlarge the community of free-market democracies around the globe.[4]

These goals underscore that the only responsible strategy for the U.S. is one of international engagement. Isolationism in any form would reduce U.S. security by undercutting the ability of the U.S. to influence events abroad that can affect the well-being of Americans. This does not mean that the U.S. seeks the role of global policeman. But it does mean that America must be ready and willing to protect its interests, both now and in the future.

As the U.S. moves into the next century, military readiness means that U.S. forces must be prepared to conduct a broad range of military missions without being spread too thin. This will require sustaining a high level of training, morale and maintaining modern reliable equipment and facilities.

The Administration has also argued for balance between defense and domestic priorities. While these priorities may compete for resources in the short term, they are wholly complementary in the longer term. The U.S. cannot be prosperous if its major trade and security partners are threatened by aggression or intimidation. Also, it cannot be secure if international economic cooperation is breaking down because the health of the U.S. economy is interwoven with the global economy. Thus prudence dictates that U.S. strategy strikes a balance. The overall U.S. budget must invest in future prosperity and productivity, while avoiding the instabilities and risks that would accompany attempts to withdraw from its security responsibilities in critical regions.

The forces and programs developed in the 1993 Bottom-Up Review and the Nuclear Posture Review, as outlined in this chapter, will provide the capabilities needed to support this ambitious strategy. United States forces today are without question the best in the world. The Administration's defense program hopes to keep them that way.

Regional Security Strategies

The security relationships established by the U.S., its allies and friends during the Cold War are essential to advancing America's post-Cold War agenda. To meet the unique challenges of the post-Cold War era, the U.S. seeks to further strengthen and adapt these partnerships and to establish new security relationships in support of U.S. interests.

In Europe, the end of the Cold War has brought new opportunities and new challenges. Hand in hand with its North Atlantic Treaty Organization (NATO) allies, the U.S. has sought to promote a free and undivided Europe that will work with the U.S. to keep the peace and promote prosperity. In the new security architecture of an integrated Europe, NATO is the central pillar. It is complemented by the European Union and a strengthened Organization for Security and Cooperation in Europe. NATO's Partnership for Peace (PFP) (unveiled at the January 1994 NATO Summit) has provided a means for expanding and intensifying political and military cooperation throughout Europe. NATO members and partners have participated in more than a dozen

PFP exercises and hundreds of other training, planning, and consultation activities. Partnership for Peace serves as a pathway for nations to qualify for NATO membership. For those partners that do not join NATO, PFP will constitute a strong link to Europe's preeminent security organization and concrete proof that the alliance is concerned about their security. Partnership for Peace and the gradual NATO enlargement bolster efforts by Central and Eastern European nations and the New Independent States to build democratic societies and strengthen region stability. Other efforts (including U.S. military programs such as the European Command's Joint Contact Team Program and Marshall Center) similarly advance U.S. defense engagement with Central and Eastern Europe and the New Independent States.

The Office of the Secretary of Defense has made building cooperative defense and military ties with Russia, Ukraine, and the other New Independent States one of the Department of Defense's (DOD's) highest priorities. Moving away from the hostility of the Cold War and reducing its lethal nuclear legacy will be neither instantaneous nor easy. Steady, continued engagement that focuses on mutual security interests will be the cornerstone in building constructive relationships with the New Independent States. Through the pursuit of a pragmatic partnership, the U.S. will strive to manage differences with Russia to ensure that shared security interests and objectives take priority. A central objective is to encourage Russia to play a constructive role in the new European security architecture through the development of NATO-Russia relations and through Russia's active participation in PFP.

The East Asian-Pacific region continues to grow in importance to U.S. security and prosperity. This region has experienced unprecedented economic growth—growth that in 1997 increased U.S. trade in the region to $479 billion and supported 3.1 million American jobs. The security and stability provided by the presence of U.S. military forces in the East Asian-Pacific area over the past 43 years, has created the conditions for such tremendous economic growth. Security, open markets, and democracy (the three strands of the President's National Security Strategy) are thoroughly intertwined in this region.

Today, the U.S. retains its central role as a force for stability in East Asia-Pacific, but it has begun to share greater responsibility for regional security with its friends and allies via a multinational force. The U.S. constructively participates in and supports regional security dialogues.

It actively encourages efforts by East Asian-Pacific nations to provide host-nation support for U.S. forces, contributes to United Nations (UN) peace operations, and participates in international assistance efforts throughout the world. While these regional initiatives are important, there is no substitute for a forward-stationed U.S. military presence (essential to both regional security and America's global military posture), or, for U.S. leadership similar to that which brought together the broad multinational coalition that convinced North Korea to relinquish its nuclear weapons program. The U.S. will remain active in this vital region.

The U.S. has enduring interests in the Middle East, especially by pursuing a comprehensive Middle East peace, ensuring the security of Israel and U.S. principal Arab partners, and, maintaining the free flow of oil at reasonable prices. The U.S. will continue to work to extend the range of Middle East peace and stability. Integral to that effort is the Administration's strategy of dual containment of Iraq and Iran for as long as those states pose a threat to U.S. interests, to other states in the region, and to their own citizens. Maintaining the long-standing U.S. military presence in Southwest Asia is critical to protecting the vital interests America shares with others in the region.

The U.S. will stay engaged in the security of South Asia militarily, as well as diplomatically and economically. Defense relationships with India and Pakistan can support broader U.S. interests and objectives, including nuclear and missile nonproliferation and global peacekeeping. The challenge DOD faces is to develop defense relationships in ways that reduce tensions in South Asia and protect U.S. vital interests in the adjacent areas. United States bilateral relationships with individual South Asian nations (even in the face of recent nuclear hostility between India and Pakistan) can advance and flourish without diminishing or tilting U.S. ties to other nations in the region.

The overarching U.S. objectives in the western hemisphere are to sustain regional stability and to increase regional cooperation. A more stable and cooperative environment would help ensure that recent strides in democracy, free markets, and sustainable development can continue, and that further progress can be made by the nations of the region. As in other regions, DOD is working to enhance the sharing of responsibility for mutual security interests with its friends and allies in the western hemisphere. Contributions might include cost-sharing for U.S. deployments, the provision of non-U.S. forces to multinational coalition operations, support for international development and de-mocratization, and, the contribution of personnel or resources to UN peace operations.

Although, at present, the U.S. has no permanent or significant military presence in Africa. The U.S. does desire access to facilities and strengthened relations with African nations through initiatives that have been or might be especially important in the event of contingencies or evacuations. The U.S. has significant interests in Africa in countering state-sponsored terrorism, narcotics trafficking, proliferation of conventional weapons, fissile materials, and related technology such as satellite encryption. The U.S. must continue to work with the continent's nations to help secure U.S. interests.

Africa also provides fertile ground for promoting democracy, sustaining development, and resolving conflict. The U.S. does not seek to resolve Africa's many conflicts, but rather to empower African states and organizations to do so themselves. It also supports the democratization and economic growth that are necessary for the long-term stability of the region. The U.S. actively participates in efforts to address the root causes of conflicts and disasters that affect U.S. national interests before they erupt. Such efforts include support for demobilization of oversized militaries, demining, effective peace operations, and strong indigenous conflict resolution facilities, including those of the Organization of African Unity and subregional organizations.

In all these regions, nations contribute to global and regional security in a wide variety of ways. The notion of responsibility sharing reflects the broad range of such contributions. In addition to providing host-nation support for U.S. forces, states can contribute to international security by maintaining capable military forces, assigning those forces to multinational coalition missions such as Operation Desert Storm; NATO's Implementation Force (IFOR) in Bosnia and Herzegovina or to UN peacekeeping operations, and providing political and financial support for such shared objectives as international economic development or the dismantlement of North Korea's nuclear program. Since the end of the Cold War, U.S. friends and allies have taken on increased shares of the burden for international security providing, for example, over 256,000 troops to Operation Desert Storm, and $80 billion to the U.S. and other multinational coalition members to help defray their expenses in the war. Yet, room for more equitable and cost-effective responsibiloity sharing remains. The Department of

Defense is committed to working with Congress and with U.S. friends and alies toward this goal.

U.S. Military Missions

As stated in the preceding National Security Strategy just discussed, the Bottom-Up Review, the National Military Strategy, and the Department of Defense will field and sustain the military capabilities needed to protect the U.S. and advance its interests. The U.S. is the only nation capable of unilaterally conducting effective, large-scale military operations far beyond its borders. There is and will continue to be a great need for U.S. forces with such capabilities, not only to protect the U.S. from direct threats, but also to shape the international environment in favorable ways. This would be particularly true in regions critical to U.S. interests, for example, the support of multinational coalition efforts to ameliorate human suffering and bring peace to regions torn by ethnic, tribal, or religious conflicts. Supporting the National Security Strategy of Engagement and Enlargement requires that the U.S. maintain robust and versatile military forces that can accomplish a wide variety of missions as delineated in the Bottom-Up Review:

First, U.S. forces must be able to offset the military power of regional states with interests opposed to those of the U.S. and its allies. To do this, the U.S. must be able to credibly deter (if required) and decisively defeat aggression. All of this would be done in concert with regional allies by projecting and sustaining U.S. power in two nearly simultaneous major regional conflicts (MRCs).

Second, U.S. forces must be forward deployed or stationed in key overseas regions in peacetime to deter aggression. They must also demonstrate U.S. commitment to allies and friends; underwrite regional stability, gain familiarity with overseas operating environments, promote joint and combined training among friendly forces, and, provide initial capabilities for timely response to crises.

Third, the U.S. must be prepared for a wide range of contingency operations in support of U.S. interests. These operations include, among others, smaller-scale combat operations, multilateral peace operations, noncombatant evacuations, and humanitarian and disaster relief operations.

Fourth, while the U.S. is redoubling its efforts to prevent the prolifer-ation of weapons of mass destruction (WMD) and associated delivery systems, it must at the same time improve its military capabilities to deter and prevent the effective use of these weapons. It must also defend against them, and fight more effectively in an environment in which such weapons are used.

Finally, to meet all these requirements successfully, U.S. forces must be capable of responding quickly and operating effectively. That is, they must be ready to fight. This demands highly qual-ified and motivated people, modern, well-maintained equipment, vi-able joint doctrine, realistic training, strategic mobility, and sufficient support and sustainment capabilities.

Deterring and Defecting Aggression

The focus of U.S. planning for a major regional conflict is based on the need to be able to project power and to deter, defend against, and defeat aggression by potentially hostile regional powers. Today, such states are capable of fielding sizable military forces that can cause serious imbalances in military power within regions important to the U.S. with allied or friendly states often finding it difficult to match the power of a potentially aggressive neighbor. Such aggressive states (e.g., India and Pakistan) may also possess WMD. Hence, to deter aggression, prevent coercion of allied or friendly governments, and ultimately defeat aggression should it occur, the U.S. must prepare its forces to assist its friends and allies in confronting this scale of threat. United States planning for fighting and winning these MRCs envisages an operational strategy that, in general, unfolds as follows (recognizing that in practice some portions of these phases may overlap):

  ♦ Halt the invasion.
  ♦ Build up U.S. and allied/coalition combat power in the theater while reducing that of the enemy.
  ♦ Decisively defeat the enemy.
  ♦ Provide for post-war stability.[5]

The U.S. will never know with certainty who the next opponent will be, how that opponent will fight, or how the conflict might unfold. Moreover, the contributions of allies to the multinational coalition's overall capabilities will vary from place to place and over time. Thus, balanced U.S. forces are needed in order to provide a wide range of complementary capabilities and to cope with the unpredictable and unexpected.

United States military strategy calls for the capability (in concert with regional allies) to fight and decisively win two MRCs that occur nearly simultaneously. This is the principal determinant of the size and composition of U.S. conventional forces. A force with such capabilities is required to avoid a situation in which an aggressor in one region might be tempted to take advantage of a perceived vulnerability when substantial numbers of U.S. forces are committed elsewhere. More fundamentally, maintaining a two-MRC force helps ensure that the U.S. will have sufficient military capabilities to defend against a multinational coalition of hostile powers or a larger, more capable adversary than is foreseen today.

United States forces fighting alongside their regional allies are capable of fighting and winning two nearly simultaneous MRCs today. Wih the capability to program enhancements to U.S. mobility/prepositioning assets, as well as improvements to encrypted satellite surveillance assets, accelerated acquisition of more effective munitions and other key improvements by U.S. military forces will forever be maintained and improved.

Stability Through Overseas Presence

The need to deploy or station U.S. military forces abroad in peacetime is also an important factor in determining overall U.S. force structure. In an increasingly interdependent world, U.S. forces must sustain credible military presence in several critical regions in order to shape the international security environment in favorable ways. Toward this end, U.S. forces permanently stationed and rotationally or periodically deployed overseas serve a broad range of U.S. interests. Specifically, these forces:

> ♦ Help to deter aggression, adventurism, and coercion against U.S. allies, friends, and interests in critical regions.
> ♦ Improve U.S. forces' ability to respond quickly and effectively in crises.
> ♦ Increase the likelihood that U.S. forces will have access to the facilities they need in theater and enroute.

- ♦ Improve the ability of U.S. forces to operate effectively with the forces of other nations.
- ♦ Underwrite regional stability by dampening pressures for competition among regional powers and by encouraging the development of democratic institutions and civilian control of the military.[6]

Through foreign military interactions, including training programs, multinational exercises, military-to-military contacts, and security assistance programs that include judicious foreign military sales, the U.S. can strengthen the self-defense capabilities of its friends and allies. Through military-to-military contacts and other exchanges the U.S. can reduce regional tensions, increase transparency, and improve its bilateral and multilateral cooperation.

By improving the defense capabilities of its friends and demonstrating its commitment to defend common interests, U.S. forces abroad enhance deterrence and raise the odds that U.S. forces will find a relatively favorable situation should a conflict arise. The stabilizing presence of U.S. forces also helps to prevent conflicts from escalating to the point where they threaten greater U.S. interests at higher costs.

Contingency Operations

United States defense strategy also requires that military forces be prepared for a wide range of contingency operations in support of U.S. interests. Contingency operations are military operations that go beyond the routine deployment or stationing of U.S. forces abroad but fall short of large-scale theater warfare. Such operations range from smaller-scale combat operations to peace operations and noncombatant evacuations. They are an important component of U.S. strategy and, when undertaken selectively and effectively, can protect and advance U.S. interests.

The U.S. will always retain the capability to intervene unilaterally when its interests are threatened. The U.S. also will advance its interests and fulfill its leadership responsibilities by providing military forces to selected allied/coalition operations, some of which may support UN Security Council (UNSC) resolutions (U.S. participation in multinational coalition sanctions enforcement and no-fly zone enforcement in Southwest Asia). Further, the U.S. will continue to participate directly in UN peace operations when it serves U.S. interests. United Nations and multinational peace operations can help prevent, contain, and resolve conflicts that affect U.S. interests. When it is appropriate to support a multinational peace operation, participating U.S. forces benefit from the authority and support of the international community and from sharing costs and risks with other nations.

Smaller-Scale Combat Operations

The U.S. will maintain the capability to conduct smaller-scale combat operations unilaterally, or in concert with others when important U.S. interests are at stake. These operations generally are undertaken to provide for regional stability (U.S. operations in Grenada), promote democracy (U.S. operations in Panama and Haiti), or otherwise respond to conflicts that affect U.S. interests.

Peace Operations

Peace operations include peacekeeping and peace enforcement. Peacekeeping involves deployment of military and/or civilian personnel with the consent of all major belligerent parties in order to preserve or maintain the peace. Such operations are normally undertaken to monitor and facilitate implementation of an existing truce agreement, and the support of diplomatic efforts to achieve a lasting political settlement. Peace enforcement is the application of military force or the threat of its use to compel compliance with resolutions or sanctions to maintain or restore international peace and security or to address breaches of the peace or acts of aggression. Such operations do not require the consent of involved states or of other parties to the conflict. These operations are authorized by the United Nations Security Council (UNSC) or a regional organization. They may be conducted by the UN, by a multinational coalition led by a member state or alliance, or by a regional organization.

The U.S. has an interest in supporting UN peace operations as a means of sharing the burdens of protecting international peace and security. Of the approximately 80,000 personnel serving in UN peace operations, about 6% are American. Previously, the U.S. was assessed 30.4% of the annual cost of UN peace operations. In FY 1999, the U.S. will be assessed only 22% of these costs. The price (in manpower and money) to protect America's interests around the world would be much greater without the burdensharing of the UN and its member states.

Members of the U.S. armed forces have been involved in UN peace-keeping missions since 1948. At the end of 1995, 3305 U.S. military personnel were participating in UN operations. During the year, significant U.S. participation was limited to three of 17 missions—Croatia (UNCRO), the former Yugoslav Republic of Macedonia (UNPREDEP), and in Haiti (UNMIH). A small number of U.S. armed forces also served as military observers or headquarters staff in other UN peace operations. The U.S. also continues to support non-UN peace operations, such as the Multinational Force, observers in the Sinai, and the Military Observer Mission in the border region between Peru and Ecuador.

Recent experiences in multilateral peace operations demonstrate that the UN, regional organizations, and member states have much to learn about how to conduct these types of operations effectively. First, the increasing size and complexity of peace operations (including the significant differences between peacekeeping and peace enforcement) and the sheer number of operations currently underway severely challenge the current capabilities of the international community to respond effectively. Second, any large-scale peace operation likely to involve combat should be conducted by a capable multinational coalition or regional organization. Recent experience also has demonstrated the need to fully integrate (at the national and international levels) political, military, economic, and humanitarian actions in peace operations, thus ensuring that military forces are adequately supported by nonmilitary efforts. Finally, DOD and other relevant agencies have also learned and applied important lessons about planning a smooth transition from a multinational coalition operation to a UN-led peace operation.

With the certainty that U.S. and allied interests will continue to be challenged by conflict, DOD has taken steps to establish more capable institutions and procedures to conduct peace operations. For example, DOD is working with the UN to improve its peacekeeping capabilities on issues ranging from encrypted satellite communications and information architecture to contracted service and material support. The U.S. military helped train the staffs of two UN peace operations that began in 1995—Haiti (UNMIH) and Angola (UNAVEM III). In both cases, this contributed significantly to the potential success of the missions.

In addition, U.S. forces continue to enhance their capabilities for conducting these operations, especially in the areas of doctrine development and training. The Joint Chiefs of Staff has recently issued *Joint Publication 3-07, Joint Doctrine for Military Operations Other Than War (MOOTW)*—providing guidance to all Services and combatant commands on the conduct of peace operations and other types of MOOTW. The U.S. Army has published *Field Manual 100-23*—a comprehensive manual on peace operations. Also, the U.S. Army Infantry School is completing a training support package that will guide brigades and battalions in the conduct of peace enforcement operations. The Marine Corps is completing the MOOTW Supplements to its Small Wars Manual. The Air Force has drafted *Air Force Doctrine Document 3, Military Operations Other Than War*, which addresses air and space power involvement in all types of MOOTW. Finally, *the Joint Task Force Commander's Handbook for Peace Operations* and the *Joint Electronic Library of Peace Operations* reference materials are also available.

As the peace operations doctrine emerged, training has also focused more directly on peace operations. The Chairman of the Joint Chiefs of Staff conducted a peace operations wargame for potential Joint Task Force commanders from the unified commands in June 1995. United States forces conducted several peace operations rotations at the Joint Readiness Training Center, preparing units for service in Haiti and hosting a pioneering exercise (Cooperative Nugget) with more than a dozen other member nations of the Partnership for Peace program. At the Combat Maneuver Training Center in Germany, U.S. and Dutch forces have trained for deployments to MOOTW environments. Many PFP (and in-the-spirit-of-PFP) exercises focus on peace operations-related training, from maritime embargoes to contingent battalions controlled by multinational headquarters. In Hawaii, the U.S. Pacific Command conducted a peace operations seminar in June 1995 that fostered dialogue between many Pacific rim nations. Also, U.S. forces have conducted an array of significant wargames and training, including multiphased exercises on MOOTW for civilian and military leaders and their staffs (such as the U.S. Marine Corps' Emerald Express). Lessons learned from past operations, discussions with other militaries, information gained from joint exercises and peace operations training have given U.S. military forces a more detailed understanding of how better to tailor training for the requirements of peace operations.

Other Key Missions

United States military forces and assets will also be called upon to perform a wide range of other important missions. Some of these can be accomplished by conventional forces fielded primarily for theater operations. Often, however, these missions call for specialized units and capabilities to deal with humanitarian and refugee assistance.

United States military forces and assets are frequently called upon to meet urgent humanitarian needs created by manmade or natural disasters including food shortages, migrant and refugee problems, and the indiscriminate use of landmines. Assisting countries with such needs, and thereby promoting good will, is integral to the U.S. strategy of engagement and enlargement. Humanitarian assistance not only provides relief, but also helps victims of violence and disaster return to the path of recovery and sustainable development. These programs support the regional unified commanders-in-chief's peacetime engagement strategy of promoting political and economic stability in their respective areas of responsibility.

During FY 1995, 105 countries benefited from DOD humanitarian assistance. The U.S. conducted several major humanitarian operations, including:

♦ Bosnia Relief, U.S. forces flew over 7700 humanitarian missions into Sarajevo and airdrops over Bosnia and Herzegovina since July 1992, contributing to the multinational coalition effort by delivering over 84,000 metric tons of humanitarian supplies.

♦ Cuban and Haitian Migrants. Operations undertaken by the U.S. armed forces facilitated refugee and migrant processing, refugee camp construction, and camp management in response to the Haitian and Cuban migration emergencies. The migrant camps in Guantanamo Bay, Cuba, closed in January 1996.

♦ Northern Iraq Relief. During most of 1995, DOD funded and managed a relief program for the population of northern Iraq, including the provision of food, basic construction materials for resettlement of villages, and medical, winterization, and agricultural supplies. On October 1, 1995, responsibility for funding the program was transferred to the U.S. Agency for International Development.[7]

In support of the Federal Emergency Management Agency, DOD has also helped provide assistance to victims of domestic disasters. Disaster responses to midwest floods, the Oklahoma City bombing, and

U.S. law. However, there has been little success in this area. Drugs still continue to flow into this country at an alarming rate. Nevertheless, the DOD supports the counterdrug mission in five key areas:

♦ Support to source nations. DOD provides training and other operational support to source-nation counterdrug units to enable them to interdict drug operations, seize deliveries, and arrest traffickers.

♦ Dismantling cartels. DOD continues to enhance its support for the Drug Enforcement Administration's strategy of dismantling the cocaine cartels and the cocaine business.

♦ Detection and monitoring the transport of illegal drugs. DOD operates detection and monitoring assets that cover the 2.6 million square mile source and transit zone stretching from South America to U.S. borders.

♦ Direct support to drug LEAs in the U.S. Active, Reserve, and Guard forces provide unique support to domestic drug LEAs in 11 categories including transportation, maintenance, training, and intelligence.

♦ Demand reduction. The Department of Defense provides community awareness and community outreach programs, as well as internal drug testing, education and training, and treatment programs.[8]

Countering the Spread and Use of WMD

Beyond the five declared nuclear weapons states, at least 22 other nations (e.g., India and Pakistan) have acquired or are attempting to acquire WMD (nuclear, biological, or chemical weapons) and the means to deliver them. In fact, many of America's most likely adversaries already possess chemical or biological weapons, and some appear determined to acquire nuclear weapons. Weapons of mass destruction in the hands of a hostile power threaten not only American lives and interests, but also the ability of the U.S. to project power to key regions of the world. The U.S. will retain the capacity to retaliate against those who might contemplate the use of WMD, so that the costs of such use will be seen as outweighing the gains.

Addressing the threat of WMD proliferation is no small challenge. The U.S. has a balanced, multitiered approach to counterproliferation, including enhancing U.S. capabilities in the following areas:

- Deterrence. Continual assessments of the strategic personality of countries with nuclear, biological, or chemical weapons to better understand their leaders' intentions and what particular combination of declaratory policy, force posture, and other political, diplomatic, and military signals can best dissuade them.
- Intelligence. Overall threat assessment and timely intelligence and detection for combat operations and management.
- Ballistic and cruise missile defense. Systems that can intercept missiles with a high degree of confidence and reliability, and prevent or limit contamination should the incoming missile be carrying a nuclear, biological, or chemical munitions.
- Passive defenses. Battlefield detection, decontamination, and individual and collective protection against chemical and biological warfare agents.
- Counterforce. Capabilities to seize, disable, or destroy WMD arsenals and their delivery means prior to their use without unacceptable collateral effects.
- Effective power projection. Reassessment of U.S. approaches to power projection to minimize the vulnerability of U.S. forces to attacks by WMD.
- Defense against covert threats. Improved capabilities to detect and disarm WMD that may be brought covertly into the U.S.[9]

The U.S. also continues to face potential nuclear threats from the New Independent States. Russia maintains a large and modern arsenal of strategic and nonstrategic nuclear weapons. Even after the Strategic Arms Reduction Treaty (START) II was ratified and entered into force, Russia still retains a formidable strategic nuclear arsenal of up to 3600 deployed warheads as well as several thousand nonstrategic nuclear weapons, which are not subject to START II. Moreover, strategic nuclear weapons from the former Soviet Union still lie outside of Russia. Perhaps more threatening is the risk that the materials, equipment, and know-how needed to make nuclear weapons will leak out of the New Independent States and into potentially hostile nations.

The U.S. seeks Russia's full implementation of the START accords. In addition, the U.S. will continue to press for the elimination of all nuclear weapons and strategic offensive arms in Ukraine, Belarus, and Kazakstan as pledged by the leaders of those countries in accordance with START I and the Nuclear Non-Proliferation Treaty. The U.S. will continue to provide assistance under the Nunn-Lugar program for the destruction of WMD and removal of all nuclear weapons from Ukraine and Belarus, ensure the safe and secure storage of nuclear weapons and, materials, and, help prevent the proliferation of WMD, their components, related technology, and expertise within and beyond national borders. These counterproliferation goals require a strong relationship with Russia and all the New Independent States.

United States nuclear forces remain an important deterrent. In order to deter any hostile nuclear state and to convince potential aggressors that seeking a nuclear advantage would be futile, the U.S. will retain strategic nuclear forces sufficient to hold at risk a broad range of assets valued by potentially hostile political and military leaders. This requirement is fully consistent with meeting America's current arms control obligations.

Summary

America's defense strategy aims first and foremost to protect the life, property, and way of life of its citizens. Its success ultimately relies on a combination of the nation's superior military capabilities, its unique position as the preferred security partner of important regional states, and its determination to influence events beyond its borders. By providing leadership and shaping the international security arena, the U.S., along with its allies and friends, can promote the continued spread of peace and prosperity. Only by maintaining its military wherewithal to defend and advance its interests and underwrite its commitments can the U.S. retain its preeminent position in the world.

From Here

This chapter covered America's defense strategy and the growing tendency for U.S. forces to engage in multinational coalitions, coupled with the transition to computer-based information systems. Chapter 11 further suggested that the ability to transmit encrypted information among dissimilar systems will be a growing concern in such missions. Additionally, the chapter reviewed tactical intelligence coordination that has always been a problem in coalition organizations.

Chapter 12 shows how the exploitation and control of space will enable U.S. forces to establish information dominance over an area of operations. The chapter also shows how this is the key to achieving success in future crises or conflicts. Chapter 12 will also show how a combination of easy access to data from U.S.-operated systems (surveillance, navigation, meteorology, and early warning), coupled with a standby ability to deny such flows in an emergency, may forestall the proliferation of uncontrolled satellite encryption systems.

Endnotes

1   "Key U.S. Security Policy Issues," National Defense University and the Institute for National Strategic Studies, Fort Lesley J. McNair, Washington, DC 20319-6000, 1995, p. 7.

2   Office of the Executive Secretary, Department of Defense, 1000 Defense Pentagon, Washington, DC, 20301-1000, 1996.

3   Ibid.

4   Ibid.

5   Ibid.

6   Ibid.

7   Ibid.

8   Ibid.

9   Ibid.

12

Managing Satellite Encryption Dimensions Over the Longer Run

The DOD's 1998 Annual Report holds that *the exploitation and control of space will enable U.S. military space forces to establish information dominance over an area of operations — the key to achieving success in future crises or conflicts.* All agree that the U.S. will enjoy considerable information superiority over any potential foe even five years hence — but how well will this superiority translate into dominance on the ground? How well will it hold up beyond the short term?

Events of the last few years suggest that the rest of the world has come to appreciate the importance of space and information, and many nations are striving to improve their capabilities in these media. In the interim, a combination of easy access to encrypted data from U.S.-operated systems, for example, surveillance, navigation, meteorology, and early warning coupled with a standby ability to deny such flows in an emergency, may forestall the proliferation of uncontrolled systems.

However, to the extent that others can piggyback on commercial satellites, third-party, or even U.S. military encrypted communication capabilities, the U.S. may be limited in its ability to deny others access to such assets. Therefore, armed with encryption technology, others may be able to do considerable damage to U.S. interests, even if they cannot prevail in a conventional military sense. If the U.S. cannot ultimately prevent the diffusion and use of such assets (encrypted satellite communications technology). U.S. military space forces will have to learn to operate in a transparent environment that it could hitherto impose upon others, but could avoid for themselves.

Now, let us take a look at the environment that U.S. space forces operate in now and in the near future. The remainder of this chapter covers how access to space systems (such as satellite communication encryption technology) by U.S. military space forces provide force multipliers that are increasingly important for sustaining an effective level of defense capability.

Space Forces Above and Beyond

The U.S. conducts activities in space in support of national security objectives. The main goals established by the President's National Security Strategy of Engagement and Enlargement in this area include:

♦ Freedom of access to and use of space.
♦ Maintaining the U.S. position as the major economic, political, military, and technological power in space.
♦ Deterring threats to U.S. interests in space and defeating aggression if deterrence fails.
♦ Preventing the spread of weapons of mass destruction (WMD) to space.
♦ Enhancing global partnerships with other space-faring nations across the spectrum of economic, political, and security issues.[1]

Department of Defense space forces will provide the means to exploit and, if required, control space to assist in the successful execution of national security strategy and national military strategy.

Space systems such as satellite encryption technology provide force multipliers that are increasingly important for sustaining an effective level of defense capability as overall U.S. force structure is downsized and restructured. Space forces meet a wide range of requirements critical to the National Command Authority (NCA), combatant commanders, and operational forces. The global coverage, high readiness, nonintrusive forward presence, rapid responsiveness, and inherent flexibility of space forces enable them to provide real-time and near-real-time support for military operations in peacetime, crisis, and across the entire spectrum of conflict. In recognition of the leverage to be gained by fully utilizing space capabilities. DOD is working to normalize space across the DOD by integrating space forces with land, sea, air, and special operations forces.

Space Forces and National Defense

Space forces are fundamental to modern military operations. They are playing a central role in the ongoing revolution in warfare because of their unique capabilities for gathering, processing, and disseminating encrypted information from satellites. As demonstrated during the Persion Gulf War of 1991, space systems can directly influence the course and outcome of war. For example, space systems helped confer a decisive advantage upon U.S. and friendly forces in terms of combat timing, operational tempo, synchronization, maneuver, and the integrated application of firepower. These inherent strengths of space forces will contribute directly to the deterrent effectiveness of U.S. armed forces.

Space Systems and C⁴I

Space forces provide key capabilities to integrate and deliver command, control, satellite communications, computer, and intelligence (C⁴I) support to land, sea, air, and special operations forces. In the planning phase of military operations, space forces provide enemy order of battle, precise geographical references and elevations, threat locations and characteristics, and, accurate cartography and geodesy. Command and control is enhanced by instantaneous satellite communications and coordination of forces, near-real-time surveillance and reconnaissance by satellites, meteorological conditions, and situational awareness of the battlefield.

Additionally, space forces provide encrypted data that are essential to military forces during the employment phase of military operations. Encrypted information provided by space systems may enable precision weapons to strike targets more effectively in any weather, day or night. Forces enroute have access to precise encrypted navigation, location, and timing information as well as continuous encrypted satellite communications with the command element and other employed forces for coordinated strikes. In addition, space assets enable secure encrypted communications among all functions in a military operation. The net result is the ability to efficiently and effectively employ forces to achieve desired objectives with a minimum of casualties and collateral damage.

In short, space-based force multipliers help to improve operational effectiveness, efficiency, and interoperability, maintain high technological superiority, and support worldwide deployment, sustainment, and operations of U.S. land, sea, air, and special operations forces. By providing almost global coverage, space forces help to compensate for reductions of forward-positioned infrastructure and provide ready, in-place capabilities to support U.S. forces worldwide.

Space Power and Deterrence

Space forces are an integral element of the overall deterrent posture of the U.S. armed forces. Any nation contemplating an action inimical to U.S. national security interests must be concerned about American space capabilities. Space systems provide the NCA, combatant commanders, and operational forces with unprecedented global situational awareness to identify and react to threats. As the U.S. draws down forces from overseas bases, space systems continue to provide nonintrusive presence because of their near-global coverage. Space forces thus help ensure that hostile actions will be discovered by the U.S. and introduce an element of uncertainty into the minds of potential adversaries.

More specifically, space forces provide unique capabilities for collecting and disseminating encrypted satellite information for determining other nations' capabilities and intentions. This includes information for indications, warning, and responding to the threat or use of force against the U.S., its armed forces, allies, and friends. Space systems perform global monitoring and are often the first to spot impending conflicts, allowing diplomatic actions to avert war. Space systems thus are critical to the ability of the U.S. to sustain a credible deterrent posture, which will continue to ensure that the costs of the threat or use of force are unacceptable to a potential adversary.

Furthermore, space forces also are essential for ensuring that U.S. land, sea, air, and special operations forces are capable of conducting operations against adversaries armed with WMD and missile systems. Space systems collect and disseminate information necessary for detecting, identifying, and characterizing threats. This includes nuclear material production, weapons systems transfers, and movements. Space systems support:

◆ Military planning, mission rehearsal, and targeting.
◆ Detect nuclear detonations.
◆ Provide launch point determination.
◆ Ensure command, control, and satellite communications.
◆ Enable precise navigation, maneuver, and weapons delivery.
◆ Facilitate smart weapons selection and force coordination.
◆ Support mapping, charting, geodesy, and terrain analysis.[2]

The force multipliers provided by space forces will enhance the effectiveness of military operations to seize, disable, or destroy WMD and missile systems, as well as provide for the alerting, survival, and protection of U.S. forces against hostile missile launches. Furthermore, space forces improve the effectiveness of active and passive defenses measures. Space systems will support the operations of active defenses that can intercept nonstrategic ballistic and cruise missiles and prevent or limit contamination should the missile be carrying a nuclear, biological, or chemical weapon. Space systems technologies are being investigated to allow cueing of missile defense forces to attacks by cruise missiles. They also will support civil defense of populations and passive defenses of operational forces. Space systems can provide strategic ballistic missile launch detection, limited theater ballistic missile launch detection, approximate impact area prediction, potential target acquisition sensor cueing, battle management, command, control, and encrypted communications, and, intelligence and missile warning dissemination.

Space Systems and the U.S. Contribution to Global Security

Space forces are a comparative national advantage of the U.S. and are an area within coalition strategy that can contribute unique capabilities for global security. In particular, space systems are capable of performing missions that place a premium on interoperability and the capacity to operate with other nations' forces. Space systems will enable U.S. and allied land, sea, and air forces to operate jointly in a more efficient and effective manner. They may also provide a means to support political commitments without putting U.S. forces at risk. Moreover, certain space systems provide dual-use capabilities employed by U.S. as well as international civil and commercial users in peacetime.

The exploitation and control of space will help enable the U.S. to achieve encrypted information warfare objectives in a military theater of operation. This could greatly enhance U.S. and allied ability to fight on more favorable terms. The ability to provide C^{4}I support to U.S. forces, and deny such support to an adversary, will enable combatant commanders and operational forces to plan and react faster than an adversary and thereby dictate the timing and tempo of operations. The responsiveness of in-theater exploitation and dissemination of space-sensor encrypted information is a key factor. Numerous countries in regions around the world are acquiring or accessing space systems, encryption technologies, and products. Foreign nations and subnational groups are obtaining space capabilities through indigenous efforts, purchases of goods and services, and cooperative activities. The spread of indigenous military and intelligence space systems, civil space systems with military and intelligence utility, and commercial space services with military and intelligence applications, poses a significant challenge to U.S. defense strategy and military operations. The spread of space capabilities compounds the dangers to U.S. national security posed by the proliferation of nuclear, biological, and chemical weapons, missile systems for their delivery, and advanced conventional weapon systems.

Consequently, the DOD must be able to ensure freedom of action in space for friendly forces and, when directed, limit or deny an adversary's ability to use the medium for hostile purposes. To ensure space control, DOD must sustain and improve capabilities to surveillance and monitor all militarily significant activities in space. The DOD also will continue to design, develop, and operate space systems with ensured survivability and endurability of their critical functions. Moreover, DOD must have capabilities to deny an adversary's use of space systems to support hostile military forces. In addition to military countermeasures, DOD's strategy to deal with the threat posed by the proliferation of space capabilities with military and intelligence applications includes:

♦ Actions to strengthen U.S. competitiveness in foreign markets.
♦ Measures to protect technologies, methodologies, and overall system capabilities which sustain U.S. advantage in space capabilities and promote continued U.S. technological advancements.
♦ Maintaining controls over significant capabilities which can be sold or transferred to foreign recipients.
♦ Government-to-government relationships with friendly states involving the sharing of space technology, products, and data.
♦ Agreements or arrangements which limit or deny foreign acquisition of, or access to, space systems, encryption technology, products, and data which could provide support to hostile forces.[3]

Major DOD Space Programs

One of the first major DOD space programs is called Space Launch. Space Launch is a key enabling capability for DOD to exploit space. Current U.S. space launch systems, however, do not meet all DOD needs and are becoming increasingly costly to use. A basic question for the past several years has been what level of DOD investment is appropriate to maintain existing capabilities and to provide for future space launch capability given current and expected fiscal constraints.

The President's National Space Transportation Policy, approved on August 5, 1994, seeks to balance efforts to sustain and modernize existing launch capabilities with the need to invest in the development of improved future capabilities. In that policy, the DOD is designated as the lead agency for improvement and evolution of the current expendable launch vehicle (ELV) fleet, including appropriate technology development. The DOD objective for this effort is to reduce costs while improving reliability, operability, responsiveness, and safety.

In order to implement this guidance, DOD is initiating an evolutionary ELV program. This program will eventually replace the medium and heavy-lift launch systems currently in the inventory. The program is defining a new relationship with the launch industry, emphasizing a measured development effort. The DOD seeks to use innovative methods to allow U.S. industry a greater leadership role in free market access to space via communication satellites. The current medium launch vehicle class will be phased out as early as 2002, and the heavy launch vehicle as early as 2005.

The Department of Defense (DOD) recently completed an assessment of the defense-related space launch industrial base. The basic conclusion of the assessment was that the industrial base has sufficient capability to meet today's defense needs. However, there is significant overcapacity in some portions of the base, which will require industry consolidation. The relatively stable commercial/defense demand, the predominantly dual-use nature of the base, and specific actions to meet DOD requirements, such as the ELV initiative, will ensure an adequate industrial base.

Space-Based Infrared Mission Area

After the cancellation of the Follow-on Early Warning System (FEWS), the DOD embarked on an intensive study to review the space-based infrared (SBIR) mission area. The goals of the SBIR study were to review infrared requirements needed to protect the U.S. within the context of two major regional contingencies and intelligence community needs, develop architectures to satisfy those requirements, and make a programmatic recommendation for system acquisition.

The SBIR study was notable in two areas: the process for conducting the study, and the results the study produced. The process brought together the various military and intelligence disciplines that use infrared data and developed a comprehensive set of requirements categorized into four areas: (1) missile warning (strategic and theater); (2) missile defense (national and theater); (3) technical intelligence; and (4) battle space characterization. Battle space characterization reflected the needs of the combatant commanders for situational awareness. Previously generated requirements for SBIR systems, new requirements, and those developed by the intelligence community were reviewed, analyzed, and adjusted to reflect current guidance.

The consolidated set of requirements was then used to develop a range of candidate architectures. These included satellite constellations in highly elliptical orbits, geosynchronous orbits, low-earth orbits, and various combinations of these orbits. In several cases, requirements that were driving the architectural design were revisited to ensure the validity of the requirement.

Based on the SBIR study, the DOD is proceeding with the development of a new high-altitude constellation of infrared detection satellites consisting of both highly elliptical and geosynchronous elements. The planned first launch of this new system is 2003. A flight demonstration of low-earth orbit satellites will be conducted to advance this technology and to investigate further phenomonologies in additional infrared frequencies. Furthermore, the high-altitude system will be designed to include the capability to integrate a low-orbit component if the need arises. Deployment of the low-altitude component may also permit the size of the high-altitude constellation to be reduced.

Military Encrypted Satellite Communications

The U.S. Army operate and mans the Defense Satellite Communications System (DSCS) for DOD through the Army Space Command at remote sites throughout the world. To update this capability, DOD's primary effort in encrypted satellite communications is the Milstar program. Conceived during the Cold War, the program was significantly restructured following the Bottom-Up Review (BUR) to reflect the increased tactical needs of current defense planning. The emphasis of the Milstar program has shifted from the provision of low data-rate, highly survivable encrypted communications to medium-encrypted data-rate communications that will provide survivable, difficult to detect, jam-resistant communications to tactical forces worldwide without reliance on foreign-based ground relays. This new emphasis was embodied in a redesign of the Milstar II system.

The BUR not only addressed the system requirements but also the affordability of the program. As a result, the constellation size for the system was reduced from six to four satellites with a determination to seek a less expensive alternative to the current design beginning early in next decade. The Milstar III program will seek to provide an advanced Extremely High-Frequency (EHF) communication system with capabilities similar to the current system on a platform that can be launched on a future medium-lift vehicle. The technological refinement required for that design are presently being pursued in an intensive investment program.

Despite the decision to pursue this advanced EHF alternative, there remain questions as to what direction military encrypted satellite communications (MILESATCOM) should take in the future. Communications are currently spread among four frequency bands on as many as seven satellite systems. All these systems will be due for replacement in the latter part of the next decade. With affordability a key concern, the Department of Defense has initiated an intensive architecture study to determine the best mix of capabilities, including commercial alternatives to support military encrypted satellite communications needs for the next century. The FY 1998 budget reflects a consolidated MILESATCOM strategy to reduce cost and improve operability.

Meteorological Satellite Convergence

The DOD, the National Oceanic and Atmospheric Administration (NOAA), and the National Aeronautics and Space Administration (NASA) completed a study in March 1994 that examined the feasibility of merging the DOD and NOAA operational polar-orbiting environmental satellite programs (the Defense Meteorological Satellite Program and the Polar-Orbiting Operational Environmental Satellite (POES) Program) while capitalizing on NASA's Earth Observing System technologies. This study culminated in the President's May 5, 1994, decision to converge U.S. polar-orbiting operational environmental satellite systems. An Integrated Program Office (IPO) has been created for the planning, development, acquisition, management, technology transition, launch, and operations of the National Polar-Orbiting Operational Environmental Satellite System (NPOESS). The DOD is the lead agency responsible for supporting the IPO in NPOESS system acquisitions. The NPOESS program also carries out a National Performance Review objective of reducing the cost of acquiring and operating polar-orbiting environmental satellite systems, while continuing to satisfy military and civil operational requirements.

The NPOESS will consist of a four-satellite constellation. The need date for the first satellite could be as early as 2005. The preferred architectural option includes a European satellite as one of the four satellites, provided this satellite meets specified U.S. conditions, including the capability to selectively deny critical encrypted data to an adversary during crisis or war, yet ensure the use of such encrypted data by U.S. and Allied military forces. A NOAA-led team, which includes DOD and NASA, is negotiating with the European Organization for the !Exploitation of Meteorological Satellites for provision of the mid-morning satellite of the four-satellite converged constellation. The DOD is working closely with NOAA and NASA to ensure that NPOESS satisfies national security requirements.

Space Support to the Warfighter

Over the past year, space forces have played important roles in every contingency where U.S. forces were engaged. In the former Yugoslavia, for example, multispectral imagery products provide support to U.S. forces that can be used for search and rescue. In Haiti, the UHF Follow-On and Milstar I military satellite communications systems provide operational support to U.S. forces for command and control as well as other functions.

To enhance the contributions of space forces to U.S. military operations, space forces also have been integrated into the Joint and Service exercise schedule. United States Space Command (USSPACECOM) components are actively engaged in supporting each combatant commander. Space systems directly supported exercises including Ulchi Focus Lens in Korea, Keen Edge in Japan, Atlantic Resolve in Europe, and Bulwark Bronze with U.S. Strategic Command and North American Aerospace Command. By fully integrating space capabilities into military operations, combatant commanders are better able to tailor their campaign planning and operations by effectively employing more available forces and achieving objectives at the least risk and cost.

To enhance the contributions of space systems to joint warfighting capabilities, USSPACECOM is proposing to establish a Joint Space and Missile Defense Warfare Center at Falcon Air Force Base, CO. It will coordinate the efforts of the Services with respect to:

- ◆ Space applications.
- ◆ Integration of joint space operations into doctrine.
- ◆ Innovation and application of joint space capabilities.
- ◆ Focused space support to the warfighter.[4]

The DOD is also actively pursuing advanced applications of space forces through Tactical Exploitation of National Capabilities (TENCAP) programs. The U.S. Army's TENCAP program, for example, is currently providing robust, in-theater space support to operational forces. The U.S. Army will continue by fielding more advanced and mobile capabilities with direct, in-theater immediate response to the warfighter. The U.S. Air Force and U.S. Navy sensor-to-shooter efforts currently underway to integrate space system-derived encrypted information into aircraft, are an additional example of ongoing activities to better exploit the force multipliers provided by space forces. These and other initiatives will improve the exploitation of space capabilities in the planning and conduct of military operations.

Summary

Space forces are essential for the successful execution of U.S. national security strategy and national military strategy. Space systems provide force multipliers that complement and enhance the capabilities of U.S. land, sea, air, and special operations forces. The organizational, operational, and modernization initiatives planned for the coming years will ensure that DOD space forces will retain the capability and versatility to accomplish their missions effectively and efficiently in support of U.S. national security objectives.

From Here

Chapter 12 showed how the exploitation and control of space will enable U.S. space forces to establish information dominance over an area of operations. The chapter also showed how this is the key to achieving success in future crises or conflicts. Chapter 12 also showed how a combination of easy access to data from U.S.-operated systems (surveillance, navigation, meteorology, and early warning) coupled with a standby ability to deny such flows in an emergency, may forestall the proliferation of uncontrolled satellite encryption systems.

Chapter 13 addresses escrowed satellite encryption, an approach aggressively promoted by the federal government as a technique for balancing national needs for information security with those of law enforcement and national security. The chapter also addresses the following: Secure Public Networks Act; the risks of key recovery, key escrow, and trusted third-party satellite encryption; satellite encryption for the rest of us; establishing identity without certification authorities; and, export criteria for key-escrow satellite encryption.

Endnotes

1 Office of the Executive Secretary, Department of Defense, 1000 Defense Pentagon, Washington, DC, 20301-1000, 1998.

2 Ibid.

3 Ibid.

4 Ibid.

13

Escrowed Satellite Encryption and Related Issues

The term *escrow*, as used conventionally, implies that some item of value (a trust deed, money, real property, other physical object) is delivered to an independent trusted party that might be a person or an organization (an escrow agent) for safekeeping, and is accompanied by a set of rules provided by the parties involved in the transaction governing the actions of the escrow agent. Such rules typically specify what is to be done with the item, the schedule to be followed, and the list of other events that have to occur. The underlying notion is that the escrow agent is a secure haven for temporary ownership or possession of the item, is legally bound to comply with the set of rules for its disposition, functions as a disinterested extra transaction party, and bears legal liability for malfeasance or mistakes.

Usually, the rules stipulate that with all conditions set forth in the escrow rules having been fulfilled, there will eventually be delivery of the item to a specified party (possibly the original depositing party, an estate, a judicial officer for custody, one or more individuals or organizations). In any event, the salient point is that all terms and conditions and functioning of an escrow process are, or can be, visible to the parties involved. Moreover, the behavior and performance of formal escrow agents are governed by legally established obligations.

As it applies to satellite encryption, the term *escrow* was introduced by the government's April 1993 Clipper initiative in the context of encryption keys. Prior to this time, the term escrow had not been widely associated with satellite encryption, although the underlying concepts had been known for some time (as described later in the chapter). The Clipper initiative promoting escrowed satellite encryption was intended *to improve the security and privacy of telephone satellite communications via satellite transmissions while meeting the legitimate needs of law enforcement.* In this original context, the term *escrowed satellite encryption* had a very specific and narrow meaning: escrowed satellite encryption was a mechanism that would assure law enforcement access to the voice communications via satellite transmissions underlying encrypted intercepts from wiretaps.

> **Note:** In the more general meaning of escrowed satellite encryption, exceptional access refers to access to plaintext by a party other than the originator and the recipient of encrypted satellite communications. For the case of stored information, exceptional access may refer to access to the plaintext of an encrypted file by someone not designated by the original encryptor of the file to decrypt it or even by persons so designated who have forgotten how to do so (see also Chapter 4). Contrast the meaning of third-party access in the original Clipper context, in which third-party access refers to assured access, under proper court authorization, by law enforcement to the plaintext of an encrypted voice conversation. The Clipper initiative was intended to support a system that provided a technically convenient means to assure fulfillment of such a requirement. Note that this meaning is much narrower than the use of the more general term *exceptional access* described previously.

However, during years of public debate and dialogue, *escrow, key escrow*, and *escrowed satellite encryption* have become terms with a much broader meaning. Indeed, many different schemes for *escrowed satellite encryption* are quite different from escrowed satellite encryption as the term was used in the Clipper initiative.

As is so often the case in computer-related matters, terminology for escrowed systems is today not clearly established and can be confusing or misleading. While new terminology could be introduced in an effort to clarify meaning, the fact is that the present policy and public and technical dialogues all use *escrow* and *escrowed satellite encryption* in a very generic and broad sense. It is no longer the very precise restricted concept embodied in the Clipper initiative and described later in this chapter. Escrow as a concept now applies not only to the initial purpose of assuring law enforcement access to encrypted materials or satellite transmissions, but also to possible end-user or organizational requirements for a mechanism to protect against lost, corrupted, or unavailable keys. It can also mean that some process such as authority to decrypt a header containing a session key is escrowed with a trusted party, or it can mean that a corporation is ready to cooperate with law enforcement to access encrypted materials or intercept encrypted satellite transmissions.

This chapter conforms to current usage, considering escrowed satellite encryption as a broad concept that can be implemented in many ways. It addresses forms of escrowed satellite encryption other than that described in the Clipper initiative. Also, escrowed satellite encryption is only one of several approaches to providing exceptional access to encrypted information. Nonescrow approaches to providing exceptional access are discussed in Chapter 26, "Satellite Encryption Technology Policy Options, Analysis and Forecasts." This chapter also describes a tool-escrowed satellite encryption that responds to the needs described in Chapter 4 for exceptional access to encrypted information. Escrowed satellite encryption is the basis for a number of Administration proposals that seek to reconcile needs for information security with the needs of law enforcement and to a lesser extent national security. As in the case of export controls, escrowed satellite encryption generates considerable controversy.

Finally, the relationship between *strong encryption* and *escrowed satellite encryption* should be noted. As previously stated, escrowed satellite encryption refers to an approach to encryption that enables exceptional access to plaintext without requiring a third party (government acting with legal authorization, a corporation acting in accordance with its contractual rights vis-à-vis its employees, an individual who has lost an encryption key) to perform a cryptanalytic attack. At the same time, escrowed satellite encryption can involve cryptographic algorithms that are strong or weak and keys that are long or short. Some participants in the public debate appear to believe that escrowed satellite encryption is necessarily equivalent to weak encryption because it does not prevent third parties from having access to the relevant plaintext. However, this is a mischaracterization of the intent behind escrowed satellite encryption, since all escrowed satellite encryption schemes proposed to date are intended to provide very strong cryptographic confidentiality (strong algorithms, relatively long keys) for users against unauthorized third parties, but no confidentiality at all against third parties who have authorized exceptional access.

Administration Initiatives Supporting Escrowed Satellite Encryption

Since inheriting the problem of providing law enforcement access to encrypted telephony from the outgoing Bush Administration in late 1992, Clinton Administration officials have said that as they considered the not-so-distant future of information technology and information security along with the stated needs of law enforcement and national security for access to information, they saw three alternatives. The first alternative was to do nothing, resulting in the possible proliferation of products with encryption capabilities that would seriously weaken, if not wholly negate, the authority to wiretap embodied in the Wiretap Act of 1968 (Title III) and damage intelligence collection for national security and foreign policy reasons. The second alternative was to support an approach based on weak encryption, likely resulting in poor security and cryptographic confidentiality for important personal and business information.

Finally, the third alternative was to support an approach based on strong but escrowed satellite encryption. If widely adopted and properly implemented, escrowed satellite encryption could provide legitimate users with high degrees of assurance that their sensitive information would remain secure but nevertheless enable law enforcement and national security authorities to obtain access to escrow-encrypted data in specific instances when authorized under law. Moreover, the Administration hoped that by meeting legitimate demands for better information security, escrowed encryption would dampen the market for unescrowed encryption products that would deny access to law enforcement and national security authorities even when they sought access for legitimate and lawfully authorized purposes.

The Administration chose the last alternative, and since April 1993, the U.S. government has advanced a number of initiatives to support the insertion of key escrow features into products with encryption capabilities that will become available in the future. These include the Clipper initiative and the Escrowed Encryption Standard, the Capstone/Fortezza initiative, and, the proposal to liberalize export controls on products using escrowed satellite encryption. These initiatives raise a number of important issues that will be discussed later in the chapter.

The Clipper Initiative and the Escrowed Satellite Encryption Standard

As previously noted, the Clipper initiative was conceived as a way for providing legal access by law enforcement authorities to encrypted telephony. The Escrowed Encryption Standard (EES; a Federal Information Processing Standard, FIPS-185) was promulgated in February 1994 as the key technological component of the Clipper initiative (see Sidebar, "Key Technical Attributes of the Clipper Initiative"). Specifically, the EES called for the integration of special microelectronic integrated circuit chips (called *Clipper chips*) into devices used for voice satellite communications. These chips, as one part of an overall system, provide voice confidentiality for the user and exceptional access to law enforcement authorities. To provide these functions, the Clipper chip was designed with a number of essential characteristics:

First, confidentiality would be provided by a classified algorithm known as Skipjack. Using an 80-bit key, the Skipjack algorithm would offer considerably more protection against brute-force attacks than the 56-bit DES algorithm (FIPS 46-1). The Skipjack algorithm was reviewed by several independent experts, all with the necessary security clearances. In the course of an investigation limited by time and resources, they reported that they did not find shortcuts that would significantly reduce the time to perform a cryptanalytic attack below what would be required by brute force.

Second, the chip would be protected against reverse engineering. It would also be protected against other attempts to access its technical details.

Third, the chip would be factory-programmed with a chip-unique secret key, the *unit key* or *device key*, at the time of fabrication. Possession of this key would enable one to decrypt all satellite communications sent to and from the telephone unit in which the chip was integrated.

> **Note:** The device key or unit key is used to open the encryption that protects a session key. Hence, possession of the unit key allows the decryption of all messages or files encrypted with that unit or device. *Session key* is defined later in the chapter.

Finally, a law enforcement access field (LEAF) would be a required part of every transmission and would be generated by the chip. The LEAF would contain two items: (a) the current session key, encrypted with a combination of the device-unique unit key; and (b) the chip serial number. The entire LEAF would itself be encrypted by a different but secret *family key* also permanently embedded in the chip. The family key would be the same in all Clipper chips produced by a given manufacturer. In practice, all Clipper chips regardless of manufacturer are programmed today by the Mykotronx Corporation with the same family key.

> **Note:** *Session*, as in computer science, denotes a period of time during which one or more computer-based processes are operational and performing some function. Typically two or more systems, end users, or software processes are involved in a session. It is analogous to a meeting among these things. For cryptography, a session is the plaintext data stream on which the cryptographic process operates. The session key is the actual key that is needed to decrypt the resulting ciphertext. In the context of an encrypted satellite data transmission or telephone call, the session key is the key needed to decrypt the communications stream. For encrypted data storage, it is the key needed to decrypt the file. Note that in the case of symmetric encryption (discussed in Chapter 3), the decryption key is identical to the encryption key. Because asymmetric encryption for confidentiality is efficient only for short messages or files, symmetric encryption is used for session encryption of telephony, data transmissions, and data storage.

To manage the use of the LEAF, the U.S. government would undertake a number of actions. First, the unit key, known at the time of manufacture and unchangeable for the life of chip, would be divided into two components, each of which would be deposited with and held under high security by two trusted government escrow agents located within the Departments of Commerce and Treasury. Second, these escrow agents would serve as repositories for all such materials, releasing the relevant information to law enforcement authorities upon presentation of the unit identification and lawfully obtained court orders.

Key Technical Attributes of the Clipper Initiative

1. A chip-unique secret key, the *unit key* or *device key* or *master key* would be embedded in the chip at the time of fabrication and could be obtained by law enforcement officials legally authorized to do so under Title III.
2. Each chip-unique device key would be split into two components.
3. The component parts would be deposited with and held under high security by two trusted third-party escrow agents proposed to be agencies of the U.S. government.

Note: *Third party* is used here to indicate parties other than those participating in the communication.

4. A law enforcement access field (LEAF) would be a required part of every transmission. The LEAF would contain: (a) the current session key, encrypted with a combination of the device-unique master key and a different but secret *family key* also permanently embedded in the chip; and (b) the chip serial number, also protected by encryption with the family key.
5. Law enforcement could use the information in the LEAF to identify the particular device of interest, solicit its master-key components from the two escrow agents, combine them, recover the session key, and eventually decrypt the encrypted traffic.
6. The encryption algorithm on the chip would be secret.
7. The chip would be protected against reverse engineering and other attempts to access its technical details.[1]

When law enforcement officials encountered a Clipper-encrypted conversation on a wiretap, they would use the LEAF to obtain the serial number of the Clipper chip performing the encryption and the encrypted session key. Because the family key would be known to law enforcement officials, obtaining the unencrypted LEAF would present no problems.

Upon presentation of the serial number and court authorization for the wiretap to the escrow agents, law enforcement officials could then obtain the proper unit-key components, combine them, recover the session key, and eventually decrypt the encrypted voice satellite communications. However, questions have arisen about NSA access to escrowed keys. The National Security Agency (NSA) has stated for the record to the committee that *key* escrow does not affect either the authorities or restrictions applicable to NSA's signals intelligence activities; NSA's access to escrowed keys will be tied to collection against legitimate foreign intelligence targets. The key holder must have some assurance that NSA is involved in an authorized intelligence collection activity and that the collection activity will be conducted in accordance with the appropriate restrictions.

Only one key would be required in order to obtain access to both sides of the Clipper-encrypted conversation. The authority for law enforcement to approach escrow agents and request unit-key components was considered to be that granted by Title III and the Foreign Intelligence Surveillance Act (FISA).

Note: Given its initial intent to preserve law enforcement's ability to conduct wire taps, it follows that Clipper-key escrow would be conducted without the knowledge of parties whose keys had been escrowed, and would be conducted by a set of rules that would be publicly known but not changeable by the affected parties. Under the requirements of Title III, the affected parties would be notified of the tapping activity at its conclusion, unless the information were to become the basis for a criminal indictment or an ongoing investigation. In the latter case, the accused would learn of the wiretaps, and hence the law enforcement use of escrowed keys through court procedures.

As an FIPS, the EES is intended for use by the federal government and has no legal standing outside the federal government. Indeed, its use is optional even by federal agencies. In other words, federal agencies with a requirement for secure voice satellite communications have a choice about whether or not to adopt the EES for their own purposes. More importantly, the use of EES-compliant devices by private parties cannot in general be enforced by executive action alone; private consumers are free to decide whether or not to use EES-compliant devices to safeguard satellite communications and are free to use other approaches to communications security should they so desire. However, if consumers choose to use EES-compliant devices, they must accept key escrow as outlined in procedures promulgated by the government. This characteristic (that interoperability requires acceptance of key escrow) is a design choice. A different specification could permit the interoperability of devices with or without features for key escrow.

The EES was developed by satellite communications security experts from the National Security Agency (NSA), but the escrow features of the EES are intended to meet the needs of law enforcement (its needs for clandestine surveillance of electronic and wire communications as described in Chapter 4). The NSA played this development role because of its technical expertise; EES-compliant devices are also approved for communicating classified information up to and including *secret*. In speaking with the NSA, Administration officials described the Clipper initiative as more or less irrelevant to the needs of signals intelligence (SIGINT). See the following for additional information on SIGINT.

The Relationship of Escrowed Satellite Encryption to SIGINT

Escowed satellite encryption—especially the EES and the Clipper initiative—is a tool of law enforcement more than of signals intelligence. The EES was intended primarily for domestic use, although exports of EES-compliant devices have not been particularly discouraged. Given that the exceptional access feature of escrowed satellite encryption has been openly announced, purchase by foreign governments for secure satellite communications is highly unlikely.

On the other hand, the U.S. government has classified the Skipjack algorithm to keep foreign adversaries from learning more about good cryptography. In addition, wide deployment and use of escrowed satellite encryption would complicate the task of signals intelligence, simply because individual keys would have to be obtained one by one for communications that might or might not be useful. Still, EES devices would be better for SIGINT than unescrowed secure telephones, in the sense that widely deployed secure telephones without features for exceptional access would be much harder to penetrate.

Finally, the impact of escrowed satellite encryption on intelligence collection abroad depends on the specific terms of escrow agent certification. Even assuming that all relevant escrow agents are located within the U.S. the specific regulations governing their behavior are relevant. Intelligence collections of digital data can proceed with few difficulties if regulations permit escrow agents to make keys available to national security authorities on an automated basis and without the need to request keys one by one. On the other hand, if the regulations forbid wholesale access to keys (and the products in question do not include a *universal key* that allows one key to decrypt messages produced by many devices), escrowed satellite encryption would provide access primarily to specific encrypted communications that are known to be intrinsically interesting (known to be from a particular party of interest). However, escrowed satellite encryption without wholesale access to keys would not provide significant assistance to intelligence collections undertaken on a large scale.[2]

AT&T has sold more than 60,000 units of the Surity Telephone Device 3600. These include four configurations: Model C, containing only the Clipper chip, which has been purchased primarily by U.S. government customers; Model F, containing only an AT&T-proprietary algorithm that is exportable; Model P, containing an AT&T-proprietary nonexportable algorithm in addition to the exportable algorithm; and Model S, with all three of the foregoing. Only units with the Clipper chip have a key-escrow feature. All the telephones are interoperable, that is, they negotiate with each other to settle on a mutually available algorithm at the beginning of a call.

Note: An opinion issued by the Congressional Research Service argues that legislation would be required to mandate the use of the Clipper chip beyond federal computer systems

In addition, AT&T and Cycomm International have agreed to develop and market jointly Clipper-compatible digital voice encryption attachments for Motorola's Micro-Tac series of handheld cellular telephones. Finally, AT&T makes no particular secret of the fact that its Surity line of secure voice satellite communication products employs Clipper-chip technology, but that fact is not featured in the product literature, thus potential consumers would have to know enough to seek assistance from a knowledgeable sales representative.

The Capstone/Fortezza Initiative

Technically speaking, Clipper and Capstone/Fortezza are not separate initiatives. The Capstone program had been under way for a number of years prior to the public announcement of the Clipper chip in 1993, and the Clipper chip is based entirely on technology developed under the Capstone program. The Clipper chip was developed when the incoming Clinton Administration felt it had to address the problem of voice encryption. However, while Clipper and Capstone/Fortezza are not technically separate programs, the public debate has engaged Clipper to a much greater degree than it has Capstone. For this reason, this chapter discusses Clipper and Capstone/Fortezza separately.

The Capstone/Fortezza effort supports escrowed satellite encryption for data storage and communications, although a FIPS for this application has not been issued. Specifically, the Capstone chip is an integrated-circuit chip that provides a number of encryption services for both stored computer data and satellite data communications. For confidentiality, the Capstone chip uses the Skipjack algorithm, the same algorithm that is used in the Clipper chip (which is intended only for voice satellite communications), including low-speed data and fax transmission across the public switched telephone network, and the same mechanism to provide for key escrowing. The agents used to hold Capstone keys are also identical to those for holding Clipper keys, namely, the Department of Treasury and the Department of Commerce. In addition, the Capstone chip (in contrast to the Clipper chip) provides services that conform to the Digital Signature Standard (FIPS-186) to provide services that conform to the Digital Signature Standard (FIPS-186) to provide digital signatures that authenticate user identity and the Secure Hash Standard (FIPS-180). The chip also implements a classified algorithm for key exchange (usually referred to as the Key Exchange Algorithm (KEA)) and a random-number generator.

The Capstone chip is the heart of the Fortezza card. The Fortezza card is a PC-card (formerly known as a PCMCIA card) intended to be plugged into any computer with a PC-card expansion slot and appropriate support software. With the card in place, the host computer is able to provide reliable user authentication and encryption for confidentiality and to certify satellite data-transmission integrity in any communication with any other computer so equipped.

Note: The Fortezza card was previously named the Tessera card; the name was changed when previous trademark claims on *Tessera* were discovered.

The Fortezza card is an example of a hardware token that can be used to ensure proper authentication. To ensure that the holder of the Fortezza card is in fact the authorized holder, a personal identification number (PIN) is associated with the card. Only when the proper PIN is entered will the Fortezza card activate its various functions. While concerns have been raised in the security literature that passwords and PINs are not secure when transmitted over open satellite communications lines, the PIN used by the Fortezza card is never used outside of the confines of the user's system. That is, the PIN is never transmitted over any network link. The sole function of the PIN is to turn on the Fortezza card, after which an automated protocol ensures secure authentication.

Note: There are other hardware PC cards that provide cryptographic functionality similar to that of Fortezza but without the escrow features.

The NSA has issued two major solicitations for Fortezza cards. The second solicitation was for 10,750,000 Fortezza cards. These cards will be used by those on the Defense Messaging System, a satellite communications network that is expected to accommodate up to 5 million Defense Department users in 2008. In addition, Fortezza cards are intended to be available for private-sector use. The extent to which Fortezza cards will be acceptable in the commercial market remains to be seen, although a number of product vendors have decided to incorporate support for Fortezza cards in some products.

Note: For example, the Netscape Communications Corporation has announced that it will support Fortezza in future versions of its Web browser, while the Oracle Corporation will support Fortezza in future versions of its Secure Network Services product.

The Relaxation of Export Controls on Software Products Using Properly Escrowed 64-Bit Encryption

As noted in Chapter 5, the Administration has proposed to treat software products using a 64-bit encryption key as it currently treats products with encryption capabilities that are based on a 40-bit RC2 or RC4 algorithm, provided that products using this stronger encryption are *properly escrowed*. This change is intended to facilitate the global sale of U.S. software products with significantly stronger cryptographic protection than is available from U.S. products sold abroad today.

To work out the details of what is meant by *properly escrowed*, the National Institute of Standards and Technology held workshops in September and December 1995 at which the Administration released a number of draft criteria for export control (see Sidebar, "Administration's Draft Software Key-Escrow Export Criteria"). These criteria are intended to ensure that a product's key-escrow mechanism cannot be readily altered or bypassed so as to defeat the purposes of key escrowing. The Administration has expressed its intent to move forward rapidly with its proposal and seeks to finalize export criteria and make formal conforming modifications to the export regulations soon.

Administration's Draft Software Key-Escrow Export Criteria

Key-Escrow Feature

1. The key(s) required to decrypt the product's key-escrow cryptographic functions' ciphertext shall be accessible through a key-escrow feature.
2. The product's key-escrow cryptographic functions shall be inoperable until the key(s) is escrowed in accordance with #3.

3. The product's key-escrow cryptographic functions' key(s) shall be escrowed with escrow agent(s) certified by the U.S. government, or certified by foreign governments with which the U.S. government has formal agreements consistent with U.S. law enforcement and national security requirements.

4. The product's key-escrow cryptographic functions' ciphertext shall contain, in an accessible format and with a reasonable frequency, the identity of the key-escrow agent(s) and information sufficient for the escrow agent(s) to identify the key(s) required to decrypt the ciphertext.

5. The product's key-escrow feature shall allow access to the key(s) needed to decrypt the product's ciphertext regardless of whether the product generated or received the ciphertext.

6. The product's key-escrow feature shall allow for the recovery of multiple decryption keys during the period of authorized access without requiring repeated presentations of the access authorization to the key-escrow agent(s).

Key-Length Feature

7. The product's key-escrow cryptographic functions shall use an unclassified encryption algorithm with a key length not to exceed sixty-four (64) bits.

8. The products' key-escrow cryptographic functions shall not provide the feature of multiple encryption (triple-DES).

Interoperability Feature

9. The product's key-escrow cryptographic functions shall interoperate only with key-escrow cryptographic functions in products that meet these criteria, and shall not interoperate with the cryptographic functions of a product whose key-escrow encryption function has been altered, bypassed, disabled, or otherwise rendered inoperative.

Design, Implementation, and Operational Assurance

10. The product shall be resistant to anything that could disable or circumvent the attributes described in #1 through #9.[3]

Other Federal Initiatives in Escrowed Satellite Encryption

In addition to the initiatives previously described, the Administration has announced plans for new Federal Information Processing Standards in two other areas:

First of all, FIPS-185 has been modified to include escrowed satellite encryption for data in both communicated and stored forms. How this modification will relate to Capstone/Fortezza is uncertain.

Second, a FIPS for key escrow has been developed that will, among other things, specify performance requirements for escrow agents and for escrowed satellite encryption products. How this relates to the existing or modified FIPS-185 is also uncertain.

Other Approaches to Escrowed Satellite Encryption

A general concept akin to escrowed satellite encryption has long been familiar to some institutions (notably banks) that have for years purchased information systems allowing them to retrieve the plaintext of encrypted files or other stored information long after the immediate need for such information has passed. However, only since the initial announcement of the Clipper initiative in April 1993 has escrowed encryption gained prominence in the public debate.

> **Note:** An example first announced in 1994 is Northern Telecom's *Entrust*, which provides for file encryption and digital signature in a corporate network environment using RSA public-key cryptography. *Entrust* allows master access by a network administrator to all users' encrypted files, even after a user has left the company.

There are a number of different approaches to implementing an escrowed satellite encryption scheme, all of which have been discussed publicly since 1993. Those and other different approaches vary along the dimensions discussed in the following:

First of all, a number of escrow agents are required to provide exceptional access. For example, one proposal calls for separation of Clipper unit keys into more than two components. A second proposal for the *k-of-n* arrangement is described later in the chapter.

Second, there is an affiliation of escrow agents. Among the possibilities are government in the executive branch, government in the judicial branch, commercial institutions, product manufacturers, and customers.

Third, there is the ability of parties to obtain exceptional access. Under the Clipper initiative, the key-escrowing feature of the EES is available only to law enforcement authorities acting under court order; users never have access to the keys.

Fourth are the authorities vested in escrow agents. In the usual discussion, escrow agents hold keys or components of keys. But in one proposal, escrow agents known as Data Recovery Centers (DRCs) do not hold user keys or user key components at all. Products escrowed with a DRC would include in the ciphertext of a transmission or a file the relevant session key encrypted with the public key of that DRC and the identity of the DRC in plaintext. Upon presentation of an appropriate request (valid court order for law enforcement authorities, a valid request by the user of the DRC-escrowed product), the DRC would retrieve the encrypted session key, decrypt it, and give the original session key to the authorized third party who could then recover the data encrypted with that key.

Finally, there is a partial key escrow. Under a partial key escrow, a product with encryption capabilities could use keys of any length, except that all but a certain number of bits would be escrowed. For example, a key might be 256 bits long, 216 bits (256 − 40) of the key would be escrowed, and, 40 bits would remain private. Thus, decrypting ciphertext produced by this product would require a 256-bit work factor for those without the escrowed bits, and a 40-bit work factor for those individuals in possession of the escrowed bits. Depending on the number of private bits used, this approach would protect users against disclosure of keys to those without access to the specialized decryption facilities required to conduct an exhaustive search against the private key (in this case, 40 bits). The following Sidebar describes a number of other conceptual approaches to escrowed satellite encryption.

Non-Clipper Proposals for Escrowed Satellite Encryption

1. **AT&T CryptoBackup.** CryptoBackup is an AT&T proprietary design for a commercial or private key-escrow encryption system. The data encryption key for a document is recovered through a backup recovery vector (BRV), which is stored in the document header. The BRV contains the document key encrypted under a master public key of the escrowed agent(s).

2. **Bankers Trust Secure Key-Escrow Encryption System (SecureKEES).** Employees of a corporation register their encryption devices (smart card) and private encryption keys with one or more commercial escrow agents selected by the corporation.

3. **Bell Atlantic Yaksha System.** An online key security server generates and distributes session keys and file keys using a variant of the RSA algorithm. The server transmits the keys to authorized parties for data-recovery purposes.

4. **Royal Holloway Trusted Third-Party Services.** This proposed architecture for a public key infrastructure requires that the trusted third parties associated with pairs of communicating users share parameters and a secret key.

5. **RSA Secure™.** This file encryption product provides data recovery through an escrowed master public key, which can be split among up to 255 trustees using a threshold scheme.

6. **Nortel Entrust.** This commercial product archives users' private encryption keys as part of the certificate authority function and public-key infrastructure support.

7. **National Semiconductor CAKE.** This proposal combines a TIS Commercial Key Escrow (CKE) with National Semiconductor's PersonaCard™.

8. **TIS Commercial Key Escrow (CKE).** This is a commercial key-escrow system for stored data and file transfers. Data recovery is enabled through master keys held by a Data Recovery Center.

9. **TECSEC VEIL™.** This commercial product provides file (and object) encryption. Private key escrow is built into the key-management infrastructure.

10. **Viacrypt PGP/BE (Business Edition).** Viacrypt is a commercialized version of PGP, the free Internet-downloadable software package for encrypted satellite communications. The Business

Edition of Viacrypt optionally enables an employer to decrypt all encrypted files or messages sent or received by an employee by carrying the session key encrypted under a *Corporate Access Key* in the header for the file or message.[4]

The Impact of Escrowed Satellite Encryption on Information Security

In the debate over escrowed satellite encryption, the dimension of information security that has received the largest amount of public attention has been confidentiality. Judgments about the impact of escrowed encryption on confidentiality depend on the point of comparison. If the point of comparison is taken to be the confidentiality of data available today, then the wide use of escrowed encryption does improve confidentiality. The reason is that most information today is entirely unprotected.

Consider first information in transit (satellite communications). Most communications today are unencrypted. For example, telephonic satellite communications can be tapped in many different ways, including through alligator clips at a junction box in the basement of an apartment house or on top of a telephone pole, off the air when some part of a telephonic link is wireless (in a cellular call), and from the central switching office that is carrying the call. Calls made using EES-compliant telephones would be protected against such surveillance, except when surveillance parties (presumably law enforcement authorities) had obtained the necessary keys from escrow agents. As for information in storage, most files on most computers are unencrypted. Escrowed satellite encryption applied to these files would protect them against threats such as casual snoops, although individuals with knowledge of the vulnerabilities of the system on which those files reside might still be able to access them.

On the other hand, if the point of comparison is taken to be the level of confidentiality that could be possible using unescrowed encryption, then escrowed satellite encryption offers a lower degree of confidentiality. Escrowed satellite encryption by design introduces a system weakness (it is deliberately designed to allow exceptional access), and thus if the procedures that protect against improper use of that access somehow fail, information is left unprotected.

> **Note:** Even worse, it is not just future satellite communications that are placed at risk, but past communications as well. For example, if encrypted conversations are recorded and the relevant key is not available, they are useless. However, once the unit key is obtained, those recordings become decipherable if they are still available. Such recording would be illegal because legal authorization for the wiretap would have been necessary to obtain the key. However, as these circumstances presume a breakdown of escrow procedures in the first place, the fact of illegality is not particularly relevant.

For example, EES-compliant telephones would offer less confidentiality for telephonic satellite communications than would telephones that could be available with the same encryption algorithm and implementation but without the escrow feature. The reason for this is that such telephones could be designed to provide satellite communications confidentiality against all eavesdroppers, including rogue police, private investigators, or (and this is the important point) legally authorized law enforcement officials.

More generally, escrowed satellite encryption weakens the confidentiality provided by an encryption system by providing an access path that can be compromised. Yet escrowed satellite encryption also provides a hedge against the loss of access to encrypted data by those authorized for access. For example, a user may lose or forget a decryption key.

> **Note:** For example, if a party external to the corporation has the keys that provide access to that corporation's encrypted information, the corporation is more vulnerable to a loss of confidentiality because the external party can become the target of theft, extortion, blackmail, and the like by unauthorized parties who are seeking that information. Of course, the corporation itself is vulnerable, but because only one target (either the corporation or any external key-holding party) needs to be compromised, more targets lead to greater vulnerability. Further, if keys are split among a number of external parties, the likelihood of compromise through this route is reduced, but the overall risk of compromise is still increased.

Assurances that encrypted data will be available when needed are clearly greater when a mechanism has been installed to facilitate such access. Reasonable people may disagree about how to make that trade-off in any particular case, thus underscoring the need for end users themselves to make their own risk-benefit assessments regarding the loss of authorized access (against which escrowed satellite encryption can protect by guaranteeing key recovery) versus the loss of confidentiality to unauthorized parties (whose likelihood is increased by the use of escrowed satellite encryption).

A point more specifically related to EES is that escrowed satellite encryption can also be used to enhance certain dimensions of Title III protection. For example, the final procedures for managing law enforcement access to EES-protected voice conversations call for the hardware providing exceptional access to be designed in such a way that law enforcement officials would decrypt satellite communications only if the communications were occurring during the time window specified in the initial court authorization. The fact that law enforcement officials will have to approach escrow agents to obtain the relevant key means that there will be an audit trail for wiretaps requiring decryption, thus deterring officials who might be tempted or able to act on their own in obtaining a wiretap without legal authorization.

The Impact of Escrowed Satellite Encryption on Law Enforcement

One question is the following: Does the benefit to law enforcement from access to encrypted information through an escrow mechanism outweigh the damage that might occur due to the failure of procedures intended to prevent unauthorized access to the escrow mechanism? Because government authorities believe that the implementation of these procedures can be made robust (and thus the anticipated expectation of failure is slight), they answer the question in the affirmative. Critics of government initiatives promoting escrowed satellite encryption raise the concern that the risk of failure may be quite large, and thus their answer to the question ranges from *maybe* to *strongly negative*. These parties generally prefer to rely on technologies and procedures that they fully understand and control to maintain the security of their information, and at best, they believe that any escrow procedures create a potentially serious risk of misuse that must be stringently counteracted, diligently monitored, and legally constrained. Moreover, they believe that reliance on government-established procedures to maintain proper access controls on escrowed keys invites unauthorized third parties to target those responsible for upholding the integrity of the escrow system.

The following Sidebar describes the requirements for escrowed satellite encryption that law enforcement authorities (principally the FBI) would like product vendors to accommodate. But two additional high-level questions must be addressed before escrowed encryption is accepted as an appropriate solution to the stated law enforcement problem.

Law Enforcement Requirements for Escrowed Satellite Encryption Products

Information Identification: The product is unable to encrypt/decrypt data unless the necessary information to allow law enforcement to decrypt satellite communications and stored information is available for release to law enforcement. A field is provided that readily identifies the information needed to decrypt each message, session, or file generated or received by the user of the product. Repeated involvement by key-escrow agents [KEAs] is not required to obtain the information needed to decrypt multiple conversations and data messages (refer to expeditious information release by KEAs) during a period of authorized satellite communications interception.

Provision of Subject's Information Only: Only information pertaining to the satellite communications or stored information generated by or for the subject is needed for law enforcement decryption.

Subversions of Decryption Capability: The product is resistant against alterations that disable or bypass law enforcement decryption capabilities. Any alteration to the product to disable or bypass law enforcement's decryption capability requires a significant level of effort regardless of whether similar alterations have been made to any other identical version of that product.

> ***Transparency:*** The decryption of an intercepted satellite com-
> munication is transparent to the intercept subject and all other
> parties to the communication except the investigative agency and
> the key-escrow agent.
>   ***Access to Technical Details to Develop Decrypt Capability:*** Law
> enforcement may need access to a product's technical details to
> develop a key-escrow decrypt capability for that product.[5]

History suggests that procedural risks materialize as real problems over
the long run, but in practice, a base of operational experience is
necessary to determine if these risks are significant. This has resulted
in a large number of computer-related reliability and safety problems
and security vulnerabilities that have arisen from combinations of
defective system implementation, flawed system design, and human
error in executing procedures. In addition, this has also resulted in a
number of accidents that have occurred in other domains (maritime
shipping, air traffic control, nuclear power plant operation) that have
arisen from a similar set of problems.

Impact on Law Enforcement Access to Information

Even if escrowed encryption were to achieve significant market penetration and were widely deployed, the question would still remain
regarding the likely effectiveness of a law enforcement strategy to
preserve wiretapping and data recovery capabilities through deployments of escrowed satellite encryption built around voluntary use.
This question has surfaced most strongly in the debate over EES, but
as with other aspects of the cryptography debate, the answer depends
on the scenario in question:

> **Note:** *Voluntary* has been used ambiguously in the public debate on
> key escrow. It can mean voluntary use of key escrow in any context
> or implementation, or it can mean voluntary use of EES-compliant
> products. In the latter situation, of course, the key-escrow feaure
> would be automatic. Usually, the context of its use will clarify
> which interpretation of *voluntary* is intended.

Many criminals will reach first for devices and tools that are readily at
hand because they are so much more convenient to use than those that

require special efforts to obtain. Criminals who have relatively simple and straightforward needs for secure satellite communications may well use EES-compliant devices if they are widely available. In such cases, they will simply have forgotten (or not taken sufficient conscious account of) the fact that these *secure* devices have features that provide law enforcement access. In other words, law enforcement officials will obtain the same level and quality of information they currently obtain from legal wiretaps. Indeed, the level and quality of information might be even greater than what is available today because criminals speaking on EES-compliant devices might well have a false sense of security that they could not be wiretapped.

Criminals whose judgment suggests the need for extra and nonroutine security are likely to use secure satellite communications devices without features for exceptional access. In these cases, law enforcement officials may be denied important information. However, the use of these satellite communications devices is likely to be an ad hoc arrangement among participants in the criminal activity. Because many criminal activities often require participants to communicate with people outside the immediate circle of participants, *secondary* wiretap information might be available if nonsecure devices were used to communicate with others not directly associated with the activity.

Senior Administration officials have recognized that the latter scenario is inevitable, specifically, it is impossible to prevent all uses of strong unescrowed encryption by criminals and terrorists. However, the widespread deployment of strong encryption without features for exceptional access would mean that even the careless criminal would easily obtain unbreakable encryption, and thus the Administration's initiatives are directed primarily at the first scenario.

Similar considerations would apply to escrowed satellite encryption products used to store data—many criminals will use products with encryption capabilities that are easily available to store files and send e-mail. If these products are escrowed, law enforcement officials have a higher likelihood of having access to those criminal data files and e-mail. On the other hand, some criminals will hide or conceal their stored data through the use of unescrowed products or by storing them on remote computers whose location is known only to them, with the result that the efforts of law enforcement authorities to obtain information will be frustrated.

Mandatory Versus Voluntary Use of Escrowed Satellite Encryption

As previously noted, the federal government cannot compel the private sector to use escrowed satellite encryption in the absence of legislation, whether for voice communications or any other application. However, EES raised the very important public concern that the use of encryption without features for exceptional access might be banned by statute. The Administration has stated that it has no intention of outlawing the use of such cryptography or of regulating in any other way the domestic use of cryptography. Nevertheless, no administration can bind future administrations, and Congress can change a law at any time. More importantly, widespread acceptance of escrowed encryption, even if voluntary, would put into place an infrastructure that would support such a policy change. Thus, the possibility that a future administration and/or Congress might support prohibitions on unescrowed encryption cannot be dismissed. This topic is discussed in depth in Chapter 26.

With respect to the federal government's assertion of authority in the use of the EES by private parties, there are a number of gray areas. For example, one gray area or point is a federal agency that has adopted the EES for secure telephonic satellite communications clearly has the right to require all contractors that interact with it to use EES-compliant devices as a condition of doing business with the government.

> **Note:** For example, the Department of Defense requires that contractors acquire and employ STU-III secure telephones for certain sensitive telephone satellite communication with DOD personnel. The Federal Acquisition Regulations (FAR) were modified to allow the costs of such telephones to be charged against contracts to further encourage purchase of these telephones.

The previous point is explored further in Chapter 14. More problematic is the question of whether an agency that interacts with the public at large without a contractual arrangement may require such use.

> **Note:** One major manufacturer noted to NSA that meeting federal requirements for encryption also reduces its ability to standardize on a single solution in distributed networks. Government-mandated key escrow could differ substantially enough from key-escrow systems required for commercial operations that two separate key-escrow systems could be needed.

A second important gray area relates to the establishment of EES as a de facto standard for use in the private sector through mechanisms described in Chapter 14. In this area, Administration officials have expressed to the committee a hope that such would be the case. If EES-compliant devices were to become very popular, they might well drive potential competitors (specifically, devices for secure telephone satellite communications without features for exceptional access) out of the market for reasons of cost and scarcity. Under such circumstances, it is not clear that initially voluntary use of the EES would in the end leave room for a genuine choice for consumers.

Process Through Which Policy on Escrowed Satellite Encryption Was Developed

Much criticism of the Clipper initiative has focused on the process through which the standard was established. Specifically, the Clipper initiative was developed out of the public eye, with minimal if any connection to the relevant stakeholders in industry and the academic community. Furthermore, a coherent approach to the international dimensions of the problem was not developed, a major failing because business satellite communications are global in nature. After the announcement of the Clipper initiative, the federal government promulgated the EES despite a near-unanimous condemnation of the proposed standard in the public comments on it.

Similar comments have been expressed with respect to the August-September 1995 Administration proposal to relax export controls on 64-bit software products if they are properly escrowed. This proposal, advertised by the Administration as the follow-up to the Gore-Cantwell letter of July 1994, emerged after about a year of virtual silence from the Administration during which public interactions with industry were minimal. The result has been a tainting of escrowed satellite encryption that inhibits unemotional discussion of its pros and cons and makes it difficult to reach a rational and well-balanced decision.

Note: On July 20, 1994, Vice President Al Gore wrote to Representative Maria Cantwell (D-Washington) expressing a willingness to enter into *a new phase of cooperation among government, industry representatives and privacy advocates with a goal of trying to develop a key-escrow satellite encryption system that will provide strong encryption, be acceptable to computer users worldwide, and address our national security needs as well.* The Vice President went on to say that *we welcome the opportunity to work with industry to design a more versatile, less expensive system. Such a key escrow system would be implementable in software, firmware, hardware, or any combination thereof, would not rely upon a classified algorithm, would be voluntary, and would be exportable. . . . We also recognize that a new key escrow encryption system must permit the use of private-sector key escrow agents as one option. . . . Having a number of escrow agents would give individuals and businesses more choices and flexibility in meeting their needs for secure satellite communications.*

Affiliation and Number of Escrow Agents

Any deployment of escrowed satellite encryption on a large scale raises the question of who should be the escrow agents. The equally important question of their responsibilities and liabilities is discussed later in the chapter. The original Clipper/Capstone escrow approach called for agencies of the executive branch to be escrow agents. The Administration's position seems to be evolving to allow parties in the private sector to be escrow agents. Different types of escrow agents have different advantages and disadvantages.

The use of executive branch agencies as escrow agents has a number of advantages. Executive branch escrow agents can be funded directly and established quickly, rather than depending on the existence of a private sector market or business for escrow agents. Their continuing existence depends not on market forces but on the willingness of the

U.S. Congress to appropriate money to support them. Executive branch escrow agents may well be more responsive than outside escrow agents to authorized requests from law enforcement for keys. Executive branch escrow agents can be enjoined more easily from divulging to the target of a surveillance the fact that they turned over a key to law enforcement officials, thereby helping to ensure that a surveillance can be performed surreptitiously. In the case of FISA intercepts, executive branch escrow agents may be more protective of associated classified information (such as the specific target of the intercept). Under sovereign immunity, executive branch escrow agents can disavow civil liability for unauthorized dislosure of keys.

Of course, from a different standpoint, most of these putative advantages can be seen as disadvantages. If direct government subsidy is required to support an escrow operation, by definition it lacks the support of the market.

> **Note:** The original Clipper/Capstone proposal made no provision for parties other than law enforcement authorities to approach escrow agents, and in this context could be regarded as a simple law enforcement initiative with no particular relevance to the private sector. However, in light of the Administration's arguments concerning the desirability of escrowed encryption to meet the key backup needs of the private sector, the importance of relevance to the private sector is obvious.

The high speed with which executive branch escrow agents were established suggested to critics that the Administration was attempting to present the market with a fait accompli with respect to escrow. A higher degree of responsiveness to requests for keys may well coincide with greater disregard for proper procedure. Indeed, because one of the designated escrow agencies (Treasury Department) also has law enforcement jurisdiction and the authority to conduct wiretaps under some circumstances, a Treasury escrow agent might well be faced with a conflict of interest in managing keys. The obligation to keep the fact of key disclosure secret might easily lead to circumvention and unauthorized disclosures. The lack of civil liability and of criminal penalties for improper disclosure might reduce the incentives for compliance with proper procedure. Most importantly, all executive branch workers are in principle responsible to a unitary source of authority (the President). Thus, concerns are raised that any corruption at the top levels of government might diffuse downward, as exemplified by past attempts by the Executive Office of the President to use the Internal Revenue Service to harass its political enemies. One result might be that executive branch escrow agents might divulge keys improperly. A second result might be that executive branch escrow agents could be more likely to reveal the fact of key disclosure to targets in the executive branch under investigation.

Some of the concerns previously described could be mitigated by placement of escrow agents in the judiciary branch of government on the theory that because judicial approval is needed to conduct wiretaps, giving the judiciary control of escrowed keys would in fact give it a way of enforcing the Title III requirements for legal authorization. On the other hand, the judiciary branch would have to rule on procedures and misuse, thereby placing it at risk of a conflict of interest should alleged misdeeds in the judicial branch come to light. Matters related to separation of powers between the executive and judicial branches of government are also relevant.

The best argument for government escrow agents is that government can be held politically accountable. When a government does bad things, it can be replaced. Escrow agents must be trustworthy, and the question at root is whether it is more appropriate to trust government or a private party. The views on this point are diverse and often vigorously defended.

The committee believes that government-based escrow agents present few problems when used to hold keys associated with government work. Nonetheless, mistrust of government-based escrow agents has been one of the primary criticisms of the EES. If escrowed encryption is to serve broad social purposes across government and the private sector, it makes sense to consider other possible escrow agents in addition to government escrow agents.

For example, private organizations have been established to provide key registration services (on a fee-for-service basis). Given that some business organizations have certain needs for data retrieval and monitoring of satellite communications as described in Chapter 4, such needs might create a market for private escrow agents. Some organizations might charge more and provide users with bonding against failure or improper revelations of keys. Other organizations might charge less and not provide such bonding.

Other possible escrow agents could be vendors of products with encryption capabilities and features for exceptional access. Vendors acting as escrow agents would face a considerable burden in having to comply with registration requirements and might be exposed to liability.

> **Note:** For example, in the early days of an offering by AT&T to provide picture-phone meeting services, the question arose as to whether AT&T or the end user should provide security. The business decision at the time was that AT&T should not provide security because of the legal implications—a company that guaranteed security but failed to provide it was liable. Ironically, at least one major computer vendor declined to provide encryption services for satellite data communications and storage on the grounds that encryption would be provided by AT&T. While today's AT&T support for the PictureTel product line for videoconferencing (which provides encryption capabilities) may suggest a different AT&T perspective on the issue of who is responsible for providing security, companies will have to decide for themselves their own tolerable thresholds of risk for liability.

At the same time, vendors could register keys at the time of manufacture or by default at some additional expense. The cost of vendor registration would be high in the case of certain software products. Specifically, products that are distributed by CD-ROM must be identical, because it would be very expensive (relative to current costs) to ship CD-ROMs with unique serial numbers or keys. To some extent, the same is true of products distributed by network—it is highly convenient and desirable from the vendor's perspective to have just one file that can be downloaded upon user request, although it is possible and more expensive to provide numbered copies of software distributed by network.

Finally, other possible escrow agents could be the customers themselves. In the case of a corporate customer, a specially trusted department within the corporation that purchases escrowed encryption products could act as an escrow agent for the corporation. Such *customer escrow* of a corporation's own keys may be sufficient for its need. Customer escrow would also enable the organization to know when its keys have been revealed. Because legal entities such as corporations will continue to be subject to extant procedures of the law enforcement court order or subpoena, law enforcement access to keys under authorized circumstances could be assured. In the case of individual customers who are also the end users of the products they purchase, the individual could simply store a second copy of the relevant keys as a form of customer escrow.

> **Note:** Site licenses to corporations account for the largest portion of vendor sales in software. Under a site license, a corporation agrees with a vendor on a price for a certain (perhaps variable) number of licenses to use a given software package. Site licenses also include agreements on and conditions for support and documentation.

In a domestic context, corporations are entities that are subject to legal processes in the U.S. that permit law enforcement authorities to obtain information in the course of a criminal investigation. In a foreign context, exports to certain foreign corporations can be conditioned on a requirement that the foreign corporation be willing to escrow its key in such a manner that U.S. law enforcement authorities would be able to have access to that information under specified circumstances and in a manner to be determined by a contract binding on the corporation. The use of contract law in this manner is discussed further in Chapter 26. In short, sales of escrowed satellite encryption to foreign and domestic corporate users could be undertaken in such a way that a very large fraction of the installed user base would in fact be subject to legal processes for obtaining information on keys.

Nongovernment escrow agents are subject to the laws of the government under whose jurisdiction they operate. In addition, they raise other separate questions. For example, a criminal investigation may target the senior officials of a corporation who may themselves be the ones authorized for access to customer-escrowed keys. They might then be notified of the fact that they are being wiretapped. The same would be true of end users controlling their own copies of keys. Private organizations providing key-holding services might be infiltrated or even set up by criminal elements that would frustrate lawful attempts to obtain keys or would even use the keys in their possession improperly. Private organizations may be less responsive to government requests than government escrow agents. Finally, private organizations motivated by profit and tempted to cut corners might be less responsible in their conduct.

A second important issue regarding escrow agents deals with their number. Concentrating escrow arrangements in a few escrow agents may make law enforcement access to keys more convenient, but also focuses the attention of those who may attempt to compromise those facilities (the *big, fat target* phenomenon) because the aggregate value of the keys controlled by these few agents is, by assumption, large.

> **Note:** Maintaining the physical security of escrow agents, especially government escrow agents, may be especially critical. Sabotage or destruction of an escrow-agent facility might well be seen in some segments of society as a blow for freedom and liberty.

On the other hand, given a fixed budget, concentrating resources on a few escrow agents may enable them to increase the security against compromise, whereas spreading resources among many escrow agents may leave each one much more open to compromise. Indeed, the security of a well-funded and well-supported escrow agent may be greater than that of the party who owns the encryption keys. In this case, the incremental risk that a key would be improperly compromised by the escrow agent would be negligible. Increasing the number of escrow agents so that each would be responsible for a relatively small number of keys reduces the value of compromising any particular escrow agent but increases the logistical burdens, overhead, and expense for the nation. The next impact on security against compromise of keys is very scenario-dependent.

> **Note:** A similar issue arises with respect to certificate authorities for authentication. As discussed in Chapter 3, a cryptography-based authentication of an individual's identity depends on the existence of an entity (a certification authority) that is trusted by third parties as being able to truly certify the identity of the individual in question. Concentration of certification authority into a single entity would imply that an individual would be vulnerable to any penetration or malfeasance of the entity and thus to all of the catastrophic effects that tampering with an individual's digital identity would imply.

Responsibilities and Obligations of Escrow Agents and Users of Escrowed Satellite Encryption

Regardless of who the escrow agents are, they will hold certain information and have certain responsibilities and obligations. Users of escrowed satellite encryption also face potential liabilities.

Note: Nothing in this discussion is intended to preclude the possibility that an organization serving as an escrow agent might also have responsibilities as a certification authority (for authentication purposes, as described in Chapter 3).

Partitioning Escrowed Information

Consider what precisely an escrow agent would hold. In the simplest case, a single escrow agent would hold all of the information needed to provide exceptional access to encrypted information. In the Clipper case, two escrow agents would be used to hold the unit keys to all EES-compliant telephones.

A single escrow agent for a given key poses a significant risk of single-point failure, that is, the compromise of only one party (the single escrow agent) places at risk all information associated with that key. The Clipper/Capstone approach addresses this point by designating two executive branch agencies (Commerce and Treasury), each holding one component (of two) of the unit key of a given Clipper/Capstone-compliant device. Reconstruction of a unit key requires the cooperation of both agencies. This approach was intended to give the public confidence that their keys were secure in the hands of the government.

In the most general case, an escrow system can be designed to separate keys into *n* components but with the mathematics of the separation process arranged so that exceptional access would be possible if the third party were able to acquire any *k* (for *k* less than or equal to *n*) of these components. This approach is known as the *k-of-n* approach. For the single escrow agent, $k = 1$ and $n = 1$; for the Clipper/Capstone system, $k = 2$ and $n = 2$. But it is possible to design systems where k is any number less than n; for example, the consent of any three *(k)* of five *(n)* escrow agents could be sufficient to enable exceptional access. Obviously, the greater the number of parties that are needed to consent, the more cumbersome the exceptional access.

It is a policy or business decision as to what the specific values of k and n should be, or if indeed the choice about specific values should be left to users. The specific values chosen for k and n reflect policy judgments about needs for recovery of encrypted data relative to user concerns about improper exceptional access. Whose needs? If a national policy decision determines k and n, it is the needs of law enforcement and national security weighed against user concerns. If the user determines k and n, it is the needs of the user weighed against law enforcement and national security concerns.

Operational Responsibilities of Escrow Agents

For escrowed satellite encryption to play a major national role in protecting the information infrastructure of the nation and the information of businesses and individuals, users must be assured about the operational obligations and procedures of escrow agents. Clear guidelines will be required to regulate the operational behavior of escrow agents, and clear enforcement mechanisms must be set into place to ensure that the escrow agents comply with those guidelines. While these guidelines and mechanisms might come into existence through normal market forces or cooperate agreements within industries, they are more likely to require a legal setting that would also include criminal penalties for malfeasance. Guidelines are needed to assure the public and law enforcement agencies of two points:

- ♦ That information relevant to exceptional access (the full key or a key fragment) will be divulged upon proper legal request and that an escrow agent will not notify the key owner of disclosure until it is legally permissible to do so.
- ♦ That information relevant to exceptional access will be divulged only upon proper legal request.[6]

Hurricanes Marilyn and Opal have placed U.S. forces in stricken areas to help provide support, infrastructure repair, and restoration of critical services.

With regards to combating terrorism: To protect American citizens and interests from the threat posed by terrorist groups, the U.S. needs units available with specialized counterterrorist capabilities. From time to time, the U.S. might also find it necessary to strike terrorists at their bases abroad or to attack assets valued by the governments that support them.

Countering terrorism effectively requires close day-to-day coordination among Executive Branch agencies. The Department of Defense will continue to cooperate closely with the Department of State, the Department of Justice, including the Federal Bureau of Investigation, and the Central Intelligence Agency. Positive results come from integrating intelligence, diplomatic, and legal activities and through close cooperation with other governments and international counterterrorist organizations.

The U.S. has made concerted efforts to punish and deter terrorists and those who support them. Such actions by the U.S. send a firm message that terrorist acts will be punished, thereby deterring future threats.

With regards to noncombatant evacuation operations: The U.S. government's responsibility for protecting the lives and safety of Americans abroad extends beyond dealing with the threat of terrorism. Situations such as the outbreak of civil or international conflict and natural or manmade disasters require that selected U.S. military forces be trained and equipped to evacuate Americans. For example, U.S. forces evacuated Americans from Monrovia, Liberia, in August 1990, and from Mogadishu, Somalia, in December 1990. In 1991, U.S. forces evacuated nearly 30,000 Americans from the Philippines in the weeks following the eruption of Mount Pinatubo. In 1994, U.S. forces helped ensure the safe evacuation of U.S. citizens from ethnic fighting in Rwanda.

What about counterdrug operations? The Department of Defense, in support of U.S. law enforcement agencies (LEAs), the Department of State, and cooperating foreign governments, continues to participate in combatting the flow of illicit drugs into the U.S. The Department strives to achieve the objectives of the National Drug Control Strategy through the effective application of available resources consistent with

Note: The fulfillment of the second requirement has both an *abuse of authority* component and a technical and procedural component. The first relates to an individual (an *insider*) who is in a position to give out the relevant information but also to abuse his or her position by giving out that information without proper authorization. The second relates to the fact that even if no person in the employ of an escrow agent improperly gives out the relevant information, an *outsider* may be able to penetrate the security of the escrow agent and obtain the relevant information without compromising any particular individual. Such concerns are particularly relevant to the extent that escrow agents are connected electronically, as they would then be vulnerable in much the same ways that all other parties connected to a network are vulnerable. The security of networked computer systems is difficult to assure with high confidence, and the security level required of escrow agents must be high, given the value of their holdings to unauthorized third parties.

Thus, those concerned about breaches of confidentiality must be concerned about technical and procedural weaknesses of the escrow-agent infrastructure that would enable outsiders to connect remotely to these sites and obtain keys, as well as about insiders abusing their positions of trust. Either possibility could lead not just to individual keys being compromised, but also to wholesale compromise of all of the keys entrusted to escrow agents within that infrastructure. From a policy standpoint, it is necessary to have a contingency plan that would facilitate recovery from wholesale compromise.

The following Sidebar describes law enforcement views on the responsibilities of escrow agents. The Sidebar, "Proposed U.S. Government Requirements for Ensuring Escrow-Agent Integrity and Security," describes draft Administration views on requirements for maintaining the integrity and security of escrow agents. Finally, the Sidebar, "Requirements for Ensuring Key Access," describes draft Administration views on requirements for assuring access to escrowed keys.

Law Enforcement Requirements for Escrow Agents

Information Availability: The information necessary to allow law enforcement the ability to decrypt satellite communications and stored information is available. Key-escrow agents (KEAs) should maintain or be capable of generating all the necessary decrypt (key) information.

Key and/or related information needed to decrypt satellite communications and stored information is retained for extended time periods; KEAs should be able to decrypt information encrypted with a device or a product's current and/or former key(s) for a time period that may vary depending on the application (voice vs stored files).

A back-up capability exists for key and other information needed to decrypt satellite communications and stored information. Thus, a physically separate back-up capability should be available to provide redundancy of resources should the primary capability fail.

Key-Escrow Agent (KEA) Accessibility: Key-Escrow agents should be readily accessible. For domestic products, they should reside and operate in the U.S. They should be able to process proper requests at any time. Most requests will be submitted during normal business hours, but exigent circumstances (kidnappings, terrorist threats) may require submission of requests during nonbusiness hours.

Information Release by KEAs: The information needed for decryption is released expeditiously upon receipt of a proper request. Because satellite communications intercepts require the ability to decrypt multiple conversations and data messages sent to or from the subject (access to each session or message key) during the entire intercept period, only one initial affirmative action should be needed to obtain the relevant information. Exigent circumstances (kidnappings, terrorist threats) will require the release of decrypt information within a matter of hours.

Confidentiality and Safeguarding of Information: Key-Escrow agents should safeguard and maintain the confidentiality of information pertaining to the request for and the release of decrypt information; KEAs should protect the confidentiality of the person or persons for whom a key-escrow agent holds keys or components thereof, and protect the confidentiality of the identity of the agency requesting decrypt information or components thereof and all information concerning such agency's access to and use of encryption keys or components thereof.

For law enforcement requests, KEA personnel knowledgeable of an interception or decryption should be of good character and have not been convicted of crimes of moral turpitude or otherwise bearing on their trustworthiness. For national security requests, KEA personnel viewing and or storing classified requests must meet the applicable U.S. government requirements for accessing and or storing classified information. Efforts are ongoing to examine unclassified alternatives.

Key-escrow agents should be legitimate organizations without ties to criminal enterprises, and licensed to conduct business in the U.S.; KEAs for domestic products should not be a foreign corporation, a foreign country, or an entity thereof.[6]

Proposed U.S. Government Requirements for Ensuring Escrow-Agent Integrity and Security

1. Escrow-agent entities shall device and institutionalize policies, procedures, and mechanisms to ensure the confidentiality, integrity, and availability of key-escrow related information.
   a. Escrow-agent entities shall be designed and operated so that a failure by a single person, procedure, or mechanism does not compromise the confidentiality, integrity or availability of the key and/or key components (two-person control of keys, split keys, etc.).
   b. Unencrypted escrowed key and or key components that are stored and or transmitted electronically shall be protected (via encryption) using approved means.
   c. Unencrypted escrowed key and or key components stored and/or transferred via other media/methods shall be protected using approved means (safes).
2. Escrow-agent entities shall ensure due form of escrowed-key access requests and authenticate the requests for escrowed key and or key components.
3. Escrow-agent entities shall protect against disclosure of information regarding the identity of the person/organization whose key

and/or key components is requested, and the fact that a key and/or key component was requested or provided.

4. Escrow-agent entities shall enter keys/key components into the escrowed key database immediately upon receipt.

5. Escrow-agent entities shall ensure at least two copies of any key and/or key component in independent locations to help ensure the availability of such key and or key components due to unforeseen circumstances.

6. Escrow-agent entities that are certified by the U.S. government shall work with developers of key-escrow satellite encryption products and support a feature that allows products to verify to one another than the products' keys have been escrowed with a U.S.-certified agent.[7]

Requirements for Ensuring Key Access

7. An escrow-agent entity shall employ one or more persons who possess a SECRET clearance for purposes of processing classified (FISA) requests to obtain keys and/or key components.

8. Escrow-agent entities shall protect against unauthorized disclosure of information regarding the identity of the organization requesting the key or key components.

9. Escrow-agent entities shall maintain data regarding all key-escrow requests received, key-escrow components released, database changes, system-administration accesses, and dates of such events, for purposes of audit by appropriate government officials or others.

10. Escrow-agent entities shall maintain escrowed keys and/or key components for as long as such keys may be required to decrypt information relevant to a law enforcement investigation.

11. Escrow-agent entities shall provide key/key components to authenticated requests in a timely fashion and shall maintain a capability to respond more rapidly to emergency requirements for access.

12. Escrow-agent entities shall possess and maintain a Certificate of Good Standing from the State of incorporation (or similar local/national authority).

13. Escrow-agent entities shall provide to the U.S. government a Dun & Bradstreet/TRW number or similar credit report pointer and authorization.

14. Escrow-agent entities shall possess and maintain an Errors & Omissions insurance policy.

15. Escrow-agent entities shall provide to the U.S. government a written copy of, or a certification of the existence of a corporate security policy governing the key-escrow-agent entity's operation.

16. Escrow-agent entities shall provide to the U.S. government a certification that the escrow agent will comply with all applicable federal, state, and local laws concerning the provisions of escrow-agent entity services.

17. Escrow-agent entities shall provide to the U.S. government a certification that the escrow-agent entity will transfer to another approved escrow agent the escrow-agent entity's equipment and data in the event of any dissolution or other cessation of escrow-agent entity operations.

18. Escrow-agent entities for products sold in the U.S. shall not be a foreign country or entity thereof, a national of a foreign country, or a corporation of which an alien is an officer or more than one-fourth of the stock that is owned by aliens or which is directly or indirectly controlled by such a corporation. Foreign escrow-agent entities for products exported from the U.S. will be approved on a case-by-case basis as law enforcement and national security agreements can be negotiated.

19. Escrow-agent entities shall provide to the U.S. government a certification that the escrow-agent entity will notify the U.S. government in writing of any changes in the forgoing information.

20. Fulfillment of these and the other criteria are subject to periodic recertification.[8]

Liabilities of Escrow Agents

In order to assure users that key information entrusted to escrow agents remains secure and authorized third parties that they will be able to obtain exceptional access to satellite encrypted data when necessary, escrow agents and their employees must be held accountable for improper behavior and for the use of security procedures and practices that are appropriate to the task of protection. Liabilities can thus be criminal or civil (or both).

For example, criminal penalties could be established for the disclosure of keys or key fragments to unauthorized parties or for the refusal to disclose such information to appropriately authorized parties. It is worth noting that the implementing regulations accompanying the EES proposal run counter to this position, in the sense that they do not provide specific penalties for failure to adhere to the procedures for obtaining keys (which only legislation could do). The implementing regulations state specifically that *these procedures do not create, and are not intended to create, any substantive rights for individuals intercepted through electronic surveillance, and, noncompliance with these procedures shall not provide the basis for any motion to suppress or other objection to the introduction of electronic surveillance evidence lawfully acquired.*

Questions of civil liability are more complex. Ideally, levels of civil liability for improper disclosure of keys would be commensurate with the loss that would be incurred by the damaged party. For unauthorized disclosure of keys that encrypt large financial transactions, this level is potentially very large.

Note: Even if these transactions are authenticated (as most large transactions would be), large transactions that are compromised could lead to loss of bids and the like by the firms involved in the transaction.

On the other hand, as a matter of public policy, it is probably inappropriate to allow such levels of damages. More plausible may be a construct that provides what society, as expressed through the U.S. Congress, thinks is reasonable (see the following Sidebar). Users of escrow agents might also be able to buy their own insurance against unauthorized disclosure.

Note: Holding government agencies liable for civil damages might require an explicit change in the Federal Tort Claims Act that waives sovereign immunity in certain specified instances, or other legislative changes.

Statutory Limitations on Liability

Government can promote the use of specific services and products by assuming some of the civil liability risks associated with them. Three examples follow:

The Atomic Energy Damages Act, also called the Price-Anderson Act, limits the liability of nuclear power plant operators for harm caused by a nuclear incident (such as an explosion or radioactive release). To operate a nuclear power plant, a licensee must show the U.S. Nuclear Regulatory Commission (U.S. NRC) that it maintains financial protection (such as private insurance, self-insurance, or other proof of financial responsibility) equal to the maximum amount of insurance available at reasonable cost and reasonable terms from private sources, unless the U.S. NRC sets a lower requirement on a case-specific basis. The U.S. NRC indemnifies licensees from all legal liability arising from a nuclear incident, including a precautionary evacuation, which is in excess of the required financial protection, up to a maximum combined licensee-and-government liability of $670 million. Incidents that cause more than $670 million in damage will trigger review by the Congress to determine the best means to compensate the public, including appropriating funds.

The Commercial Space Launch Act provides similar protection to parties licensed to launch space vehicles or operate launch sites, but with a limit on the total liability the U.S. accepts. The licensee must obtain financial protection sufficient to compensate the maximum probable loss that third parties could claim for harm or damage, as determined by the Secretary of Transportation. The most that can be required is $600 million or the maximum liability insurance available from private sources, whichever is lower. The U.S. is obliged to pay successful claims by third parties in excess of the required protection, up to $2.6 billion, unless the loss is related to the licensee's willful misconduct. The law also requires licensees to enter into reciprocal waivers of claims with their contractors and customers, under which each party agrees to be responsible for losses it sustains.

The swine flu vaccination program of 1976 provides an example in which the U.S. accepted open-ended liability and paid much more than expected. Doctors predicted a swine flu epidemic, and Congress appropriated money for the Department of Health, Education, and Welfare (HEW) to pay four pharmaceutical manufacturers for vaccines to be distributed nationwide. The manufacturers' inability to obtain liability insurance delayed the program until Congress passed legislation (P.L. 94-380) in which the U.S. assumed all liability other than manufacturer negligence. The government's liability could thus include, for example, harmful side effects. Claims against the U.S. would be processed under the Federal Tort Claims Act (which provides for trial by judge rather than jury and no punitive damages, among other distinctions). Some of the 45 million people who were immunized developed complications, such as Guillain-Barre syndrome, consequently, the program was cancelled. By September 1977, 815 claims had been filed. The U.S. ultimately paid more than $200 million to settle claims, and some litigation is still pending today. Manufacturers, who by law were liable only for negligence, were not used.[9]

Furthermore, the amount of liability associated with compromising information related to satellite data communications is likely to dwarf the analogous volume for voice satellite communications. If escrowed satellite encryption is adopted widely in satellite data communications, compromise of escrow agents holding keys relevant to network encryption may be catastrophic, and may become easier as the number of access points that can be penetrated becomes larger.

Note: Liability of escrow agents may be related to the voluntary use of escrow. A party concerned about large potential losses would have alternatives to escrowed satellite encryption (namely, unescrowed encryption) that would protect the user against the consequences of improper key disclosure. Under these circumstances, a user whose key was compromised could be held responsible for his or her loss because he or she did not choose to use unescrowed encryption. An escrow agent's exposure to liability would be limited to the risks associated with parties that use its services. On the other hand, if escrowed satellite encryption were the only cryptography permitted to be used, then by assumption the user would have no alternatives, and so in that case, an escrow agent would shoulder a larger liability.

Another aspect of liability could arise if the escrow agents were also charged with the responsibilities of certificate authorities. Under some circumstances, it might be desirable for the functions of escrow agents and certificate authorities to be carried out by the same organization. Thus, these dual-purpose organizations would have all of the liabilities carried by those who must certify the authenticity of a given party.

The Role of Secrecy in Ensuring Product Security

The fact that EES and the Fortezza card involve classified algorithms has raised the general question of the relationship between secrecy and the maintenance of a product's trustworthiness in providing security. Specifically, the Clipper/Capstone approach is based on a secret (classified) encryption algorithm known as Skipjack. In addition, the algorithm is implemented in hardware (a chip) whose design is classified. The shroud of secrecy surrounding the hardware and algorithms needed to implement EES and Fortezza makes skeptics suspect that encrypted satellite communications could be decrypted through some secret *back door* (without having the escrowded key).

Note: A kind of de facto secret back door can result from the fact that vendors of security products employing Clipper or Capstone technology are not likely to advertise the fact that the relevant encryption keys are escrowed with the U.S. government. Thus, even if the escrowing capability is *open* in the sense that no one involved makes any attempt to hide that fact, a user who does not know enough to ask about the presence or absence of escrowing features may well purchase such products without realizing their presence. Functionally, escrowing, of which the use is ignorant, is equivalent for that user to a *secret* back door.

Logically, secrecy can be applied to two aspects of a satellite encryption system: the algorithms used, and the nature of the implementation of these algorithms. Each is addressed in turn. The following Sidebar describes a historical perspective on cryptography and secrecy that is still valid today.

Perspectives on Secrecy and System Security

The distinction between the general system (a product) and the specific key (of an encrypted message) was first articulated by Auguste Kerckhoffs in his historic book: *La Crytographic Militaire*, published in 1883.

Kerckhoffs deduced that compromise of the system should not inconvenience the correspondents. Perhaps the most startling requirement, at first glance, was Kerckhoffs' explanation of *system*. He meant *the material part of the system; tableaux, code books, or whatever mechanical apparatus may be necessary*, and not *the key proper*. Kerckhoffs here makes for the first time the distinction, now basic to cryptology, between the general system and the specific key. Why must the general system *not require secrecy*? Because, Kerckhoffs said, *it is not necessary to conjure up imaginary phantoms and to suspect the incorruptibility of employees or subalterns to understand that, if a system requiring secrecy were in the hands of too large a number of individuals, it could be compromised at each engagement. This has proved to be true, and Kerckhoffs' second requirement has become widely accepted under a form that is sometimes called the fundamental assumption of military cryptography: that the enemy knows the general system. But he or she must still be unable to solve messages in it without knowing the specific key. In its modern formulation, the Kerckhoffs doctrine states that secrecy must reside solely in the keys.*

A more modern expression of this sentiment is that the security of a cryptosystem should depend only on the secrecy of the keys and not on the secrecy of the algorithms. This requirement implies that the algorithms must be inherently strong. That is, it should not be possible to break a cipher simply by knowing the method of encipherment. This requirement is needed because the algorithms may be in the public domain, or known to a cryptanalyst.

Algorithm Secrecy

The use of secret algorithms for satellite encryption has advantages and disadvantages. From an information security standpoint, a third party who knows the algorithm associated with a given piece of ciphertext has an enormous advantage over one who does not (if the algorithm is unknown, cryptanalysis is much more difficult). Thus, the use of a secret algorithm by those concerned about information security presents an additional (and substantial) barrier to those who might be eavesdropping. From a signals intelligence (SIGINT) standpoint, it is advantageous to keep knowledge of good satellite encryption out of the hands of potential SIGINT targets. Thus, if an algorithm provides good cryptographic security, keeping the algorithm secret prevents the SIGINT target from implementing it. In addition, if an algorithm is known to be good, studying it in detail can reveal a great deal about what makes any algorithm good or bad. Algorithm secrecy thus helps to keep such information out of the public domain.

Note: Of course, if other strong algorithms are known publicly, the force of this argument is weakened from a practical standpoint. For example, it is not clear that the disclosure of Skipjack would be harmful from the standpoint of making strong algorithms public, because triple-DES is already publicly known, and triple-DES is quite strong.

On the other hand, algorithm secrecy entails a number of disadvantages as well. One is that independent analysis of a secret algorithm by the larger community is not possible. Without such analysis, flaws may remain in the algorithm that compromise the security it purports to provide. If these flaws are kept secret, users of the algorithm may unknowingly compromise themselves. Even worse, sophisticated users who need high assurances of security are unable to certify for themselves the security it provides (and thus have no sense of the risks they are taking if they use it). In most cases, the real issue is whether the user chooses to rely on members of the academic cryptography communities publishing in the open literature, members of the classified military community, or members of the commercial cryptography community who are unable to fully disclose what they know about a subject because it is classified or proprietary.

A second disadvantage of algorithm secrecy is the fact that if a cryptographic infrastructure is based on the assumption of secrecy, public discovery of those secrets can compromise the ends to be served by that infrastructure. For example, if a cryptographic infrastructure based on a secret algorithm were widely deployed, and if that algorithm contained a secret and unannounced *back door* that allowed those with knowledge of this back door easy access to encrypted data, that infrastructure would be highly vulnerable and could be rendered untrustworthy in short order by the public disclosure of the back door.

A third disadvantage is that a secret algorithm cannot be implemented in software with any degree of assurance that it will remain secret.

Software, as it exists ready for actual installation on a computer (so-called object code or executable code), can usually be manipulated with special software tools to yield an alternate form (namely, source code) reflecting the way the creating programmer designed it, and therefore revealing many, even most, of its operational details, including any algorithm embedded within it. This process is known as *decompling* or *disassembly* and is a standard technique in the repertoire of software engineers.

> **Note:** As one example, the RC2 encryption algorithm, nominally a trade secret owned by RSA Data Security Inc. was posted to the Internet in early 1996, apparently as the result of an apparent *disassembly* of a product embedding that algorithm.

All of the previous comments apply to secrecy whether it is the result of government classification decisions or vendor choices to treat an algorithm as a trade secret. In addition, vendors may well choose to treat an algorithm as a trade secret to obtain the market advantages that proprietary algorithms often bring. Indeed, many applications of cryptography for confidentiality in use today are based on trade-secret algorithms such as RC2 and RC4.

Product Design and Implementation Secrecy

Product design and implementation secrecy has a number of advantages. For example, secrecy in how a product has been designed makes it more difficult for an outsider to reverse-engineer the product in such a way that he or she could understand it better or, even worse, modify it in some way. Because vulnerabilities sometimes arise in implementation, keeping the implementation secret makes it harder for an attacker to discover and then exploit those vulnerabilities. Design and implementation secrecy thus protects any secrets that may be embedded in the product for a longer time than if they were to be published openly.

On the other hand, it is taken as an axiom by those in the security community that it is essentially impossible to maintain design or implementation secrecy indefinitely. Thus, the question of the time scale of reverse engineering is relevant, that is, given the necessary motivation, how long will it take and how much in resources will be needed to reverse-engineer a chip or a product?

For software, reverse engineering is based on decompilation or disassembly (as described previously). The larger the software product, the longer it take to understand the original program. Even a small one can be difficult to understand, especially if special techniques have been used to obscure its functionality. Modification of the original program can present additional technical difficulties (the product may be designed in such a way that disassembling or decompiling the entire product is necessary to isolate critical features that one might wish to modify). Certain techniques can be used to increase the difficulty of making such modifications, but there is virtual unanimity in the computer community that modification cannot be prevented forever.

> **Note:** For example, Trusted Information Systems Inc. of Glenwood, MD, has advocated an approach to preventing modification that relies on the placement of integrity locks at strategic locations. With such an approach, a change to the disassembled source code would have to be reflected properly in all relevant integrity locks. Doing so might well involve disassembly of an entire product rather than of just one module of the product. Nevertheless, such an approach cannot prevent modification, although it can make modification more difficult. Such anti-reverse-engineering features may also increase the difficulty of vendor maintenance of a product. Increased difficulty may be a price vendors must pay in order to have secure software implementations.

How robust must these anti-reverse-engineering features be? The answer is that they must be robust enough that the effort needed to overcome them is greater than the effort needed to develop a satellite encryption system from scratch.

For hardware, reverse engineering takes the form of physical disassembly and/or probing with x-rays of the relevant integrated circuit chips. Such chips can be designed to resist reverse engineering in a way that makes it difficult to understand the functions of various components on the chip. For example, the coating on a die used to fabricate a chip may be designed so that removal of the coating results in removal of one or more layers of the chip, thus destroying portions of what was to be reverse-engineered. The chip may also be fabricated with decoy or superfluous elements that would distract a reverse engineer. For all of these reasons, reverse engineering for understanding a chip's functions is difficult (though not impossible), and under some circumstances it is possible to modify a chip. In general, reverse engineering of the circuits and devices inside a chip requires significant expertise and access to expensive tools.

> **Note:** Estimates of the cost to reverse-engineer the Clipper chip nondestructively cover a wide range, from *doable in university laboratories with bright graduate students and traditions of reverse engineering* (as estimated by a number of electrical engineers in academia with extensive experience in reverse engineering), to as much as $40 million to $60 million (as estimated in informal conversations between JASON members and DOD engineers). The cost may well be lower if large numbers of chips are available for destructive inspection.

An important factor that works against implementation secrecy is the wide distribution of devices or products whose implementation is secret. It is difficult to protect a device against reverse engineering when millions of those devices are distributed around the world without any physical barriers (except those on the implementation itself) to control access to them. Everyone with an EES-compliant telephone or a Foretzza card, for example, will have access to the chip that provides satellite encryption and key-escrow services.

The previous comments refer to the feasibility of maintaining implementation secrecy. But there are issues related to its desirability as well. For example, implementation secrecy implies that only a limited number of vendors can be trusted to produce a given implementation. Thus, foreign production of Clipper/Capstone-compliant devices under classification guidelines raises problems unless foreign producers are willing to abide by U.S. security requirements.

A more important point is that implementation secrecy also demands trust between user and supplier/vendor. Users within government agencies generally trust other parts of the government to provide adequate services as a supplier. But in the private sector, such trust is not necessarily warranted. Users who are unable to determine for themselves what algorithms are embedded in computer and satellite communications products used must trust the vendor to have provided algorithms that do what the user wants done. And, the vast majority of users fall into this category. Such opacity functions as a de facto mechanism of secrecy that also impedes user knowledge about the inner workings and that is exploited by the distributors of computer viruses and worms. As a result, choosing between the use of a self-implemented source code and a pre-packaged program to perform certain functions is in many ways analogous to choosing between the use of unclassified and classified algorithms.

An information security manager with very high security needs must make trade-offs of assurance versus cost. In general, the only way to be certain that the algorithms used are the ones claimed to be used is to implement them on one's own. Yet if the manager lacks the necessary knowledge and experience, a self-implementation may not be as secure or as capable as one developed by a trusted vendor. A self-implementer also carries the considerable burden of development costs that a commercial vendor can amortize over many sales.

As a result, security-conscious users of products whose inner workings are kept secret must: (1) trust the vendor implicity (based on factors such as reputation); or (2) face the possibility of various extreme scenarios. Two such scenarios are as follows:

First, the hardware of a secret device can be dynamically modified. For example, electrically erasable read-only memories can direct the operation of a processor. One possible scenario with secret hardware is that a chip that initially provides Clipper-chip functionality might be reprogrammed when it first contacts a Clipper/Capstone-compliant device to allow a nonescrowed but unauthorized access to it. Such a means of *infection* is common with computer viruses. In other words, the Skipjack algorithm may have been embedded in the chip when it was first shipped, but after the initial contact, the algorithm controlling the chip is no longer Skipjack.

Second, an algorithm that is not Skipjack is embedded by the manufacturer in chips purporting to be Clipper or Capstone chips. Because the utility of a vector test depends on the availability of an independent implementation of the algorithm, it is impossible for the user to perform this test independently if the user has no reference point. As a result, the user has no access to an independent test of the chip that is in the user's *Clipper/Capstone-compliant* device, and so any algorithm might have been embedded.

Note: The review team for Skipjack (previously discussed) compared the output from Clipper chips with output from the software version of Skipjack that the review team obtained to verify that the algorithm on the chips was the same as the software version.

Any technically trained person can invent many other such scenarios. Thus, public trust in the technical desirability of the EES and Fortezza for exceptional access depends on a high degree of trust in the government, entirely apart from any fears about compromising escrow agents wherever they are situated.

Of course, some of the same considerations go beyond the Skipjack algorithm and the Clipper/Capstone approach. In general, users need confidence that a given product with satellite encryption capabilities indeed implements a given algorithm. Labeling a box with the letters *DES* does not ensure that the product inside really implements DES. In this case, the fact that the DES algorithm is publicly known facilitates testing to verify that the algorithm is implemented correctly.

Note: As described in Chapter 5, the product tester can use the product to encrypt a randomly chosen set of values with a randomly chosen key, and then compare the encrypted output to the known correct result obtained through the use of a product known to implement the algorithm correctly. This is known as a vector test.

If the algorithm's source code is available for inspection, other security-relevant aspects of a software product can be examined to a certain extent, at least up to the limits of the expertise of the person checking the source code. But for software products without source code, and especially for hardware products that cannot easily be disassembled (and even more so for hardware products that are specifically designed to resist disassembly), confidence in the nonalgorithm security aspects of the product is more a matter of trusting the vendor than of the user making an independent technical verification of an implementation. In some sectors (banking, classified military applications), however, independent technical verification is regarded as essential.

Note: The previous discussion is not meant to preclude the possibility of an independent certifying authority, a kind of *Consumers Reports* for crypto equipment and products. Such organizations have been proposed to evaluate and certify computer security.

Finally, a given product may properly implement an algorithm but still be vulnerable to attacks that target the part of the product surrounding the implementation of the algorithm. Such vulnerabilities are most common in the initial releases of products that have not been exposed to public test and scrutiny. For example, a security problem with the Netscape Navigator's key-generation facility could have been found had the implementation in which the key generator was embedded been available for public examination prior to its release, even though the encryption algorithm itself was properly implemented.

Note: This security problem is referenced in Chapter 3.

The Hardware-Software Choice in Product Implementation

After the Clipper initiative was announced, and as the debate over escrowed satellite encryption broadened to include the protection of stored data from satellite data communications, the mass market software industry emphasized that a hardware solution to cryptographic security (as exemplified by the Clipper chip) would not be satisfactory. The industry argued with some force that only a software-based approach would encourage the widespread use of satellite encryption envisioned for the world's electronic future. The industry made several points:

First, customers have a strong preference for using integrated cryptographic products. While stand-alone products with satellite encryption capabilities could be made to work, in general, they lack operational convenience for the applications that software and systems vendors address.

Second, compared to software, hardware is expensive to manufacture. In particular, the relevant cost is not simply the cost of the hardware satellite encryption device compared to a software encryption package, but also the cost of any modifications to the hardware environment needed to accept the hardware satellite encryption device. For example, one major company noted to NSA that the adoption of the Fortezza card (a card that fits into the PC-card slots available on most laptop computers) would be very expensive in their desktop computing environment. Why? Because most of their desktop computers do not have a PC-card slot and would have to be modified to accept a Fortezza card. By contrast, a software satellite encryption product can simply be loaded via common media (a CD-ROM or a floppy disk) or downloaded via a network.

Third, the fact that hardware is difficult to change means that problems found subsequent to deployment are more difficult to fix. For example, most users would prefer to install a software fix by loading a CD-ROM into their computers than to open up their machines to install a new chip with a hardware fix.

Fourth, hardware-based security products have a history of being market-unfriendly. Hardware will, in general, be used only to the extent that the required hardware (and its specific configuration) is found in user installations. Moreover, hardware requirements can be specified for software only when that hardware is widely deployed. For example, a technical approach to the software piracy problem has been known for many years. The approach requires the installation of special-purpose hardware that is available only to those who obtain the software legitimately. This *solution* has failed utterly in the marketplace, and software piracy remains a multibillion-dollar-per-year problem.

Finally, hardware for security consumes physical space and power in products. For example, a hardware-based satellite encryption card that fits into an expansion slot on a computer takes up a slot permanently, unless the user is willing to install and deinstall the card for every use. It also creates an additional power demand on electronic devices where power and battery life are limited.

In general, products with satellite encryption capabilities today use software or hardware or both to help ensure security. The crux of the hardware-software debate is what is good enough to ensure security.

Note: The dividing line between hardware and software is not always clear. In particular, product designers use the term *firmware* to refer to a design approach that enters software into a special computer memory (an integrated circuit chip) that usually is subsequently unchangeable (read-only memory; ROM). Sometimes an alternate form of memory is used that does permit changes under controlled conditions (electrically programmable ROM; EPROM). Such software-controlled hardware (microprogrammed hardware) has the convenience that the functionality of the item can be updated or changed without redesign of the hardware portion.

The security needed to manage electronic cash in the international banking system needs to be much stronger than the security to protect wordprocessing files created by private individuals. Thus, software-based cryptography might work for the latter, while hardware-based cryptography might be essential for the former.

Products with satellite encryption capabilities must be capable of resisting attack. But as such products are often embedded in operating environments that are themselves insecure, an attacker may well choose to attack the environment rather than the product itself. For example, a product with satellite encryption capabilities may be hardware-based, but the operating environment may leave the encryption keys or the unencrypted text exposed. More generally, in an insecure environment, system security may well not depend very much on whether the cryptography per se is implemented in hardware or software or whether it is weak or strong.

In the context of escrowed satellite encryption, a second security concern arises—a user of an escrowed satellite encryption product may wish to defeat the escrow mechanism built into the product. Thus, the escrow features of the product must be bound to the product in a way that cannot be bypassed by some reverse-engineered modification to the product. This particular problem is known as binding or, more explicitly, escrow binding. Escrow binding is an essential element of any escrow scheme that is intended to provide exceptional access.

Concern over how to solve the escrow-binding problem was the primary motivation for the choice of a hardware approach to the Clipper initiative. As suggested previously, the functionality of a

hardware system designed to resist change is indeed difficult to change, and so hardware implementations have undeniable
advantages for solving the escrow-binding problem. An EES-compliant
device would be a telephone without software accessible to the user,
and would provide high assurance that the features for exceptional
access would not be bypassed.

> **Note:** A device controlled by software stored in a programmable
> read-only memory is for all intents and purposes the same as *pure
> hardware* in this context.

As the debate has progressed, ideas for software-based escrow processes have been proposed. The primary concern of the U.S. government about software implementations is that once a change has been
designed and developed that can bypass the escrow features (*break the
escrow binding*), such a change can easily be propagated through many
different channels and installed with relatively little difficulty. In the
U.S. Congressional Committee of privacy and technology's view, the
important question is whether software solutions to the escrow-binding problem can provide an acceptable level of protection against the
reverse engineer. Whether an escrowed satellite encryption product is
implemented in software (or hardware for that matter), the critical
threshold is the difficulty of breaking the escrow binding (bypassing
the escrowing features) compared to the effort necessary to set up an
independent unescrowed encryption system (perhaps as part of an
integrated product). If it is more difficult to bypass the escrow features
than to build an unescrowed system, then *rogues* who want to defeat
exceptional access will simply build an unescrowed system. The
bottom line is that an escrowed satellite encryption product does not
have to be perfectly resistant to breaking the escrow binding.

A possible mitigating factor is that even if a software *patch* is developed that would break the escrow binding of an escrowed satellite
encryption software product, it may not achieve wide distribution even
among the criminals that would have the most to gain from such a
change. Experience with widely deployed software products (operating
systems) indicates that even when a software fix is made available for
a problem in a product, it may not be implemented unless the
anomalous or incorrect software behavior is particularly significant to
an end user. If this is the case for products that are as critical as

operating systems, it may well be true for products with more specialized applications. On the other side of the coin, many parties (criminals) may be greatly concerned about the presence of escrowing and
thus be highly motivated to find *fixes* that eliminate them.

Responsibility for Generation of Unit Keys

Key generation is the process by which cryptographic keys are generated. Two types of keys are relevant:

First, a session key is required for each encryption of plaintext into
ciphertext. This is true whether the information is to be stored or
communicated. Ultimately, the intended recipients of this information
(those who retrieve it from storage or those who receive it at the other
end of a satellite communications channel) must have the same session
key. For maximum information security, a new session key is used
with every encryption (see previous discussions on this subject for
more information).

Second, a unit key is a cryptographic key associated with a particular
product or device owned or controlled by a specific individual. Unit
keys are often used to protect session keys from casual observation in
escrowed satellite encryption products, but precisely how they are
used depends on the specifics of a given product.

In the most general case, the session key is a random number, and a
different one is generated anew for each encryption. But the unit key
is a cryptographic variable that typically changes on a much longer
time scale than does the session key. In many escrowed satellite
encryption schemes, knowledge of the unit key enables a third party
to obtain the session key associated with any given encryption.

The Clipper/Capstone approach requires that the unit key be generated
by the manufacturer at the time of manufacture (*at birth*) and then
registered prior to sale with escrow agents in accordance with established procedures. Such an approach has one major advantage from the
standpoint of those who may require exceptional access in the future — it guarantees registration of keys because users need not take
any action to ensure registration.

At the same time, because the Clipper/Capstone approach is based on a hardware-based implementation that is not user-modifiable, a given device has only one unit key for its entire lifetime, although at some cost the user may change the Clipper chip embedded in the device. If the unit key is compromised, the user's only recourse is to change the chip. A user who does not do so violates one basic principle of information security—frequent changing of keys (or passwords). In addition, the fact that all unit keys are known at the time of manufacture raises concerns that all keys could be kept (perhaps surreptitiously) in some master databank that would be accessible without going to the designated escrow agents.

> **Note:** However, because the Skipjack algorithm is classified, simple knowledge of the unit key (or the session key) would enable only those with knowledge of the algorithm to decrypt the session key (or the session).

The implication is that the user is forced to trust several organizations and individuals involved with the manufacturing process. Such trust becomes an implicit aspect of the secrecy associated with EES-compliant devices.

One alternative to unit key generation at birth is the generation (or input) of a new unit key at the user's request. This approach has the advantage that the user can be confident that no one else retains a copy of the new key without his or her knowledge. The disadvantage is that escrow of that key would require explicit action on the user's part for that purpose.

An alternative that has some of the advantages of each approach is to install and register a unit key at birth, but to design the product to allow the user to change the unit key later. Thus, all products designed in this manner would have *default* unit keys installed by the manufacturer and recorded with some escrow agent. Each of these keys would be different. Users who took the trouble to install a new unit key would have to take an explicit action to escrow it, but in many cases the inconvenience and bother of changing the unit key would result in no action being taken. Thus, valid unit keys would be held by escrow agents in two cases — for products owned by users who did not change the unit key, and for products owned by users who chose to register their new keys with escrow agents.

Who is responsible for the collection of unit keys? Under the Clipper/Capstone approach, the responsible party is the U.S. government. But if nongovernment agencies were to be responsible for escrowing keys (see previous discussion in this chapter), a large market with many vendors producing many different types of satellite encryption products in large volume could result in a large administrative burden on these vendors.

The specific implementation of EES also raises an additional point. As proposed, EES requires that unit keys be given to government authorities upon presentation of legal authorization. If these keys are still available to the authorities after the period of legal authorization has expired, the EES device is forever open to government surveillance. To guard against this possibility. Administration plans for the final Clipper key-escrow system provide for automatic key deletion from the decrypting equipment upon expiration of the authorized period. Key deletion is to be implemented on the tamper-resistant device that law enforcement authorities will use to decrypt Clipper-encrypted traffic. However, the deployed interim key-escrow system has not been upgraded to include that feature.

Issue Related to the Administration Proposal to Exempt 64-Bit Escrowed Satellite Encryption in Software

As noted in Chapter 5, the Administration has proposed to treat software products with 64-bit encryption using any algorithm as it currently treats products that are based on 40-bit RC2/RC4 algorithms, provided that products using this stronger encryption are *properly escrowed*. This change is intended to make available to foreign customers of U.S. software products stronger cryptographic protection than they have at present. This proposal has raised several issues that we will examine next.

The Definition of Proper Escrowing

The definition of *proper escrowing* (as the phrase is used in the Administration's proposed new export rules in Sidebar, "Administration's Draft Software Key-Escrow Export Criteria") is that keys should be escrowed only with *escrow agent(s) certified by the U.S. government, or certified by foreign governments with which the U.S. government has formal agreements consistent with U.S. law enforcement and national security requirements.* These agents would not necessarily be government agencies, although they could be in principle.

The obvious question is whether foreign consumers will be willing to purchase U.S. products with satellite encryption capabilities when it is openly announced that the information security of those products could be compromised by or with the assistance of escrow agents certified by the U.S. government. While the draft definition does envision the possibility that escrow agents could be certified by foreign governments (those in the country of sale), formal agreements often take a long time to negotiate, during which time U.S. escrow agents would hold the keys, or the market for such products would fail to develop.

For some applications (U.S. companies doing business with foreign suppliers), interim U.S. control of escrow agents may prove acceptable. But it is easy to imagine other applications for which it would not, and in any case a larger question is begged: What would be the incentive for foreign users to purchase such products from U.S. vendors if comparably strong but unescrowed foreign products with satellite encryption capabilities were available? As the discussion in Chapter 3 points out, integrated products with satellite encryption capabilities are generally available today from U.S. vendors. However, how long the U.S. monopoly in this market will last is an open question.

The Proposed Limitation of Key Lengths to 64 Bits or Less

The most important question raised by the 64-bit limitation is this: If the keys are escrowed and available to law enforcement and national security authorities, why does the length of the key matter? In response to this question, senior Administration officials have said that the limitation to 64 bits is a way of hedging against the possibility of finding easily proliferated ways to break the escrow binding built into software, with the result that U.S. software products without effective key escrow would become available worldwide. Paraphrasing the remarks of a senior Administration official at the 1995 International Cryptography Institute: *The 64-bit limit is there because we might have a chance of dealing with a breakdown of software key escrow 10 to 15 years down the line. But, if the key length implied a work factor of something like triple-DES, we would never (emphasis in original) be able to do it.*

Two factors must be considered in this argument. One is the likelihood that software key escrow can in fact be compromised (this subject was previously discussed). But a second point is the fact that the 64-bit limit is easily circumvented by multiple encryption under some circumstances. Specifically, consider a stand-alone security-specific product for file encryption that is based on DES and is escrowed. Such a product (in its unaltered state) meets all of the proposed draft criteria for export. But disassembly of the object code of the program (to defeat the escrow binding) may also reveal the code for DES encryption in the product. Once the source code for the DES encryption is available, it is a technically straightforward exercise to implement a package that will use the product to implement a triple-DES encryption on a file.

Summary

Escrowed satellite encryption is one of several approaches to providing exceptional access to encrypted information. The U.S. government has advanced a number of initiatives to support the insertion of escrow features into products with satellite encryption capabilities that will become available in the future, including the Escrowed Satellite Encryption Standard, the Capstone/Fortezza initiative, and a proposal to liberalize export controls on products using escrowed satellite encryption. Its support of escrowed satellite encryption embodies the government's belief that the benefit to law enforcement and national security from exceptional access to encrypted information outweighs the damage owing to loss of confidentiality that might occur with the failure of procedures intended to prevent unauthorized access to the escrow mechanism.

Escrowed satellite encryption provides *more* confidentiality than leaving information unprotected (as most information is today), but less confidentiality than what could be provided by good implementations of unescrowed cryptography. On the other hand, escrowed satellite encryption provides more capability for exceptional access under circumstances of key loss or unavailability than does unescrowed encryption. All users will have to address this trade-off between level of confidentiality and key unavailability. The central questions with respect to escrowed satellite encryption are the following:

1. With what degree of confidence is it possible to ensure that third parties will have access to encrypted information only under lawfully authorized circumstances?
2. What is the trade-off for the user between potentially lower levels of confidentiality and higher degrees of confidence that encrypted data wil be available when necessary?

From Here

Chapter 13 addressed escrowed satellite encryption, an approach aggressively promoted by the federal government as a technique for balancing national needs for information security with those of law enforcement and national security. The chapter also addressed the following: Secure Public Networks Act; the risks of key recovery, key escrow, and trusted third-party satellite encryption; satellite encryption for the rest of us; establishing identity without certification authorities; and export criteria for key-escrow satellite encryption.

Chapter 14 discusses other dimensions of the national satellite encryption policy, including the Digital Telephony Act of 1995 (aka the Communications Assistance for Law Enforcement Act) and a variety of other levers used in national cryptography policy that do not often receive much attention in the debate.

Endnotes

1   Dorothy Denning and Miles Smid, "Key Escrowing Today," *IEEE Communications*, Volume 32(9), September 1994, IEEE Communications. All Rights reserved. © 1994 by IEEE. 3 Park Avenue, New York, NY, 10016-5997, Volume 32, Number 9, pp. 58–68.

2 David L. Sobel, Legal Counsel, "Comments on Draft Export Criteria for Key Escrow Encryption," Electronic Privacy Information Center, National Institute of Standards of Technology, 666 Pennsylvania Ave., S.E., Suite 301, Washington, DC 20003, 1995.

3 National Institute of Standards and Technology, "Draft Software Key Escrow Encryption Export Criteria, November 6, 1995.

4 Dorothy Denning and Miles Smid, "Key Escrowing Today."

5 Federal Bureau of Investigation, International Cryptography Institute 1995, September 22, 1995.

6 © 1996 by the National Academy of Sciences. Courtesy of the National Academy Press, 2101 Constitution Avenue, NW, Washington, D.C., 20418.

7 Federal Bureau of Investigation, International Cryptography Institute 1995, September 22, 1995.

8 National Institute of Standards and Technology, "Draft Software Key Escrow Encryption Export Criteria."

9 Ibid.

10 U.S. Senate, "Secure Public Networks Act," 105th Congress, Washington, DC, 1997.

14

Other Dimensions of National Satellite Encryption Policy

In addition to export controls and escrowed satellite encryption current national policy on satellite encryption is affected by government use of a large number of levers available to it, including the Communications Assistance for Law Enforcement Act, the standards-setting process, R&D funding, procurement practices, education and public jawboning, licenses and certification, and arrangements both formal and informal with various other governments (state, local, and foreign) and organizations (specific private companies). All of these are controversial because they embody judgments about how the interests of law enforcement and national security should be reconciled against the needs of the private sector. In addition, the international dimensions of satellite encryption are both critical (because cryptography affects satellite communications and communications are fundamentally international) and enormously difficult (because national interests differ from government to government).

The Communications Assistance for Law Enforcement Act

The Communications Assistance for Law Enforcement Act (CALEA) was widely known as the *digital telephony* bill before its formal passage. The CALEA is not explicitly connected to the national satellite encryption policy, but it is an important aspect of the political context in which the national satellite encryption policy has been discussed and debated.

Brief Description of and Stated Rationale for the CALEA

The Communications Assistance for Law Enforcement Act (CALEA) was passed in October 1994. The act imposes on telecommunications carriers four requirements in connection with those services or facilities that allow customers to originate, terminate, or direct satellite communications:

The first requirement is to enable the government to isolate and to intercept expeditiously pursuant to court order or other lawful authorization, all wire and electronic communications in the carrier's control to or from the equipment, facilities, or services of a subscriber, in real time or at any later time acceptable to the government. Carriers are not responsible for decrypting encrypted satellite communications that are the subject of court-ordered wiretaps, unless the carrier provided the encryption and can decrypt it. Moreover, carriers are not prohibited from deploying a satellite encryption service for which it does not retain the ability to decrypt communications for law enforcement access.

The second requirement is to enable the government to isolate and access expeditiously pursuant to court order or other lawful authorization, reasonably available call-identifying information about the origin and destination of satellite communications. Access must be provided in such a manner that the information may be associated with the communication to which it pertains and is provided to the government before, during, or immediately after the satellite communication's transmission to or from the subscriber.

The third requirement is to make intercepted communications and call-identifying information available to the government, pursuant to court order or other lawful authorization, so that they may be transmitted over lines or facilities leased or procured by law enforcement to a location away from the carrier's premises.

Finally, the fourth requirement is to meet foregoing requirements with a minimum of interference with the subscriber's service and in such a way that protects the privacy of communications and call-identifying information that are not targeted by electronic surveillance orders, and that maintains the confidentiality of the government's interceptions.

The CALEA also authorizes federal money for retrofitting common carrier systems to comply with these requirements. As of this writing, however, no funds have been appropriated for this task.

The CALEA requirements apply only to those services or facilities that enable a subscriber to make, receive, or direct calls. They do not apply to information services, namely, the services of electronic mail providers, on-line services such as Compuserve or America Online, Internet access providers, or to private networks or services whose sole purpose is to interconnect carriers. Furthermore, the CALEA requires law enforcement authorities to use carrier employees or personnel to activate a surveillance. The CALEA also provides that a warrant be required to tap a cordless telephone. Wiretaps on cellular telephones are already governed by Title III or the Foreign Intelligence Surveillance Act.

The Stated Rationale for the CALEA

Historically, telecommunuications service providers have cooperated with law enforcement officials in allowing access to communications upon legal authorization. New telecommunications services (call forwarding, paging, cellular calls) and others expected in the future have diminished the ability of law enforcement agencies to carry out legally authorized electronic surveillance. The primary impact of the CALEA is to ensure that within the next decade telecommunications service providers will still be able to provide the assistance necessary to law enforcement officials to conduct surveillance of wire and electronic communications (both content and call-identifying information) controlled by the carrier, regardless of the nature of the particular services being offered.

Reducing Resource Requirements for Wiretaps

Once a court-approved surveillance order has been granted, it must be implemented. In practice, the implementation of a surveillance order requires the presence of at least two agents around the clock. Such a presence is required if real-time minimization requirements are to be met.

> **Note:** Minimization refers to the practice, required by Title III, of monitoring only those portions of a conversion that are relevant to the crime under investigation. If a subject discusses matters that are strictly personal, such discussions are not subject to monitoring. In practice, a team of agents operate a tape recorder on the wiretapped line. Minimization requires agents to turn off the tape recorder and to cease monitoring the conversation for a short period of time if they overhear nonrelevant discussions. At the end of that time period, they are permitted to resume monitoring. For obvious reasons, this practice is conducted in real time. When agents encounter a foreign language with which they are unfamiliar, they are allowed to record the entire conversation. The tape is then *minimized* after the fact of wiretapping.

As a result, personnel requirements are the most expensive aspect of electronic surveillance. The average cost of a wiretap order is $68,000, or approximately one-half of a full-time-equivalent agent-year. Such costs are not incurred lightly by law enforcement agencies.

Under these circumstances, procedures and/or technologies that could reduce the labor required to conduct wiretaps pose a potential problem for individuals concerned about excessive use of wiretaps. Specifically, these individuals are concerned that the ability to route wiretapped calls to a central location would enable a single team of agents to monitor multiple conversations.

Note: For example, such a concern was raised at the Fifth Conference on Computers, Freedom, and Privacy held in San Francisco in March 1995. The argument goes as follows: While the CALEA authorizes $600 million to pay for existing in-place telephone switch conversions to implement the capabilities desired by law enforcement, this amount is intended as a one-time cost. Upgrades of switching systems are expected to implement these capabilities without government subsidy. Moreover, the Congress has not yet appropriated this money. The point is that additional wiretap orders would not pose an additional incremental cost (though the original cost of $68,000 would still obtain), and the barrier of incremental cost would not impede more wiretap orders. In short, critics argue that it would make good economic sense to make additional use of resources if such use can *piggyback* on an already-made investment.

Such time sharing among monitoring teams could lower wiretap costs significantly. From the standpoint of law enforcement, these savings could be used for other law enforcement purposes, and they would have the additional effect of eliminating an operational constraint on the frequency with which wiretap authority is sought today.

At present, technologies that would enable minimization without human assistance are in their infancy. For example, the technology of speech recognition for the most part cannot cope with speech that is speaker-independent and continuous. And, artificial intelligence programs today and for the foreseeable future will be unable to distinguish between the criminally relevant and nonrelevant parts of a conversation. Human agents are an essential component of a wiretap, and law enforcement officials have made three key points in response to the concern previously raised.

Most importantly, today's wiretaps are performed generally with law enforcement agencies paying telecommunications service providers for delivering the intercepted satellite communications to a point selected by the law enforcement entity. From an operational standpoint, the real-time minimization of wiretapped conversations requires agents who are personally familiar with the details of the case under investigation. In this way they know when the subjects are engaged in conversations related to the case. Agents exceed their authority if they monitor unrelated conversations.

Procedural rules require that all evidence be maintained through a proper chain of custody and in a manner such that the authenticity of evidence can be established. Law enforcement officals believe that the use of one team to monitor different conversations could call into question the ability to establish a clear chain of custody.

Obtaining Access to Digital Streams in the Future

In the conduct of any wiretap, the first technical problem is simply gaining access to the relevant traffic itself, whether or not it is encrypted. For law enforcement, products with satellite encryption capabilities and features that allow exceptional access are useless without access to the traffic in question. The CALEA was an initiative spearheaded by law enforcement to deal with the access problem created by new telecommunications services.

The problems addressed by the CALEA will inevitably resurface as newer satellite communications services are developed and deployed for use by common carriers and private entities (corporations) alike. It is axiomatic that the complexity of interactions among satellite communications systems will continually increase, both as a result of increased functionality and the need to make more efficient use of available bandwidth. Consequently, isolation of the digital streams associated with the party or parties targeted by law enforcement will become increasingly difficult if the cooperation of the service provider is not forthcoming for all of the reasons described in Chapter 3. It is for this reason that the CALEA applies to parties that are not at present common carriers upon appropriate designation by the Federal Communications Commission.

Moreover, even when access to the digital stream of an application is assured, the structure of the digital stream may be so complex that it would be extremely costly to determine all of the information present without the assistance of the application developer. Tools designed to isolate the relevant portions of a given digital stream transmitted on open systems will generally be less expensive than tools for proprietary systems. However, as both open and proprietary systems will be present in any future telecommunications environment, law enforcement authorities will need tools for both. The development of such tools will require considerable technical skill, such skill that is most likely possessed by the application developers. Cooperation with product developers may decrease the cost of developing these tools.

Finally, as the telecommunications system becomes more and more heterogeneous, even the term *common carrier* will become harder to define or apply. The routing of an individual data communication through the *network* will be dynamic and may take any one of a number of paths, decisions about which are not under the user's control. While only one link in a given route need be a common carrier for CALEA purposes, identifying that common carrier in practice may be quite difficult.

The CALEA Exemption of Information Service Providers and Distinctions Between Voice and Data Services

At present, users of data communications services access networks such as the Internet either through private networks (via their employers) or through Internet service providers that provide connections for a variety of individuals and organizations. Both typically make use of lines owned and operated by telecommunications service providers. In the former case, law enforcement access to the digital stream is more or less the same problem as it is for the employer (and law enforcement has access through the legal process to the employer). In the latter case, the CALEA requires the telephone service provider to provide to law enforcement authorities a copy of the digital stream being transported.

The CALEA exempts online information service providers such as America Online and Compuserve from its requirements. In the future, other CALEA issues may arise as the capabilities provided by advanced information technologies grow more sophisticated. For example, the technological capability exists to use Internet-based services to supply real-time satellite voice communications. Even today, a number of Internet and network service providers are capable of supporting (or are planning to support) real-time *push-to-talk satellite* voice communications. The CALEA provides that a party providing satellite communications services, that in the judgment of the FCC are *a replacement for a substantial portion of the local telephone exchange service*, may be deemed a carrier subject to the requirements of the CALEA. Thus, one possible path along which telecommunications services may evolve could lead to the imposition of CALEA requirements on information service providers, even though they were exempted as an essential element of a legislative compromise that enabled the CALEA to pass in the first place.

These possibilities are indicative of a more general problem: the fact that lines between *voice* and *data* services are being increasingly blurred. This issue is addressed in greater detail in Chapter 26.

Other Levers Used in National Satellite Encryption Policy

The government has a number of tools to influence the possession and use of satellite encryption domestically and abroad. How the government uses these tools in the context of the national satellite encryption policy reflects the government's view of how to balance the interests of the various stakeholders affected by satellite encryption.

Federal Information Processing Standards

Federal Information Processing Standards (FIPs) are an important element of national cryptography policy, and all federal agencies are encouraged to cite FIPs in their procurement specifications. The following Sidebar contains a brief description of all FIPs related to satellite encryption. The National Institute of Standards and Technology (NIST) is responsible for issuing FIPs.

Satellite Encryption-Related Federal Information Processing Standards

FIPS 46, 46-1 and 46-2—Data Encryption Standard (DES)

This standard specifies the DES algorithm and rules for implementing DES in hardware; FIPS 46-1 recertifies DES and extends it for software implementation; FIPS 46-2 reaffirms the Data Encryption Standard algorithm until 1999 and allows for its implementation in software, firmware or hardware. Several other FIPSs address interoperability and security requirements for using DES in the physical layer of data communications (FIPS 139) and in fax machines (FIPS 141), guidelines for implementing and using DES (FIPS 74), modes of operation of DES (FIPS 81), and use of DES for authentication purposes (FIPS 113).

FIPS 180-1—Secure Hash Standard

The standard specifies a Secure Hash Algorithm (SHA) that can be used to generate a condensed representation of a message called a message digest. The SHA is required for use with the Digital Signature Algorithm (DSA) as specified in the Digital Signature Standard (DSS) and whenever a secure hash algorithm is required for federal applications. The SHA is used by both the transmitter and intended receiver of a message in computing and verifying a digital signature.

FIPS 186—Digital Signature Standard

This standard specifies a Digital Signature Algorithm (DSA) appropriate for applications requiring a digital rather than a written signature. The DSA digital signature is a pair of large numbers represented in a computer as strings of binary digits. The digital signature is computed using a set of rules (the DSA) and a set of parameters such that the identity of the signatory and integrity of the data can be verified. The DSA provides the capability to generate and verify signatures.

FIPS 140—Security Requirements for Cryptographic Modules

This standard provides specifications for cryptographic modules that can be used within computer and telecommunications systems to protect unclassified information in a variety of different applications.

FIPS 185—Escrowed Encryption Standard

This standard is covered thoroughly throughout the chapter.

FIPS 171—Key Management Using ANSI X9.17

This standard specifies a selection of options for the automated distribution of keying material by the federal government when using the protocols of ANSI X9.17. The standard defines procedures for the manual automated management of keying materials and contains a number of options. The selected options will allow the development of cost-effective systems that will increase the likelihood of interoperability.

Other FIPSs address matters related more generally to computer security. For example:

- FIPS 48: Guidelines on Evaluation of Techniques for Automated Personal Identification.
- FIPS 83: Guidelines on User Authentication Techniques for Computer Network Access Control.
- FIPS 112: Password Usage.
- FIPS 113: Computer Data Authentication.
- FIPS 73: Guidelines for Security of Computer Applications.[1]

The FIPSs can have enormous significance to the private sector as well, despite the fact that the existence of a FIPS does not legally compel a private party to adopt it. One reason is that to the extent that a FIPS is based on existing private-sector standards (which it often is), it codifies standards of existing practice and contribute to a planning environment of greater certainty. A second reason is that a FIPS is often taken as a government endorsement of the procedures, practices, and algorithms contained therein, and thus a FIPS may set a de facto *best practices* standard for the private sector. A third reason is related to procurements that are FIPS-compliant as discussed in the next section of this chapter.

The NIST has traditionally relied on private sector standards-setting processes when developing FIPSs. Such practice reflects NIST's recognition of the fact that the standards it sets will be more likely to succeed (in terms of reducing procurement costs, raising quality, and influencing the direction of information technology market development) if they are supported by private producers and users.

The existence of widely accepted standards is often an enormous boon to interoperability of computers and communication devices, and the converse is generally true as well. The absence of widely accepted standards often impedes the growth of a market.

In the domain of satellite encryption, FIPSs have had a mixed result. The promulgation of FIPS 46-1, the Data Encryption Standard (DES) algorithm for encrypting data, was a boon to satellite encryption and vendors of cryptographic products. On the other hand, the two satellite encryption-related FIPSs most recently produced by NIST (FIPS-185, the Escrowed Satellite Encryption Standard (EES), and FIPS-186, the Digital Signature Standard (DSS)) have met with a less favorable response. Neither was consistent with existing de facto industry standards or practice, and both met with significant negative response from private industry and users.

The promulgation of the EES and the DSS, as well as current Administration plans to promulgate a modification of the EES to accommodate escrowed satellite encryption for data storage and communications. Another FIPS for key escrow to performance requirements for escrow agents and for escrowed encryption products, has generated a mixed market reaction. Some companies see the promulgation of these standards as a market opportunity, while others see these standards as creating yet more confusion and uncertainty in pushing escrowed satellite encryption on a resistant market.

The Government Procurement Process

Government procurement occurs in two domains. One domain is special-purpose equipment and products, for which government is the only consumer. Such products are generally classified in certain ways; weapons and military-grade satellite encryption are two examples. The other domain is procurement of products that are useful in both the private and public sectors.

Where equipment and products serve both government and private sector needs, in some instances the ability of the government to buy in bulk guarantees vendors a large enough market to take advantage of mass production, thereby driving down the unit costs of a product for all consumers. Through its market power, government has some ability to affect the price of products that are offered for sale on the open market. Furthermore, acceptance by the government is often taken as a *seal of approval* for a given product that reassures potential buyers in the private sector.

History offers examples with variable success in promoting the widespread public use of specific information technologies through the use of government standards. The DES was highly successful; DES was first adopted as a cryptographic standard for federal use in 1975. Since then, its use has become commonplace in cryptographic applications around the world, and many implementations of DES now exist worldwide.

A less successful standard is GOSIP, the Government OSI Profile, FIPS-146. The GOSIP was intended to specify the details of an OSI configuration for use in the government so that interoperable OSI network products could be procured from commercial vendors and to encourage the market development of products. The GOSIP has largely failed in this effort, and network products based on the TCP/IP protocols now dominate the market.

> **Note:** OSI refers to Open Systems Interconnect, a standardized suite of international networking protocols developed and promulgated in the early 1980s.

In the case of the EES, the government chose not to seek legislation outlawing satellite encryption without features for exceptional access, but chose instead to use the EES to influence the marketplace for encryption. This point was acknowledged by Administration officials to NSA on a number of occasions. Specifically, the government hoped that the adoption of the EES to ensure secure satellite communications within the government and for communications of other parties with the federal government would lead to a significant demand for EES-compliant devices, thus making possible production in larger quantities and thereby driving unit costs down and making EES-compliant devices more attractive to other users. A secondary effect would be the fact that two nongovernmental parties wishing to engage in secure communications would be most likely to use EES-compliant devices if they already own them rather than purchase other devices. As part of this strategy to influence the market, the government persuaded AT&T in 1992 to base a secure telephone on the EES.

In the case of the Fortezza card, the large government procurement for use with the Defense Messaging System may well lower unit costs sufficiently so that vendors of products intended solely for the commercial nondefense market will build support for the Fortezza card into their products. Given the wide availability of PC-Card slots on essentially all notebook and laptop computers, it is not inconceivable that the security advantages offered by hardware-based authentication would find a wide commercial market. At the same time, the disadvantages of hardware-based cryptographic functionality discussed in Chapter 13 would remain as well.

Implementation of Policy: Fear, Uncertainty, Doubt, Delay, Complexity

The implementation of policy contributes to how those affected by policy will respond to it. This important element is often unstated. It refers to the role of government in creating a climate of predictability. A government that speaks with multiple voices on a question of policy, or one that articulates isolated elements of policy in a piecemeal fashion, or one that leaves the stakeholders uncertain about what is or is not permissible, creates an environment of fear, uncertainty, and doubt that can inhibit action. Such an environment can result from a deliberate choice on the part of policymakers, or it can be inadvertent, resulting from overlapping and/or multiple sources of authority that may have at least partial responsibility for the policy area in question. Decisions made behind closed doors and protected by government security classifications tend to reinforce the concerns of those who believe that fear, uncertainty, and doubt are created deliberately rather than inadvertently.

The National Security Agency observes that the satellite encryption policy has indeed been shrouded in secrecy for many years and that many agencies have partial responsibility in this area. It also believes that fear, uncertainty, and doubt are common in the marketplace. For example, the introduction of nonmarket-driven standards such as the DSS and the EES may have created market uncertainty that impeded the rapid proliferation of high-quality products with encryption capabilities both internationally and domestically. Uncertainty over whether or not the federal government would recertify the DES as a FIPS has plagued the marketplace in recent years, because withdrawal of the DES as a FIPS could cause considerable consternation among some potential buyers who might suddenly be using products based on a decertified standard, although in fact the government has recertified the DES in each case. On the other hand, the DES is also a standard of the American National Standards Institute and the American Banking Association, and if these organizations retain their endorsement of the DES, the DES will arguably represent a viable algorithm for a wide range of products.

Many parties in industry believe that the complexity and opacity of the decision making process with respect to satellite encryption are major contributors to this air of uncertainty. Of course, the creation of uncertainty may be desirable from the perspective of policymakers if their goal is to retard action in a given area. Impeding the spread of high-quality poducts with satellite encryption capabilities internationally is the stated and explicit goals of export controls. On the domestic front, impeding the spread of high-quality products with encryption capabilities has been a desirable outcome from the standpoint of senior officials in the law enforcement community.

A very good example of the impact of fear, uncertainty, and doubt on the marketplace for satellite encryption can be found in the impact of government action (or more precisely, inaction) with respect to authentication. As noted in Chapter 3, satellite encryption supports digital signatures, a technology that provides high assurance for both data integrity and user authentication. However, federal actions in this area have led to considerable controversy. One example is that the federal government failed to adopt what was (and still is) the de facto commercial standard algorithm on digital signatures, namely, Rives-Shamir-Adleman (RSA) algorithm. Government sources told NSA that the fact that the RSA algorithm is capable of providing strong confidentiality as well as digital signatures was one reason that the government deemed it inappropriate for promulgation as a FIPS.

> **Note:** The specific concern was that widespread adoption of RSA as a signature standard would result in an infrastructure that could support the easy and convenient distribution of DES keys. The two other reasons for the government's rejection of RSA were the desire to promulgate an appoach to digital signatures that would be royalty-free (RSA is a patented algorithm) and the desire to reduce overall system costs for digital signatures.

Furthermore, the government's adoption of the Digital Signature Standard (DSS) in 1993 occurred despite widespread opposition from industry to the specifics of that standard. In other words, the DSS is based on an unclassified algorithm known as the Digital Signature Algorithm (DSA) that does not explicitly support confidentiality.

However, the DSS and its supporting documentation do amount to U.S. government endorsement of a particular one-way hash fuction, and document in detail how to generate the appropriate number-theoretic constants needed to implement it. Given this standard, it is possible to design a confidentiality standard that is as secure as the DSS. Actually, the DSS is a road map to a confidentiality standard, although it is not such a standard explicitly. Whether an ersatz confidentiality standard would pass muster in the commercial market remains to be seen.

R&D Funding

An agency that supports research (and/or conducts such research on its own in-house) in a given area of technology is often able to shape the future options from which the private sector and policymakers will choose. For example, an agency that wishes to maintain a monopoly of expertise in a given area may not fund promising research proposals that originate from outside. Multiple agencies active in funding a given area may thus yield a broader range of options for future policymakers.

In the context of satellite encryption and computer and satellite communications security, it is relevant that the National Security Agency (NSA) has been the main supporter and performer of R&D in this area. The NSA's R&D orientation has been, quite properly, on technologies that would help it to perform more effectively and efficiently its two basic missions: (1) defending national security by designing and deploying strong encryption to protect classified information; and (2) performing signals intelligence against potential foreign adversaries. In the information security side of the operation, NSA-developed technology has extraordinary strengths that have proven well suited to the protection of classified information relevant to defense or foreign policy needs.

Note: It is important to distinguish between R&D undertaken internally and externally to NSA. Internal R&D work can be controlled and kept private to NSA. By contrast, it is much more difficult to control the extent to which external R&D work is disseminated. Thus, decisions regarding specific external satellite encryption-related R&D projects could promote or inhibit public knowledge of cryptography.

How useful such technologies will prove for corporate information security remains to be seen. Increasing needs for information security in the private sector suggest that NSA technology may have much to offer, especially if such technology can be made available to the private sector without limitation. At the same time, the environment in which private-sector information-security needs are manifested may be different enough from the defense and foreign policy worlds that these technologies may not be particularly relevant in practice to the private sector. Furthermore, the rapid pace of commercial developments in information technology may make it difficult for the private sector to use technologies developed for national security purposes in a less rapidly changing environment.

These observations suggest that commercial needs for satellite encryption technology may be able to draw on NSA technologies for certain applications, and most certainly will draw on nonclassified R&D work in cryptography (both in the U.S. and abroad). Even the latter will have a high degree of sophistication. Precisely how the private sector will draw on these two sources of technology will depend on policy decisions to be made in the future. Finally, it is worth noting that nonclassified research on satellite encryption appearing in the open literature has been one of the most important factors leading to the dilemma that policymakers face today with respect to cryptography.

Patents and Intellectual Property

A number of patents involving satellite encryption have been issued. Patents affect satellite encryption because patent protection can be used by both vendors and governments to keep various patented approaches to cryptography out of broad use in the public domain.

The DES, first issued in 1977, is an open standard, and the algorithm it uses is widely known. According to NIST, devices implementing the DES may be covered by U.S. and foreign patents issued to IBM (although the original patents have by now expired). However, IBM granted nonexclusive, royalty-free licences under the patents to make, use, and sell apparatus that complies with the standard.

RSA Data Security Inc. (RSA) holds the licensing rights to RC2, RC4, and RC5, which are variable-key-length ciphers developed by Ronald

Rivest (the R in RSA); RC2 and RC4 are not patented, but rather are protected as trade secrets (although their algorithms have been published on the Internet without RSA's approval). Moreover, RSA has applied for a patent for RC5 and has proposed it as a security standard for the Internet. Another alternative for satellite data encryption is IDEA, a block cipher developed by James Massey and Xueija Lai of the Swiss Federal Institute of Technology (ETH), Zurich. The patent rights to IDEA are held by Ascom Systec AG, a Swiss firm; IDEA is implemented in the software application, PGP.

In addition to the previously discussed patents (which address symmetric-key encryption technologies), there are several important patent issues related to public-key satellite encryption. The concept of public-key satellite encryption, as well as some specific implementing methods are covered by U.S. Patents 4,200,770 (M. Hellman, W. Diffie, and R. Merkle, 1980) and 4,218,582 (M. Hellman and R. Merkle, 1980), both of which are owned by Stanford University. The basic patent for the RSA public-key cryptosystem, U.S. Patent 4,405,829 (R. Rivest, A. Shamir, and L. Adleman, 1983), is owned by the Massachusetts Institute of Technology. U.S. Patent 4,218,582 has counterparts in several other countries. These basic public-key patents and related ones have been licensed to many vendor worldwide. With the breakup of the partnership that administered the licensing of Stanford University's and MIT's patents, the validity of the various patents has become the subject of current litigation. In any event, the terms expired in 1997 for the first two of the previously discussed patents and the third is due to expire in 2001.

Note: In 1994, Congress changed patent terms from 17 years after issuance to 20 years from the date of filing the patent application. However, applications for these patents were filed in or before 1977, and so they will not be affected.

In 1994, NIST issued the Digital Signature Standard, FIPS 186. The DSS uses the NIST-developed Digital Signature Algorithm, which according to NIST is available for use without a licence. However, during the DSS's development, concern arose about whether the DSS might infringe on the public-key patents cited previously, as well as a patent related to signature verification held by Claus Schnorr of Goethe University in Frankfurt, Germany. The NIST asserts that the DSS does not infringe on any of these patents At the least, U.S. government users have the right to use public-key satellite encryption without paying a licence fee for the Stanford and MIT patents because the concepts were developed at these universities with federal reseach support. However, there remains some disagreement about whether commercial uses of the DSS (for example, in a public-key infrastructure) will require a license from one or more of the various patent holders.

A potential patent dispute regarding the key-escrow features of the EES may have been headed off by NIST's negotiation of a nonexclusive licensing agreement with Silvio Micali in 1994. Micali has patents that are relevant to dividing a key into components that can be separately safeguarded (by escrow agents) and later combined to recover the original key.

A provision of the U.S. Code (Title 35, US Code 181) allows the Patent and Trademark Office (PTO) to withold a patent and order that the invention be kept secret if publication of the patent is detrimental to national security. Relevant to satellite encryption is the fact that a patent application for the Skipjack encryption algorithm was filed on February 7, 1994. This application was examined and all of the claims allowed, and notification of the algorithm's patentability was issued on March 28, 1995. Based on a determination by NSA, the Armed Services Patent Advisory Board issued a secrecy order for the Skipjack patent application. The effect of the secrecy order is that even though Skipjack can be patented, a patent will not be issued until the secrecy order is rescinded. Because applications are kept in confidence until a patent is issued, no uninvolved party can obtain any information concerning the application. In this way, the patentability of the algorithm has been established without having to disclose the detailed information publicly. As Title 35 USC 181 also provides that the PTO can rescind the secrecy order upon notification that publication is no longer detrimental to national security, compromise ad subsequent public revelation of the Skipjack algorithm (through reverse-engineering of a Clipper chip) might well cause a patent to be issued for Skipjack that would give the U.S. government control over its subsequent use in products.

Formal and Informal Arrangements with Various Other Governments and Organizations

International agreements can be an important part of national policy. For example, for many years the CoCom nations cooperated in establishing a common export control policy on militarily significant items with civilian purposes, including encryption.

International agreements can take a variety of different forms. The most formal type of agreement is a treaty between (or among) nations that specifies the permissible, required, and prohibited actions of the various nations. Treaties require ratification by the relevant national political bodies as well as signature before entry into force. In the U.S. treaties must be approved by the Senate by a two-thirds vote. Sometimes treaties are self-executing, but often they need to be followed by implementing legislation enacted by the Congress in the normal manner for legislation.

Another type of agreement is an executive agreement. In the U.S., executive agreements are, as the name implies, entered into by the executive branch. Unlike the treaty, no Senate ratification is involved, but the executive branch has frequently sought approval by a majority of both houses of Congress. For all practical purposes, executive agreements with other countries bind the U.S. in international law just as firmly as treaties, although the treaty may carry greater weight internally due to the concurrence by a two-thirds vote of the Senate. Executive agreements can also be changed with much greater flexibility than treaties.

Finally, nations can agree to cooperate through diplomacy. Even though cooperation is not legally required under such arrangements, informal understandings can work very effectively so long as relationships remain good and the countries involved continue to have common goals. In fact, informal understanding is the main product of much diplomacy and is the form that most of the world's business between governments takes place. For example, although the U.S. maintains formal mutual legal assistance treaties with a number of nations, U.S. law enforcement agencies cooperate (sometimes extensively) with foreign counterparts in a much larger number of nations. Indeed, in some instances, such cooperation is stronger, more reliable, and more extensive than is the case with nations that are a party to a formal mutual legal assistance treaty with the U.S.

Note: The more formal the agreement, the more public is the substance of the agreement. Such publicity often leads to attention that may compromise important and very sensitive matters, such as the extent to which a nation supports a given policy position or the scope and nature of a nation's capabilities. When informal arrangements are negotiated and entered into force, they may not be known by all citizens or even by all parts of the governments involved. Because they are less public, informal arrangements also allow more latitude for governments to make decisions on a case-by-case basis. In conducting negotiations that may involve sensitive matters or agreements that may require considerable flexibility, governments are often inclined to pursue more informal avenues of approach.

Certification and Evaluation

Analogous to *Good Housekeeping* seals of approval or *check ratings* for products reviewed in *Consumer Reports*, independent testing and certification of products can provide assurance in the commercial marketplace that a product can indeed deliver the services and functionality that it purports to deliver. For example, the results of government crash tests of automobiles are widely circulated as data relevant to consumer purchases of automobiles. Government certification that a commercial airplane is safe to fly provides significant reassurance to the public about flight safety. At the same time, while evaluation and certification would in principle help users to avoid products that implement a sound algorithm in a way that undermines the security offered by the algorithm, the actual behavior of users demonstrates that certification of a product is not necessarily a selling point. Many of the DES products in the U.S. have never been evaluated relative to FS-1027 or FIPS 140-1, and yet such products are used by many parties.

The government track record in the satellite encryption and computer security domain is mixed. For example, a number of DES products were evaluated with respect to FS-1027 (the precursor to FIPS 140-1) over several years and a number of products were certified by NSA. For a time, government agencies purchased DES hardware only if it met FS-1027, or FIPS-140. Commercial clients often required compliance because it provided the only assurance that a product embodying DES was secure in a broader sense. In this case, the alignment between government and commercial security requirements seems to have been reasonably good and thus this program had some success. Two problems with this evaluation program were that it addressed only hardware and that it lagged in allowing use of public-key management technology in products (in the absence of suitable standards).

A second attempt to provide product evaluation was represented by the National Computer Security Center (NCSC), which was established by the Department of Defense for the purpose of certifying various computer systems for security. The theory underlying the center was that the government needed secure systems but could not afford to build them. The quid pro quo was that industry would design and implement secure operating systems that the government would test and evaluate at no cost to industry. Operating systems meeting government requirements would receive a seal of approval.

Although the NCSC still exists, the security evaluation program it sponsors, the Trusted Product Evaluation Program (TPEP), has more or less lapsed into disuse. In the judgment of many, the TPEP was a relative failure because of an underlying premise that the information security problems of the civil government and the private sector were identical to those of the defense establishment. In fact, the private sector has for the most part found a military approach to computer security inadequate for its needs. A second major problem was that the time scale of the evaluation process was much longer than the private sector could tolerate. Products that depended on NCSC evaluation would reach market already on the road to obsolescence—perhaps superseded by a new version to which a given evaluation would not necessarily apply. In late 1995, articles in the trade press reported that the Department of Defense was attempting to revive the evaluation program in a way that would involve private contractors.

A recent attempt to provide certification services is the Cryptographic Module Validation Program (CMVP) to test products for conformance to FIPS 140-1, *Security Requirements for Cryptographic Modules*; FIPS 140-1 provides a broad framework for all NIST satellite encryption standards, specifying design, function, and documentation requirements for cryptographic modules (including hardware, software, *firmware*, and combinations thereof) used to protect sensitive, unclassified information in computer and telecommunication systems. The CMVP was established in July 1995 by NIST and the Communications Security Establishment of the government of Canada.

The validation program is currently optional: agencies may purchase products based on the vendor's written assurance of compliance with the standard. However, U.S. federal procurement now requires cryptographic products to be validated by an independent, third party. Under the program, vendors submit their product for testing by an independent, NIST-accredited laboratory.

> **Note:** As of September 1995, the National Institute of Standards and Technology's National Voluntary Laboratory Accreditation Program had accredited three U.S. companies as competent to perform the necessary procedures: CygnaCom Solutions Laboratory (McLean, VA), DOMUS Software Limited (Ottawa, Canada), and InfoGard Laboratories (San Luis Obispo, CA.)

Such a laboratory evaluates both the product and its associated documentation against the requirements in FIPS 140-1. The NIST has also specified test procedures for all aspects of the standard. Examples include attempting to penetrate tamper-resistant coatings and casings, inspecting software source code and documentation, attempting to bypass protection of stored secret keys, and statistically verifying the performance of random number generators. The vendor sends the results of independent tests to NIST, which determines whether these results show that the tested product complies with the standard and then issues validation certificates for those products. Time will tell whether the CMVP will prove more successful than the NCSC.

Nonstatutory Influence

By virtue of its size and role in society, government has considerable ability to influence public opinion and to build support for policies. In many cases, this ability is not based on specific legislative authority, but rather on the use of the *bully pulpit*. For example, the government can act in a convening role to bring focus and to stimulate the private sector to work on a problem.

Note: One advantage of government's acting in this way is that it may provide some assurance to the private sector that any coordinated action they may take in response to government calls for action will be less likely to be interpreted by government as a violation of antitrust provisions.

The bully pulpit can be used to convey a sense of urgency that is tremendously important in how the private sector reacts, especially large companies that try to be good corporate citizens and responsive to informal persuasion by senior government officals. Both vendors and users can be influenced by such authority.

Note: For example, in responding favorably to a request by President Clinton for a particular action in a labor dispute, the chairman of American Airlines noted: *He [President Clinton] is elected leader of the country. For any citizen or any company or any union to say 'No, I won't do that' to the President requires an awfully good reason.*

In the security domain, the Clinton Administration has sponsored several widely publicized public meetings to address security dimensions of the national information infrastructure (NII). These were meetings of the NII Security Issues Forum, held in 1994 and 1995. They were announced in the *Federal Register* and were intended to provide a forum in which members of the interested public could air their concerns about security.

Note: Office of Management and Budget press release, National Information Infrastructure Security Issues Forum Releases NII Security: The Federal Role, Washington, DC, June 14, 1995. The subjects of these meetings were Commercial Security on the NII, which focused on the need for intellectual property rights protection in the entertainment, software, and computer industries; Security of Insurance and Financial Information; Security of Health and Education Information; *Security of the Electronic Delivery of Government Services and Information; Security for Intelligent Transportation Systems and Trade Information*; and *The NII: Will It Be There When You Need It?* These all address the availability and reliability of the Internet, the public switched telecommunications network, and cable, wireless, and satellite communications services.

In the satellite encryption domain, the U.S. government has used its convening authority to seek comments on various proposed cryptographic standards and to hold a number of workshops related to key escrow (discussed in Chapter 13). Many in the affected communities believe that these attempts at outreach were too few and too late to influence anything more than the details of a policy outline upon which government had already decided. A second example demonstrating government's nonstatutory influence was the successful government request to AT&T to base the 3600 Secure Telephone Unit on the Clipper chip instead of an unescrowed DES chip.

Interagency Agreements Within the Executive Branch

Given that one government agency may have expertise or personnel that would assist another agency in doing its job better, government agencies often conclude agreements between them that specify the terms and nature of their cooperative efforts. In the domain of cryptography policy, NSA's technical expertise in the field has led to memoranda of understanding with NIST and with the FBI.

The memorandum of understanding (MOU) between NIST and NSA outlines several areas of cooperation between the two agencies that are intended to implement the Computer Security Act of 1987. Joint NIST-NSA activities are described in the following Sidebar. This MOU has been the subject of some controversy, with critics believing that the MOU and its implementation cede too much authority to NSA and defenders believing that the MOU is faithful to both the spirit and letter of the Computer Security Act of 1987.

Overview of Joint NIST-NSA Activities

The National Security Agency provides technical advice and assistance to the National Institute of Standards and Technology in accordance with Public Law 100-235, the Computer Security Act of 1987. An overview of NIST-NSA activities follows.

National conference
The NIST and NSA jointly sponsor, organize, and chair the prestigious National Computer Security Conference, held yearly for the

past 20 years. The conference is attended by over 5000 people from government and private industry.

Common criteria

The NSA is providing technical assistance to NIST for the development of computer security criteria that would be used by both the civilian and defense sides of the government. Representatives from Canada and Europe are joining the U.S. in the development of the criteria.

Product evaluations

The NIST and NSA are working together to perform evaluations of computer security products. In the Trusted Technology Assessment Program, evaluations of some computer security products will be performed by NIST and its laboratories, while others will be performed by NSA. The NIST and NSA engineers routinely exchange information and experiences to ensure uniformity of evaluations.

Standards development

The NSA supports NIST in the development of standards that promote interoperability among security products. Sample standards include security protocol standards, digital signature standards, key management standards, and encryption algorithm standards (the DES, Skipjack).

Research and development

Under the Joint R&D Technology Exchange Program, NIST and NSA hold periodic technical exchanges to share information on new and ongoing programs. Research and development are performed in areas such as security architectures, labeling standards, privilege management, and identification and authentication. Test-bed activities are conducted in areas related to electronic mail, certificate exchange and management, protocol conformity, and satellite encryption technologies.[2]

The MOU between the FBI and NSA, declassified for the National Research Council, states that the NSA will provide assistance to the FBI upon request, when the assistance is consistence with NSA policy (including protection of sources and methods), and in accordance with certain administrative requirements. Furthermore, if the assistance requested is for the support of an activity that may be conducted only pursuant to a court order or with the authorization of the Attorney General, the FBI request to the NSA must include a copy of that order or authorization.

In 1995, the National Security Agency, the Advanced Research Projects Agency, and the Defense Information Systems Agency signed a memorandum of agreement (MOA) to coordinate research and development efforts in system security. This MOA provides for the establishment of the Information System Security Research Joint Technology Office (ISSR-JTO). The role of the ISSR-JTO is *to optimize use of the limited research funds available, and strengthen the responsiveness of the programs to DISA, expediting delivery of technologies that meet DISA's requirements to safeguard the confidentiality, integrity, authenticity, and availability of data in Department of Defense information sytems, provide a robust first line of defense for defensive information warfare, and permit electronic commerce between the Department of Defense and its contractors.*

Organization of the Federal Government with Respect to Information Security

The extent to which the traditional national security model is appropriate for an information infrastructure supporting both civilian and military applications is a major point of contention in the public debate. There are two schools of thought on this subject:

First, the traditional national security model should be applied to the national information infrastructure, because protecting those networks also protects services that are essential to the military. And, the role of the defense establishment is indeed to protect important components of the national infrastructure that private citizens and businesses depend upon. For example, the Joint Security Commission recommended that *policy formulation for information systems security be consolidated under a joint DOD/DCI security executive committee, and that the committee oversee development of a coherent network-oriented information systems security policy for the DOD and the Intelligence Community that could also serve the entire government.*

The second school of thought is that the traditional national security model should not be applied to the national information infrastructure, because the needs of civilian activities are so different from those of the military, and, the imposition of a national security model would impose an unacceptable burden on the civilian sector. Proponents of this view argue that the traditional national security model of information security (a top-down approach to information security management) would be very difficult to scale up to a highly heterogeneous private sector involving hundreds of millions of people and tens of millions of computers in the U.S. alone.

Role of National Security vis-à-vis Civilian Information Infrastructures

There is essential unanimity that the world of classified information (both military and nonmilitary) is properly a domain in which the DOD and NSA can and should exercise considerable influence. But moving outside this domain raises many questions that have a high profile in the public debate. Specifically, what should the DOD and NSA role be in dealing with the following categories of information:

1. Unclassified government information that is military in nature.
2. Unclassified government information that is nonmilitary in nature.
3. Nongovernment information.[3]

Policy decisions have been made that give the DOD jurisdiction in information security policy for category 1. For categories 2 and 3, the debate continues. It is clear that the security needs for business and for national security purposes are both similar (see Sidebar, "Similarities in Commercial Security Needs and National Security Needs) and different (see Sidebar, "Differences in Commercial Security Needs and National Security Needs"). In category 2, the argument is made that DOD and NSA have a great deal of expertise in protecting information, and that the government should draw on an enormous historical investment in NSA expertise to protect all government information. At the same time, NIST has the responsibility for protecting such information under the Computer Security Act of 1987, with NSA's role being one of providing technical assistance. Some commentators believe that NIST has not received resources adequate to support its role in this area.

Note: For example, the Office of Technology Assessment stated that *the current state of government security practice for unclassified information has been depressed by the chronic shortage of resources for NIST's computer security activities in fulfillment of its government-wide responsibilities under the Computer Security Act of 1987. Since enactment of the Computer Security Act, there has been no serious (adequately funded and properly staffed), sustained effort to establish a center of information-security expertise and leadership outside the defense/intelligence communities.* A similar conclusion was reached by the Board on Assessment of NIST Programs of the National Research Council, which wrote that *the Computer Security Division is severely understaffed and underfunded given its statutory security responsibilities, the growing national recognition of the need to protect unclassified but sensitive information, and the unique role the division can play in fostering security in commercial arthitectures, hardware, and software.*

Similarities in Commercial Security Needs and National Security Needs

1. Strong aversion to public discussion of security breaches. Information about threats is regarded as highly sensitive. Such a classification makes it very difficult to conduct effective user education, because security awareness depends on an understanding of the true scope and nature of a threat.
2. Need to make cost-benefit trade-offs in using security technology. Neither party can afford the resources to protect against an arbitrary threat model.
3. Strong preference for self-reliance (government relying on government, industry relying on industry) to meet security needs.
4. Strong need for high security. Both government and industry need strong encryption with no limitations for certain applications. However, the best technology and tools are often reserved for government and military use because commercial deployment cannot be adequately controlled, resulting in opportunities for adversaries to obtain and examine the systems so that they can plan how to exploit them.
5. Increasing reliance on commercial products in many domains (business, Third-World nations).

6. Increasing scale and sophistication of the security threat for businesses, which is now approaching that posed by foreign intelligence services and foreign governments.
7. Possibility that exceptional access to satellite-encrypted information and data may become important to commercial entities.[3]

Differences in Commercial Security Needs and National Security Needs

Business wants market-driven cryptographic technology. Government is apprehensive about such technology. For example, standards are a critical element of market-driven satellite encryption. Market forces and the need to respond to rapidly evolving dynamic new markets demand an approach to establishing satellite encryption standards. Businesses want standards for interoperability, and they want to create market critical mass in order to lower the cost of satellite encryption.

By its nature, the environment of business must include potential adversaries within its security perimeter. Commercial enterprises now realize that electronic delivery of their products and services to their customers will increase. They must design systems and processes explicitly so that customers can enter into transactions with considerable ease. Business strategies of today empower the customer through software and technology. Enterprise networks have value in allowing the maximum number of people to be attached to the network. Customers will choose which enterprise to enter in order to engage in electronic commerce, and making it difficult for the customer will result in loss of business. But adversaries masquerading as customers (or who indeed may be customers themselves) can enter as well. By contrast, the traditional national security model keeps potential adversaries outside the security perimeter, allowing access only to those with a real need. However, to the extent that U.S. military forces work in collaboration with forces of other nations, the security perimeter for the military may also become similarly blurred.

Business paradigms value teamwork, openness, trust, empowerment, and speed. Such values are often difficult to sustain in the national security establishment. The cultures of the two worlds are different and are reflected in, for example, the unwillingness of business to use multilevel security systems designed for military use. Such systems failed the market test, although they met Defense Department criteria for security.

National security resources (personnel with cryptographic expertise, funding) are much larger than the resources in nondefense government sectors and in private industry and universities. As a result, a great deal of cryptographic knowledge resides within the world of national security. Industry wants access to this knowledge to ensure appropriate use of protocols and strong algorithms, and development of innovative new products and services.

National security places considerable emphasis on confidentiality as well as on authentication and integrity. Today's commercial enterprises stress authentication of users and data integrity much more than they stress confidentiality (although this balance may shift in the future). For example, improperly denying a junior military officer access to a computer facility may not be particularly important in a military context, whereas improperly denying a customer access to his or her bank account because of a faulty authentication can pose enormous problems for the bank.

While both businesses and national security authorities have an interest in safeguarding secrets, the tools available to businesses to discourage individuals from disclosing secrets (generally civil suits) are less stringent than those available to national security authorities (criminal prosecution).[4]

In category 3, the same argument is made with respect to nongovernment information on the grounds that the proper role of government is to serve the needs of the entire nation. A second argument is made that the military depends critically on nongovernment information infrastructures (the public switched-telecommunications network) and that it is essential to protect those networks not just for civilian use but also for military purposes.

Note: NSA does not have broad authority to assist private industry with information security, although it does conduct for industry, upon request, unclassified briefings related to foreign information security threats; NSD-42 also gives NSA the authority to work with private industry when such work involves national security information systems used by private industry.

Other Government Entities with Influence on Information Security

As previously noted, NSA has primary responsiblity for information security in the classified domain, while NIST has primary responsibility for information security in the unclassified domain, but for government information only. No organization or entity within the federal government has the responsibility for promoting information security in the private sector.

The Security Policy Board (SPB) does have a coordination function. Specifically, the charge of the SPB is to consider, coordinate, and recommend for implementation to the President's policy directives for U.S. security policies, procedures, and practices, including those related to security for both classified and unclassified government information. The SPB is intended to be the principal mechanism for reviewing and proposing legislation and executive orders pertaining to security policy, procedures, and practices. The Security Policy Advisory Board provides a nongovernmental perspective on security policy initiatives to the SPB and independent input on such matters to the President. The SPB does not have operational responsibilities.

Other entities supported by the federal government have some influence over information security, though little actual policymaking authority. These include:

- ◆ The Computer Emergency Response Team (CERT).
- ◆ The Information Infrastructure Task Force's (IITF) National Information Infrastructure Security Issues Forum.
- ◆ The Computer System Security and Privacy Advisory Board (CSSPAB).
- ◆ The National Counterintelligence Center (NACIC).

Computer Emergency Response Team (CERT)

The CERT was formed by the Defense Advanced Research Projects Agency (DARPA) in November 1988 in response to the needs exhibited during the Internet worm incident; CERT's charge is to work with the Internet community to facilitate its response to computer security events involving Internet hosts, to take proactive steps to raise the community's awareness of computer security issues, and to conduct research targeted at improving the security of existing systems. The

CERT offers around-the-clock technical assistance for responding to computer security incidents, educates users regarding product vulnerability through technical documents and seminars, and provides tools for users to undertake their own vulnerability analyses.

Information Infrastructure Task Force's (IITF) National Information Infrastructure Security Issues Forum

The form is charged with addressing institutional, legal, and technical issues surrounding security in the NII. A draft report issued by the forum proposes federal actions to address these issues. The intent of the report, and of the Security Issues Forum, more generally, is to stimulate a dialogue on how the federal government should cooperate with other levels of government and the private sector to ensure that participants can trust the NII. The draft report proposes a number of security guidelines (proposed NII security tenets), the adoption of Organization of Economic Cooperation and Development security principles for use on the NII, and a number of federal actions to promote security.

Computer System Security and Privacy Advisory Board (CSSPAB)

The CSSPAB was created by the Computer Security Act of 1987 as a statutory federal public advisory committee. The law provides that the board shall identify emerging managerial, technical, administrative, and physical safeguard issues relative to computer systems security and privacy; advise the National Institute of Standards and Technology and the Secretary of Commerce on security and privacy issues pertaining to federal computer systems; and report its findings to the Secretary of Commerce, the Directors of the Office of Management and Budget, the National Security Agency, and the appropriate committees of the Congress. The board's scope is limited to federal computer systems or those operated by a contractor on behalf of the federal government and which process sensitive but unclassified information. The board's authority does not extend to private sector systems, systems that process classified information, or DOD unclassified systems related to military or intelligence missions as covered by the Warner Amendment (10 USC 2315). The activities of the board bring it into contact with a broad cross section of the nondefense agencies and departments. Consequently, it often deals with latent policy considerations and societal consequences of information technology.

National Counterintelligence Center (NACIC)

Established in 1994 by Presidential Decision Directive NSC-24, NACIC is primarily responsible for coordinating national-level counterintelligence activities, and it reports to the National Security Council. Operationally, the NACIC works with private industry through an industry council (consisting of senior security officials or other senior officials of major U.S. corporations) and sponsors counterintelligence training and awareness programs, seminars, and conferences for private industry. The NACIC also produces coordinated national-level, all-source, foreign intelligence threat assessments to support private sector entities having responsibility for the protection of classified, sensitive, or proprietary information, as well as such assessments for government use. In addition, a number of private organizations (trade or professional groups) are active in information security.

International Dimensions of Satellite Encryption Policy

The satellite encryption policy of the U.S. must take into account a number of international dimensions. Most importantly, the U.S. does not have the unquestioned dominance in the economic, financial, technological, and political affairs of the world as it might have had at the end of World War II. Indeed, the U.S. economy is increasingly intertwined with that of other nations. To the extent that these economically significant links are based on satellite communications that must be secure, satellite encryption is one aspect of ensuring such security. Differing national policies on satellite encryption that lead to difficulties in communicating internationally work against overall national policies that are aimed at opening markets and reducing commercial and trade barriers.

Other nations have the options to maintain some form of export controls on satellite encryption, as well as controls on imports and use of cryptography. Such controls form part of the context in which the U.S. satellite encryption policy must be formulated. Specifically, foreign export control regimes that are more liberal than that of the U.S. have the potential to undercut U.S. export control efforts to limit the spread of satellite encryption. On the other hand, foreign controls on imports and use of satellite encryption could vitiate relaxation of U.S. export control laws. Indeed, relaxation of U.S. export controls laws might well prompt a larger number of nations to impose additional barriers on the import and use of satellite encryption within their borders. Finally, a number of other nations have no explicit laws regarding the use of satellite encryption, but nevertheless have tools at their disposal to discourage its use. Such tools include laws related to the postal, telephone, and telegraph (PTT) system, laws related to content carried by electronic media, laws related to the protection of domestic industries that discourage the entry of foreign products, laws related to classification of patents, and informal arrangements related to licensing of businesses.

As a first step in harmonizing satellite encryption policies across national boundaries, the Organization for Economic Cooperation an Development (OECD) held a December 1995 meeting in France among member nations to discuss how these nations were planning to cope with the public policy problems posed by satellite encryption. What the Paris meeting made clear is that many OECD member nations are starting to come to grips with the public policy problems posed by satellite encryption, but that the dialog on harmonizing policies across national borders has not yet matured. Moreover, national policies are quite fluid at this time, with various nations considering different types of regulation regarding the use, export, and import of satellite encryption.

Summary

While export controls and escrowed encryption are fundamental pillars of the current national satellite encryption policy, many other aspects of government action also have some bearing on it: The Communications Assistance for Law Enforcement (Digital Telephony) Act calls attention to the relationship between access to a communications stream and government access to the plaintext associated with that digital stream. The former problem must be solved (and was solved by the CALEA for telephone communications) before the latter problem is relevant.

The government can influence the deployment and use of satellite encryption in many ways. Federal Information Processing Standards often set a *best practice* standard for the private sector, even though they have no official standing outside government use. By assuring large-volume sales when a product is new, government procurement practices can reduce the cost of preferred satellite encryption products to the private sector, giving these products a price advantage over possible competitors. Policy itself can be implemented in ways that instill action-inhibiting uncertainty in the private sector. Government R&D funding and patents on cryptographic algorithms can narrow technical options to some degree. Formal and informal arrangements with various other governments and organizations can promote various policies or types of cooperation. Product certification can be used to provide the information necessary for a flourishing free market in products with satellite encryption capabilities. Convening authority can help to establish the importance of a topic or approach to policy.

In some ways, the debate over the national satellite encryption policy reflects a tension in the role of the national security establishment with respect to information infrastructures that are increasingly important to civilian use. In particular, the use of satellite encryption has been the domain of national security and foreign policy for most of its history, a history that has led to a national satellite encryption policy that today has the effect of discouraging the use of satellite encryption in the private sector.

From Here

Chapter 14 discussed other dimensions of the national satellite encryption policy, including the Digital Telephony Act of 1995 (aka the Communications Assistance for Law Enforcement Act) and a variety of other levers used in national cryptography policy that do not often receive much attention in the debate. Chapter 15 discusses the following: The planning process; details; implementing the plan; and, the commanding uplink (receiving) process.

Endnotes

1 G. A. Keyworth II and David E. Colton, Esq., The Computer Revolution, Encryption and True Threats to National Security (Permission granted to reproduce as long as acknowledgment is made. Richard F. O'Donnell, Editor), The Progress & Freedom Foundation, 1301 K Street, N.W., Suite 650 West, Washington, DC 20005, 1996.

2 "Information Security and Privacy in Network Environments," National Security Agency, April 1994 (as printed in U.S. Congress, Office of Technology Assessment, OTA-TCT-606, U.S. Government Printing Office, Washington, DC, September 1994, Box 4-8, p. 165).

3 © 1996 by the National Academy of Sciences. Courtesy of the National Academy Press, 2101 Constitution Avenue, NW, Washington D.C. 20418.

4 "Information Security and Privacy in Network Environments," National Security Agency, April 1994 (as printed in U.S. Congress, Office of Technology Assessment, OTA-TCT-606, U.S. Government Printing Office, Washington, DC, September 1994, Box 4-8, p. 165).

5 Ibid.

Part Three

Implementing Satellite Encryption

End-to-End Encrypted Data Flow — The Uplink Part

This chapter, together with the next one (the downlink part), builds upon many of the topics that have been discussed in previous chapters and shows how they work together as a whole system to implement Satellite Encryption. At one end of the system are the people who plan and operate the satellites or spacecrafts. At the other end of this pipeline is the public. The results are delivered to your doorstep in the form of news and images on your TV and your computer, in schoolbooks and scientific journals, in coffee table books and technical texts as discussed in Chapter 16, "End-to-End Encrypted Data Flow—The Downlink Part."

Uplink is a general term given to a radio signal sent *up* from Earth to a spacecraft or satellite as shown in Figure 15.1.[1] The signal is a highly directional microwave radio beam, normally in frequencies around 3 billion to 7 billion hertz (3 to 7 gigahertz) and at power levels of less than 20,000 watts. The purposes for an uplink include commanding (which we we'll concentrate on in this chapter), tracking, and radio science.

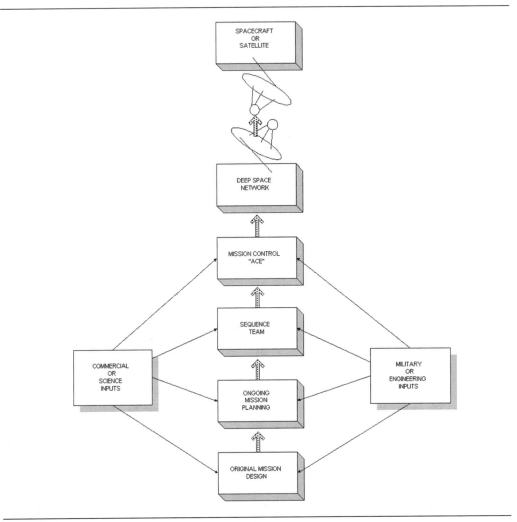

Figure 15.1 The commanding uplink process.

Downlink, as you would expect, is a general term for a radio signal sent *down* to Earth from a spacecraft or satellite. The signal is usually in the same radio-frequency bands as the uplink. Downlink serves the purposes of tracking, telemetry, and radio science.

Note: The radio signal transmitted to a spacecraft or satellite is known as uplink. The transmission from spacecraft or satellite to Earth is downlink. Uplink or downlink may consist of a pure radio frequency (RF) tone, called a carrier, or carriers may be modulated to carry encrypted information in each direction. Commands transmitted to a spacecraft or satellite are sometimes referred to as an upload. Encrypted communications with a spacecraft or satellite involving only a downlink are called one-way. When an uplink is being received by the spacecraft or satellite at the same time a downlink is being received at Earth, the communications mode is called *two-way*.

The Planning Process

On the radio-frequency uplink to the spacecraft or satellite, encrypted command signals are sent by the user that the spacecraft or satellite receives, decodes and acts upon. However, before that can be done, a bit of planning is required because it is a complex process with millions of possibilities to consider and many decisions to be made. For example, the process has to take into consideration what scientific, commercial or military observations the spacecraft or satellite is going to make, when to make them, exactly when to fire its rocket engine or thrusters, how the spacecraft or satellite must be oriented when it fires them, what kinds of measurements to send to Earth, and what encrypted data rates to use. Therefore the commanding process as shown in Figure 15.1 begins long before sequences of commands are actually placed on the uplink.

The process starts with the scientists (also called investigators or users) associated with the spacecraft's or satellite's instruments. These scientists are typically located at universities and aerospace companies worldwide and are usually supported by their own research assistants and graduate students, as well as by the military. To carry out the scientists' investigations requires that the spacecraft or satellite operate its instruments to make observations, and perform classified military experiments under just the right conditions and at just the right time. To match the desired scientific investigations with exactly the right encrypted commands requires a good amount of information about the spacecraft or satellite and about the target system (ground-based or space-based). For example, exactly where will the spacecraft or satellite be at a particular time, and what will its orientation in space be? When, and in what direction, must it turn to capture a view of the targets of interest? How long must an instrument's shutter remain open to obtain the right exposure? What other settings will the instrument need?

Determining the path, or trajectory, of the spacecraft or satellite is the job of a team of navigators. They obtain their information from the intricate process of tracking the spacecraft or satellite. They determine where the spacecraft or satellite will be at any given time in relation to the classified objects (space- or ground-based) to be studied.

Of course, to do that, they need to know where those targeted classified objects are going to be: The predicted locations of space-based targeted objects as ephemerides (singular, *ephemeris*—meaning a table giving the coordinates of a celestial body or other space-based target at a number of specific times during a given period). These are maintained by the worldwide astronomical and military communities. The spacecraft or satellite navigators take the ephemerides and the spacecraft or satellite tracking encrypted data into account in their processes, using highly developed computer programs. After lots of number-crunching, they provide the predictions necessary for planning how and when the spacecraft or satellite will be able to make observations. These programs are also made to provide information on what the spacecraft or satellite must occasionally be commanded to do in order to make small adjustments in its trajectory so it will be exactly where it needs to be at the proper time.

More Details

Now, we know where everything is. And, from the ephemeris and encrypted trajectory information, we also know what the spacecraft's or satellite's attitude (position) will be at any given time, based on the last sequence of encrypted commands sent to the spacecraft or satellite. This is needed so that the spacecraft or satellite can be commanded to point its instruments in the right direction and point the communications antenna toward Earth to beam these valuable bits of encrypted data down to the waiting scientists or military officials.

There is a huge amount of encrypted information that needs to be considered for planning operations. But there is more. We need to consider many details about the spacecraft's or satellite's onboard subsystems, too. How much power will be required to operate the desired instrument? Will that power be available at the required time, or will we first need to turn off something else? What other *consumables*, such as electrical power and propellant will be affected? If we rotate the spacecraft or satellite, will we lose communication with Earth? If so, for how long? Can we afford to do without tracking encrypted data for that long at that point in time? When the spacecraft or satellite rotates, is there any danger of sunlight entering a sensitive instrument and causing damage? Exactly where in memory should the encrypted data from this observation be stored? Do we need to do any onboard *housekeeping*? Of course, a lot of planning for the achievement of major goals is accomplished many years before launch, and has been incorporated into the spacecraft's or satellite's design, the choice of launch vehicle, and the original design of the overall mission. However, all the planning cannot be done at once; it has to be an ongoing process.

The resources to be considered during planning include special opportunities for making observations, time available from the Deep Space Network tracking system (DSN), human workforce availability, spacecraft or satellite consumables, and so on. It is not uncommon for the various scientific investigators or military personnel to desire conflicting observations, nor is it uncommon for more than one spacecraft or satellite to desire conflicting use of the DSN's precious time — it tracks *Voyager 1, Voyager 2, Galileo, Mars Global Surveyor, Mars Pathfinder, Clementine* and many others. Conflicts are resolved in high-level meetings with the principal people concerned.

Implementing the Plan

Once the planners have done their job, all the details of a plan of action for a given period are passed to another team (usually caled a sequence team) that is responsible for creating the actual encrypted commands to be uplinked. This team relies on highly advanced computer programs to help with such tasks as selecting and time-tagging the proper encrypted commands, placing them in the correct order, checking that no operating constraints are violated and making sure that all of the instructions to the spacecraft or satellite will fit into the spacecraft's or satellite's available computer memory. The encrypted commands produced for a particular time period are typically called a sequence.

The sequence is then passed to the team member (a person identified as *Ace*) who will actually send it to the spacecraft or satellite. In addition to the sequence, there may be other encrypted commands, usually called real-time encrypted commands, the Ace needs to send. These are typically much shorter than sequences, and sometimes need to be sent on short notice, such as when scientists or military personnel want to make quick adjustments to their instruments' states.

The Ace chooses or checks the proper time for transmitting the encrypted command loads and real-time encrypted commands to the spacecraft during an appropriate DSN tracking period. Next, the encrypted data commands are formatted for transmission and sent electronically using the ground communications facility (GCF) to the proper site in the DSN, where they are then loaded onto disk in the remote command computer. The GCF uses a combination of communications satellites, conventional surface means and undersea cables to electronically link the remote DSN sites.

Ready to Go

At this point, the sequence of encrypted data commands is finally ready to go to the spacecraft or satellite. The Ace makes sure the DSN's transmitter is on, radiating a carrier signal to the spacecraft or satellite. Then he or she manipulates the command computer under remote control from the Jet Propulsion Laboratory, causing the sequence to be modulated (*vocalized*, if you will) onto the uplink. The encrypted commands begin their journey to the spacecraft or satellite, which may take hours. The spacecraft or satellite reports, via telemetry, the fact that each encrypted command in the sequence has been received and properly stored on board. Once the timed encrypted commands are in the spacecraft's or satellite's memory, the onboard clock will cause each of them to be executed at the proper instant.

The Ace and others, including spacecraft or satellite engineers, instrument scientists or military personnel, will watch the downlink over the period of time covered by the encrypted command sequence, making sure all is going according to plan. That is where Chapter 16 picks up—the downlink part of end-to-end encrypted data flow.

From Here

Chapter 15 discussed the following: The planning process; details; implementing the plan; and, the commanding uplink (receiving) process. Chapter 16 analyzes the following: How we listen; Telemetry Delivery System (TDS); downlink (transmitting) process; and, how to serve a world of data consumers.

Endnotes

1 Dave Doody, "Basics of Spaceflight," The Planetary Report, The Planetary Society, Jet Propulsion Laboratory, JPL Advanced Mission Operations Section, NASA, Pasadena, CA, 1997.

16

End-to-End Data Flow — The Downlink Part

This chapter, together with Chapter 15 on uplink, offers a large-scale view of how uplink and downlink work together as a system, bringing measurements of ground- and space-based objects or military targets to our front doors and to our home computers, and to the supercomputers of the military establishment, respectively. Pretty scary, isn't it?

As previously explained, *uplink* refers to radio signals sent to a spacecraft or satellite; *downlink* is a transmission from the spacecraft to Earth. Once an uplink has been sent and its encrypted commands have been carried out by the spacecraft or satellite, two kinds of results come back on the downlink. One is telemetry, meaning onboard measurements that have been converted to encrypted radio signals. The other includes encrypted tracking, radio science, and military data.

The downlink contains engineering, military, commercial and science encrypted information. Engineering telemetry, for example, includes the temperature of nearly every spacecraft or satellite component as

449

well as voltages, currents, pressures, and other measurements. Engineers rely on these measurements in their operation of the spacecraft or satellite. Encrypted tracking and navigation data fall under the engineering heading too.

For instance, under the science heading, encrypted data is included from the science instruments and imaging equipment (which are telemetry) and the radio science data. In a radio science experiment, the downlink may yield atmospheric-sensing results or provide tracking data to determine the mass of a planet or moon or some ground-based imagery observation here on earth.

Transmission of encrypted military data, whether telemetry or radio science, enjoys the highest priority in the end-to-end system. Engineering encrypted data have a repetitive nature: If some information is lost, it is usually no big deal, because the same measurements will probably recur in a short time. Military data are the diamond amid rhinestones. Receipt of these data is the purpose of flying DOD spacecraft (like Clementine) or spy satellite missions the first place.

How We Listen

A broad definition of *downlink* encompasses receiving, deciphering (decrypting), storing, distributing, displaying, archiving, and publishing the data. The result of the end-to-end system is that encrypted science data reach the worldwide public, including the people who fund missions with their taxes.

Figure 16.1 depicts a typical Jet Propulsion Laboratory (JPL) project, such as *Galileo, Mars Global Surveyor, Clementine* (DOD/NASA Lunar Mapping), or the *Cassini mission* to Saturn.[1] Many of the components pictured are shared by several projects. Going from left to right, Figure 16.1 shows that the spacecraft's or satellite's scientific instruments collect images, spectra, and other data, which then go to a storage device aboard the spacecraft or satellite, such as a tape recorder. That device, under the control of a computer, plays back the encrypted information when Earth is listening. Hardly a bit of precious science or military encrypted data is ever lost permanently. If information gets garbled on the way down—because of a rainstorm, for example, over

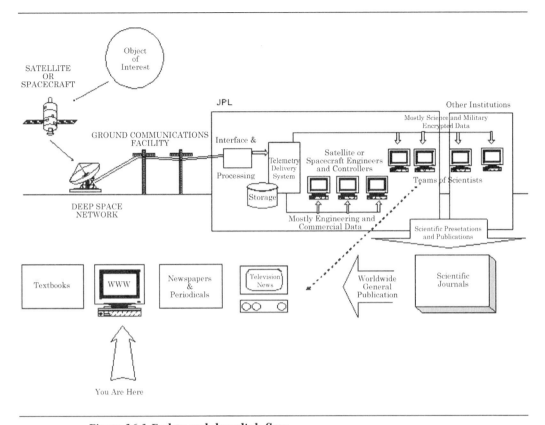

Figure 16.1 End-to-end downlink flow.

the receiving station of the Deep Space Network (DSN) — controllers can command the spacecraft to replay the encrypted data when Earth rotates a drier DSN station into view (see Sidebar, "Telecommunications Issues").

The downlink encrypted radio signal is pretty weak, having traveled a long distance. It takes large-aperture radio telescopes to collect and concentrate the signal. The antenna has to be pointed in exactly the right direction to locate the spacecraft or satellite in the sky. Once the downlink encrypted signal has been funneled in, it is boosted in the antenna with a low-noise amplifier kept at very low temperature to minimize electronic noise.

From the low-noise amplifier, the encrypted signal goes to a radio receiver that automatically tunes in the right signal. Once the receiver has locked onto the downlink, the encrypted signal goes to a telemetry processor, which then converts the signal to binary digits (bits), all ones and zeros—the same digital information originally generated by the spacecraft or satellite instruments. The DSN transmits these encrypted telemetry bits to JPL via the Ground Communications Facility (GCF), which despite its name relies on satellites for communications. In addition to telemetry, the DSN captures satellite tracking data and radio science data. In Figure 16.1 encrypted data are coming in to JPL, where preliminary processing removes GCF transport information and decommutates the channels.

Decommutates the channels! Remember, a spacecraft or satellite sends lots of military, engineering and science measurements. *Decommutating* means determining the kind of encrypted data being received. *Channels* are categories of encrypted data. For example, instead of always referring laboriously to *the temperature data on reaction wheel motor number one*, it is called something like *Channel E-1654*. C-channels are for commercial encrypted data, E-channels are for encrypted engineering data, S-channels are for encrypted science data and M-channels are for encrypted military data.

Telecommunications Issues

This sidebar gives a broad view of some telecommunications issues, including both spacecraft or satellite and Earth-based encrypted communication. This view is of abbreviated depth.

Encrypted Signal Power: Your local entertainment radio broadcast station may have a radiating power of 50 kW, and the transmitter is probably no more than 100 km away. Your portable receiver probably has a simple antenna inside its case. Spacecraft or satellites have nowhere near that amount of power available for transmitting, yet they must bridge distances measured in tens of billions of kilometers. A spacecraft or satellite might have a transmitter with no more than 20 W of radiating power. How can that be enough? One part of the solution is to employ microwave frequencies, and concentrate all available power into a narrow beam, and then to send it in one direction instead of broadcasting in all directions. This is typically done using a parabolic dish antenna on the order of 1 or 5 m in diameter. Even when these concentrated signals reach Earth, they have vanishingly small power. The rest of the solution is provided by the DSN's large aperture reflectors, cryogenically cooled low-noise amplifiers and sophisticated receivers, as well as encrypted data coding and error-correction schemes.

Uplink and Downlink: As previously stated in Chapter 15 and reiterated again here, the encrypted radio signal transmitted to a spacecraft is known as uplink. The transmission from spacecraft or satellite to Earth is downlink. Uplink or downlink may consist of a pure RF tone, called a carrier, or carriers may be modulated to carry information in each direction. Encrypted commands transmitted to a spacecraft or satellite are sometimes referred to as an upload. Communications with a spacecraft or satellite involving only a downlink are called one-way. When an uplink is being received by the spacecraft at the same time a downlink is being received at Earth, the communications mode is called *two-way*.

Modulation and Demodulation: Consider the carrier as a pure tone of, say, 3 GHz, for example. If you were to quickly turn this tone off and on at the rate of a thousand times a second, we could say it is being modulated with a frequency of 1 kHz. Spacecraft or satellite encrypted carrier signals are modulated, not by turning off and on, but by shifting each waveform's phase slightly at a given rate. One scheme is to modulate the carrier with a frequency, for example, near 1 MHz. This 1 MHz modulation is called a subcarrier. The subcarrier is in turn modulated to carry individual phase shifts which are designated to represent groups of binary 1s and 0s—the spacecraft's or satellite's encrypted telemetry data. The amount of phase shift used in modulating encrypted data onto the subcarrier is referred to as the modulation index, and is measured in degrees. The same kind of scheme is also used on the uplink.

Demodulation is the process of detecting the subcarrier and processing it separately from the carrier, detecting the individual binary phase shifts, and decoding them into encrypted digital data for further processing. The same processes of modulation and demodulation are used commonly with Earth-based computer systems and fax machines transmitting encrypted data back and forth over a telephone line. The device used for this is called a modem, short for modulator/demodulator. Modems use a familiar audio frequency carrier which the telephone system can readily handle.

Binary encrypted digital data modulated onto the uplink is called encrypted command data. It is received by the spacecraft or satellite and either acted upon immediately or stored for future use or execution. Encrypted data modulated onto the downlink is called telemetry, and includes science or military data from the spacecraft's or satellite's instruments and spacecraft health data from sensors within the various onboard subsystems.

Multiplexing: Not every instrument and sensor aboard a space-craft or satellite can transmit its data at the same time, so the encrypted data are multiplexed. In the time-division multiplexing (TDM) scheme, the spacecraft's or satellite's computer samples one measurement at a time and transmits it. On Earth, the samples are demultiplexed, that is, assigned back to the measurements which they represent. In order to maintain synchronization between multi-plexing and demultiplexing (also called mux and demux) the space-craft or satellite introduces a known binary number many digits long, called the pseudo-noise (PN) code at the beginning of every round of sampling (telemetry frame), which can be searched for by the ground data encryption system. Once recognized, it is used as a starting point, and the measurements can be demuxed since the order of muxing is known.

Newer spacecraft or satellites use packetizing rather than TDM. In the packetizing scheme, a burst or packet of encrypted data is transmitted from one instrument or sensor, followed by a packet from another, and so on, in nonspecific order. Each burst carries an identification of the measurement it represents for the ground data encryption system to recognize it and handle it properly. These schemes generally adhere to the International Standards Organi-zation (ISO)'s Open Systems Interconnection (OSI) protocol suite, which recommends how computers of various makes and models can intercommunicate. The ISO OSI is distance independent, and holds for spacecraft or satellite light-hours away as well as between workstations.

Coherence: Aside from the information modulated on the down-link as telemetry, the carrier itself is used for tracking the spacecraft or satellite, and for carrying out some types of science or military experiments. For each of these uses, an extremely stable downlink frequency is required, so that Doppler shifts on the order of fractions of a hertz may be detected out of many GHz over periods of hours. But it would be impossible for any spacecraft or satellite to carry the massive equipment on board required to generate and maintain such stability. The solution is to have the spacecraft generate a downlink which is phase-coherent to the uplink it receives.

Down in the basement of each DSN Signal Processing Center, there looms a hydrogen maser-based frequency standard in an environmentally controlled room. This is used as a reference for generating an extremely stable uplink frequency for the spacecraft or satellite to use in generating its coherent downlink.

The resulting spacecraft or satellite downlink, based on and coherent with an uplink, has the same extraordinarily high frequency stability as does the massive hydrogen maser-based system in its controlled environment in the DSN basements. It can thus be used for precisely tracking the spacecraft or satellite, and for carrying out science or military experiments. The spacecraft or satellite also carries a low-mass oscillator to use as a reference in generating its downlink for periods when an uplink is not available, but it is not highly stable, and its output frequency is affected by temperature variations on the spacecraft or satellite. Some spacecrafts or satellites carry an Ultra-Stable Oscillator (USO). Because of the stringent frequency requirements for spacecraft or satellite operations, JPL stays at the forefront of frequency and timing standards technology.

Most spacecrafts or satellites may also invoke a noncoherent mode, which does not use the uplink frequency as a downlink reference as shown in Figure 16.2.[2] Instead, the spacecraft or satellite uses its onboard oscillator as a reference for generating its downlink frequency. This mode is known as Two-Way Non-Coherent (TWNC, pronounced *twink*). When TWNC is on, the downlink is noncoherent.

Recall that *two-way* means there is an uplink and there is a downlink, and does not indicate whether the spacecraft's or satellite's downlink is coherent to that station's uplink or not. However, in common usage, operations people commonly say *two-way* to mean *coherent*, which is generally the case. Correctly stated, a spacecraft or satellite downlink is coherent when it is two-way with TWNC off. When a spacecraft or satellite is receiving an uplink from one station and its coherent downlink is being received by another station, the downlink is said to be *three-way* coherent as shown in Figure 16.2.[3]

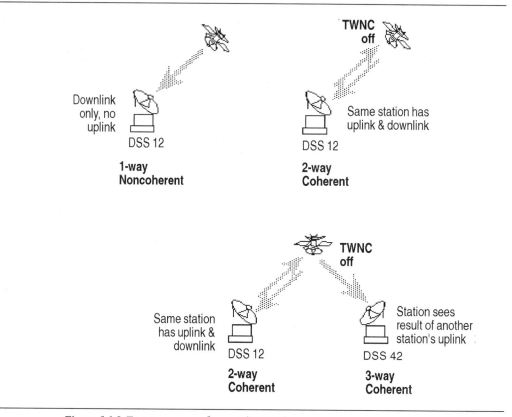

Figure 16.2 Two-way noncoherent (TWNC).

Telemetry Delivery System (TDS) Has Results Right Now

The telemetry, now *channelized* to identify its contents, is typically stored within a Telemetry Delivery System (TDS) for the entire life of the mission. This system distributes encrypted telemetry to people who should receive it right away: the real-time flight controller who is looking out for surprises, the engineer who needs to watch a propellant tank temperature, or the scientist waiting for findings from a particular instrument.

TDS users can custom-order the telemetry they need. For example, a science team analyzing encrypted data from a photopolarimeter experiment might need to look at radio-receiver performance to be sure of their results. Another supplementary data set includes spacecraft and satellite positions and instrument pointing. With this information, one can determine exactly where in space the encrypted data were gathered from and thus where they fit in relation to other collected data.

Once a team of scientists or military personnel has correlated and analyzed the encrypted telemetry and supplementary data, they will want to publish their results in a scientific journal (except for the military), but even before these scholarly articles appear, the public gets a look at the scientific data absent scholarly analysis! The policy at NASA calls for immediate publication of images returned from planetary encounters (represented by the dashed arrow in Figure 16.1); and, an unwritten policy of not publishing anything even remotely considered to be a UFO (unidentified flying object). These *first-look* images are preliminary and usually lack the high resolution scientists need for extensive analysis and reporting.

Serving a World of Data Consumers

Encrypted data from planetary missions are made available worldwide through Regional Planetary Imaging Data Facilities, operated by NASA at more than a dozen sites around the United States and overseas. Encrypted data from military missions are only made available to those in the military establishment who have a need to know (such as NSA, CIA, etc.). Each site around the world maintains a complete photographic library of images from lunar and planetary missions. These sites are open to the public by appointment for browsing and ordering materials for purchase.

Researchers funded by NASA obtain planetary imaging data via the Planetary Data System, which consists of a central on-line catalog and data retrieval nodes at various research facilities. Classroom educators can obtain videocassettes and a wide variety of other materials on NASA's flight projects through Educator Resource Centers (which until just recently were called Teacher Resource Centers). They cannot obtain materials that were classified as part of a military or DOD mission.

From Here

Chapter 16 analyzed the following: how we listen; Telemetry Delivery System (TDS); downlink (transmitting) process; and, how data consumers are served. Chapter 17 covers the following: geosynchronous satellites; low-earth orbit satellites; multiple access methods; geosynchronous systems; low-earth orbit systems; and, advantages and disadvantages of each.

Endnotes

1   Dave Doody, "Basics of Spaceflight," The Planetary Report, The Planetary Society, Jet Propulsion Laboratory, JPL Advanced Mission Operations Section, NASA, Pasadena, CA, 1997.

2   "Spaceflight Projects," Jet Propulsion Laboratory, JPL Advanced Mission Operations Section, NASA, Pasadena, CA, 1998.

3   Ibid.

17

Satellite Data and Voice Communications Encryption

A communications satellite provides encrypted communications over long distances by reflecting or relaying radio-frequency signals. The distance an encrypted radio signal can be transmitted is limited by the curvature of the Earth. Any receiving station beyond the horizon of the transmitter will not receive any transmissions because they will be blocked by the planet itself. One earth-bound solution to this limitation is to make the transmitting and/or receiving towers taller, thus placing the horizon farther away; another is to place repeating stations to receive transmissions and then rebroadcast them. A communications satellite represents an extension of these ideas into space. It can be thought of as a really tall station.

The first communications satellites were very simple devices, operating in a passive mode; they simply reflected signals transmitted to them from the ground back to the ground. The problem with passive satellites is that they require extremely powerful ground transmitters and very large ground antennas. Modern communications satellites have active systems, in which a transponder receives signals from the ground, amplifies them, then retransmits them back to the ground on a different frequency (or, in some cases, to other satellites). The use of a different frequency for retransmission is to prevent interference between the inbound and outbound signals. The inbound signal (uplink) is always a higher frequency than the outbound (downlink) signal. One frequency band used, 600 MHz wide, is divided into repeater channels of various bandwidths (located at 7 GHz for upward, or uplink, transmission and 5 GHz for downward, or downlink, transmission). A band at 15 GHz (uplink) and 13 or 14 GHz (downlink) is also much in use, mostly with fixed (nonmobile) ground stations. A 90-MHz-wide band at about 1.6 GHz (up-and-downlink) is used with small, mobile ground stations (ships, land vehicles, and aircraft). Power for reception and transmission is provided by solar energy cells mounted on large panels attached to the satellite.

In the 1960s and 1970s, technology was introduced that allowed satellites to *beam*, or concentrate, their signals only to the Earth, rather than radiating them in all directions. Multiple beam operation is now possible, and a satellite's encrypted signal can be concentrated upon a small region of the earth. Frequency reuse is a technique that allows a satellite to communicate with multiple ground stations using the same frequency by transmitting in narrow beams directed to each of the stations. Beam widths can be as small as the state of Delaware or as large as the entire North American continent. Stations far enough apart can receive different messages on the same frequency. Satellite antennas have been designed that can transmit several beams in different directions.

Geosynchronous Satellites

Most communications satellites currently in use are in geosynchronous orbits. In a geosynchronous orbit, the satellite orbits in a circular path above the equator, at a height of 46,900 kilometers (km) (29,313 miles), in the same direction as the earth's rotation. This results in the satellite completing one orbit in 24 hours, the same time it takes the earth to rotate once, thus keeping it in a fixed position above the equator and providing uninterrupted contact with ground stations in its line of sight. Three satellites in geosynchronous orbits could have the entire planet in range. There are communications problems inherent in the use of geosynchronous satellite systems, however. For example, due to the extreme altitude of the satellites, there is a need for large antenna dishes (as much as 20 meters wide) or megawatts (MW) of focused beam power. Another is a half-second time delay (propagation delay) for signals being relayed through a geosynchronous satellite. This is an annoyance for encrypted voice communications, but near eternity for computer communications.

Low-Earth Orbit Satellites

An alternative to geosynchronous satellite systems are Low- Earth Orbit (LEO) satellite constellations, in which the satellites orbit at much lower altitudes, between 600 and 1500 km above the earth, most likely in a polar or near-polar orbit. Below 600 km, the orbit would decay and the satellite would re-enter the atmosphere too soon for the satellite to have a practical lifetime. Above 1500 km, radiation from the Van Allen belts would sharply decrease the satellites' lifetimes. These LEO satellite systems will require much larger numbers of satellites than geosynchronous satellite systems for two reasons. First, because they orbit at lower altitudes, the satellites will not remain stationary relative to the earth's surface. For any station on earth, a satellite will move into view overhead, travel across the sky, and move out of view. This requires a continuous string of satellites, with a new satellite moving into view before the prior one is out of view, and the encrypted communication being *handed off* to the new satellite. In order to achieve global coverage, there will need to be several such strings, orbiting in different planes. The other reason is also as a result of the lower altitude. At a lower altitude, the *footprint* of the satellite (a circular region on earth in which the satellite would be in view) is much smaller than that of a geosynchronous satellite.

Low-Earth Orbit systems will also require ground stations to be able to track their satellites as they move across the sky. Despite these requirements, LEOs will allow smaller, less powerful, and thus less expensive satellites to be used, as well as smaller, less powerful, less expensive user terminals.

A variation on the LEO constellations previously described is a single-satellite store-and-forward LEO, for non-real-time encrypted communication. A ground station sends an encrypted digital message to the LEO satellite, which stores it in an onboard system. With the satellite in a highly inclined orbit, its motion and that of the earth rotating below it will combine to bring every point on earth within its footprint at least twice each day. When the destination station is below, the satellite then retransmits the encrypted message to its intended recipient. By using a single small satellite and low-cost communications equipment, this can provide the lowest-cost satellite communications. The drawback, of course, is that the sending ground station must wait until the satellite comes into view to upload a message, and then the satellite must wait until the destination station comes within its footprint. The delay in message delivery can be anywhere from a few seconds up to 20 hours (h).

Multiple Access Methods

Satellite communication systems also differ in the method they use to allow multiple access to the systems. One such system, Time Division Multiple Access (TDMA), gives each ground station an assigned time slot on the same channel for transmitting its encrypted messages. All stations monitor these slots and select the communications directed to them. In this way, more than one user can have access to the system at the same time. A drawback to this method, however, is that it requires the limited available frequency spectrum to be divided up ahead of time. Thus, TDMA has to be done using one computer. Two different systems cannot use the same frequency spectrum at the same time. A second method, Code Division Multiple Access (CDMA), uses codes to allow multiple users, even on different systems, to use the same spectrum at the same time. Frequency spectrums do not have to be assigned ahead of time. Instead, different users have different codes that are used to transmit their messages. Each user ignores messages not in their code.

Now let us examine some of the major satellite systems, both currently in operation and being planned for the near future, which do or soon will provide encrypted data communications. These satellite systems were initially introduced in Chapter 1.

Geosynchronous Systems

As previously stated, most communications satellites currently in use are in geosynchronous orbits. This is where the satellite orbits in a circular path above the equator. Let us take a look at some of these types of satellites:

- Intelsat.
- Hughes Network Systems.
- Inmarsat.

Intelsat

The International Telecommunications Satellite Organization (INTELSAT) is an international nonprofit cooperative of more than 140 nations (begun in 1964), which operates a global system of 24 satellites in geosynchronous orbits. It provides transoceanic, regional and domestic telephone and television services, as well as business services such as international video, teleconferencing, facsimile, and data and telex at transmissions rates up to 150 megabits per second (Mbps).

INTELSAT's two basic types of service, Channel/Carrier-Based and Leased, are not complete turnkey solutions; INTELSAT does not supply user terminals or ground stations, it merely provides access to its satellite system. The channel or carrier-based service requires subscribers to use earth stations and other equipment that meet INTELSAT technical specifications. This service provides a guaranteed, specified quality of service. The other option, leased service, places no requirements on the equipment used other than that it not interfere with other communications on the system. This alternative gives no guarantee regarding quality of service, however. Earth stations meeting INTELSAT's technical specifications would fall into one of the following categories:

- *A*: For voice, data, and TV (including IBS and IDR), using the 6/4 GHz frequency band, and having a 15–18 meter (m) antenna.
- *B*: Meeting the same criteria as for *A*, but with an 11–13 m antenna.

- *C*: For voice, data and TV (including IBS and IDR), using the 14/11 GHz frequency band, and having an 11–14 m. antenna.
- *D1*: For Vista service, using the 6/4 GHz frequency band, and having a 4.5-6 m antenna.
- *D2*: Meeting the same criteria as for *D2*, but with an 11 m antenna.
- *E1*: For IBS service, using the 14/11 and 14/12 GHz frequency band, and having a 3.5–4.5 m antenna.
- *E2*: For IBS & IDR service, using the 14/11 and 14/12 GHz frequency band, and having a 5.5–7 m antenna.
- *E3*: Meeting the same criteria as for *E2*, but with an 8–10 m antenna.
- *F1*: For IBS service, using the 6/4 GHz frequency band, and having a 4.5–7 m antenna.
- *F2*: For IBS & IDR service, using the 6/4 GHz frequency band, and having a 7–8 m antenna.
- *F3*: For voice and data (including IBS & IDR), using the 6/4 GHz frequency band, and having a 9–10 m antenna.
- *G*: For international and domestic lease service, using the 6/4 and 14/11 GHz frequency bands, and having antenna of any size.[1]

INTELSAT does offer another service, Demand Assignment Multiple Access (DAMA), for thin-route Public Switched Network (PSN) services. This digital technology provides instant dial-up connectivity between a large community of users, using dynamic satellite circuit sharing. The DAMA service offers basic telephony service for speech, fax, and voiceband data applications, and should be capable of providing on-demand n × 64 kbps connections. With the DAMA service, a central Network Management and Control Center allocates space segment resources from a pool of available channels on-demand. There is no fixed association between satellite encrypted channels and earth station channels. Customers are charged only for the duration of answered calls. The terminal, supplied by INTELSAT, is installed within an existing earth station facility.

The satellites owned by INTELSAT consist of 13 INTELSAT V/V-A series, 6 INTELSAT VI series, and 7 INTELSAT VII series. Each of the INTELSAT V satellites has 22 C-band transponders and 5 Ku-band transponders. They each have a capacity of 13,000 two-way telephone circuits and 3 TV channels. The INTELSAT V-A series has 27 C-band transponders and 7 Ku-band transponders. They have a capacity of 16,000 two-way telephone circuits and 3 TV channels. Both series have a projected life span of 8 years. Many have already been in service for over 13 years, however. The INTELSAT VI series has 39 C-band transponders and 11 Ku-band transponders, a capacity of 36,000 two-way telephone circuits and 4 TV channels, and a projected lifetime of 14 years. The INTELSAT VII series has 27 C-band transponders and 11 Ku-band transponders, a capacity of 24,000 two-way telephone circuits and 4 TV channels, and a projected lifetime of 11–16 years.

Hughes Network Systems

Hughes Network Systems, Inc. (HNS) provides interactive Very Small Aperture Terminal (VSAT) satellite networks for businesses and governments in the United States using three geosynchronous satellites. It is currently the largest supplier of VSAT networks in the world, responsible for 80% of those in operation.

A VSAT network allows transmission of encrypted data, voice, and video between geographically dispersed locations. Each site has a VSAT terminal, which is made up of an antenna between .86 and 3.5 m in diameter; electronics mounted on the antenna to receive and transmit signals in the C-band (4/6 GHz) and Ku-band frequencies; and, connections to user computers, telephones and video equipment. Encrypted data transfer rates of up to 20.3 kilobits per second (kbps) for asynchronous serial; 2.3–64 kbps for synchronous serial; 10 Mbps for Ethernet LAN; and, 4/16 Mbps for Token-Ring LAN are possible. The protocols supported by Hughes' VSAT networks are: Ethernet; Token-Ring; SDLC (PU4-PU2, PU4-PU4); X.25; Burrough's Poll; BSC 3270; TINET; Bit and Byte Transparent; HASP; Frame Transparent; X.3/X.28 PAD; Broadcast; and other specialized protocols.

Hughes offers four VSAT network configurations. Its Point-to-point VSAT network provides communications between three remote sites.

Hughes' Star network configuration enables multipoint communications between a central hub station and an unlimited number of VSATs at remote sites. This type of network has two hub options. Companies may have a Dedicated Hub—owned, staffed, and operated by the customer. If desired, Hughes will provide the staff to operate a Dedicated Hub. The other alternative is a Shared Hub, owned, staffed and operated by Hughes—providing turnkey service for customers. Hughes operates five such centers around the U.S.

The third configuration available is a Mesh network, which allows all VSATs on the network direct communications with all other sites without routing through a hub. Each VSAT communicates with other sites with a single satellite hop.

Hughes also provides a satellite service called DirecPC. A 16-bit ISA adapter card installed in a user's PC allows it to receive high-speed satellite data from DirecPC's Network Operations Center. A satellite-grade cable connects the adapter card to a 24-in dish antenna. DirecPC subscribers have access to a catalog of electronic documents, on-line news, financial, sports information, and on-line ordering of commercial software packages. Users also have a number of enhanced services available.

Turbo Internet is an additional service available to DirecPC subscribers. The user makes a request for a document or file from the Internet through a dial-up modem and public switched telephone network. The information requested is then transmitted to the user's PC by the DirecPC satellite at up to 500 kbps.

Digital Package Delivery is a DirecPC service available to companies. This service provides one-way transmission of encrypted software, computer-based training, and electronic files and documents to an unlimited number of sites in digital form. A company could send encrypted information to be transmitted (and the addresses to which it is to be sent) to the DirecPC Network Operation Center, where it is then broadcast by way of Ku-band satellite to all specified sites at 14 Mbps.

Another Hughes satellite service is in TELEconference, a multipoint videoconferencing by satellite which provides high-quality 495 kbps video. This system consists of a Network Control Center (NCC) at a site designated by the user and satellite Earth stations at remote sites. During conferences, the NCC polls remote sites to determine status, manages scheduling, processes requests for extension and termination, and transmits conference information. The Earth stations consist of a small antenna, an electronics unit for encrypted communication with the video codec, and a PC-based controller.

Hughes has plans to expand its present system into a 10-satellite system called Spaceway, eventually expanding to 18 satellites. This system would use Ka-band geosynchronous satellites with 59 communication beams and 16-year lifetimes. Standard terminals will be nonmobile 26-in VSATs able to receive at a high encrypted data rate, but normally transmitting at 16 kbps (voice/fax) and T1 for places where terrestrial infrastructure is lacking. A higher rate terminal will provide multimegabit uplinks. These uplink rates would be a 2.6 Mbps burst for VSAT terminals and a 7 Mbps burst for broadcast terminals.

Inmarsat

Inmarsat is an international organization with 80 member countries, which was set up to provide mobile satellite encrypted communications for the worldwide maritime community. It has been providing those services for over a decade.

Inmarsat uses geostationary satellites orbiting 33,334 miles above the earth. The organization owns five Inmarsat-2 satellites and leases the European Space Agency's Marecs B2 satellite; the maritime communications subsystems on the International Telecommunications Satellite Organization's Intelsat-V satellites; and, capacity on three Comsat General Marisat satellites.

Inmarsat-2 weighs 1400 kg, has 1300 W of available power, and is designed for 11 years of service. These satellites were launched between 1990 and 1992 to replace the first generation of Inmarsat satellites. Inmarsat-2 has three transponders, which provide outbound and inbound links to mobile terminals in the 6.4/1.5 and 1.6/3.6 GHz bands. Bandwidth can be dynamically allocated for communications. Inmarsat has recently launched six new generation Inmarsat-3 satellites, which will be about nine times more powerful than the Inmarsat-2.

Inmarsat currently offers a number of services. Inmarsat-A, with more than 36,000 users, is the most common. Inmarsat-A provides two-way direct dial telephone, facsimile, telex, encrypted e-mail, and data communications. Most of the Inmarsat-A terminals, which are produced by 18 manufacturers around the world, feature dynamically driven antennas less than one meter in diameter mounted onboard ocean-going ships; but, there are also transportable versions that fit into one or two suitcases. Some of these terminals can support data rates of up to 64 kbps, allow full duplex encrypted transmission, and can transmit still and compressed encrypted video, broadcast-quality audio, and video conferencing.

Encrypted transmissions from Inmarsat-A terminals are received by the satellite constellation and routed to land Earth stations, which deliver the transmissions to national and international telecommunications and data networks. Inmarsat-A terminals use the L-band (1.5/1.6 Ghz) to receive and transmit encrypted signals. The land Earth stations use the C-band (4/6 Ghz).

Inmarsat-B service, which was introduced in 1993, offers services similar to Inmarsat-A, but is all-encrypted digital. This allows the system to be smaller, lighter, and less expensive than Inmarsat-A. It also results in lower encrypted transmission charges for users.

Inmarsat-M is a telephone service that provides direct-dial telephone, facsimile, or 2.4 kbps encrypted data transmission. The shipboard terminals for this service are only one-eighth the size of the A/B terminals. The mobile terminals for Inmarsat-M are the size of a briefcase. The cost of terminals and user charges are also much lower than for the A/B systems.

Inmarsat-C service provides the smallest terminals of all. They weigh only 3 to 4 kg and have small, omnidirectional atennas that can be mounted to a vehicle or ship easily. Users can get terminals with their own preparation and display facilities or with RS232 interfaces to their own personal computers. The Inmarsat-C service operates at 700 bps and provides two-way, store-and-forward message, text, data reporting and polling services. Terminals of this service can be programmed to receive multiple address messages, which have special headers to indicate to which group of terminals or to which geographic area they are intended.

Inmarsat also has three aeronautical services. Aero-C provides Inmarsat-C service to aircraft. Aero-L provides 700 bps real-time encrypted data transmission for operational and administrative purposes. Aero-H supports multichannel voice, facsimile and data transmissions for aircraft at up to 11.6 kbps.

Low-Earth Orbit Systems

As previously discussed, Low-Earth Orbit (LEO) satellite constellations are satellites which orbit at much lower altitudes. Their orbits are between 600 and 1,500 km above the Earth, most likely in a polar or near-polar orbit. Now, let us take a look at some of these types of satellites:

♦ Iridium.
♦ ORBCOMM.
♦ Globalstar.
♦ Teledesic.
♦ Odyssey.

Iridium

Iridium is a company founded by Motorola, intended to establish a global cellular satellite network. The plan is for a network of 77 Low-Earth Orbit (LEO) satellites, orbiting at a height of 853 km (533 miles) in 12 nearly polar orbits (tilted 5° off from a true polar orbit). The system will provide voice, data, fax, paging, messaging, and position location encrypted communications. Voice transmission will be at 2.4/4.8 kbps, data transmission at 2.4 kbps. The life expectancy of the satellites is 4 to 11 years.

Unlike most other proposed LEO systems, whose satellites will be simple bent-pipe repeaters (sending the handset's encrypted signal back down to a ground station that feeds into land lines), Iridium's satellites will communicate with one another to form a network in space, passing on conversations and handing them off when they drift out of range. Because of these crosslinks between satellites, once a link is established, calls will be able to be carried on without the intervention of any ground stations. This capability comes with a cost, however. The satellites will have to have onboard computers to handle system networking. This will increase the initial cost of the satellites and increase power consumption in orbit. Furthermore, those computers and their memory will also have to have backups, further raising the cost. Additionally, the satellites will need an extra set of antennas to communicate with one another, and will need to be positioned much more precisely than satellites that communicate only with the ground. All of this also results in a larger, heavier satellite that costs more to launch.

Another difference between Iridium and most other proposed LEO systems is in their use of the available spectrum. Iridium plans to use time division multiple access (TDMA) as its access method. This method allows operators to divide a signal up into tiny fractions of a second, allowing more than one person to use the same frequency spectrum at the same time. Most other proposed LEO systems plan to use code division multiple Access (CDMA), which allows multiple users of the same frequency spectrum by use of multiple encrypted codes.

User uplinks and downlinks will be in the L-band (1616–1616.5 MHz). Feeder uplink (19.4-19.6 GHz) and downlink (29.1–29.3 GHz) will be in the Ka-band. One-way propagation delay is expected to be 2.6–8.22 milliseconds (ms). Each satellite will have 1200 voice circuits, 49 beams, and a *footprint* diameter of 5800 km (3625 miles).

User terminals will be hand-held. They will also have an ordinary cellular telephone mode, and should cost $2,200–$2,700. Call rates are projected to be $2 per minute; and, Iridium hopes to have its system operational in 1999, at a cost of $4.8 billion.

ORBCOMM

ORBCOMM is a joint venture of Orbital Sciences Corporation (OSC), a Virginia-based aerospace company and Teleglobe Inc., a Canadian global telecommunications provider. When completed, ORBCOMM will provide digital encrypted transmission of direct two-way messages, e-mail and data packets, complete connectivity to the Internet, and e-mail networks.

ORBCOMM will be based on a constellation of 39 Low-Earth Orbit (LEO) satellites in orbits 600 to 1100 miles above the earth. This constellation should place a satellite continuously in view for a U.S. user over 96% of the time, with less than a two minute wait the remainder of the time. Satellites can be repositioned to compensate for outages.

An ORBCOMM satellite, weighing only 96 pounds, uses 137–138 MHz for downlink and 148—150.05 MHz for uplink. It has 170 W of solar array Power, three VHF user link transmitters, two VHF feeder link transmitters, two UHF transmitters, eight user link receivers and three feeder links receivers. ORBCOMM uses Dynamic Channel Activity Assignment to optimize spectrum sharing with existing users of the radio frequencies. Three satellites have been placed in orbit and tested so far, with the remainder planned to be in place by the end of 1999.

ORBCOMM's Subscriber Communicators (SCs) are lightweight, pocket-sized, and support a variety of interfaces—such as RS-232 and analog/digital input/output, to provide connection to PCs, control equipment, sensors, etc. Each country that uses the ORBCOMM System will have a Network Control Center (NCC) and at least two unmanned Gateway Earth Stations (GES). The U.S. is to have five GESs, located in five places across the country. The GES will transmit to the LEOs at 600 W and receive 4-W encrypted transmissions from them, using three steerable high-gain VHF antennas that track the LEO satellites as they move across the sky. The GES also transmits encrypted data to and receives encrypted data from the NCC.

An encrypted message transmitted from a SC will be received by a LEO satellite and relayed down to a GES. The GES will then relay the message to the NCC, either by satellite or by a dedicated terrestrial line. The NCC is responsible for message processing, network management, and satellite control. It is connected to X.400 and X.25 networks, has dedicated lines to other networks, and supplies dial-in access. The NCC will determine the location of the recipient and relay the message to the appropriate GES for transmission to a satellite, and from there down to the recipient. The message transit time from SC to SC is six seconds. ORBCOMM is making agreements with service providers which will make messaging to almost 80 countries possible.

Globalstar

Globalstar is a limited partnership, founded by Loral Corporation and QUALCOMM, Inc., to develop a Low-Earth Orbit (LEO) digital satellite communications system made up of 59 satellites orbiting 1400 km (874 miles) above the earth. This system will provide voice, data, fax, paging, short message service, and position location. Encrypted voice transmissions will be at 3.5 to 10.7 kbps, encrypted data transmissions at 8.3 kbps.

The satellites will be small repeater types that will act as simple bent pipes to relay encrypted signals received directly to the ground. No onboard processing will be needed. The three-axis, body-stabilized satellites will weigh 450 kg, have a total transponder power of 1000 W, and an estimated lifetime of 8.6 years. The satellites will provide 3000–4000 voice links per satellite, 17 beams, and have a *footprint* diameter of 6960 km (4350 miles). User links will use the S-band (2483.5–2500 MHz) for downlink and the L-band (1610–1626.5 MHz) for uplink. Feeder links will use the C-band (5.091–5.25 MHz) for uplink and for downlink (6.875–7.055 MHz). One-way propagation delay is estimated to be 4.63–11.5 ms.

On the ground, there will be three satellite operations control centers (SOCCs), which will track and control the satellite constellation, using command and telemetry equipment located in three telemetry stations and in gateway stations around the world. The gateway stations, which will be located at the site of an existing cellular (or some other kind of telecommunications) switch, will be Globalstar's access points to land-based telecommunications systems, and will have as many as four antennas to track satellites in their view. Each gateway station should be able to provide fixed coverage for an area larger than Saudi Arabia and Egypt together, and mobile coverage for an area slightly smaller than Eastern Europe.

There will also be three ground operations control centers (GOCCs), which will plan and control satellite utilization by gateway terminals, coordinate information from the SOCCs, monitor system performance, collect billing information, ensure that gateways do not exceed allocated system capacity, and allocate system capacity dynamically among nearby regions. Here CDMA will be used as the multiple access method.

User terminals will be hand-held and capable of normal cellular telephone functions, and are expected to cost about $640. The call rate is expected to be $.34–$.54 per minute. Globalstar expects the system to go into operation in 1999, at a cost of $3.6 billion.

Teledesic

The Teledesic Corporation was formed in 1994 by William Gates of Microsoft, and Craig McCaw of McCaw Cellular Communications. Teledesic is planned to be a satellite network made up of 950 Low-Earth Orbit (LEO) satellites orbiting at a height of 848 km (530 miles), providing encrypted voice, data, fax, paging and video communications. Alone among the planned LEO systems, Teledesic is fully focused on extending computer networks to a broadband satellite system.

Teledesic is planned to transmit voice communications at up to 16 kbps and data communications at up to 2.048 Mbps. Here TDMA will be used as the multiple access method. The satellites will be about 4 m long and 1 m wide, have 200,000 voice circuits, onboard processing (which is necessary for TDMA), and an estimated lifetime of 11 years. Teledesic will be using the Ka-band (30/20 GHz), each satellite will have 65 beams, and one-way propagation delay is estimated at 2.32-3.4 ms. The *footprint* diameter for the satellites will be 2523 km (1577 miles).

User terminals will be portable, but not hand-held. Teledesic plans to begin operations in 2003, at a projected cost of $10 billion.

Odyssey

Odyssey is a planned global mobile personal communications system (founded by TRW Inc. and Teleglobe Canada Inc.), which will offer encrypted voice, data, fax, paging, messaging, and position location communications. The system will be based on a satellite constellation of 23 satellites in Medium- Earth Orbit (MEO) at an altitude of 12,072 km (7545 miles). Encrypted voice transmissions will be at 4.8 kbps and data transmissions will be at 9.6 kbps.

The satellites will be simple repeaters, with no onboard processing, having 3400 voice circuits and 62 beams. User downlink will be in the S-band (2483.5–2500 MHz) and uplink will be in the L-band (1610–1626.5 MHz). Feeder uplink (29.5–30 GHz) and downlink (19.7–20.2 GHz) will be in the Ka-band. One-way propagation delay is predicted to be 34.6–44.3 ms. Here CDMA will be the multiple access method.

The satellite's *footprint* will be 21,650 km (13,531 miles), and its expected lifetime is 11 years.

User terminals will be hand-held, have a normal cellular telephone mode, and cost about $200; the expected call rate is $.64 per minute. Odyssey is expected to become operational in the year 2002, at an expected cost of $4.3 billion.

Advantages and Disadvantages

What are the relative strengths and weaknesses of these competing systems in the real world? Hughes, INTELSAT and Inmarsat have the advantage of satellites already in operation and a proven record with them. Hughes and Inmarsat have the advantage of significantly higher encrypted data transmission rates than most of the competition. They also have the advantage that under current law, geosynchronous systems have absolute priority over LEO systems in bandwidth assignment. The disadvantages they all have are those inherent with geosynchronous satellites, especially propagation delay.

All LEO systems have an advantage over the geosynchronous systems in their much smaller propagation delays, and smaller, cheaper satellites and terminals. They also share several disadvantages—they require a much greater number of satellites than the geosynchronous systems, they are unproven, and they are at least three years away from being operational (if everything goes right).

Teledesic has an advantage over the other LEOs of a high encrypted data transmission rate, but it is the costliest of the LEO systems and one of the most complex. On the other hand, Iridium has an advantage over other LEOs in being completely independent of ground stations, but it also is one of the most complex, and has a higher call rate than its competition.

A disadvantage in common for Teledesic and Iridium is the use of TDMA as their access method, which does not allow different systems to share the same frequency spectrum and requires that limited spectrum available to be assigned ahead of time. However, Motorola has the very real real-world advantage of having a great deal of influence in regulatory circles.

Globalstar has the advantages of the smallest number of satellites required among the LEOs, the lowest calling rate, and its choice of CDMA as its access method. This will allow the sharing of the frequency spectrum among different systems and does not require allocation of the spectrum ahead of time. On the other hand, Odyssey is the odd one out here because, with its MEO, it falls between the geosynchronous and LEO systems.

So, which has the greatest chance of success for the future? Most of the experts are qualifying their opinions with a great many *ifs* and *coulds*. In other words, *only time*, as the saying goes, *will tell.*

From Here

Chapter 17 covered the following: geosynchronous satellites; Low-Earth Orbit satellites; multiple access methods; geosynchronous systems; Low-Earth Orbit systems; and, advantages and disadvantages of each. Chapter 18, focuses on the following: Channel Control System and AMS software intricacies; DEC's MicroVax II computers and VMS computer operating system; the exact, technical nature of video and audio baseband systems; the idiosyncrasies of satellite uplink equipment; the technical characteristics of current generation satellites; the wide variety of satellite downlink systems; the technical liaison with General Instruments and DEC; the interface with the uplink operator to ensure transmission quality control; support of cable headend technicians who have intranet reception troubles; loading of DBS/IPPV program epochs from printed schedules; creation of all DBS tier-addressed messages; routine tape backups of DEC computer system hard disks; spot monitoring of signal quality with our test equipment; software development to enhance encryption systems; full-time and occasional leasing of VideoCipher encryption systems; immediate utilization of all features—absolutely no learning curve; quick identification and resolution of operational problems; very low cost to achieve and maintain complete operational competence; security of the Videocrypt system; ISO card protocol; Videocrypt protocol; and VCL File Format.

Endnotes

1 Joseph A. Mercer, "Satellite Voice and Data Communications," MIS 468, Southern Illinois University, Edwardsville, Illinois, 2208 Rock Road, Granite City, IL. 62040, 1996, pp. 3–4.

18

Satellite Video And Audio Encryption

When HBO first introduced scrambling to the marketplace in 1986, the technology used did not provide adequate protection against piracy. While the actual encryption algorithm was never broken, the decoder module hardware of the original system (VideoCipher II by General Instruments) was easily defeated. A very active pirate community sprung up almost overnight. Since the advent of the VideoCipher II Plus (VCII) scrambling system from General Instrument, the problem of satellite signal theft has been brought substantially under control, but not entirely. Today, the majority of people watching satellite television subscribe legally and pay for the programming they receive. More importantly, the majority of satellite dealers now recognize that they are doing both themselves and their customers a great disservice by supporting satellite signal theft.

One of the aftereffects of the intense pirate activity during the late 1980s is that many of the pirates sorely miss the days of being able to make a quick, albeit illegal, buck by selling *chipped* VCII decoders. Unfortunately, individuals are now attempting to repeat history. They are resorting to some very creative marketing practices to lure satellite dealers and consumers into parting with their money.

An entity that appears to be based in the Cayman Islands, BWI, placed an advertisement that first appeared in the October 1995 issue of *Satellite Watch News.* The ad describes a product called the *Cyber-1 MIP.* It is hailed as a *small integrated test device that installs easily into the VCII Plus board.*

Facts about the Cyber-1 MIP Device

The MIP device does allow manipulation of the *non-secured* areas of memory contained on the VCII Plus board. The device can manipulate some functions of the internal character generator in the decoder module and can alter the manner in which diagnostic information is displayed on the screen. Specifically, *The MIP* allows the user to:

- Blank the zip code coordinates stored in decoder RAM, thereby allowing the customer to view programs, such as sporting events, that might otherwise be blacked-out in his/her area.
- Prevent the display of advertising messages (TAMS) used by programmers to solicit new subscribers.
- Change the rating ceiling in the decoder to allow the customer to view programs that would otherwise be unavailable because of a rating restriction.
- Delete passwords stored in the decoder RAM that are normally used to restrict user functions.[1]

Note: The MIP does not interact in any way with the secure very large scale integration (VLSI) chip contained within each VCII Plus module. VCII Plus security has not been entirely broken by this device. *The MIP* cannot easily decode the audio for fully scrambled services and it does not have the complete ability to decrypt VCII Plus keys. While people may be thinking that this is a *break* in VCII Plus security, it is only an attempt. VCII Plus security remains intact for the most part (see the Sidebar that follows).

Debunking Cyber-1 Marketing Claims

McKibben Communications (a California-based supporter of encryption systems for a variety of satellite programming suppliers) did an independent analysis of the claims made in the various marketing materials produced by Cyber-1. Here are a few excerpts from its ad followed by McKibben's comments in parenthesis. *(Disclaimer: The following comments or opinions do not necessarily represent those of the author, but are simply documented here as information for comparison purposes only.)*

- SUPERBINARY ENCIPHERMENT routines for peak ECM protection. (*Superbinary encipherment* is a phrase coined by Cyber-1. It is pure gibberish.)
- A state of the art custom microprocessor governs the software operation and handles all security routines. (Marketing hype — *The MIP* uses a processor that costs less than $1.00.)
- Exceptionally technical KILL ROUTINES. (Gibberish aimed at a non-technical person.)
- Revenue generating technology. (Appeals to the greed of unscrupulous dealers.)
- Performance that exceeds expectations. (Implies that the MIP does more than just provide a means to *test* signals.)
- We designed the MIP to be used as a legal enhancement device within North America and as a test device to allow dealers outside of North America the ability to test subscription programming they do not subscribe to. (Test or no test, this still sounds like signal piracy to us.)
- This device is not designed to allow the user to watch programs without paying for them. (What other purpose could it have? Most people have no need to *test* signals.)
- The international feature of the device has a great potential for abuse within North America. (This is the message that Cyber-1 wants to get to unscrupulous dealers in the U.S. that could be potentially duped into buying the Cyber-1 products.)
- Designed by a team of industry leading partners that are as old as the VCII itself. (This conjures up an image of a bunch of 10-year old kids doing the design work for Cyber-1 since

VCII was introduced in 1986. Maybe we should be taking this phrase literally.)

The Cyber-1 *press release* does some major backpedaling on the delivery date of *The MIP*. Again, McKibben Communications excerpted verbatim some phrases from the *press release* and have provided their comments in parenthesis:

+ The delay has been attributed to a number of unforeseen difficulties in the security design of the product. (To stall for more time, Cyber-1 needs to make excuses. It's obvious that this is just the first in a long series of such excuses.)
+ We wanted to release the product before the Christmas season with fully functional features including the advanced international Security Code Integration (SCI) routines. (More meaningless gibberish aimed at the non-technical person.)
+ These routines allowed our international clientele the ability to test services they could not subscribe to such as HBO, Cinemax, Showtime and the Movie Channel just to name a few. (By naming specific channels, Cyber-1 holds out the promise of being able to receive premium channels without a subscription.)
+ We realize that the delay will come as a great disappointment to all of the dealers who have ordered the MIP. However, a product with a limited life span is of no use to anyone. (Meant to keep suspicious dealers at bay for as long as possible.)
+ We have decided to ship each and every person who ordered our product, whether by prepaying or by COD, a free package on the MIP. This will include an extensive video tape, a special pre-release MIP processor and complete instructions. (If you include the cost of the videotape, this offer will cost Cyber-1 about $10.00 plus shipping. Dealers may think they're getting something of value, but the *pre-release* MIP bears no resemblance to what is implied in the Cyber-1 marketing materials. Dealers will have either pre-paid the $199 or will have to pony-up when the COD package shows up. A very creative interpretation of the word *free*.)[2]

Caveat Emptor: Let the Buyer Beware

The Cyber-1 MIP is a simple device that only manipulates some of the ancillary functions of a VCII Plus decoder. Some of these same functions can be manipulated through the remote control on any VCII Plus-equipped system. The only significant security problem that *The MIP* presents is its ability to defeat regional sports blackouts. Even though blackouts can be defeated, programming cannot be received in the absence of a proper subscription for the channel. There is no access to fully scrambled signals and audio signals are not available on channels for which no subscription has been purchased.

The parts used in *The MIP* cost less than $5.00. The device is being marketed at a price of $199 to satellite dealers. At that price, a dealer is promised a videotape containing a *MIP-Tutorial*, some software, and an instruction manual. Cyber-1 is offering a special *promotional* package containing 10 MIP modules, the tutorial videotape, and some software utilities for $1492.50. Estimated cost of goods in this package is less than $50. The suggested retail price to consumers for *The MIP* could be well in excess of $400.

Because of the history of piracy in the Direct To Home (DTH) market, dealers and consumers are likely to believe that a breach of VCII Plus security has actually occurred. Conventional logic dictates that if it happened once, it could happen again. The Cyber-1 folks are counting on people remembering the *good old VCII days* when a *chipped* decoder worked quite well. Cyber-1 portrays *The MIP* to hold that promise. It is a form of the old *bait and switch* sales technique with no recourse for the consumer or dealer who falls for it. The truth is that the device does not provide the kind of breach in security that people remember from the days before VCII Plus. By the time dealers and consumers wise up to the game, a lot of money will have changed hands.

Many satellite television programmers are inadvertently helping the Cyber-1 people to perpetrate their false claims. A number of channels intentionally run their VCII Plus scrambling system in *Fixed-Key* (FK) or *Universal Access* (UA) modes when running promotions. In these modes, the use of tier-addressed messages (TAMS) encourages consumers to become subscribers. Because *The MIP* does allow TAMS to be blanked out, someone observing a channel operating in FK or UA modes might be convinced that *The MIP* device is actually decoding the signal. The reality is, in this instance, *The MIP* is only preventing the TAM from being displayed. It does, however, allow for a rather convincing *demonstration* of its alleged capabilities.

While *The MIP* is very easy to install on a VCII Plus board, the holographic seals will be broken in the process, thus voiding the warranty. A consumer who allows this to be done to his/her decoder module will ultimately end up having to purchase a new VCII Plus module to *undo* the damage.

Next, let us take a look at another satellite video and audio encryption pay-TV access control and scrambling system known as Videocrypt. Consider this part of the chapter as some kind of frequently asked questions list with answers about the system.

Videocrypt

Videocrypt encodes the TV image by cutting each line of the image in two pieces at some cut point and then exchanges these two line fragments in the broadcasted pictures. For example, if a line such as:

0123456789

passes the encoder, the output might look like:

4567890123

where the digits represent the pixels of the image. There are 256 possible cut points and no cut points directly near the image border (the minimum distance from the margin is about 12–15% of the image width), which is the reason why you sometimes still can see vertical patterns even on an encrypted image (video). The sound (audio) is currently not encrypted.

Several times per second, a computer at a broadcasting station generates a 32-byte long message, which when broadcasted is encoded together with forward error correction information in the first invisible lines of the TV signal similar to teletext. About every 2.5 seconds, one of these 32-byte messages is processed in the encoder by a secret hash algorithm, which transforms the 32-byte message into a 60-bit value.

These 60 bits are then used by a second algorithm in order to determine the 8-bit cut point coordinates for each line for the next 2.5 seconds. No details about this second algorithm are known, but think of it just as some kind of 60-bit pseudo random number generator (PRNG) where the 60-bit output from the secret hash function is used as a start value (seed).

The decoder receives the 32-byte messages and other data together with the TV signal, applies some error correction algorithms, and passes all 32-byte packets to the smart card in the decoder's card slot. The smart card implements the same secret hash function and answers with the same 60-bit value (like the one used in the encoder). By using this 60-bit answer from the card, the decoder hardware can generate with the same PRNG the same cut point sequence as the encoder. It can also reconstruct the original image by exchanging (again) the two line fragments. The secret hash function is a cryptographically strong system designed in such a manner as to make it extremely difficult for one to guess the algorithm of the function even if looking at many pairs of 32-byte/60-bit values.

Apart from being the source for the generation of the 60-bit PRNG seed, the 32-byte messages from the broadcasting station contain card numbers so that individual cards can be addressed. They also contain commands like activation, deactivation, and pay-per-view account modification. In addition, the 32-byte packets contain a digital signature (currently 4 bytes) that allows the card to test whether the 32-byte messages really originate from the encoder and have not been generated by someone analyzing the card. Again, this digital signature, like the hash function, has been designed so that it is difficult to find out how to generate a correct signature even by looking at enough examples. This prevents chosen-text attacks — where someone tries to probe the secret hash function with very carefully selected 32-byte messages. It also prevents hackers from generating new activation commands for the card.

For instance, in early 1993, someone managed to get access to the secret hash functions of several stations that use Videocrypt (British Sky Broadcasting (BSkyB), Adult Channel, JSTV, BOB, Red Hot TV). Most of these systems used the same hash and signature algorithm. The only difference between the stations was a 32-byte secret key table. It is not known how it was possible to get this information. Either someone from the company who manufactured the cards (News

Datacom Ltd.) released this information or it was possible for someone to read out the EEPROM contents of the card processor (very difficult, but theoretically possible). With this knowledge, it was then quite easily possible for the original hackers to produce *clone cards.* These are simple PCBs with a cheap microcontroller (one of Microchip's PIC family), which implements only the secret hash function and serial 1/0 procedures in its EPROM and answers with the correct 60-bit values to 32-byte messages just as the real cards do. For several channels, clone cards are still available. But, BSkyB distributed new 09 series cards in the spring 1994 and switched on 1994-05-18 to a now secret hash and signature function. Consequently, all clone cards stopped working.

The clone cards did not implement any interpretation procedures for card activation, deactivation, and pay-per-view functions. Thus, the clone card's software is considerably simpler than the one in real cards. This resulted in some tiny differences between the reaction of the clone card software and the original card software on pathological 32-byte messages. These differences were used in countermeasures (commonly referred to as ECHN) against clone cards several times in 1993 and 1994 by BSkyS and News Datacom, in order to deactivate clone cards. But, it was quite easy each time to find the tiny bugs in the clone card software and correct them.

There are two microprocessors in a typical Videocrypt decoder. An Intel 8052 microcontroller manages the communication between the smart card and the rest of the system. Since the software of this processor is not read protected, it was also possible to reprogram this chip (by using the EPROM version 8752BH) so that the hash algorithm is performed inside the decoder. Thus, no external card is needed at all for the channels in which the hash algorithm was implemented in the 8752. The second processor is a Motorola 6805 variant. Its internal ROM :contents cannot be read out easily. The Motorola decodes the data that comes with the TV signal, applies error correction algorithms to this data, exchanges the 32-byte messages and 8-byte answers with the Intel processor, and controls the PRNG and the on-screen display hardware.

There are also Videocrypt II decoders available. These work almost like the Videocrypt decoders, with the only important difference being a new software in the Intel and Motorola processor. Videocrypt II decoders get their data from other invisible TV lines rather than from Videocrypt. And, it is possible to broadcast a signal encrypted in a way that allows both Videocrypt and Videocrypt II to decode it with different smart cards.

More detailed basic information about Videocrypt has been published in the European patent BP 0 428 252 A2 (a system for controlling access to broadcast transmissions). You can order a copy for approximately 10 Deutsche marks from the European Patent office (Schottenweldgasse 29, A-1072 Wien, Austria) if you are interested.

Videocrypt System Security

The Videocrpyt system is very secure because all the secret parts that are essential to a successful decryption are located in the smart card. And, if the card's secret hash algorithm/key becomes known, it can easily be replaced by just sending new cards to the subscribers. This card exchange can also be used if details about the format of the commands hidden in the 32-byte sequences sent to the card become known. Together with the knowledge of the signature algorithm, this allows generation of new activation messages and filtering of deactivation messages. There are, however, at least two obvious security flaws of the system that cannot be removed by new smart card generations.

The first security flaw has to do with the fact that the dialog between the card and the decoder is the same synchronously for all Videocript decoders switched to this channel. (the decoder does not add any card specific or decoder specific information to the traffic). This makes it possible to use one card for several decoders. In other words, it is possible to record the 32-byte messages broadcast by a station during an evening with a PC, then to send these messages to someone else with an original card who asks his or her card for the 60-bit answers to all recorded messages. If this person then sends the 60-bit answers back, you can then use this data to descramble the VCR recorded program of a particular evening (delayed data transfer). However, decoding VHS recorded encrypted signals produces minor color distortions. Also, a few VCRs do not preserve the Videocrypt data stream in the first invisible lines that accompany their V signal. It is also possible to distribute the 60-bit answers from one card in real time with cables, to many decoders in a house or with radio signals to many decoders in a larger region.

The second security flaw is concerned with the simple cut-and-exchange encryption method. It is also concerned with the fact that two consecutive lines in an image are almost always nearly identical—making it possible to try all 256 possible cut points and to select the one that causes both lines to fit together best. This method has already been implemented on fast PCs with framegrabbers—which load the image into the memory and display it corrected on the computer screen (many seconds per frame); on parallel supercomputers which allow almost real-time decryption; and, with special hardware that achieves real-time decryption. However, with this decoding method, there are severe image quality losses and many additional problems—which, when put together with the high hardware costs required (much higher than a regular subscription), do not make this approach very practical for every day usage.

Both of these security gaps in the Videocrypt systems do not allow for a very high-quality decryption method that is used in a card. However, the described methods have already been successfully used by a few technically skilled people who like to analyze encrypted programs.

ISO Card Protocol

The card and the protocol used to communicate with it conform exactly to the International Standard ISO 7816. The options used from this standard are: $T = 0$ asynchronous halfduplex character transmission protocol, active low reset, and inverse convention. Only a few basic principles of the ISO protocol will be explained here. For additional detailed information, please read the ISO standard, which you can order from your national standards body (DIN, ANSI, APNOR, BSI, *DS*, etc.). There are three parts of the standard: ISO 7816-1 describes physical characteristics of the card and quality tests a card has to survive; ISO 7816-2 describes the location and meaning of the contacts; and ISO 7816-3 (most important) describes the electrical characteristics, the answer-to-reset message, and the protocol.

The data format is an asynchronous 9600 bit/a serial format similar to that used on RS-232 lines with 8 data bits, 1 parity bit and 2 stop bits. The parity is even (but if inverse bit meaning convention is used, a RS-232 interface has to be programmed for odd parity in order to produce the correct bit). There is also an error detection and character repetition mechanism in the protocol, which is not supported by

RS-232 interfaces. If the receiving device (decoder) detects a parity error, it sends an impulse during the stop bit time. This will tell the sender to retransmit 1 byte.

After a reset impulse to the card, the card answers with an answer-to-reset message with some information about the card. If the first byte is 3fh, then this means that in order to read the bytes with an RS-232 interface, you will have to invert and reverse all bits. A typical answer-to-reset looks like the following one shown in Table 18.1.[3]

The answer-to-reset message has a variable length format. Some bits specify whether certain bytes are present or not. If the lowest bit in the high nibble of the second byte is 1, then the previously shown third byte is present and determines the relation between the bit rate and the clock frequency after the reset answer.

Table 18.1 Answer-to-Reset Message

3f	fa	11	25	05	00	01 b0 02 00 4d 59 00 81 80

						'historic characters' with
						information about chip and
						software version, etc.
				±	low nibble: protocol type T = 0
					high nibble: end of ISO part
			±	requests 5 additional stop bits
		±	encodes programming voltage and max. programming
		current (here: SV, 50 mA)
	±	clock freq.: 11h = 3.5 MHz, 31h-7 MHz
±	the 0ah low nibble means: 10 'historic characters, which are not defined in the ISO standard, are appended to the reset answer

Note: 11h means that 372 clock cycles are 1-bit duration (default). With a clock frequency of 3.5712 Mhz, the bit frequency is 9600 Hz.

In the Videocrypt system, the bit rate is always 9600 bits/s, but a value of 31h (=factor 744) in the third byte requests a doubled clock frequency (−7MHz) from the decoder. Other values are not supported by the Videocrypt decoder.

The Videocrypt decoder supports several programming voltages (5 V, 12.5 V, 15 V and 21 V, max. 50 mA current) and different numbers of stop bits ($\geq$ 5) sent to the card. All these parameters can be selected in the answer-to-reset. Of the *historic characters* part, the decoder only verifies that it is at least 7 characters long and that the values 4dh and 59h are at the positions as in the example, otherwise the card in rejected. No more details about the information in the historic characters part of a Videocrypt card is currently known. For the detailed format of the answer-to-reset message, please consult ISO 7816-3.

The T = 0 protocol is a half duplex master slave protocol. The decoder can send commands to the card followed by a data transmission either to or from the card. The card can do some limited flow control and can request or deactivate the programming voltage VPP selected in the answer-to-reset using *procedure bytes*. If the decoder initiates a command, it sends five header bytes to the card as follows:

53 78 00 00 08

The first byte (CLA) is the command class code, and is always 53h in the Videocrypt system. The second byte (INS) is the instruction code. Its lowest bit is always 0, and the instruction codes have never had a 6 or 9 high nibble (you will see why). The following 2 bytes (Pl and P2) are a reference (an address) completing the instruction code, and a Videocrypt decoder sets them always to 00 00. The final byte (P3) codes the number of data bytes to be transmitted during the command. P3 = 0 has a special meaning: in data transfers from the card, it indicates 256 data bytes. In data transfers from the decoder, P3 = 0 indicates 0 bytes. The direction of the data transfer is determined by CLA and INS and must be known in advance by both the card and the decoder.

After transmission of such a 5-byte header, the decoder waits for a *procedure byte* from the card. The following procedure bytes are possible:

- *60h*: Please wait, I will send another procedure byte soon, do not timeout.
- *INS*: Now let us transfer all (remaining) data bytes, I do not need programming voltage.
- *INS + L*: Now let us transfer all (remaining) data bytes and please activate VPP.
- *INS xor ffh*: Now let us transfer another single data byte, I do not need programming voltage.
- *(INS + 1) xor ffh*: Now let us transfer another single data byte, and please activate VPP.
- *6Xh or 9Xh*: This byte (SWI) indicates an end of the data transfer and requests to deactivate VPP. A second status byte (SW2) follows from the card. SW1 SW2 = 90 00 indicates a normal termination, other values report an error.[4]

After each data transfer, the decoder waits for another procedure byte. A typical decoder⟨-⟩card dialog looks like this (command 78h requests the 60-bit answer as 8 bytes from the card):

- Decoder mends header:

    78 00 00 08

- Card sends procedure byte (all at once, no VPP):

    78

- Card sends P3 data bytes:

    80 52 02 79 f5 39 7c 0a

- Card closes with SWl and SW2:

    90 00[5]

Videocrypt Protocol

The newer Videocrypt smart cards do not require any programming voltage (the VPP pin isn't even connected). Although, the ISO standard requires only 2 stop bits after each transfered byte, Videocrypt decoders seem to require more than 5 stop bits. Because PC serial ports do not support more than 2 stop bits directly, a card emulator software has to wait for about 0.5–1.5 ms after each byte. Cards can announce in the answer-to-reset message how many stop bits they require. And

Videocrypt cards also require more than 2 stop bits. A videocrypt decoder knows the 10 commands (all with CLA = 53h and Pl-P2 = 00h) as shown in Table 18.2.[6]

Next, let us take a look at some of the things that are known about the data bytes of the commands in Table 18.2.

70h

In BSkyB cards, the 70h data contains the card issue number (07 or 09) in the low nibble of the first byte. The high nibble of the first byte seems to be always 2. The next 4 bytes form a 32-bit bigendian integer value, which corresponds to the decimal card number without the final digit of the card number (which is perhaps a check digit, algorithm unknown). The meaning of the final byte is unknown.

72h and 7ch

Several times per second, the decoder requests with 7ch 16 bytes from the card. If a card is removed and a new card is inserted in the decoder without switching off the power of the decoder, then shortly after the card reset, the decoder sends the latest 7ch data bytes from the previous card in a 72h message to the new card. In this way, 16 bytes

Table 18.2 The 10 Commands

INS	Length (P3)	Direction	Purpose
70h	6	from card	serial number, etc.
72h	16	to card	message from previous card
74h	32	to card	message from station
76h	1	to card	authorize button pressed
78h	8	from card	60-bit answer
7ah	25	from card	onscreen message
7ch	16	from card	message to next card
7eh	64	from card	??? \
80h	1	to card	??? > perhaps Fiat-Shamir
82h	64	from card	??? / authentication?

information (the status of a pay-per-view account or a list of activated channels?) can be transferred from one card to the next.

74h and 78h

The 74h command transfers the 32-byte messages from the broadcasting station to the card. If the third bit (value 8) in the first byte is set, then the decoder will ask with a 78h command for the 60-bit :answer. This happens about every 5th 74h packet every 2.5 seconds. The high nibble of the final byte in the 78h data is ignored by the decoder (only 60 bits are needed). The high nibble of the first 74h byte seems to have the value *oh* or *fh* in normal encrypted operation, and *ch* or *dh* in the *soft scrambled* mode where the decoder can descramble the image even without any card.

The following information is valid for the 07 and 09 BSkyB card and need not necessarily be true for future smart cards, because these data bytes do not seem to be interpreted in the decoder and so their meaning can be exchanged. A typical BSkyB 74h packet for the 09 series card looks like this:
e843 OaSS8261 Oc 29e4O3f6
20202020202020202020202020202020 fb54acO2 51

The second byte indicates the current date and counts the months since January 1989. In the 07 card, this month code selects one of several 32-byte secret key tables that are used by the hash function. When the switch from the 07 hash algorithm to the new 09 algorithm happened on 1994-05-18, this value jumped from 40h (1994-05) to 43h (1994-08) which might indicate that the activation of the 09 algorithm was originally planned for August. In the 07 card, this value was only interpreted to find an offset into a table with various 32-byte secret keys.

The third byte seems to be a random number. This byte together with the month code is used to generate with a quite simple algorithm four XOR bytes that are necessary to decode the command byte and the card number prefix (described next). If you XOR these four bytes with bytes 8 to 11; and, if you XOR only the first of the four bytes with byte 4, then you have decrypted the card number and the command code.

The fourth byte is an encrypted command code. Some decrypted known values:

0x00   Deactivate whole card (message: 'PLEASE CALL 0506 4847771)

0x01   Deactivate Sky Movies (message: 'THIS C L IS BLOCKED')

0x02   Deactivate Movie Channel

0x03   Deactivate Sky Movies Gold

0x06   Deactivate Sky Sports

0x08   Deactivate TV Asia

0x0c   Deactivate Multichannels

0x20   Activate whole card (remove 'PLEASE CALL 0506 484 777')

0x2l   Activate Sky Movies (remove 'THIS CHANNEL IS BLOCKED')

0x22   Activate Movie Channel

...

0x2c   Activate Multichannels

0x40   Pay-per-view account management command

0x80   \

0x81   \ perhaps 09 card ECM

0xf0   / commands

0xf1   /[7]

Packets with incorrect command bytes and correct signatures can irreversibly kill a card (it does not even answer the reset). The fifth and sixth bytes seem to be parameters for pay-per-view account management (program number and number of tokens) and do not seem to have a meaning for enabling and disabling commands. The lower 7 bits of the seventh byte contain a channel ID.

A card number is represented by a 5-byte card address consisting of a 4-byte prefix and a 1 byte suffix. The five bytes for a card are identical to the first 5 bytes of the 70h answer. Only the high nibble of the first address byte seems to have a different purpose (unknown). Up to 16 cards with the same card address prefix can be addressed with one single 32-byte 74h message. The bytes 8-11 might contain the common prefix to the addressed cards and the bytes 12-27 the various suffixes. If there are fewer than 16 different cards to be addressed, then the same suffix byte is repeated several times in order to fill the space. The A-byte prefix is encrypted like the command byte by XORing it with the four bytes generated using the bytes 2 and 3.

The 4 bytes 28-31 contain the digital signature, which is simply an intermediate result of the iterations of the hash algorithm. If the checksum, the digital signature, or some of the values in the first 7 bytes of a 74h command are not correct, then the 78h answer will only contain 8 00 bytes or in some cases 01 00 00 00 00 00 00 00. The final byte 32 is a simple checksum that makes the sum of all 32 bytes a multiple of 256.

The 07 card (and also cards used by Sky New Zealand) has an interesting security hole: The card sends to the decoder as many data bytes as specified in P3. By sending a higher length value in the command header to the card, one can get up to 256 data bytes back—which seem to be values from the card's RAM that allow some insight into the internal data structures of the card software.

76h And 7ah

If the authorize button on the decoder is pressed for a few seconds, then the decoder will send a single 76h message with a 00 data byte to the card. The 7ah on the other hand requests from the card an ASCII text, which is then displayed on the TV screen. The display field is 12 characters wide, one or two lines high and no lowercase letters are supported. The lower 5 bits in the first byte indicate the length of the text that is to be displayed: 0 for no display, 12 for a single line, and 24 for 2 lines. The highest 3 bits of the first byte seem to be some kind of display priority. The number there (0-3) must be high enough if standard decoder messages have to be suppressed. The remaining 24 bytes contain the ASCII test.

The meaning of the other commands is unknown, some of them are never used currently. Perhaps these commands are used for the Fiat-Shamir identification exchange described in the patent. Some cards understand also additional instruction codes which cannot be issued by a normal decoder. A BSkyB 09 card also understands 12h, 86h, 88h, 8ah and 8ch. These commands are perhaps used in order to test or configure the card at the factory, etc.

VCL File Format

The Videocrypt Card Logfile format (VCL) is used by some for performing the delayed data transfer procedure described earlier. Person *A* with a valid card can record the dialog between the decoder and the card for a certain program *P* and transmits this information as a VCL file to person *B* who has no card and has recorded with a VCR only the encrypted signal of program *P*. Person *B* now connects the Videocrypt decoder between the VCR and the TV set and connects the card slot of the decoder to a PC. Using the information in the VCL file, *B's* computer can now also decrypt program *P*. This is of course only possible for the few hours that are covered by the information in the VCL file.

Not all of the information exchanged between the card and the decoder is necessary for descrambling the TV signal. The VCL format uses this fact in order to save a lot of storage space. Only 12 bytes of high entropy (that means: almost uncompressible) are stored every 2.5 seconds. Thus, a VCL file of a 1 hour program is only about 17 kbytes large. In addition, VCL files do not contain any information about the card owner (especially not the card serial number), which usually appears in normal full log files. The only potential security hole is the remaining nibble in the 78h data. Consequently, the data should be cleared in order to avoid card specific information to leak into the VCL file.

The VCL files have a very simple binary format consisting of a 128-byte header and arbitrarily many 12-byte records. At the end, VCL files may be padded with zero bytes to a multiple of the operating system's disk sector size so that no RAM contents can leak in there out of an unsecure system like MS-DOS. Do not forget to use a binary mode if you transfer VCL files or their contents will be rendered unusable. The 128-byte header has the following format as shown in Table 18.3.[8]

After the first 128-bytes, just as many 12-byte records follow as initially announced in bytes 4-7. Each record represents a 74h/78h Videocrypt protocol pair and consists of two fields: The first 4 bytes are the final 4 bytes of the 74h data part. The remaining 8 bytes are the data part of the corresponding 78h command. Four bytes of each 74h packet are enough to allow a card emulator to quickly and reliably synchronize with the queries of the decoder. The final four bytes of the 74h commands have been selected because of their high entropy (signature and checksum).

Table 18.3 The VCL File 128-Byte Header

Byte number	Purpose
0–3	ASCII String 'VCL1' which identifies the tile type and version of the format.
4–7	The number of 12-byte records stored in this file encoded as a bigendian (most significant byte first) 32-bit unsigned integer value
8–23	Date and time when the recording started. Format: yyyymmddThhumssZ, where yyyymmdd are year, month and day ('19940618'), hhmmss are hour, minute and second ('235959'), T is just the ASCII letter T, and Z is the ASCII letter Z if the time is in UTC or a zero byte, if the time is local time. The digits are ASCII characters.
24–25	Name of the satellite or cable system from which the recording was done. This is a zero terminated ASCII string with only characters between 20h and 7eh. As many zero bytes are appended as necessary for filling up the 32 bytes. The same format is also used for the next two text fields. Example: 'Astra.'
56–63	Name/number of the transponder from which the recording was done. Example: '08' for Sky One on Astra.
64–127	Description of what has been recorded. Example: 'Star Trek: TNG, episode 123.'

Summary

The C-Band, DTH satellite industry is under intense competitive pressure and needs to be concerned about Cyber-1 activities. Consumers were burned once before by pirate dealers and, as a result, the industry suffered severely. Any disruption to consumer confidence that now exists in the DTH marketplace could be very detrimental in view of the competitive pressures from the new operators in the Direct Broadcast Satellite band.

The most insidious aspect of the Cyber-1 product is the manner in which it is being marketed. The illusory claims contained in the Cyber-1 advertisement and *press release* are intended to do only one thing: entice dealers and consumers into sending in their money. Once *The MIP* product fails to meet the implied promises that are being made to dealers, the result will be more marketplace confusion and consumer aggravation. Someone expects to make a lot of money with this *product*. Most of the money will end up being sent to a post office box in the Cayman Islands and will just disappear from there.

This harmful activity can be short-circuited only through a concerted educational effort by programmers, manufacturers, packagers, distributors and retail dealers. Responsible organizations in the DTH satellite television industry are strongly encouraged to take steps to educate their customers (particularly the consuming public), so that the damage being done to the satellite industry by the Cyber-1 marketing ploy is minimized.

Videocrypt on the other hand is a pay-tv scrambling system jointly developed by Thomson Consumer Electronics and News Datacom. Over 5 million users receive Videocrypt encrypted signals; and this system, as of this writing, has remained secure from illicit decoder manufacturers through the protection of revenue from Videocrypted television channels.

What is most interesting about VideoCrypt is that while the product appears to be more secure than VideoCipher, its cost is appreciably less. A direct cost comparison is hard to come by. However, integrated receiver descramblers in the U.S. sell for between $300—$1100 compared to $220 in England. And, a U.S. satellite dish owner upgrading to the VideoCipher RS will plunk down between $260 and $360, while the decoder portion of the Sky Television receiver amount is under $110.

From Here

Chapter 18 focused on the following: satellite encryption security; security of the Videocrypt system; ISO card protocol; Videocrypt protocol; and VCL File Format. Chapter 19 presents a framework for dealing with the following: satellite encryption policy; protection of personal data and privacy; security of information systems; intellectual property protection; international instruments; and rising demand for hardware-based data security;

Endnotes

1   W. Mark McKibben, "VideoCipher II Plus Satellite Security," McKibben Communications, 20640 Bahama Street, Chatsworth, CA 91311, 1996, p.2.

2   Ibid., pp.3-4.

3   Markus G. Kuhn, "Pay TV Access Control Systems," University of Cambridge, Computer Laboratory, TG1, New Museums Site, Pembroke Street, Cambridge CB2 3QG, United Kingdom, 1994, p.4.

4   Ibid., p. 5.

5   Ibid., p. 6.

6   Ibid.

7   Ibid., p. 7.

8   Ibid., p. 9.

19

Encrypting Intellectual Property Distribution and Electronic Commerce

Emerging information and communications networks and technologies raise questions about the nature and scope of national sovereignty and the role of government in the new electronic environment. These networks and technologies (and their projected rates of development and use) create conditions that require reconsideration of existing national rules and global consensus on new rules. It is widely recognized that individual national solutions may be inefficient and potentially anticompetitive.

Work in the Organization for Economic Cooperation and Development (OECD) on protection of personal data and privacy, security of information systems, satellite encryption policy, and protection of intellectual property is directed toward worldwide harmonization. This goal is reflected in the international instruments of the OECD in these fields.

The OECD member countries as well as nonmember countries adopt legislation based on the OECD Privacy Guidelines and Security Guidelines. In addition to legislation, they also establish standards, technical criteria, and other regulations. Hundreds of private sector entities have adopted the Guidelines and use them as a basis for codes of conduct and internal management, administrative, and technical procedures. It is to be anticipated that additional OECD work in these areas will be taken up in the same practical manner.

Note: OECD is made up of 29 member countries from North America, Western Europe, and the Pacific. The membership more or less corresponds to the world's most wealthy industrialized countries.

Satellite Encryption Policy

In recent years, OECD member countries have undertaken to develop and implement policies and laws relating to satellite encryption. In many countries these are still in the process of being developed. Disparities in policy may create obstacles to the evolution of national and global information and communications networks and hinder the development of international trade. Member governments have recognized the need for an internationally coordinated approach to facilitate the smooth development of an efficient, secure information infrastructure. The OECD is playing a role in this regard by developing consensus about specific policy and regulatory issues relating to information and communications networks and technologies, including satellite encryption issues.

In early 1996 the OECD initiated a project on satellite encryption policy by forming the Ad Hoc Group of Experts on Satellite Encryption Policy Guidelines under the auspices of the Committee for Information, Computer and Communications Policy (ICCP). The ad hoc group was charged with drafting guidelines for satellite encryption policy to identify the issues that should be taken into consideration in the formulation of satellite encryption policies at the national and international level. In addition to OECD member government officials, private sector representatives and experts on privacy, data protection and consumer protection participated in the drafting process. The ad hoc group had a 1-year mandate to accomplish this task and it completed its work in December 1996. The Council of the OECD adopted the recommendation of the council concerning guidelines for satellite encryption policy on 27 March 1997 (see the Sidebar that follows).

Satellite Encryption Policy Guidelines

The Organization for Economic Cooperation and Development (OECD) adopted guidelines for satellite encryption policy for its 29 member countries. These guidelines are nonbinding for the OECD member nations, and are merely suggestions to guide the individual countries in their efforts to formulate national and international policies and legislation regarding satellite encryption. These guidelines are summarized by the OECD in eight basic principles:

1. Satellite encryption methods should be trustworthy in order to generate confidence in the use of information and satellite communications systems.
2. Users should have a right to choose any satellite encryption method, subject to applicable law.
3. Satellite encryption methods should be developed in response to the needs, demands and responsibilities of individuals, businesses and governments.
4. Technical standards, criteria, and protocols for satellite encryption methods should be developed and promulgated at the national and international level.
5. The fundamental rights of individuals to privacy, including secrecy of communications and protection of personal data, should be respected in national satellite encryption policies and in the implementation and use of cryptographic methods.
6. National satellite encryption policies may allow lawful access to plaintext, or cryptographic keys, of encrypted data. These policies must respect the other principles contained in the guidelines to the greatest extent possible.
7. Whether established by contract or legislation, the liability of individuals and entities that offer satellite encryption services or hold or access cryptographic keys should be clearly stated.

8. Governments should co-operate to co-ordinate satellite encryption policies. As part of this effort, governments should remove, or avoid creating in the name of cryptography policy, unjustified obstacles to trade.

The OECD recommends that these eight guidelines *be taken as a whole in an effort to balance the various interests at stake.* The OECD formed an ad hoc group of experts to draft these guidelines early in 1996. The ad hoc group was chaired by Norman Reaburn (Australian Attorney General's Department). The ad hoc group included over 100 representatives from OECD member nations. According to an OECD press release, this group of representatives comprised *government officials from commerce, industry, telecommunications and foreign ministries, law enforcement and security agencies, privacy and data protection commissions, as well as representatives of private sector.*

The OECD also received input from its Business and Industry Advisory Committee, as well as experts in the fields of *privacy, data protection, and consumer protection.* While the OECD views these guidelines as recommendations for national policies, it hopes that *in the future, the Guidelines could form a basis for agreements on specific issues related to international satellite encryption policy.* The member nations of the OECD are:

- Australia
- Austria
- Belgium
- Canada
- Czech Republic
- Denmark
- Finland
- France
- Germany
- Greece
- Hungary
- Iceland
- Ireland
- Italy

- Japan
- Korea
- Luxembourg
- Mexico
- Netherlands
- New Zealand
- Norway
- Poland
- Portugal
- Spain
- Sweden
- Switzerland
- Turkey
- United Kingdom
- United States[1]

Protection of Personal Data and Privacy

The 1980 OECD Guidelines on the Protection of Privacy and Transborder Flows of Personal Data have been adopted by all OECD Member countries. They were followed by the 1985 Declaration on Transborder Data Flows.

A fundamental question relates to where responsibility for protection of personal data and privacy should lie. It should be possible, via a combination of legal, technological and other measures, return to and preserve for the individual direct control over personal data and its use. Current OECD efforts are directed toward the establishment of global consensus and an international framework for privacy and individual autonomy (freedom of movement, freedom of assembly, and fundamental human rights) in the information infrastructure.

Security of Information Systems

In 1988, the OECD Member countries asked the OECD Secretariat to prepare a comprehensive analytical report on security of information systems, covering technological, management, administrative, and legal matters. Following from this work and recognizing that explosive growth in the use of information systems for all types of applications in all parts of life had made provision of proper security essential, the OECD member countries negotiated and adopted the I 992 OECD Guidelines for the Security of Information Systems. The Guidelines provide an international framework for the development and implementation of coherent security measures, practices, and procedures in the public and private sectors.

The Recommendation of the Council of the OECD Concerning Guidelines for the Security of Information Systems recommends that OECD member countries review the Guidelines every five years with a view to improving international co-operation on issues relating to the security of information systems. The Guidelines were adopted in 1992. In accordance with the Recommendation of the Council, OECD member countries reviewed the Guidelines again in 1997.

Intellectual Property Protection

In contrast to data protection and privacy, security and the satellite encryption policy, there is already a long-established international foundation for the protection of intellectual property in numerous international conventions and the international organizations that administer them: the World Intellectual Property Organization (WIPO), the World Trade Organization (WTO), and the United Nations Educational, Scientific and Cultural Organization (UNESCO). The OECD concentrates its efforts in the field of intellectual property on issues related to new technologies such as information and communications technologies—and biotechnology in cases where there are questions of the applicability of existing rules or the need for new rules.

For example, one of these new intellectual property protection technologies is VLSI Technology, Inc.'s VLSI VMS230 GhostRider Security Chip for PC-based satellite encryption and intellectual property distribution. *Disclaimer: The following comments or opinions with regard to the VMS230 GhostRider Security Chip do not necessarily represent those of the author. Nor are they included in this book for promotional purposes of said product, but are simply documented here as information for comparison purposes only.*

The VMS230 GhostRider Security Chip

VLSI Technology, Inc.'s VMS230 GhostRider Security Chip promises to make PCI-based (peripheral component interconnect) computers, modems, web browsers, and set-top-boxes safer for intellectual property distribution and electronic commerce through the hardware implementation of its new PCI bus-compatible real-time satellite encryption/decryption chip solution. Already identified as a key element of Microsoft's interactive PC/TV Platform, the GhostRider security processor enables a full slate of functions including file encryption, secure satellite communications, safe distribution of high-value intellectual property, secure electronic commerce, private e-mail and user authentication with minimal impact on system speed or data throughput.

Establishing a high confidence level in the security of electronic commerce is impossible without the protected transmission of intellectual property. The GhostRider data security processor converts the PC into a platform for electronic commerce and content distribution by entertainment companies, publishers, software houses, and other information providers.

Rising Demand for Hardware-Based Data Security

While traditional approaches to data security stress software solutions, software-only data security has proven vulnerable to attack. Furthermore, software security algorithms typically run on the same microprocessors used to manage other data.

This often slows down overall computer performance, particularly when using strong, mathematically intense cryptographic algorithms. The GhostRider chip via VLSI's security integrated circuit (IC) architecture vastly strengthens data security and privacy by performing all satellite security encryption/decryption within a dedicated chip, closed to outside attack. In addition, ICs integrate an on-chip RISC processor with encryption/decryption engine functional system blocks (FSB™) that accelerate encryption processing and relieve main CPUs of data security processing overhead.

Designed into Microsoft PC/TV Platform

The GhostRider chip has already been designated as the standard satellite encryption engine for the Microsoft interactive PC/TV platform. The PC/TV services deliver a wide range of electronic commerce products and services via satellite to end users including digital video, software downloads, home shopping, and financial services. The PC/TV lead customers and partners for the GhostRider chip include, ComStream, DIRECTV(R), Hughes Network Systems, and Samsung.

VLSI and Bell Labs Codevelopment

The GhostRider security chip results from cooperative development between VLSI and Bell Laboratories—the research and development arm of Lucent Technologies. Bell Labs provided much of the encryption/decryption firmware embedded in the GhostRider chip.

GhostRider Chip Security Features and Benefits

The GhostRider real-time satellite encryption/decryption chip is an application-specific standard product (ASSP) designed as a PCI-based cryptographic processor. Main elements of the chip include a PCI bus master-slave interface block, VKIIO Security FSB, secure vROM^{TM} OTP (one time programmable) memory, Advanced RISC Machines (ARM) Reduced Instruction Set Coding/Computer (RISC) processor FSB, an I^{2}C serial interface block and a generic 8-bit input interface block. The VK110 security block supports data encryption standard (DES), Triple DES, and random number generation. The ARM RISC processor core manages operations of the GhostRider satellite encryption chip— accelerating encryption/decryption functions. The VMS230 chip's PCI block acts as the gateway between the crypto subsystem and the PC's main data bus (see Figure 19.1).[2]

All satellite encryption operations are performed on the GhostRider chip. A Bell Labs-developed proprietarily on-chip operating system manages the functions of performing RSA public key management as well as controlling data flows to and from the high-speed DES engine.

PCI-based, DES and RSA Satellite Encryption Processor

As previously mentioned, VMS230 is a PCI cryptographic system on a single chip that meets Microsoft's Broadcast Data Network (MSBDN)

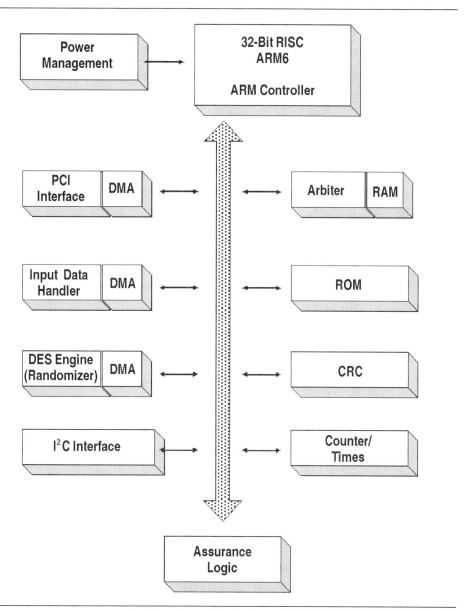

Figure 19.1 VMS230 Architecture.

DSS Satellite receiver conditional access hardware requirements. The VMS230 MSBDN conditional access chip incorporates RSA key management, triple DES for payload encryption and data interface, and, services via a packet handling and a master/stave PCI bus.

As well as providing MSBDN security capabilities, the VMS230 is an excellent vehicle for PC-based security applications such as disk encryption, high-speed secure satellite communications, cable modem applications, and public key encryption. The VMS230's DES Engine consists of DES Core FSB™. This block performs satellite data encryption and decryption, key generation (performed via Randomizer), key storage (in fast access cache memory), and secure secret key storage using vROM™ technology.

A high performance PCI interface links the VMS230 to a host processor. The VMS230 PCI supports both master and target operations as defined in the PCI Local Bus specification for a 32-bit Master/slave configuration.

A 32-bit embedded ARM processor on the VMS230 provides system control and secure function processing. Cryptographic processes occur such that keys and other sensitive data never leave the chip in the clear. The ARM and the various DMA logic handle all data movement throughout the VMS230. The DMA logic provides fast PCI transfers to and from the DES Security block, on-chip memory (RAM, ROM), and external memory. The VMS230's compliant PCI interface is the main interface with the host processor, while the IDH (Input Data Handler) is used for byte-wide data reception and the I^2C is used to communicate with other components in the system.

ARM

A 32-bit embedded ARM RISC processor block is used for system control and cryptofunction processing. All cryptographic processes execute on the ARM and occur under a secure mode, where keys and other sensitive data never leave the chip in the clear. The ARM controls the data flow, implements multiple DES encryption and decryption algorithms, and provides arbitrated shared memory communications. The CPU provides 11 basic instruction types and has a simple and consistent programming model. The design offers a high MIPS/watt ratio, a cost-effective die size, and a multistage pipeline to allow the many parts of the CPU to operate concurrently.

The ARM performs all on-chip cryptographic functions. These functions are accessed through a special Software Interrupt (SWI) instruction. Once the special SWI instruction is processed, the cryptographic functions are designed to operate entirely from instructions located in on-chip memory, requiring external access to data only. Utilizing the cryptographic services that are available on the VMS230, users can customize private cryptographic functions to meet specific security requirements.

DMA

Multiple data paths through the VMS230 and multitasking necessitate direct memory access (DMA) to alleviate potential bottlenecks and move data efficiently as shown in Figure 19.2.[3] The VMS230 utilizes DMAs on the IDR, PCI and the DES security block. These DMAs ensure that incoming packet data are quickly moved into internal memory, and then sent to the host via the PCI bus simultaneous with the host sending data to the VMS230 via the PCI bus for satellite encryption or decryption.

The VMS230 has three logical DMAs while supporting four different DMA channels—two for use with the IDH to PC Host connection and two for the PC Host to the DES block. The PCI DMA is for data transfer to and from the PC Host memory. The VMS230 utilizes on-chip SRAM to stage data that is moving in or out of the chip. The PCI DMA simultaneously controls the flow of data to and from the PCI bus and the IDH and DES blocks. Data transfers from the IDH to SRAM are handled by the IDH DMA, while the DES DMA is used to move data from the SRAM to the DES block for satellite encryption. Upon completion, the DES DMA also moves data from the DES block back to SRAM.

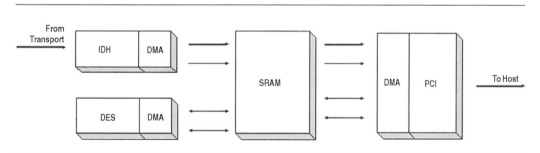

Figure 19.2 VMS230 DMA.

PCI

The PCI bus interface is used to communicate with the external host processor for control and data transfer. The DMA logic provides fast PCI transfers to and from the DES Security block, on-chip memory (RAM, ROM), and external memory. The VMS230 PCI bus interface is compliant and has been deemed compatible with industry leading PC chipsets by XXCAL, Inc.

IDH

The custom IDH interfaces to any 8-bit wide interface. The IDH interface accepts byte wide data and formats the data to the ARM 32-bit bus for preprocessing. The ARM and DMA controls all the processing activities of the IDH interface.

I^2C

The I^2C interface allows the VMS230 to communicate upstream with other components in the system. During start-up, the I^2C interface is used to read an external EEPROM and initialize the PCI configuration registers.

Assurance Logic

As previously mentioned, the GhostRider Chip provides physical security features to thwart attacks that attempt to alter the voltage or clock frequency of the chip. When under- or overvoltage is detected or the clock frequency is out of range, the chip issues a reset. Several resets result in the chip locking in reset and requiring power to be cycled.

CRC

The CRC FSB is designed for a parallel CRC computation of up to four (4) bytes and an accumulation of the results. In addition, the FSB also computes a CRC for a 3-, 2- or 1-byte size while preserving the 32-bit CRC accumulated value. The CRC FSB computes a CRC in one clock and can also be loaded with a 32-bit start value. For example, the CRC FSB check algorithm (in hardware) per the Ethernet polynomial is:

$$1 + D + D2 + D4 + D5 + D7 + D8 + DIO + DlI + DI2 + DI6$$
$$+ D22 + D23 + D26 + D32$$

DES Engine and Satellite Encryption Primitives

The VMS230 performs on-chip cryptographic functions that are accessible to the host. The cryptographic functions are designed to operate entirely from instructions located in on-chip ROM, requiring external access to data only. All cryptographic processes occur in a secure mode where keys and other sensitive data never leave the chip in the clear.

The DES core performs triple DES at a sustained rate of 15 Mbits/s in MSBDN mode. When operating in non-MSBDN mode, the DES engine can encrypt at the rate of 45 Mbits/s in single DES. The hardware randomizer within this block is all digital and fully nondeterministic. And, the patented one-time programmable vROM is included for unique variable storage. A key cache is available for high-speed cryptographic context switching. Also, on-chip cryptographic primitives provide users the flexibility to perform RSA key management functions and DES encryption operations

VMS230's IDH Interface

As previously mentioned, the VMS230's Transport interface is an 8-bit interface capable of loading byte words at 5.5 MHz or greater. The Transport is selected using a chip select (SCS) input as the qualifier, while the clock input is used to load the byte data on the rising edge. This interface receives two types of packets per the MSBDN requirements, 127-byte MPT packet, and the 130-byte DSS packet in test mode. The IDH interface is designed to receive any packet type up to 4096 bytes.

The example in Figure 19.3 shows one way the VMS230 can interface with a Transport chip.[4] In this example, the VMS230 interfaces with Transport via a high-speed port and Parallel I/0 port.

VMS230 In MSBDN and PC Crypto Applications

As an MSBDN cryptoprocessor, the VMS230 performs several tasks simultaneously. While receiving packets at 30 Mbits/s via the IDH Interface, the VMS230 also performs RSA decryption of 1024-bits in less than 5 s and can carry out a triple DES encryption/decryption. In MSBDN mode, the VMS230's triple DES throughput is 15 Mbits/s. However, acting as a general purpose PC crypto chip and utilizing only the DES engine, the VMS230's throughput for single DES is 45 Mbits/s. As a PC Crypto processor, the VMS230 offers the ability to securely store data and secure high-speed satellite communication links such as e-mail and cable modem applications as shown in Figure 19.4.[5]

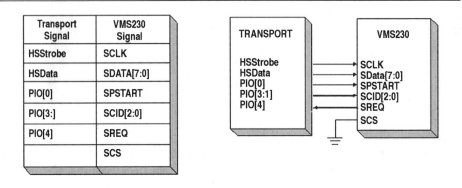

Figure 19.3 VMS230 IDH Interface.

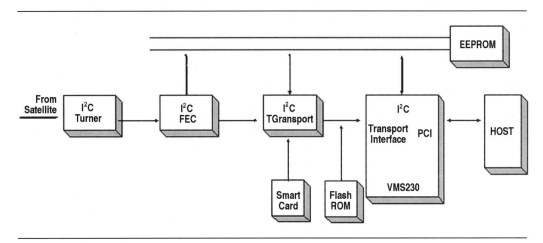

Figure 19.4 VMS230 n MSBDN applications.

From Here

Chapter 19 presented a framework for dealing with the following: satellite encryption policy, protection of personal data and privacy, security of information systems, intellectual property protection, international instruments, and rising demand for hardware-based data security. Chapter 20 discusses the following: mobile tracking, satellite encryption systems planning, design, and operation at C and Ku Bands, principles of satellite encryption systems, mobile satellite communications (MSAT), land mobile satellite encryption systems, MSAT propagation, modulation and coding for MSAT, space segment technology, mobile segment, vehicle antenna technology, aeronautical mobile satellite encryption systems, maritime mobile communications and surveillance, radio determination satellite encryption services, Inmarsat systems, mobile satellite encryption integration, current and future mobile systems, and trends in mobile satellite encryption systems.

Endnotes

1   Abraham J. Mund, "OECD Adopts Encryption Guidelines," Communications Media Center at New York Law School, 57 Worth Street, New York, NY 10013, 1997, pp. 1–3.

2   VLSI Technologies, Inc. 1109 McKay Drive, San Jose, CA 95131, 1997.

3   Ibid.

4   Ibid.

5   Ibid.

20

Satellite Encryption with Emphasis on Mobile Applications

Mobile tracking services with emphasis on satellite encryption provide the ability to track mobile units (trucks, ships, and planes) via satellite, collect data from mobile platforms, and provide message storage and retrieval. The mobile tracking function collects data transmitted from the mobile units, logs the information, plots location tracks on a world map, and computes the distance and arrival times to specified destinations. Alarms are generated when there is a deviation from the plan or when the status of a unit changes.

The mobile tracking function may also provide remote monitoring, including the collection and distribution of data such as meteorological readings or fishing catch reports. This information can be transmitted along with a position report to the customer's facilities to support global weather modeling or fishing limit verification. The system may also forward messages from mobile units to a public-switched data network. In applications where public access to the data is desired,

Figure 20.1 The system development division (SDD) has supported COMSAT Mobile Communications in developing a maritime message referral, tracking and display system, which was used to monitor the BOC Challenge. With COMSAT-supplied software on their PCs, users could display the current position and historical track of the racers superimposed on the world map.

such as the BOC Challenge (see the Sidebar that follows), the system can make this information available in response to dial-in inquiries as shown in Figure 20.1.[1]

Tracking the BOC Challenge

The BOC Challenge was a solo, around-the-world sailboat race with approximately 35 racers which started in Charleston, South Carolina, in September 1994 and ended in May 1995. This race and its sister race, the BOC Transatlantic Challenge, were organized by

The BOC Group, a British company that was the primary sponsor. The CMC, also a sponsor of the races, provided satellite communications and information services to the racers and race management, and tracked the racers to provide the news media with constantly updated information on each boat's location, and other race news.

All racers were equipped with Inmarsat-C mobile earth stations having integrated global positioning (GPS) satellite receivers. This automated equipment reports its position periodically without manual intervention. The race position reports passed through one of the Inmarsat land earth stations and were collected at COMSAT's server in Clarksburg, Maryland. The server, an IBM-compatible PC, published the reports on a computer bulletin board, allowing simultaneous access by users around the world.[2]

Mobile Tracking with Mobile Satellite Communications (MSAT)

The North American mobile satellite system (MSAT) began providing the United States and Canada with an unprecedented range of innovative mobile satellite encryption services in 1995. The MSAT is the first dedicated mobile tracking system in North America for mobile telephone, radio, facsimile, paging, position location, and encrypted data communications for users on land, at sea, and in the air (see Figure 20.2).

Canada-based TMI Communications & Company Ltd. of Ottawa, Ontario, and American Mobile Satellite Corporation (AMSC) in Reston, Va., signed contracts with Hughes and Spar Aerospace Ltd. of Canada in 1990 to build their respective satellites for the initial systems as shown in Figure 20.3.[4] TMI and AMSC each own and operate identical spacecraft. Both will provide complementary mobile tracking services, and each will provide backup and restoration capacity for the other. MSAT-1 operates at long. 106.5°W and AMSC-1 is at long. 101°W.

Hughes was the prime contractor for AMSC's satellite, called AMSC-1, and Spar was the prime for TMI's spacecraft, called MSAT-1. Hughes Space and Communications Company in El Segundo, CA is managing the program and providing the HS 601 satellite buses. Spar's Space

Figure 20.2 MSAT.

Systems Group is providing the high-power payloads and conducting spacecraft integration and testing at the Canadian Space Agency's David Florida Laboratory in Ottawa. The payload is the result of a 13-year mobile payload technical development program supported by the Canadian federal government and Spar investment.

Each satellite has the capacity to support up to 2000 simultaneous radio channels, depending on the type of antenna used and bandwidth allocated. Encrypted communications between the mobile users and the satellites are accomplished in L-band; terrestrial feeder stations use Ku-band (see the Sidebar that follows) to communicate with the satellite and with one another.

Wideband Systems

CSS100 Series C- and Ku-Band Transportable Satellite Systems

Commercial Satellite System's CSS100 Series Transportable Satellite Communications Systems operate in C- or Ku-band over domestic and international satellites providing encrypted voice, facsimile,

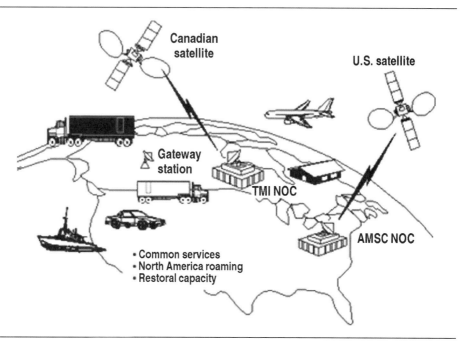

Figure 20.3 The MSAT/AMSC spacecraft will provide the U.S. and Canada with a full range of mobile satellite services.

high-speed data, and video teleconferencing solutions for an organization's communications requirements. The systems are integrated into a self-contained trailer-mounted package which can be towed and/or air transported to the desired operational location as shown in Figure 20.4.

Features

♦ Operates in C- or Ku-bands over domestic and international satellites.
♦ Voice, data, image transmission, and video teleconferencing in virtually any combination.
♦ Data rates from 64 Kbps through 1544 Kbps, half or full-duplex.
♦ Rugged 16-foot tandem axle trailer—C130 air transportable.
♦ Generator/UPS backup power system.
♦ Rapid deployment.

Mobile Applications

♦ Full period high-data rate circuit.
♦ Remote interactive video training systems.
♦ Multiplexed multiple voice and data circuits on a single satellite circuit.
♦ Remote database access and reporting.
♦ Remote electronic mail and Internet access.
♦ Image transfer, facsimile transmission, and video teleconferencing.
♦ Morale, Welfare and Recreation (MWR) voice and e-mail systems.

Typical Components

♦ Ku- or C-band transceiver.
♦ High-speed error-correcting modem.
♦ 2.4-m dish antenna.
♦ Backup generator.
♦ UPS.
♦ Equipment shelter with rack and environmental control unit (ECU).

Figure 20.4 CSS 100 series C- and Ku-band transportable satellite systems.

Available Options

♦ Lifting rings for sling loading.
♦ Antenna tracking controller.
♦ Antenna deicing system.
♦ Redundancy.
♦ Multiplexer (voice/data).
♦ Spectrum analyzer.

Services

♦ System design and integration.
♦ Foreign and domestic half-circuit lease arrangements for customers.
♦ Arrange for domestic teleport service and public switch telephone system interconnect.
♦ Training and postinstallation support.
♦ Leasing arrangements for any duration.

CSS200 Series C- and Ku-Band Flyaway Satellite Systems

Commercial Satellite System's CSS200 Series Flyaway Satellite Communications Systems operate in C- or Ku-band over domestic and international satellites providing encrypted voice, facsimile, high-speed data, and video teleconferencing solutions for an organization's communications requirements (see Figure 20.5). The systems are configured in ruggedized transport cases, allowing vehicular or air transport. Snap-on front and back covers allow rapid deployment on-site. All required cables are included.

Features

- Operates in C- or Ku-bands over domestic and international satellites.
- Voice, data, image transmission, and video teleconferencing in virtually any combination.
- Data rates from 64 Kbps through 1544 Kbps, half or full-duplex.
- Systems configured for unique customer requirements.

Mobile Applications

- Full period high-data rate circuit.
- Remote interactive video training systems.
- Multiplexed multiple voice and data circuits on a single satellite circuit.
- Remote database access and reporting.
- Remote electronic mail and Internet access.
- Image transfer, facsimile transmission, and video teleconferencing.
- Morale, Welfare and Recreation (MWR) voice and e-mail systems.

Typical Components

- Ku- or C-band transceiver.
- High-speed error-correcting modem.
- Voice and data multiplexer (if required).
- 1.2 or 2.4-m dish antenna with nonpenetrating mount.
- Interconnect cabling.

Figure 20.5 CSS 200 series C- and Ku-band flyaway satellite systems.

Services

♦ System design and integration.
♦ Foreign and domestic half circuit lease arrangements for customers.
♦ Arrange for domestic teleport service and public switch telephone system interconnect.
♦ Training and postinstallation support.
♦ Leasing arrangements for any duration.

CSS300 Series Fixed Satellite Earth Stations

The availability of commercial satellite communications offers worldwide connectivity to users with communications requirements in remote operating regions. The CSS300 Series Fixed Satellite Earth Stations provide encrypted voice, video, facsimile, and high-speed data communications in a fixed earth station configuration as shown in Figure 20.6.

Features

♦ Operates in C- or Ku-bands over domestic and international satellites.

- Voice, data, fax, and video teleconferencing in virtually any combination.
- Data rates from 64 Kbps through 45 Mbps (DS3) available for half or full-duplex.
  - Redundant systems.
  - Permanent or nonpenetrating antenna mounts.
- Available in a variety of configurations to suit any application.

Mobile Applications

- Full period high-data rate circuit.
- Remote interactive video training systems.
- Multiplexed multiple voice and data circuits on a single satellite circuit.
  - Remote database access and reporting.
  - Satellite-based Internet access.
- Image transfer, facsimile transmission, and video teleconferencing.

Typical Components

- Ku- or C-band transceiver.
- High-speed error-correcting modem.
- Voice and data multiplexer (if required).
- 1.2 to 9.3-m dish antenna with mount (permanent or nonpenetrating).
- Interconnect cabling.
- Optional antenna tracking controller.
- Optional antenna deicer.
- Optional UPS.

Services

- System design and integration.
- Turnkey service, including space segment.
- Domestic teleport service and terrestrial connectivity.
- On-site training and support.
- System installation and startup.
- Systems available for purchase or lease of any duration.[5]

Figure 20.6 CSS 300 series fixed satellite earth stations.

The L-band effective isotropic radiated power (EIRP) is 57.3 dBW. A 30-in (.75-m) shaped reflector antenna connects the Earth stations in Ku-band. Its EIRP is 36 dBW. Such high signal amplification by the satellite permits the use of small, low-power mobile and portable antennas. The Ku-band is driven by one powerful traveling wave tube (TWT) amplifier (with two spares).

Each satellite uses four spot beams at L-band to cover North America and 200 miles (300 k) of coastal waters. Another beam serves Alaska and Hawaii. The Caribbean beam includes Puerto Rico, the U.S. Virgin Islands, and Mexico. Each beam transponder is equipped with eight surface acoustic wave filters covering the 29 MHz L-band allocation, allowing selection of filters to match traffic needs and to coordinate with other international users. The spacecraft were designed with the capability for frequency reuse between the North American east and west beams. The beams are combined into two L-band power pools, one covering the east and central beams, and the other covering the remaining service areas. Each power pool is generated by a hybrid matrix amplifier assembly. The satellites have 16 active and 4 backup Spar-designed solid-state power amplifiers for L-band, each operating in a linear mode nominally at 38 W.

TMI's MSAT-1 was carried by an Ariane 4 booster from Kourou, French Guiana, on April 20, 1996. The AMSC-1 satellite was launched on an Atlas IIA rocket from Cape Canaveral, FL, on April 7, 1995.

Like others in the HS 601 series, the MSAT satellites consist of a cube-shaped center payload section, with the solar panel wings extending from the north and south sides, and an antenna array. The HS 601 is composed of two modules: the primary structure, which carries all launch vehicle loads and contains the propulsion system, bus electronics, and battery packs; and a payload module, which holds communications equipment and isothermal heat pipes. Reflectors, antenna feeds, and solar arrays mount directly to the primary module, and antenna configurations can be placed on three faces of the bus. Such a modular approach allows work to proceed in parallel on both structures, thereby shortening the manufacturing schedule and test time.

The HS 601 spacecraft was introduced in 1987 to meet anticipated requirements for high-power, multiple-payload satellites for such applications as direct television broadcasting to small terminals, private business networks, and mobile encrypted communications.

Each MSAT satellite measures approximately 62 feet (18.9 m) across with its two antennas deployed, and 68 feet, 9 inches (21 m) long from the tip of one three-panel solar wing to the tip of the other as shown in Figure 20.7.[6] Each stowed MSAT satellite measures approximately 25 feet (7.6 m) in length (height) and 12 feet and 1 inch (3.7 m) in width, as shown in Figure 20.8.[7] These arrays generate more than 3 kW, backed up by a 28-cell nickel-hydrogen battery for power during eclipse.

The MSAT spacecraft are the first to use Hughes' innovative Springback antennas. Weighing 45 pounds (20 kg) each, they are made of graphite in a 22.3-foot-by-17-foot (6.8-m-by-5.25-m) elliptical shape. The unique design not only provides a lightweight antenna, it also takes advantage of normally unused space in the top of the rocket

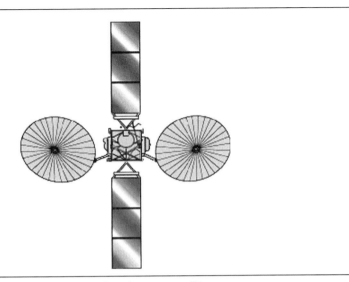

Figure 20.7 Deployed MSAT satellite.

Figure 20.8 Stowed MSAT satellite.

fairing. Instead of being folded against the spacecraft body for launch as conventional antennas are, two Springback antennas are rolled together into a 16-foot (4.9-m)-high cone shape atop the satellite. The cone is about 5 feet (1.5 m) in diameter at the top and about 10 feet (3 m) in diameter at the bottom.

For launch, the solar arrays fold accordion-style against the sides of the spacecraft. A flight-proven bipropellant propulsion system with an integral 110-lbf liquid apogee motor and 12 5-lbf thrusters afford a minimum 10-year service life. At the beginning of life in orbit, the spacecraft will weigh approximately 3783 pounds (1716 kg). Furthermore, Hughes/Spar as a team has successfully built and launched seven spacecraft: three Anik Cs and two Anik Ds for Canada and two SBTS satellites (Brasilat A series) for Brazil.

From Here

Chapter 20 discussed the following: mobile tracking, satellite encryption systems planning, design, and operation at C and Ku bands, mobile satellite communications (MSAT), land mobile satellite encryption systems, and MSAT propagation, modulation and coding for MSAT. Chapter 21 discusses the following: getting ready for the next war, putting the electronic battlefield to the test, the new terror fear—biological weapons, advances in encrypted signal processing technology for electronic warfare, radar target identification, advanced electronic warfare principles, theater missile defense, soldier identification encryption system utilizing low probability of intercept (LPI) techniques, battle management systems, and IRCM and IRCCM (Infrared Countermeasures and Counter-Countermeasures).

Endnotes

1   Eric Timmons, "Satellite System Management," COMSAT Laboratories, 22300 COMSAT Drive, Clarksburg, MD 20871, USA, 1998, p.3.

2   Ibid.

3   "Mobile Satellite System for Canada, U.S.," Hughes Space and Communications Company, P.O. Box 92919, Los Angeles, CA. 90009, 1996, p. 1.

4 Ibid., p. 3.

5 Commercial Satellite Systems, Inc., 835 Crosskeys Office Park, Fairport, NY 14450, 1998.

6 Hughes Space and Communications Company, P.O. Box 92919, Los Angeles, CA 90009, 1996, p. 1.

7 Ibid.

21

Satellite Encryption for the Electronic Battlefield of Today

In 1961, President John F. Kennedy issued a breathtaking challenge to the nation's scientists, engineers, and taxpayers: to land a man on the moon in under a decade. In 1983, President Reagan just as stunningly announced the Strategic Defense Initiative (SDI), which was to provide a *space shield* over the United States against intercontinental ballistic missiles by the end of the century. In the next few years, in the face of enormous political and technological debate, SDI almost, but not quite, withered away. Even its proponents acknowledged that much of it was not so much technologically audacious as outright impossible.

The SDI went underground for a while and retooled its weapons proposals, with its R&D funding cut off or maintained at drastically lower levels. But then came the 1991 Gulf War, whose televised missile battles were like rain on SDI's parched landscape. Ballistic missiles reentered public consciousness in a way not seen since the 1950s. No longer considered some vaguely imagined, doomsday weapons, they had become real and present dangers to U.S. lives.

The Ballistic Missile Defense (BMD) program, since 1993 the new incarnation of SDI, does resemble the manned space program in terms of technological audacity, funding required, and national interest (criteria matched as well only by the Manhattan Project, which of course was carried out in secrecy). But the BMD program of today and NASA's effort in the 1960s differ in important ways. Today's political climate is vastly different from that during the Cold War, when NASA's critics were quickly silenced by the echo of Sputnik. The current political arguments are far more deeply interwoven with claims about state-of-the-art and even as yet unknown engineering issues, whereas space technology was proven in slow steps. And, above all, mass death was not an issue.

Responsibility for BMD is shared among the nation's military services and some allies, mainly the United Kingdom and Israel. The Clinton Administration expects to spend $34 billion to develop and deploy battlefield (theater) missile defenses through the year 2004; should national (strategic) missile defenses be deployed that year, add at least $6 billion. The funding would not stop at these dates, only the projections do. In addition, the Pentagon is investigating space-based lasers and interceptors, which if deployed could raise costs to a total of some $64 billion, according to the nonpartisan Congressional Budget Office (CBO).

National defenses include missiles that will take down incoming ones by hitting them, using only their combined kinetic energy for destruction—what the military calls a hit-to-kill or kinetic-energy method. The warhead of a traditional surface-to-air missile explodes near its target, giving it a greater margin of error. Strategic defense would also be carried out in space, by orbiting lasers and by kinetic energy destruction of enemy satellites.

Hit-to-kill systems also hold sway in theater-level warfare, the type seen in the Patriot-versus-Scud encounters of the Gulf War. In addition to defenses against midcourse or reentering tactical missiles and warheads, the Pentagon is funding airborne laser and kinetic-energy weapons to destroy missiles on launch.

When it comes down to it, distinguishing a tactical missile from a strategic one is not very easy. For example, if North Korea lobs a missile at South Korea, is that a tactical move to slow down its troops or a strategic threat against the nation of South Korea? A definition agreed to at the March, 1997 Helsinki Summit by presidents Bill Clinton and Boris Yeltsin bases the distinction on physical criteria: the speed and range of the missiles against which the defense is tested. Strategic weapons are those traveling at more than 6 km/s and which have a range of 4600 km or more.

The issue is far more than nominal. Testing of strategic antiballistic systems, including radar and other sensors, is severely constrained by the 1972 Antiballistic Missile Treaty (signed by the United States and the Soviet Union). Similarly, the militarization of space is regulated by several treaties.

But what of, say, a laser weapon, still in development, that in principle could take down either a tactical or strategic missile? The ramifications of this kind of question are legion, as much for engineers as for diplomats and the public they represent. Suffice it to say that BMD has little precedent in that so many competing claims of weapon performance (actual and potential) are so widely held, even within engineering circles. Even rarer is for these claims to take such a leading role in debates on the U.S. military force structure and on national and international politics.

To that end, this chapter report begins with an overview of the grand plans of BMD. In BMD's layered defense concept, there are different types of antimissile missiles shield areas that depend on the distance a warhead has penetrated. Cueing and tracking encrypted data for these systems, as well as for strategic defenses, would be handled by picket fences of new satellites and ground radars.

An even better defense against warheads is to prevent them from reentering in the first place. Of the two boost-phase weapons in the works, the big-ticket one for now is the Air Force's proposed Airborne Laser (ABL). In the cold reaches of the upper atmosphere, this converted Boeing 747 would pin a megawatt beam on a boosting missile hundreds of kilometers away.

The ABL, as well as other laser weapons being independently developed by the Army and Navy, are founded on a notion of great antiquity. In the Punic Wars in the third century B.C., not only did Hannibal have his elephants, but also (it is said) Archimedes had his mirror weapon, which focused sunlight into burning heat rays. Fast forward to 1966, six years after the invention of the laser, when Pentagon weaponeers first notice this new energy device.

Now, the ultimate heat-ray weapon finally may be in the offing: that Reagan favorite, a space-based directed-energy weapon. This chapter also examines the population explosion planned for BMD electronics in space, notably concentric rings of early warning and targeting satellites.

Finally, the chapter examines BMD and enemy missiles that fight back. An attacking missile will no doubt attempt to outmaneuver, confuse, or hide from the defense—actions known as countermeasures. Furthermore, as with most complex technologies, there is likely to be a divide between the public perception of BMD and its reality—or its reality in the making.

Now, let us look at the proposed (in development, and in test) new U.S. ballistic missile defenses. This would range from tiers of hit-to-kill missiles to novel laser weapons and satellites that transmit encrypted data (uplink and downlink).

Theater Missile Defense Plans

Under cover of darkness, hostile forces cross the borders of a country of strategic importance to the U.S. and overrun it. United States forces are poised to rush to the region. But the opponent has ballistic missiles that may be armed with chemical, biological, or nuclear warheads. What is more, the missiles may have ranges long enough to strike the airfields and ports that U.S. forces must use to deploy to the region and to attack the population centers of regional allies. What should the United States do?

A futuristic scenario? Not at all. It happened when Iraq invaded Kuwait in 1990. Early the next year Iraq launched 99 missiles against targets in Saudi Arabia and Israel (the exact number is in dispute), but all carried only conventional warheads. Postwar analysis revealed that in point of fact Iraq did have chemical warheads for its missiles. More serious still, it was also building biological weapons and was closer than experts elsewhere had expected to being able to build the nuclear weapons.

The Gulf War illustrated how essential ballistic missile defenses have become to the U.S. Today, some 30 countries in the developing world have short-range ballistic missiles and the number is likely to rise in future years. Many of those same countries also currently possess, or are developing, chemical or biological weapons and may be converting them into warheads that can be delivered by ballistic missiles. In fact, a few (notably North Korea, Iraq, India, Pakistan and Iran) have made marked progress toward developing nuclear weapons and are trying to develop longer-range missiles.

United States military leaders worry that in future regional (or so-called theater) conflicts their forces may be vulnerable to attack or denied access to bases in allied countries within range of enemy ballistic missiles, and have proposed an impressive plan of layered defenses as shown in Figure 21.1.[1] In addition, the Clinton administration has restarted U.S. efforts to develop a National Missile Defense program, prompted by concerns that so-called rogue states might some day develop missiles with ranges long enough to reach U.S. soil.

Note: The term *rogue states*, in current U.S. usage, applies to countries whose behavior does not conform to international norms and may not be deterred by the threat of conventional or nuclear retaliation.

Yet defenses against ballistic missiles, despite the seeming imperative to build them, remain controversial. They present a unique set of technical challenges that must be overcome before they can be effective. They are also expensive and place additional demands on resources at a time when the U. S. Department of Defense (DOD) is struggling to afford the forces and equipment it already has. Moreover, there is some uncertainty about the types of threats that the United States and its troops will face, just when those threats will come to pass, and how much the country should worry about them.

Into the bargain are international arms control treaties, in particular the Antiballistic Missile (ABM) Treaty. Unlike in other areas, treaty limitations can influence the technology. Conversely, attempts to define treaty-compliant technical solutions run the risk of undermining specific provisions intended to prevent large defenses from being deployed (see the Sidebar that follows).

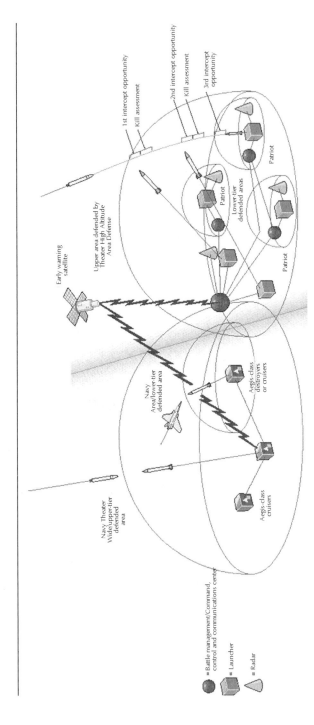

Figure 21.1 The U.S. plans to use a layered approach to protect its forces and allies against short-range ballistic missiles, which have ranges of less than 3500 km. Area defenses such as the land-based Theater High-Altitude Area Defense (Thaad) and the sea-based upper-tier defenses will cover broad areas. Patriot PAC-2 and PAC-3 and the sea-based lower-tier defenses will be a backup for targets missed by the upper tier and protect high-value targets such as airfields, ports, headquarters, and population centers. The Airborne Laser (not shown) will aim to destroy missiles during the boost phase, when their rocket motors are burning and before they can deploy warheads, decoys, or submunitions. A theater battle management system will coordinate the system's many components.

Treaty Complaints and Constraints

Few technology development efforts are influenced by treaty limitations. For the United States, ballistic missile defenses are a notable exception. So far, only one treaty formally constrains U.S. efforts to build ballistic missile defenses: the Antiballistic Missile (ABM) Treaty between the U.S. and the former Soviet Union. It applies only to defenses against strategic ballistic missiles; theater defenses are not even mentioned. Nor does it establish a clear, formal definition of what constitutes a *strategic ballistic missile*, a fact that has generated much debate in recent years.

The treaty was written to allow very limited defenses of only two regions in each country: the national capital, because it is home to the national command authority, and one field of ballistic missiles that are part of the strategic retaliatory force. It was also designed to make it difficult for either side to build large defenses quickly and covertly. A 1974 protocol reduced the number of allowed sites to one. The USSR chose to defend its capital, while the U.S. opted to defend a missile field near Grand Forks, ND. The treaty was negotiated so that the superpowers could limit the size of their nuclear arsenals without having to worry about the other side building defenses that would call into question the effectiveness of its deterrent.

The treaty restricts the universe of acceptable technical solutions to a defense problem in a number of ways. For defenses deployed against strategic missiles, the treaty limits each side to a defense at one site consisting of no more than 100 ground-based interceptors. It prohibits defenses that are mobile or based in the air, at sea, or in space. It also limits where ABM radars can be located and prohibits either side from building a defense of its national territory.

While not explicitly limiting theater missile defenses, the treaty does prohibit giving non-ABM missiles, launchers, and radars the ability to intercept long-range (strategic) ballistic missiles. This issue raises a prickly question: when does a theater system have such a capability? The struggle over the answer dates back to at least the early 1980s, when the U.S. argued that the Soviet Union had given its SA-12 air defense system (the Soviet version of the Patriot) the ability to counter strategic missiles. More recently, debate has focused on Russian concerns that new U.S. theater systems such as the Thaad, the Navy Upper Tier system, and the Airborne Laser would be able to intercept Russian strategic missiles.

Negotiating a demarcation between theater and strategic defenses has occupied both countries for the past four years. It may have only been resolved in April, 1997 at the Helsinki summit, when Russia agreed in principle to accept the U.S. position that a defense system will not be considered strategic if it is tested only against targets moving slower than 6 km/s and with ranges of less than 4600 km.

Why is this document, signed in 1972 by Presidents Richard Nixon and Leonid Brezhnev, still the subject of such fervent debate? Because it is at the core of a broader debate about how the U.S. should respond to the challenges of the post–Cold War world and manage its relationship with Russia.

Supporters of the treaty, including the leadership of both the U.S. and Russia today, see it as *the cornerstone of nuclear arms control*. They contend that none of the arms limitations or reduction treaties signed to date would have been possible without it. Neither side would have been willing to limit its nuclear weapons if it thought that the other side could build large missile defenses that would jeopardize the effectiveness of its own deterrent.

It was just such logic that required the treaty to be included in the first set of accords that limited nuclear weapons. It has been cited by both sides as a key element of all subsequent nuclear arms reduction treaties. To underscore its relevance today, Presidents Bill Clinton and Boris Yeltsin reaffirmed their commitment to it at the Helsinki summit by signing agreements that clarified the distinction between theater and national defenses and taking steps to ease passage of the Strategic Arms Reduction Talks (Start) II treaty in Russia's parliament.

Opponents of the ABM treaty are fundamentally opposed to limiting the U.S. in its ability to defend itself or its troops against ballistic missiles. In their view, the lengthy negotiations about which systems or technologies are allowed is an example of exactly what the United States should not be doing. They believe that the ABM treaty is a Cold War relic that has outlived its usefulness in a world where the more probable threats are posed by proliferation in the so-called rogue states—countries that may not be deterred by U.S. nuclear forces.

The Clinton administration has tried to forge a middle path. It agrees with opponents of limits that new threats must be addressed. But it argues that the best approach is to clarify or modify the treaty to address those threats while still retaining the fundamental constraints embodied in the treaty. The Administration plans to finalize the agreement with Russia about theater missile defenses and may try to negotiate enough modifications to the treaty that the United States can deploy a limited National Missile Defense that will protect all 50 states against a small attack.

There are two principal critiques of this plan. Supporters of the ABM treaty argue that the Administration's actions are likely to undermine the ability of the treaty to keep defenses limited. They also suggest that the Administration and others are overstating the threat posed by North Korea and other hostile states. Opponents of the treaty believe that the Administration is ignoring the possible threat posed by North Korea and others and dragging its feet on deploying a defense. They tend to chafe against the fetters of the treaty.[2]

In fact, no other defense issue today entwines cutting-edge technology, arms control, and international relations in such a tight knot, one that is virtually impossible to separate. It is futile for analysts to discuss the ABM treaty and other arms control measures in a vacuum. They must acknowledge the revolutionary technical changes that over the past decade have brought effective missile defenses closer to reality — and, at the same time, made the proliferation of ballistic missiles and weapons of mass destruction more likely.

Finally, policymakers cannot discuss missile defenses without noting that the bipolar, relatively stable world of the Cold War has given way to a less certain one where regional powers, subnational groups, and terrorist organizations are no longer constrained by the superpowers. And the technology and know-how to build weapons of mass destruction are more available than ever before.

Threats Small and Large

Any consideration of missile defenses must start with the basic question: Why should the United States build such defenses? The answer can be found, in part, in the types of threats that the country and its allies face these days and will face in the future. Unfortunately, projecting future threats is, at best, an uncertain business.

A substantial threat today, however, comes from theater missiles, that is, tactical missiles developed for shorter-range encounters than strategic missiles. United States forces faced them in the Gulf War and will continue to be threatened by them. Since that war, for instance, North Korea has deployed, but not fully tested, several missiles that might be able to reach Japan.

Also to be factored in are the other methods available to protect U.S. forces. These include using passive defenses to protect troops and civilians from the effects of chemical and biological weapons; destroying ballistic missiles and warheads before their launch; slowing the spread of these weapons in the first place; and deterrence by the implicit or explicit threat of retaliation, either conventional or nuclear.

An additional question is whether or when potential regional opponents will develop chemical, biological or nuclear warheads to ride atop their missiles. Would they use them or threaten to use them? Should the U.S. expect them to develop submunitions or countermeasures that could significantly complicate the problem for U.S. defenses? Challenges to ballistic missile defense are covered later in the chapter.

> **Note:** Submunitions are small warheads containing explosives, chemicals, or biological agents. By dispersing dozens or even hundreds of them, a missile can spread its effects over a larger area than it could with a single warhead.

Currently, the United States itself faces no threats of ballistic missile attack from rogue countries. That situation is likely to continue, according to a 1995 report that was released by the intelligence community. Called the National Intelligence Estimate, it projected the ballistic missile threat to the U.S. for a period of 16 years.

The report found that it would take a developing country at least 16 years before it could threaten the continental U.S. with long-range missiles that it had developed indigenously. The intelligence community qualified this finding by noting that North Korea might be able to develop missiles after 2001 that could strike parts of Alaska and Hawaii. Recent press reports indicate that North Korea is in the early stages of developing a missile called the Taepo Dong, which has a range of some 3000 km or more. A 3500-km range missile could reach the western tip of the Aleutian Island chain; a 6000-km range one could reach most of Alaska and the western third of the Hawaiian chain. It would take a range of 7500 km to reach all of the five main Hawaiian islands as shown in Figure 21.2.[3]

Note: Although the U.S. itself faces no threats of ballistic missile attack from rogue countries, it does face a very real and serious threat from terrorist groups who would bring nuclear weapons to our major cities and then set them off remotely via encrypted satellite signal from the relative comfort of their own countries. It would be almost virtually impossible to detect that kind of threat in advance; unless of course, our SDI satellites have the capability of detecting nuclear material from space as it carried from place to place. Makes you wonder, doesn't it?

Nevertheless, the Clinton administration interpreted the National Intelligence Estimate to support its position that the United States should develop a limited national missile defense called the 3-Plus-3 system

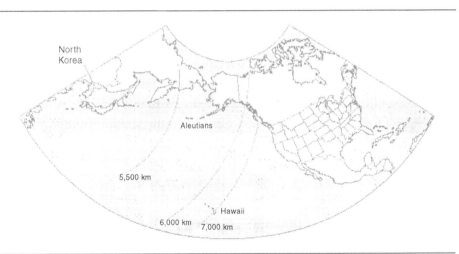

Figure 21.2 North Korea reportedly is in the early stages of developing a missile called the Taepo Dong, whose range is 3000 km or more. A North Korean missile would need a range of 3500 km to reach the tip of the Aleutian island chain in Alaska. A range of 6000 km would reach most of Alaska and the western third of the Hawaiian islands, and a 7500-km range would reach the rest of Hawaii.

(discussed next), to be deployed only when warranted by a projected threat.

But advocates of building a national missile defense as soon as possible were quick to point out that the report ignored the possibility that Russia or China could sell intercontinental ballistic missiles to hostile states. They also argued that the potential threats to Alaska and Hawaii identified in the estimate make it urgent that the United States deploy a National Missile Defense as soon as possible.

An independent panel commissioned by the U.S. Congress to review the National Intelligence Estimate was skeptical of North Korea's ability to develop within the next few years a missile capable of reaching Alaska or Hawaii. The panel noted that every country that had developed long-range missiles had taken years (16 to 21 years in some cases) to do so. And all of them had greater resources than North Korea. But the panel felt that the report ignored the threat of long-range cruise missiles and shorter-range missiles launched from ships close to U.S. coasts.

Ballistics Basics

Most defenses are designed to destroy the warhead on a missile before it can reach its target. Theater missile defenses that have already been deployed, as well as those likely to be deployed in the near future, intercept the warhead during the terminal phase of its flight — after it has reentered the atmosphere. Other defenses under development will attempt to intercept the warhead outside the atmosphere during the midcourse phase. Still others will try to destroy the missile during its boost phase after launch while the rocket motor is still burning.

Once a ballistic missile is launched, the following defensive events occur. First, the infrared sensors on the early-warning satellites in geosynchronous orbits detect the hot exhaust plume of the missile as it rises above the clouds. The satellites alert military commanders (via encrypted signals) that a launch has taken place and indicate the general area toward which the missile is headed. That encrypted information can be used to point (cue) the defense's sensors to the right spot for tracking.

Those sensors then track the target and all of its associated clutter and decoys (if any), distinguish which object is the actual warhead (a process called discrimination), and then tell the interceptor where to head so that it will be in a position to intercept the warhead. Mid-course sensors have traditionally been ground-based radar, but in the years to come, radars will be complemented by satellites in low earth-orbits that carry a suite of infrared sensors as shown in Figures 21.3 and 21.4.[4]

Based on this information, the interceptor flies toward the estimated intercept point, receiving updated estimates along the way. At some point the interceptor's kill vehicle separates from the missile and continues on its trajectory. The kill vehicle will slam into the attacking warhead. It does have small thrusters of its own, but these are used only in the final moments before intercept. When close enough to the target, the interceptor uses its own sensors and guidance algorithms to distinguish the target from decoys and clutter, and homes in for the kill.

This intricate dance must be choreographed by a battle management system in exquisite detail and within very tight time constraints. For example, the total flight time of a 400-km range Scud on a maximum-range trajectory is 5 minutes. Longer-range theater missiles are in flight for 16 minutes or less.

This procedure applies to both theater (tactical) and national (strategic) missile defenses. The only differences between them are the flight times available, the speed of the incoming warheads, and the area that must be protected, all of which are greater for national missile defenses.

Of course, this picture of how a missile defense would work is idealized. In practice, each of the steps described presents its own technical challenges, some of which are discussed later in the chapter.

Layers and Salvos

To improve the odds of destroying warheads, a salvo of two or more interceptors can be shot at each target. A more efficient tactic is to launch interceptors only at targets surviving the first intercept at-tempt—an approach that is called shoot-look-shoot.

The same effect can be achieved by using layered defenses. Intercep-tors in each of the layers can attack a missile during a different part of

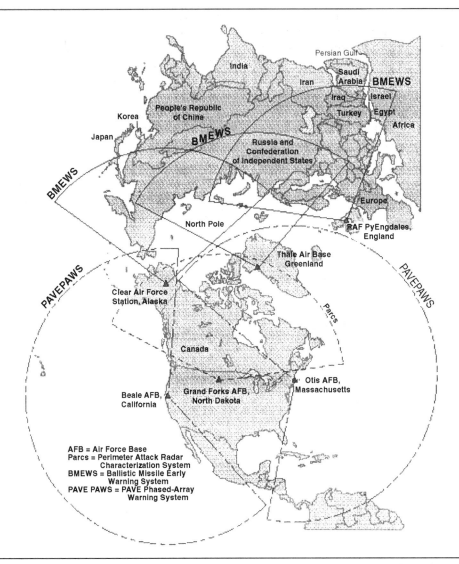

Figure 21.3 To track incoming warheads, countermeasures and debris, the radars of the so-called 3-Plus-3 National Missile Defense system would be upgraded from those shown here, part of the current U.S. early-warning radar network. A new 10-GHz radar would be deployed at Grand Forks, ND, to replace the existing one there (a phased-array radar used in Ballistic Missile Early Warning System (BMEWS) is pictured in Figure 21.4). Because of the earth's curvature, however, that radar would be unable to see any of the trajectory of a missile launched from North Korea and aimed at Alaska, Hawaii, or parts of the U.S. West Coast.

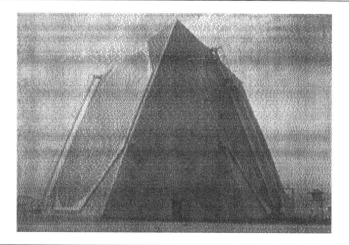

Figure 21.4 A phased-array radar used in BMEWS.

its trajectory, taking advantage of different vulnerabilities. For example, if missiles are intercepted during the boost phase, the large rocket plume is easy to find. And if the missile is attacked before it has a chance to deploy its warheads, as well as decoys and other counter-measures, the task of the terminal defenses that form the lower layer is greatly simplified.

The U.S. Army plans to complement its improved Patriot system by adding an upper layer or tier called the Theater High-Altitude Area Defense (Thaad), which will attack missiles in the upper reaches of the atmosphere or outside it altogether. The U.S. Navy plans to take a similar approach with its upper-tier and lower-tier defenses.

BMD in Action

Over the years, the mechanisms by which interceptors destroy a target have kept pace with rapid advances in missile, computer, and sensor technologies. In its first ABM system, the U.S. deployed interceptors that would rely on the blast from nuclear explosions to destroy incoming Soviet warheads. The large kill radius of a nuclear blast was necessary to overcome the limitations of the defense. Although the system was short-lived (it was operational for only a few months in 1976), an ABM system tipped with nuclear warheads continues in operation near Moscow.

The first U.S. theater missile defense system (the Patriot of Gulf War fame) used chemical explosives and steel fragments to attack warheads. It was originally developed as an air defense weapon against aircraft and cruise missiles — *soft* targets against which blast fragmentation warheads are particularly effective. They are also effective against some soft ballistic missile warheads, but destroying most other kinds of missiles requires more energy than this type of defense system can deliver.

The Reagan administration's Strategic Defense Initiative (known to its critics as *Star Wars*) introduced a new kill mechanism to the mix — directed energy from lasers. Large sums were spent on laser programs during the 1980s. The descendants of those efforts persist today in the form of chemical lasers that would be used to destroy missiles in their boost phase.

Most of the development efforts for theater and national missile defenses are now focused on hit-to-kill interceptors that use the large kinetic energy of objects moving toward each other at speeds exceeding several kilometers a second to destroy the target. When the interceptor hits the target, the results are spectacular. Both objects are obliterated in a fiery explosion.

But hitting the target is no easy task at such high speeds. The kill vehicle is essentially a heavy object such as a rod or a disk that is only tens of centimeters in diameter, and it must hit a target that is often only a meter or so across. According to the oft-used, but still accurate cliché, hit-to-kill BMD *is like hitting a bullet with a bullet* — but at least 10 times faster.

BMD's U.S. Plans

The U.S. is planning to develop six theater missile defenses: the PAC-3 version of the Army's Patriot, the Thaad, the Medium-Altitude, Extended Area Defense System (Meads), the Navy Area defense (also known as Navy Lower Tier), the Navy Theater Wide defense (also known as Navy Upper Tier), and, finally, the Airborne Laser. The land-based defenses will protect troops, ports, and airfields once they have arrived in theater. Sea-based defenses will protect ports and littoral regions until the land-based defenses arrive as shown in Figure 21.1. Two other systems have already been deployed: the Marine Corps' upgraded version of the Hawk and the Patriot PAC-2, which has recently been upgraded from its Gulf War configuration with enhancements to its radar and missile guidance system. And, as has been noted, the U.S. is developing a limited national missile defense as shown in Figure 21.5.[5]

Terminals System Defenses

While most of the focus of U.S. theater missile defense efforts is on terminal systems, that is, systems that intercept warheads after they reenter the atmosphere, there are no plans to develop a terminal defense to protect the U.S. The advantage of terminal defenses is that they can use the atmosphere to separate warheads from lightweight decoys. The disadvantages are that they can defend only a limited area, have a limited time to respond with a second or third shot, and can be overwhelmed by submunitions.

Four of the six theater systems under development are terminal defenses: the Patriot PAC-3, the Thaad, the Meads, and the Navy Area defense. The Thaad system is actually a hybrid system that can operate against the incoming warhead's midcourse and terminal phases, because it will be able to intercept targets at altitudes from 50 km to a few hundred kilometers.

Note: The sensible atmosphere ends at about 110 km.

All of these systems except the Navy Lower Tier defense will make use of hit-to-kill interceptors. Because it will retain a significant air defense role against enemy aircraft, the Navy system's interceptor will use a blast fragmentation warhead. The operationally improved Hawk and PAC-2 systems use blast fragmentation warheads for the same reason.

The Meads is an international program. It was spun off from the U.S.-developed Corps Surface-to-Air Missile program in 1995. Development now enjoys support from Germany and Italy. To date, the U.S.

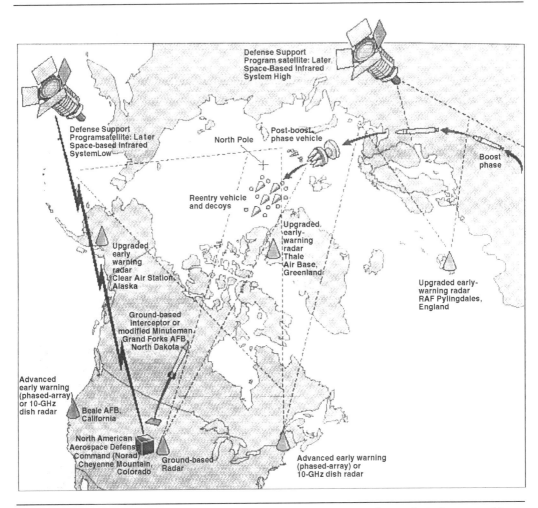

Labels within figure:

Defense Support Program satellite: Later Space-Based Infrared System High

Defense Support Programsatellite: Later Space-based Infrared SystemLow

North Pole

Post-boost phase vehicle

Boost phase

Reentry vehicle and decoys

Upgraded early-warning radar Thale Air Base, Greenland

Upgraded early warning radar Clear Air Station, Alaska

Upgraded early-warning radar RAF Pylingdales, England

Ground-based interceptor or modified Minuteman, Grand Forks AFB, North Dakota

Advanced early warning (phased-array) or 10-GHz dish radar

Beale AFB, California

North American Aerospace Defense Command (Norad) Cheyenne Mountain, Colorado

Ground-based Radar

Advanced early warning (phased-array) or 10-GHz dish radar

Figure 21.5 In an ideal defensive deployment, the 3-Plus-3 National Missile Defense would use satellites, and additional 10-GHz radars would be deployed closer to the regions from which threats are likely. In this composite of options proposed by the U.S. Army and Air Force, early-warning satellites, either the current Defense Support Program (DSP) system or the future Space-Based Infrared System, would detect the exhaust plume from the burning rocket motor of an attacking missile and other systems. Forward-based 10-GHz radars and infrared-detecting satellites in the Space and Missile Tracking System would use their ability to resolve smaller objects to try to distinguish warheads from clutter and decoys. Based on that encrypted data, the ground-based interceptor — a newly developed hit-to-kill weapon proposed by the Army, or revamped Minutemen in the Air Force's proposal — would fly toward an approximate intercept point, receiving course corrections along the way from the battle management system based on more up-to-date tracking data. As the interceptor neared the target, its own sensors would guide it to the impact point.

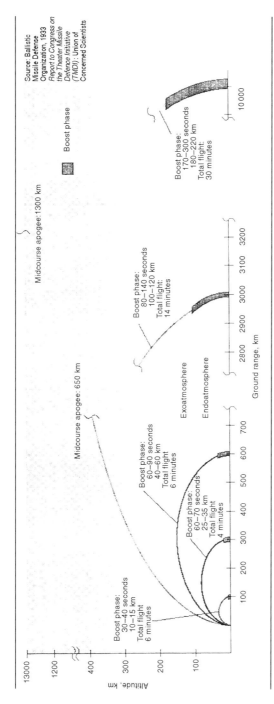

Figure 21.6 The boost phase of a ballistic missile—when the rocket engine is providing thrust—is a critical time during its flight. The plumes from the engine make it easily visible, and the missile is not yet able to release decoys or submunitions. Shown here are the time spent and altitudes reached when ballistic missiles fired on maximum-range trajectories are boosting.

has committed itself only to early stages of development for the Meads. France is in the early stages of developing a shipborne lower-tier interceptor called Aster. The most advanced (and so far successful) shared overseas program is the Arrow defense being developed in Israel with significant U.S. support.

Boost-Phase Theater Missile Defenses

Concerns about submunitions have prompted work on boost-phase theater missile defenses, which attempt to destroy the target shortly after its launch. This undertaking is simply a matter of leverage. Every missile that is destroyed before it can release its many small warheads or decoys eases the challenge facing terminal defenses as shown in Figure 21.6.[6] In fact, boost-phase defenses appear to be the only way to defend against submunitions that are released early in the mid-course.

One advantage of the boost-phase defense is that the missiles' rocket motors are spewing out hot gases that are easy to locate. But the challenge lies in being close enough to the missiles to intercept them because their motors burn for, at most, a few minutes. In that time, the defense must detect the launch, track it for long enough to get a fix on its trajectory, and then intercept it.

Currently, the Pentagon is developing in earnest one boost-phase defense against theater missiles: the Air Force's Airborne Laser. This chemical laser, carried in a 747 airplane, will be able to intercept missiles from a few hundred kilometers away, according to the Air Force. The beauty of a laser beam is that it travels at the speed of light, saving precious seconds during the short burn time of a theater ballistic missile. But several technical obstacles must be overcome before this revolutionary system can be deployed. The Airborne laser is covered in more detail later in the chapter.

Another boost-phase intercept concept is undergoing testing by the Ballistic Missile Defense Organization and Israel: kinetic energy interceptors (missiles) that would be carried by unmanned air vehicles. These drone aircraft would have to be in position and loiter close enough to enemy missile fields (on the order of some 50 km away), so that their missiles could reach their targets in time. This means that a large number of drones would be required if the opponent has been able to disperse his launchers over a large area. The Pentagon believes that this program will provide a degree of insurance in case the Airborne Laser program does not pan out as hoped. In some cases, the Navy's Theater Wide defense (also known as Navy Upper Tier) could intercept theater missiles in the early post-boost phase if the ship can position itself near the launch site — for example, near a North Korean site where a missile might be launched against Japan.

The Advantages of Midcourse Defenses

Both the Navy's Upper Tier defense and the Thaad would intercept missiles in midcourse. The Thaad should also be able to intercept missiles inside the atmosphere. The advantage of midcourse defenses is that they have a lot of time to intercept their targets, particularly intercontinental ballistic missiles that spend more than 20 minutes of their 30-minute flight in the midcourse. This allows ground-based interceptors to defend large areas. It also affords the defense enough time for what is known as a *shoot-look-shoot* strategy (that is, launch, check if the attacking missile was destroyed, and, if not, launch again).

But without an atmosphere to help separate light decoys from the real warhead, midcourse defenses can be baffled by inexpensive counter-measures. The U.S. is devoting sizable sums of money to solve the countermeasures problem by developing both radar and optical tech-niques for discriminating between warheads and decoys.

Each region of the electromagnetic spectrum has its own set of advan-tages and disadvantages, which is why current plans envision a mix of sensor types. An important part of the U.S. solution lies in a planned constellation of 27 to 35 satellites equipped with a variety of infrared sensors. A detailed discussion of these types of satellites occurs later in the chapter.

Nevertheless, using these Space and Missile Tracking System satel-lites, scientists plan to observe the encrypted signatures of warheads and decoys in space across several infrared and visible bands and to exploit the differences between them. For example, a warhead that weighs several hundred kilograms would cool at a different rate in space from a balloon that weighs only one-tenth of a kilogram. By fusing together encrypted data from ground-based radar and different space-based infrared sensors (which measure the heat emitted by an object), the battle management system will, in theory, be able to separate the wheat from the chaff. An opponent could complicate this task, however, by using a technique called antisimulation, to make the warhead itself look like something else — perhaps by encasing it in a balloon. Again, as previously mentioned, the future challenges to ballistic missile defense will be discussed in detail later in the chapter.

National Missile Defense

The Clinton administration is developing a National Missile Defense that will intercept warheads outside the atmosphere. The idea is primarily to protect the United States, including Alaska and Hawaii, against an attack by a rogue state using a handful of warheads outfitted with relatively simple countermeasures. The system would also provide protection against an accidental launch of a few warheads by Russia or China.

As much as possible, the Administration wants the system to comply with the ABM treaty. As a result, the defense will consist of no more than 200 hit-to-kill interceptors based at the old ABM site near Grand Forks, ND. It will be supported by a constellation of Space and Missile Tracking System infrared tracking satellites and several ground-based early-warning, tracking, and fire-control radars.

When the Administration will actually deploy the National Missile Defense has not yet been decided. It plans to develop the system through the year 2001 to the point where it can be deployed four years later, if warranted by a threat. If no near-term threat is on the horizon by 2001, the United States will wait to deploy the system, keep on improving it, and reassess the threat every year thereafter to see if deployment is warranted.

In the view of the Administration, the United States could deploy a defense before a threat emerges without locking itself into a specific technology prematurely. For lack of a better name, the Administration has dubbed its defense the 4-Plus-4 system: four years to develop and four years to deploy. Critics who doubt the Administration's commitment derisively call the system 4-Plus-Infinity.

Details of the system are undecided. There are two competing concepts as shown in Figure 21.5. One proposed by the Army is the Ground-Based Interceptor, a new missile built from components of existing missiles. This interceptor would carry the Exoatmospheric Kill Vehicle, an Army hit-to-kill warhead in development.

The Army would deploy a new phased-array 10-GHz radar at Grand Forks, ND., called the Ground-Based Radar, as well as similar Advanced Early-Warning Radars at several other locations. It would also upgrade several existing early-warning radars as shown in Figure 21.5. This system would cost in the region of $10 billion. Nevertheless, recent cost overruns in the 4-Plus-4 program suggest that costs would be at least $2 billion higher for the various systems described here (see the Sidebar that follows).

Competition and Costs

Costs are always an important element in the perennial debates about building missile defenses. The Clinton administration plans to spend $27 billion to develop and build missile defenses over the next seven years, the period for which detailed plans exist: $16 billion for theater missile defenses, nearly $6 billion for national missile defenses, $3 billion for early warning satellites, and more than $2 billion to develop other technologies. That total averages to about $3.8 billion a year as shown in Figures 21.7, 21.8, and 21.9. These figures exclude the cost of deploying the national missile defense, which would add at least $6 billion more to the total.

Beyond 2004, aval costs should increase as many systems enter procurement. Some critics argue that the Administration is developing too many redundant theater (tactical) missile defense systems and designing national (strategic) missile defenses well before any threat is on the horizon. Others contend that the Administration is not doing enough and should accelerate its plans to develop both theater and national missile defenses.

In one sense, costs should not really matter. If U.S. citizens or U.S. troops are threatened by ballistic missiles, particularly those armed with weapons of mass destruction, it is incumbent on the Congress and the Administration to protect them. On the other hand, the immediacy of the threats (in the case of national missile defenses) or their severity (in the case of theater defenses) are not known with certainty.

In the current era of tight budgets, the Congress and the President must allocate resources across many competing programs as they

System name	Prime contractor(s)	Type of warhead	Air defense role	Approx. radius defended area, km	Number to be bought[a]	Date of initial deployment	Acquisition cost
			Lower-tier (point defenses				
Patriot PAC-2 (Army)	Raytheon Co.	Blast fragment	Substantial	10–15	2247 missiles, modified	1991	US $0.3 billion
Patriot PAC-3 (Army)	Raytheon/Loral Corp.	Hit-to-kill	Substantial	40–50	1200 missiles, 54 fire units	1999	$6.2 billion
Navy Area Defense (a.k.a. Navy Lower Tier)	Standard Missile Co. (=Raytheon/ Hughes Aircraft Co.)	Blast fragment	Substantial	50–100	1500 missiles	2000[b]	$6.2 billion
Meads (a.k.a. Corps Sam)	Lockheed Martin Corp. vs. Hughes/ Raytheon	Hit-to-kill	Substantial	<10	*To be determined*	*To be determined*	*To be determined*
Improved Hawk (Marines)	Raytheon/ Lockheed Martin (GE)	Blast fragment	Substantial	<10	700 missiles (modified)	1995	$0.3 billion

			Upper-tier (area) defenses				
Thaad (Army	Lockheed Martin Missiles and Space Co. (Raytheon: radar)	Hit-to-kill	None	1233 missiles, 77 launchers, 11 radars	A few hundred	2006[c]	$12.8 billion
Navy Theater Wide Defense (a.k.a. Navy Upper Tier)	Standard Missiles vs. Lockheed Martin	Hit-to-kill	None	650 missiles on 22 Aegis cruisers	More than a few hundred	TBD	$5 billion
Arrow (Israel/ United States)	Israel Aircraft Industries	Blast fragment	None	*To be determined*	A few hundred	*To be determined*	*To be determined*
			Boose-phase defenses				
Airborne Laser (Air Force)	Boeing Defense and Space	Directed energy	None	7 aircraft	Possibly huge	2006	$6.1 billion
UAV Boost Phase Defense (Israel/United States)	Israel Aircraft Industries	Hit-to-kill	None	*To be determined*	Possibly huge	*To be determined*	*To be determined*

SOURCE: Ballistic Defense Organization (BMDO); Department of Defense.

[a] This number excludes missiles bought for development.

[b] By 2000 the BMDO plans to deploy prototypes of the Navy Area Defense and Thaad, called User Operational Evaluation Systems, that would be available in a crisis. Recent setbacks in the Thaad testing program will likely delay its prototype date.

Meads = Medium Extended Air Defense System

Thaad = Theater High Altitude Aerial Defense

UAV = Unmanned Air Vehicle

Figure 21.7 Table showing the capability and cost of planned theater missile defenses.

System name(s)	Prime contractor(s)	Types of infrared sensors	Earth orbit, km	No. in constellation	Initial deployment	Cost
Defense Support Program (DSP)	TRW Inc.	Short-wave	Geostationary	4–5	Early 1970s	U.S. $9 billion
Space-Based Infrared System (Sbirs High)	Lockheed Martin Missiles and Space Co.	Short- and medium-wave near visible	Geostationary .... High earth-orbit	4 2	2002	$5 billion
Space and Missile Tracking System (a.k.a. Brilliant Eyes, Sbirs Low)	TRW/Hughes vs. Boeing/Lockheed	Short-, medium- and long-wave	Low earth-orbit	24	2004	$5 billion

SOURCE: Department of Defense

Figure 21.8 Table showing the capability and cost of planned space-based sensors.

Proposed system	Interceptor					Antiballistic missile (ABM) radar			Cost
	Name	Kill vehicle	Type of booster	No. deployed	Location	Name	Type	Base radars	
Candidates for Administration's 3-Plus-3 System									
Army System	Ground-Based Interceptor (GB)	Exo-atmospheric Kill Vehicle (EKV)	New or off the shelf	100	Grand Fords, N.D.	Ground-Based Radar (GBR)	100-GHz phased array	10-GHz phased-array Advanced Early-Warning Radars (much like GBR)	$9 billion
Air Force System	Minuteman Interceptor	Kinetic Kill Vehicle	Minuteman III (modified)	20–100	Grand Forks, N.D.	Have Stare	10-GHz dish (based on existing radar)	10-GHz dish (same as ABM radar)	$4–6 billion
Other possible ground-based defenses									
4-Plus-2-Weeks System	Payload Launch Vehicle	EKV	Payload Launch Vehicle (existing)	20	Grand Forks, N.D.,— but in a crisis at sites in Alaska and Hawaii	Have Stare in Alaska	2–4-GHz and 100-GHz dish	Unknown sea-based radar (Cobra Judy?) or Have Stare in Japan	Unknown— a few billion
Multiple Site Defense	GBI	EKV	New or off the shelf	200–300	Northeast, Northwest, and perhaps Grand Forks	Ground-Based Radar	10-GHz phased array	10-GHz phased-array Advanced Early Warning Radars (much like the GBR)	$13–15 billion

SOURCE: Department of Defense; Congressional Budget Office (CBO). Costs are from the CBO. They exclude $2.3 billion recently added by the Pentagon to increase testing and reduce technical and schedule risk. This addition would raise these estimates, although probably not by the full amount, since the CBO had earlier included $1 billion for those activities.

Figure 21.9 Table showing the characteristics and costs of ground-based national missile defenses.

struggle to balance the Federal budget and cut taxes. Resources spent on missile defense must be taken away from defense programs that protect citizens in other ways. For example, in the case of national missile defenses, the Pentagon and the Congress must assess how much the nation should spend on a possible future threat when a country or terrorist group could use other less-expensive and more readily available methods for delivering weapons of mass destruction to the United States, such as ships or commercial aircraft. Furthermore, some analysts believe that rogue states may threaten the United States with cruise missiles long before intercontinental ballistic missiles will be available.

Ballistic missile defenses must also compete with other means of addressing proliferation. One means is the destruction of an opponent's weapons of mass destruction and his ability to produce and deliver them before they are used in a conflict (a tactic that the Department of Defense refers to as counterproliferation). Another is the use of passive measures to protect troops and allied populations such as gas masks or vaccinations against biological agents, at least in the case of theater missiles. And another is the use of diplomatic efforts to stop or slow down the spread of ballistic missiles and weapons of mass destruction, such as the Nuclear Nonproliferation Treaty, the Missile Technology Control Regime, and diplomatic pressure and sanctions.

In the end, it is a balancing act between a variety of systems that protect citizens from a broad spectrum of threats. Only time will show how the political process sorts out all of these competing priorities.[7]

Another concept was proposed by the Air Force, which believes that the right way to do missile defense in the long run is with weapons based in space. It argues that defending against a few warheads from North Korea in the interim does not require a brand new ground-based system.

Instead, the Air Force's candidate is a ground-based system that uses off-the-shelf components as much as possible: modified Minuteman missiles carrying an improved version of the Leap kill vehicle that the Navy is developing for the Upper Tier system,10-GHz dish radars based on a design that has already been deployed, and the Minuteman command and control system. This system would cost at least $7 billion for 200 interceptors. Recently, the service modified its proposal and now plans to use the Army's proposed Exoatmospheric Kill Vehicle instead of Leap, raising costs somewhat. Over the next year or so, the Ballistic Missile Defense Organization and the Joint National Missile Defense Program Office will decide which proposal, or combination of the two, will be developed.

A third suggested approach might be capable of protecting Alaska and Hawaii sooner and at lower cost. This approach would address the most pressing threat (a possible North Korean missile that could reach parts of Alaska and Hawaii) without requiring that the United States rush to build a national missile defense. The interceptor would add the Army's Exoatmospheric Kill Vehicle (EKV) by having it ride atop the Army's Payload Launch Vehicle (PLV). A few PLVs, unlike the EKV, are operational, and in fact were designed with EKV testing in mind.

Because the Payload Launch Vehicle was built to carry the Exoatmospheric Kill Vehicle during test flights, the integration has already largely been done and would cause little interruption to the effort to develop a national missile defense. While the EKV-and-PLV missile is too small and slow to defend Alaska and Hawaii from a base in North Dakota, the concept is to keep the missiles parked on trailers in a large garage at Grand Forks, ND so that Russia can keep track of them. Then, when a crisis emerges (say, North Korea tests its Taepo Dong missile) the 30-missile defense would be deployed to a prepared site in Alaska or Hawaii, or both. Appropriate radars would already have been built at optimal locations. Costs are not known, but would likely be lower than for either of the other systems.

This system has been dubbed the 5-Plus-3 defense: five years to develop and build and three weeks to deploy during a crisis. It would clearly violate portions of the ABM treaty, but conversely may be easier for Russia to accept than the Administration's 4-Plus-4 system — largely because it would have a very limited ability to intercept Russian missiles and would not protect the entire country.

Other BMD Approaches

All the forementioned concepts for missile defense would be ground-based and designed to deal with very small attacks. Part of the reason for this approach is that the anticipated threats are small (except for the unlikely possibility of a large accidental or deliberate attack by

Russia). Another reason is the desire by the Administration and others to keep the defenses within the numerical and geographic limits of the ABM treaty.

But several other approaches are being considered as well. One would deploy a system similar to that proposed by the Administration, but base it at two sites — one on the West Coast of the United States, which would be better suited to defending Alaska and Hawaii, and the other at Grand Forks or on the East Coast. Senator Richard Lugar (R-IN) and former Senator Sam Nunn (D-GA) have endorsed this approach.

There have also been calls to develop a sea-based National Missile Defense. A group of analysts from the Heritage Foundation, Washington, DC, have proposed using the Navy Upper Tier theater missile defense to protect the United States from ballistic missiles launched from North Korea or the Middle East. They contend that this would be the best and least costly way to provide national missile defense until space-based defenses could be deployed. This system would require that Space and Missile Tracking System (SMTS) infrared tracking satellites be deployed to ensure that missiles were intercepted shortly after they left the atmosphere. Setting up such a system would cost about $6 billion, and deploying SMTS as a part of it would cost another $6 billion.

In contrast, the Navy doubts that the Upper Tier system could provide an effective national missile defense — its missiles are unable to fly high enough or fast enough and its kill vehicle is too readily fooled by countermeasures. Instead, it should develop a new, larger missile to deploy on its Aegis cruisers. The missiles would carry the Exoatmospheric Kill Vehicle that the Army is developing for its national missile defense and rely on SMTS satellites as their primary sensor to make up for the inadequate resolution and range of the Aegis 24-GHz (S-band) radar. Preliminary estimates of the cost of this system, which as yet exists only on paper, run to about $20 billion, not including the use of SMTS and a few of the other supporting systems that the Navy thinks it would need.

Two other approaches that make use of space-based weapons have been proposed. Both would have some ability to destroy longer-range theater missiles. One would deploy 23–30 laser satellites to destroy missiles during boost phase. This program, which traces its lineage back to the original Strategic Defense Initiative program, faces several technical hurdles. The Administration plans to fund it at only $39 million a year through 2004. Congress has quadrupled funding in each of the past three years, but even if that trend continues, it will not be nearly enough for deployment by the next decade.

Yet another approach, called Brilliant Pebbles, was halted by the Clinton administration in 1993. It was to have deployed 600 or more small autonomous satellites, each carrying a hit-to-kill missile to destroy missiles during the boost and midcourse phase.

The Ballistic Missile Defense Organization (BMDO) has estimated that these two systems would cost $36 billion and $27 billion to deploy, respectively. But estimates of the cost of a space-based laser system vary widely — from a $26 billion industry estimate to a $200 billion internal Pentagon estimate.

Simulation and Testing

A critical question facing the BMDO in developing and sustaining a highly reliable and effective missile defense is the proper amount and type of testing required. Testing is essential during development. Moreover, it must continue long after the system has been deployed in order to ensure that the defense remains reliable.

The Proof That Systems Do Work

To have confidence in the outcome of the development tests used to prove the system works, enough of them must be done for the results to be statistically significant. If too few are conducted, there is a chance that successful tests will be random events and not true indicators of the system's performance.

The degree of confidence that the military requires in a system relates to its mission. For example, the Pentagon demands a very high degree of confidence for the missiles that carry nuclear weapons, and it budgets for a correspondingly high number of flight tests. But missile defenses, too, particularly national missile defenses, must meet stringent standards.

In fact, testing missile defenses is more complex than testing nuclear delivery systems. Unlike ballistic missiles, which essentially try to hit a fixed point on the earth, interceptors must be able to engage a target coming from any number of directions, at different speeds and ranges, and at many different points in targets accompanied by a wide variety of decoys and other countermeasures. What's more, the system must be able to prove that it can engage many targets at once.

Components of the system — such as interceptors, radars, surveillance and tracking satellites, and battle management systems — must each be tested enough times so that the BMDO is convinced of their individual performance and reliability. Then, those systems must all be tested together as an integrated system enough times for the BMDO to verify the performance of the total system.

Nor are flight tests inexpensive. They can cost between $20 million and $60 million apiece because, unlike aircraft or radar, interceptors are destroyed during the tests.

Early on, the BMDO recognized that it would be unable to afford extensive flight testing. It also determined that some parameters were better measured in a laboratory, where conditions could be controlled and measurements made repeatedly and more accurately. Nor could the organization test the system against all possible contingencies, even with unlimited resources. Flight tests would have to be conducted at established test ranges to avoid dropping rocket boosters from the interceptors or dummy warheads on citizens who lived along the flight paths. For similar reasons, test ranges also impose constraints on the types of trajectories allowed.

So the BMDO has devised a testing strategy (adapted from earlier programs to develop air defenses) that makes extensive use of encrypted computer simulations and laboratory tests. Some laboratory tests will use actual flight hardware and simulate the environmental conditions and target signatures that the system will see in flight, so-called hardware-in-the-loop tests.

The concept underlying the BMDO's approach is to ensure that as many components and subcomponents as possible work properly in the laboratory under close to realistic conditions before they are used in a flight test. In turn, the results from flight tests will be used to validate and improve the computer models and simulations.

Under its simulation-intensive approach, the BMDO plans an average of 20 flight tests during development for each of its four hit-to-kill systems: PAC-3, Thaad, Navy Upper Tier, and the 4-Plus-4 National Missile Defense. This is a very low number compared to the 222 flight tests that the United States conducted to develop its first ABM system, Safeguard, in the 1970s or even to the 222 tests conducted for the Advanced Medium-Range Air-to-Air Missile (Amraam).

How effective this approach will be has yet to be seen. Clearly, more simulations and laboratory tests will increase confidence in the system. But whether they can substitute for flight tests to the extent that the BMDO is proposing is unknown; no other development program has relied so heavily on computer simulation and lab testing for a system that is the first of its kind. Indeed, the ability to gain knowledge about complex systems through simulation is a new frontier.

The central question is whether a historically low number of flight tests will allow military leaders to certify to the President that a national missile defense will work with the reliability and performance expected of an intercontinental ballistic missile system. Likewise, officials must have enough confidence in theater defenses to commit troops to future conflicts in which chemical or biological weapons might be used.

Early results indicate the difficulties that the BMDO faces. Thaad, for example, is scheduled to have 22 flight tests in its demonstration and evaluation phase. With so few tests, its intercept record of five failures for five tests looms the larger, and may delay the program more than might have been the case if the program had planned 30 or 60 flight tests.

Nevertheless, if a comparison of the Department of Energy's (DOE) nuclear weapons testing program is made with BMDs, there is some reason for optimism—even with limited testing. DOE's weapons laboratories have used computer simulations and laboratory experiments extensively to design new weapons and test many of their components. The labs exploded only a handful of weapons in underground tests for each of the new weapons before deploying it.

Because so few weapons are exploded, the Department of Energy cannot present performance and reliability measures statistically. Rather, it relies on the expert judgment of weapons scientists to estimate those parameters, based on past experience and laboratory tests.

There are, all the same, important differences between the nuclear weapons program and current efforts to develop ballistic missile defenses. First, before they began to rely so heavily on simulation for the missile defenses, scientists had acquired a lot of experience designing and exploding weapons. Second, the peer review process between two competing laboratories during the nuclear weapons program (Los Alamos in New Mexico and Lawrence Livermore in California) reduced the chances that expert judgment might become wishful thinking.

System Confidence

The requirements for flight tests do not end when the defense is deployed. For the military to retain confidence that the system will perform reliably over the years, it must perform enough tests each year to measure a sample that is large enough to properly represent the characteristics of the population of deployed missiles.

Typically, U.S. intercontinental and sea-launched ballistic missiles are launched in operational tests 7 to 24 times a year for the first six years or so of a system's life to establish a sufficient sample measuring the missile's operational performance and reliability. Thereafter, missiles are flown four to seven times a year in so-called follow-on tests to ensure that their performance and reliability do not degrade with age. Follow-on tests continue for most of the missile's service life.

The BMDO will have to conduct similar tests of its systems after they have been deployed. So far, those plans have not been laid out in detail, but as with development testing, the BMDO will rely heavily on simulation.

The Bottom Lines

One of the remarkable aspects of the evolution of missile defenses is that few policymakers question the fundamental ability of missile defenses to be effective. Instead, they focus on issues of timing, costs, and quantities. This is a sharp change from the Reagan years, perhaps because the technology used is closer at hand and the threats are smaller.

Indeed, the Patriot PAC-2 (in both its original and its upgraded configurations) and the Navy Lower Tier systems have intercepted ballistic missile targets with their blast fragmentation warheads. And the hit-to-kill interceptor that will be used in the PAC-3 system has intercepted several targets.

But many more attempts have failed, particularly for hit-to-kill systems. In the last 16 years the United States has conducted 30 hit-to-kill intercepts, for the BMD programs discussed here as well in other tests. Seven intercepts were successful; 24 of those tests were done within the last six years, and among them, four intercepts succeeded. The Pentagon's premier area defense, the Thaad, has failed in all five of its intercept attempts.

Furthermore, no real attempts have been made to intercept uncooperative targets — those that make use of clutter, decoys, maneuver, antisimulation, and other countermeasures. Nor have any tests attempted to use a real battle management system that integrates data from a diverse array of actual tracking sensors and directs an interceptor to a target.

The long record of failures to intercept must be kept in perspective, however. Missing its target does not necessarily mean that a system is badly designed or a program is poorly managed. Rather, it may illuminate a more basic truth. Developing an effective missile defense indeed is challenging. If decisionmakers recognize this, they will not be surprised by schedule delays and cost increases that will probably accompany many missile defense programs as program managers endeavor to conquer those challenges.

In one recent example, the Pentagon decided to add $3.4 billion over the next four years to the National Missile Defense program to keep it on schedule — tripling the cost of the development program in those years. And internal Pentagon documents reveal that a similar increase is expected for the Thaad, according to press reports, even as the schedule continues to slip.

Next, let us move on to the high-flying megawatt laser that pinpoints a missile lifting off hundreds of kilometers away and is an alternative to terminal defenses.

The High-Flying Airborne Laser

The U.S. Air Force announced the three prime contractors for a radically new weapon: a modified Boeing 747 freighter carrying a laser beam for disabling or destroying ballistic missiles as they leave their launch pads. If the various demonstration phases meet expectations — an in-flight kill by the Airborne Laser is scheduled for 2003 — the United States proposes to fund construction of eight of the weapons. Boeing Defense and Space Group, in Seattle, WA is handling system integration of the Airborne Laser; Lockheed Martin Missiles & Space Co., Sunnyvale, CA is providing the optics and beam control system; and TRW Inc., Redondo Beach, CA will supply the high-energy laser.

In operation, each plane (estimated at $2 billion a copy) will patrol friendly airspace. If an enemy missile launch is detected by any of a variety of U.S. sensor systems, the information is relayed to the airplane. Its nose will swivel, and a 1.5-meter mirror inside the nose will focus the beam from a megawatt-class chemical laser onto the missile and keep the beam locked on that small supersonic target perhaps hundreds of kilometers away as shown in Figures 21.10, 21.11, and 21.12.[8] If the beam is able to dwell on the same spot for long enough (an interval known as its dwell time), the metal is fatally weakened.

The utility of the Airborne Laser (ABL) in future regional conflicts, and the consequent policy options the United States can exercise, will be highly dependent on its range. The ABL's range in turn is determined by the accuracy with which the primary laser beam can be pointed, the power density it can deliver, and the structural design of the missile being attacked. The following examination of these issues, including all the estimates of the ABL's capabilities, are from the author's own analyses. The findings combine information found in the open literature with the basic physics and engineering involved in propagating intense laser beams through the atmosphere (see the Sidebar that follows).

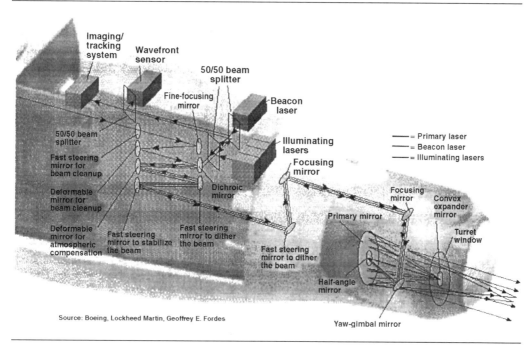

Figure 21.10 The Airborne Laser (ABL) would have three critical laser systems linked by mirrors.

Figure 21.11 The Airborne Laser (ABL

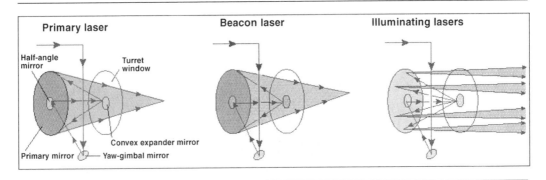

Figure 21.12 The primary mirror both gathers in light beams and focuses them outward. Other mirrors reflect the laser beams, split them, sometimes into two separate beams of different wavelengths (dichroic mirrors), steer them, and, for the killing beam, shape its wavefront (a detail of the final stages of the beams is shown). A megawatt chemical oxygen iodine laser (COIL) is the primary (killing) beam. Beams from the pulsed illuminating lasers are injected into the continuous killing beam's path through a dichroic mirror. The illuminating and killing beams, as well as pulses from the beacon laser are directed and focused by the same mirrors that aim (dither) the killing beam. The beacon and illuminating pulses are reflected from the target missile. The primary mirror focuses them on, respectively, the wavefront sensor (to optimize the killing beam) and the imaging/ tracking system, to continue the lock-on of the target.

U.S. Laser Weapons Program Highlights

1960 (May): In a nondefense research project, Theodore Maiman builds the world's first laser using a ruby rod that is pumped by a flash lamp.

1966: The U.S. Air Force creates a Weapons Laboratory task force to study possible uses of laser technology for weapons systems.

Early 1970s: General Dynamics Co. initiates the Airborne Laser Laboratory (ALL) design effort at the behest of the Air Force. Hughes Aircraft Co. receives a contract for an Airborne Pointer/Tracker.

1971: The Naval High-Energy Laser Program begins to develop carbon dioxide gas dynamic high-energy laser (HEL) technology for use against antiship missiles.

1973: TRW Inc. builds the first high-energy chemical laser, using deuterium fluoride.

1975: As part of the ALL program, a laser is fired for the first time from an aircraft.

1977: The U.S. Navy starts its Sealite project to demonstrate a laser that would be lethal against fast antiship missiles.

1978: TRW integrates its high-energy laser with the Navy Pointer/Tracker constructed by Hughes Aircraft Co. The system is successfully tested in a live fire exercise against TOW antitank missiles at TRW's Capistrano test site, near San Clemente, CA.

1980 (September): The TRW-built megawatt-class Mid-Infrared Advanced Chemical Laser (Miracl) is tested at TRW's Capistrano range. Now at the U.S. Army's High Energy Laser Test Facility (Helstf) at White Sands, NM, it is the highest-average-power laser in the United States.

1981: TRW builds a chemical oxygen iodine laser (COIL).

1981 (May): The ALL makes its first attempt at defeating an AIM-9 Sidewinder air-to-air missile at China Lake in California.

1983 (July): It is reported that the ALL manages to *destroy or defeat* five AIM-9 Sidewinder air-to-air missiles.

1987: The Miracl/Sealite Beam Director (SLBD) tracks and destroys three subsonic drones.

1989: The Miracl/SLBD is successfully tested against a supersonic Vandal missile flying in excess of Mach 2.

1990 (February)–1991 (April): Using adaptive optics and a cooperative beacon, the Massachusetts Institute of Technology's Lincoln Laboratory demonstrates near-diffraction limited imaging of a satellite along a near-vertical path.

1992: The U.S. Air Force Phillips Laboratory gives a contract to Boeing to assess how well a large existing airplane, such as a Boeing 707, 747, 767, or B-52, would perform while carrying a high-energy laser and beam control system.

1995: The Phillips Laboratory's ABLE ACE series of experiments is completed. Horizontal atmospheric turbulence effects over ranges of up to 200 km were measured at high altitudes over the United States, Japan, and South Korea.

1996 (February): A Miracl laser with a Sealite Beam Director shoots a Katyusha rocket out of the air at the U.S. Army's High-Energy Laser Systems Test Facility at the White Sands Missile Range in New Mexico (discussed in greater detail later in chapter).

1996 (August): A TRW COIL produces a beam of hundreds of kilowatts lasting for several seconds.

1996 (November): The Air Force's Airborne Laser System Program Office awards Boeing, TRW, and Lockheed Martin a $1.1 billion, 5-year contract to develop and flight-test a laser weapon system to severely damage or destroy ballistic missiles in their boost phases. A demo shootdown is expected in the year 2003.

1997 (March): TRW and Lockheed Martin announce a successful 0.5-second ground test of the Alpha laser and control systems, in work toward the half-scale Starlite demonstrator for the Space-Based Laser.[9]

Engaging Enemy Missiles While Still under Power

The well-known Patriot missile defense system and other mid-course or terminal defenses target warheads as they reenter the atmosphere. But engaging enemy missiles while they are still under power, as the ABL does, has significant advantages.

Tactically, an attack during the boost phase can destroy missiles carrying chemical or biological agents before any smaller warheads (submunitions) are released. The deadly debris could even fall over enemy territory. Technologically, rocket plumes from a missile under power are easier to track than is a warhead in its coast phase.

Structurally, too, theater ballistic missiles are quite vulnerable during their boost phase. During the last seconds of powered flight, they commonly endure compressive loads about five times their launch weight. Usually, military designers desire maximum range; yet to supply that range, the missiles must weigh as little as possible (all other things being equal). But skimping on stronger, but heavier, structural components leaves the missile with slim safety margins, so that even minimal damage to its structural integrity can result in its destruction.

Modes of Kill

The ABL attacks missiles by focusing its primary laser beam on their metal surface, heating it until the metal strength drops dramatically. Each kind of metal has its own characteristic failure point — 460 °C for steel and 182 °C for aluminum.

Given this as a starting point, the ABL would take down missiles as they were being launched, in one of two ways. The first, against missiles with liquid fuel (such as those deployed by most Third World countries), is to heat a fuel tank to the point of structural failure, whereupon it will rupture from its internal pressure. The rupture will terminate or lessen the thrust of the missile, which will then fall short of its intended target.

The other scenario is to have the laser beam cause the missile as a whole to fail catastrophically. To do this, the beam must heat a sufficiently large arc along the missile's circumference to the critical structural-failure temperature. When it does so, the aerodynamic and inertial forces acting along the missile's axis will bend the structure in half as shown in Figure 21.13.

Airborne Laser Flight

According to engineers from Lockheed Martin, the ABL while patrolling on station will fly above the clouds at an altitude of 12.9 km. Battle management officers onboard would most likely have the aircraft travel along an elongated figure-eight flight path, the long axis being perpendicular to the direction of the missile's expected launch. This flight path optimizes laser beam atmospheric propagation as much as possible.

On board the ABL are three major laser systems: the primary beam (referred to as the killing beam), which is a continuous laser; and two

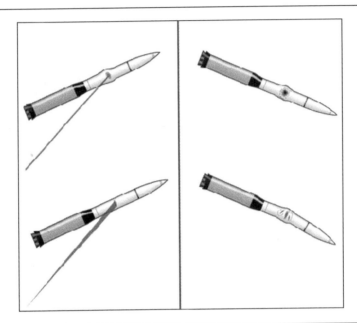

Figure 21.13 Boost-phase defense systems attack a missile while it is still under power. The Airborne Laser is intended to take down a boosting missile by heating the latter's metal skin to a failure point (rather than melting it). In one technique, it heats a spot on a missile's liquid-fuel tank until internal pressure forces a rupture at the weakened site, causing early termination of the flight (top, left and right). In the other technique, the missile is heated in a back-and-forth arc around its circumference. It then crumples in its entirety along the weakened arc from the force of its own acceleration (bottom, left and right).

pulsed lasers, one to track and keep a lock on the target and another to minutely adjust the properties of the primary beam in response to atmospheric changes (see Figures 21.10, 21.11, and 21.12).

The heart of the weapon is the primary laser. This is a chemical oxygen iodine laser (COIL) positioned in the rear of the 747 that produces a continuous infrared beam with a wavelength of 1.315 μm. The ABL will require a megawatt-class COIL, a considerably higher power than has been produced by lasers on the ground, which have peaked in the area of hundreds of kilowatts. The calculations have been based on a 3-MW COIL (see the Sidebar). By way of analogy, 30 acetylene torches have a combined power of 3 MW—but of course, they cannot be directed across hundreds of kilometers.

COIL

Electrical laser pumping, although the most common, is not the only form of laser excitation (see Sidebar, "An Array Of Lasers"). For instance, the first optical lasers of the 1960s used flash lamps to pump the laser medium, at the time a ruby crystal rod. Another option is to use a chemical reaction. Chemical generation of the laser energy has an advantage over the more common electrical mechanisms because it allows a more compact unit. These chemical methods have been used to create lasers with large efficiencies, as much as 25% overall efficiency, as opposed to less than 1% for helium-neon lasers.

In a chemical lasing process, two or more chemicals mixed together will try to form molecules with a lower total energy. The excess energy, usually observed as heat, is left in different forms in the resulting molecules. For instance, when hydrogen and fluorine are mixed at high temperatures in the so-called Miracl (Mid-Infrared Advanced Chemical Laser) system, the resulting molecule, deuterium fluoride, is usually both spinning and vibrating more than it would in its ground state.

This molecular motion by itself is not enough to create the needed population inversion for lasing. The different final states of the molecule, as characterized by the varied amounts of spinning and vibrating, have different rates of production. It is these different reaction constants that produce a greater proportion of excited states than would be present in the gas in thermal equilibrium.

In the chemical oxygen iodine laser (COIL) of the U.S. Air Force's Airborne Laser project, excited atomic iodine is used as the lasing medium. The first step involves blowing chlorine gas past a basic hydrogen peroxide solution. Chlorine migrates into the liquid and reacts to produce excited oxygen molecules. Excited oxygen then escapes from the solution and is mixed downstream with molecular iodine. The iodine molecules are broken up and individual atoms are excited by a nearly resonant reaction with the oxygen in multiple reactions. This last transfer of energy leaves atomic iodine in an inverted population, and this takes place between the mirrors of the laser's resonator.[10]

An Array of Lasers

The operation of the common helium-neon laser is a good introduction to the basic principles of lasing. An electric power supply is used to *pump* a medium (helium, in this case) until the majority of the atoms in it have been excited to a higher-than-normal energy level; the resulting state, which rarely occurs in nature, is known as a population inversion.

Eventually one atom falls to a lower energy-state by radiating a photon along the axis of the laser. This photon's electromagnetic fields are amplified by mirrors placed along this axis. Other excited atoms, stimulated by these fields, also radiate their excitation energy in the same direction, creating a wave of energy. The resulting lightwave is coherent, meaning that the phases of all the individual electromagnetic waves are the same. Individual photons' electric and magnetic fields are then added constructively inside the laser cavity.

Gas dynamic lasers and chemical lasers share some important outward similarities. Both types have a supersonic flow of gases into the laser cavity, where the population inversion needed for lasing is produced. But they differ in how they produce the inversion.

Gas dynamic lasers start with a molecular gas at elevated temperatures and pass it through supersonic nozzles to cool it. The gases expand on their way through the bell-shaped nozzles, in this way reducing the molecules' kinetic motion. Vibrational and rotational motions are, to a large extent, unaffected by this cooling, so that highly excited vibrational and rotational states are populated out of proportion to the new gas temperature. A common characteristic of all lasers utilizing rotationally and vibrationally excited states is their multiple lasing lines.

Chemical lasers, on the other hand, produce the population inversion by mixing chemicals inside the laser cavities. Reactions leading to useful lasers preferentially produce the excited states.

Some examples of a few lasers and their characteristics follow:

CO_2: An example of a gas dynamic laser, this was one of the first types of lasers to produce a beam in the hundreds-of-kilowatt range (at Avco Everett Research Laboratory, in Massachusetts). So the military invested in a long-term research project to decide on the feasibility of its use as a weapon. As things turned out, its long characteristic wavelengths, spanning the range of 9–11 μm, are not suitable for focusing small beam spots across large distances. Carbon dioxide lasers primarily serve industrial purposes such as drilling, cutting, and welding of materials.

HF: This chemical laser combines heated hydrogen (produced in a combustion chamber similar to the one in a rocket engine) with fluorine gas to produce excited hydrogen fluoride molecules. The light beam that results radiates on multiple lines between 2.7–2.9 μm. These wavelengths transmit poorly through the atmosphere. Even so, the Ballistic Missile Defense Organization is considering HF lasers for space-based defenses needing to propagate through only the upper atmosphere.

DF: The deuterium fluoride laser is chemically the same as the HF laser. However, the increased mass of heavy hydrogen, deuterium, shifts the laser wavelengths to 3.5–4.0 μm. This range is superior to the HF range for transmission through the atmosphere. Deuterium fluoride is used as the lasing medium for the Mid-Infrared Advanced Chemical Laser (Miracl). The U.S. Army Space and Strategic Defense Command has given TRW Inc., in Redondo Beach, CA, a contract to develop a tactical high-energy laser employing DF technology. It is the basis for the Nautilus project, whose goal it is to defend Israeli settlements against Katyusha and other short-range rockets.

COIL: The chemical oxygen iodine laser differs from previous chemical lasers in that the excited iodine atom, responsible for the lasing, radiates only a single line at 1.315 μm. This short wavelength reduces diffraction effects that limit the utility of other chemical lasers. The COIL laser has been chosen for the Airborne Laser missile defense system. One advantage of the COIL process is that most of the excess heat is liberated in the production of the excited oxygen. This greatly reduces the turbulence inside the laser cavity and facilitates the production of high-quality beams (see Sidebar, "COIL").[11]

After leaving the COIL, the ABL's killing beam travels inside a pipe toward the front of the aircraft. It passes stations for operators who monitor the various subsystems. Just aft of the nose turret, the killing beam enters a complex optical bench, which isolates the optical components on it as much as possible from the vibrations of normal flight. These components do the fine-grained beam control needed to point the primary beam at a missile traveling at transonic speeds hundreds of kilometers away.

Pointing the beam is done with fast, lightweight steering mirrors, which are rotated or tilted to follow the apparent motion of the target. The control loop driving the motions of these mirrors uses as input the image of the missile formed by the tracking system (of which more later).

> **Note:** Directing the primary beam is not done with lenses, as might be expected, but by mirrors, to reduce the power that otherwise would be absorbed within the optical train. But a laser beam powerful enough to melt the metal skin of a missile also would damage any ordinary, silvered mirrors inside the optical train. Therefore, each of the control mirrors will most likely be fabricated from single infrared transparent crystals tens of centimeters across. The reflecting surface is produced by a series of thin layers (films) of dielectric materials with alternating high and low indices of refraction — a measure of how much a material affects the speed of light.

Beam Correction

The atmosphere is the bane of the killing laser. Its water droplets, winds, heatable air, and other components all must be taken into account if a usable killing beam is to be formed. To counteract these atmospheric distortions, encrypted information from an additional *beacon* laser beam is used to preshape a deformable mirror that adjusts the killing beam.

Accordingly, the optical bench supporting the control elements also contains a deformable mirror, which preshapes the primary beam's wavefront (the surface, roughly perpendicular to the beam's axis, on which all the light rays have the same phase). These types of deformable mirrors are usually made from thin faceplates — thin enough for them to be distorted several micrometers across a surface distance of a few millimeters and in several thousandths of a second. Tiny pistons, or actuators, attached to the back of the mirror push a section of the surface forward or pull it back as shown in Figure 21.14.

When the primary laser beam hits different points on the deformable mirror, the effective path lengths of light rays reflected from it are changed. As they travel along the different path lengths, rays that began in phase wind up out of phase. A properly distorted mirror can

Figure 21.14 The adaptive optics system on the Airborne Laser employs deformable mirrors (left) to shape the killing beam for optimum travel through the atmosphere. Hundreds of tiny pistons attached to the mirror's back distort its faceplate by several micrometers across a surface distance of a few millimeters.

allow a ray whose phase is lagging behind to catch up by causing other rays to travel longer path lengths.

Note: All of this is possible because differences in phase arise from variations not only in the index of refraction but also in path lengths.

Each pulse of the beacon beam arrives at the missile slightly ahead of the target spot and just before the next adjustment to the killing beam; its job is to *sample and report back* on atmospheric conditions at extremely short time intervals—conditions to which the killing beam must be adapted as it passes through the same atmospheric profiles as shown in Figure 21.15.

A reflection of the beacon beam from a small spot on the missile *records* the atmospheric distortions it passes through on its return

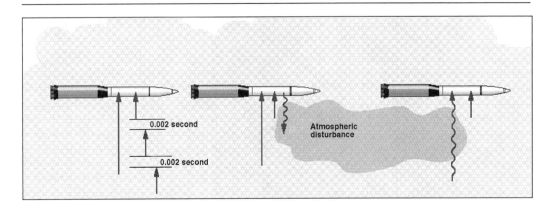

Figure 21.15 The Airborne Laser's adaptive optics molds its continuous killing beam using reference pulses emitted by a beacon laser every 0.002 second, which have the same target point as the killing beam. When a pulse is reflected from the missile to the main mirror, its optical signature bears distortions caused by the atmosphere through which it has just passed. Deformable mirrors then *predistort* **the killing beam in a way exactly complementary to the distortions the pulse has undergone.**

flight to the ABL. Because that profile is changing rapidly (both the ABL and the target missile are after all traveling at hundreds of kilometers an hour) the faster the sampling by the beacon beam, the better. This analysis employs a moderate rate of a pulse every 0.002 second.

The different optical signatures of each pulse, upon its return, are used to determine, in real time, the best shape for the deformable mirror that will form and guide the killing beam. When the killing beam is bounced off the deformable mirror, it is *predistorted* in a way exactly complementary to the distortions produced by the atmosphere (or at least to those distortions produced 0.002 second before). Similar adaptive optics technology is used in ground-based observatories to correct for the atmospheric distortion of astronomical images — to remove the *twinkle* of a star, for example, so as to get a clearer picture.

Tracking and Imaging

The missile-tracking laser system has an equal share in this process. Missile tracking is cued by input from reconnaissance assets such as satellites or Airborne Warning and Control System (AWACS) planes. The initial fix on the missile's plume, followed by coarse tracking, is performed by several small infrared telescopes mounted along the 747 body and with much wider fields of view than the primary mirror.

The turret housing the 1.5-meter diameter primary mirror at the front of the ABL is its main window to the world: a ball-shaped protuberance behind an infrared-transparent canopy, which smoothes air flow around the turret. The turret rotates and can be pointed up or down to allow the main mirror to follow the gross motions of the target, which can span several degrees of arc during an engagement.

The angular coverage of the laser extends across the front of the 747 and backward on either side until blocked by the fuselage. In principle, this coverage might span 270° horizontally, but atmospheric effects make the forward direction not particularly usable.

The main mirror not only focuses the killing and beacon beams on the target, but by imaging the target (taking light in) it is also a major player in tracking. The imaging system must get an extremely rapid fix on the missile's body if motion blur is not to distort the image. The time scales are quite daunting.

For instance, a North Korean Nodong-1 missile launched 200 km away from the ABL has a vertical speed of 290 m/s as it crosses the engagement plane. It represents the lowest point at which the ABL can start to fire its main laser. Assuming the ABL is flying perpendicular to the direction of the missile at 200 m/s, the total apparent speed is 350 m/s.

Note: This plane is an imaginary sheet spreading out horizontally from the position of the ABL.

Taking this relative motion of ABL and the target into account, a *shutter speed* of 6310^{-5} second (a pulse rate well within the means of the illuminating laser) produces a smearing of the image of 2 cm. That is an acceptable tenth of what is known as the diffraction limit (the smallest resolvable image) of the main beam.

> **Note:** Killing-beam spots greater than the diffraction limit decrease the average intensity of the beam. In order to compensate, the beam will have to dwell on the target longer, which will decrease the effective range.

Once tracking of the missile is locked on, blurring is reduced considerably because the *camera* follows the target. But the faster the shutter speed, the less time light has to enter. Just as with home cameras, the subject needs additional illumination, even in full daylight.

The additional illumination required is provided by the third major laser system, a number of ancillary infrared lasers that are also directed at the target. They are most probably of the pulsed, solid-state Nd:YAG (neodymium-doped yttrium aluminum garnet) type, with characteristic wavelengths around 1.06 m. Light from these lasers illuminates a region around the missile's nose cone, probably 5–10 m in diameter. The primary mirror gathers the 1.06-μm illuminating light reflected from the target. The *image* of the missile is formed by focusing these pulses on the imaging/tracking system.

On the optical bench, images of the target are separated from the laser beams by a dichroic mirror (in this case, a mirror whose layers of dielectric materials are selected to transmit light at 1.06 μm and to reflect the intense 1.315-μm primary beam). The features of the missile, such as the point of the nose cone, are found with the aid of pattern-recognition techniques — all of which are used by a tracking control loop to point and lock-on the primary and beacon beams.

The jitter/smearing from the effects of both motion-blurring and diffraction-limiting will increase the time required for the primary laser to dwell on the target. The calculations show, for instance, that when the ABL's pointing accuracy at the diffraction limit of 1 microradian, its maximum range against, for example, Iraq's al-Husayn missiles, decreases from 470 km to 420 km — in other words, a 20% decrease in area coverage occurs as shown in Figure 21.16.[12]

Name (country of origin)	Missile				Airborne laser	
	Range, km	Burn time, seconds	Diameter, meters	Skin metal and thickness	Range for decisive engagement[a] km	Maximum range, km
Scud-B (USSR)	300	75	0.84	Steel, 1 mm	240	320
al-Husayn (Iraq)	650	90	0.84	Steel, 1 mm	320	470
Nodong-1 (North Korea)	1000[b]	70[b]	1.2[b]	Steel, 3 mm[b]	185	320
ICBM (SS-18-like) (USSR)	10,000	324	3	Aluminum,	—	>1000

[a] Decisive engagements require a 45-degree arc of the missile circumference to be heated to the point of rupturing.
[b] Estimated.

Figure 21.16 Table showing engagement parameters of the airborne laser and various missiles.

The Killing Fields

At this point, several of the most important of these phenomena and the methods that can be employed to overcome them can be examined. The ABL cannot optically track the missile through dense clouds. Instead, it must wait until the missile has risen above any clouds that might intervene. Only then can the tracking algorithms start to lock onto the target. High-altitude balloon experiments sponsored by the U.S. Air Force's Phillips Laboratory in New Mexico suggest that reaching this lock-and-tracking stage might consume several precious seconds.

There is a well-established height below which most clouds usually occur, corresponding to the tropopause, the boundary for convective circulation in the lower atmosphere. Clouds normally stop at the tropopause because the moisture associated with them cannot be transported any higher. The nominal ABL cruising altitude of 12.9 km is above the tropopause in most latitudes, and clearly is an important factor in any battle scenario.

Even clear air weakens the laser beams. The mechanisms underlying this weakening vary with the wavelength of the light. The wavelengths associated with the ABL lasers are mostly prone to scattering by aerosols (chiefly water droplets) and to absorption.

When a water droplet scatters light, it removes energy from the beam. Scattering, though, is a relatively benign process, as it merely diminishes the delivered power density, measured in megawatts per square meter.

Absorption of the light, on the other hand, starts a much more pernicious nonlinear effect. The energy lost to the beam by absorption heats up the column of air the beam is passing through. Because the column is being heated nearly uniformly along its length, the air must expand radially outward. Radial migration of molecules along the beam produces a drop in the density of the air along the central axis as shown in Figure 21.17.[13]

Over the hundreds of kilometers that the beam must travel, a significant lensing effect can build up — the thermal blooming phenomenon. When it is encountered, the beam diverges and the power density delivered at the target is lessened.

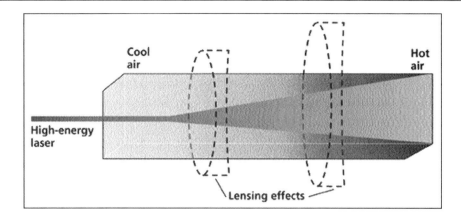

Figure 21.17 A phenomenon known as thermal blooming can degrade laser performance. When a laser beam heats cool air, the air column expands radially outward, and air density along the central axis drops. As the beam travels hundreds of kilometers through nonuniform air, it diverges, as if passing through lenses. The greater the divergence, the less power the beam can deliver.

Several factors affect thermal blooming, most importantly altitude and crosswinds. The net effect of increasing the ABL's altitude is to lessen thermal blooming. The reason is that the aerosol content of the atmosphere, and therefore the absorption of beam energy, is larger at lower altitudes.

Crosswinds blowing perpendicular to the beam (including those generated by the motion of the airplane itself) also mitigate thermal blooming, by blowing the heated, expanding gases out of the way. In fact, they can be very effective at reducing the beam-degrading effects of the phenomenon. Slewing the beam when the missile's trajectory is followed has the same desirable effect, but is less effective near the aircraft.

Crucial during an engagement is the relative positioning of the ABL and enemy missile, known as the engagement geometry. Consider a hypothetical engagement between the ABL and a North Korean Nodong-1 missile launched at Japan. On patrol, the ABL would most likely be flying parallel to the North Korean coast, about 90 km out into the Sea of Japan to avoid surface-to-air missiles.

A Nodong launched toward Japan would fly over the ABL, coming from a direction perpendicular to the ABL's flight path. In this case, the beam is fired sideways relative to the air stream of the ABL, which is moving at 200 m/s. This crosswind is fairly efficient at reducing the effect of thermal blooming to an acceptable level. By looking at the calculations in this engagement geometry, thermal blooming would reduce the peak beam intensity by roughly a factor of 10.

Were the Nodong-1 launched during the brief period when the ABL is making its turn at the end loop of its figure-eight patrol pattern, thermal blooming would be far worse. Now the air stream would be blowing the heated air not out of the beam but along it. Only the slewing motion of the beam as it follows the target introduces a sort of crosswind. This motion would do little to mitigate thermal blooming, which could reduce the peak intensity of the beam by a factor of 1000.

Turbulence is the other atmospheric effect of significance for laser beams. This turbulence is associated not with the airplane's motion through the air, but with slight fluctuations in temperature at different positions along the beam. The fluctuations generate small volumes of expanding or contracting air, known as turbulence cells—evident in the shimmering of distant objects on a blistering hot day.

The greater the altitude, the more homogeneous the atmosphere, and the larger the turbulence cells. As with the regions of air in thermal blooming, the cell regions have different indices of refraction, so that the light rays passing through them are minutely bent in random directions. If the turbulence cells are smaller than the laser beam diameter (which is 1.5 m as it leaves the main mirror), the beam is dispersed. With enough dispersion, the beam breaks into separate smaller beams. And adaptive optics based on deformable mirrors (that is, phase-only adaptive optics) cannot correct a beam to which this happens.

The model of the atmosphere used here has a completely homogeneous atmosphere above 20 km, and therefore no turbulence above that height. But there can be large fluctuations from day to day as well as possible dependencies on latitude. The Air Force Phillips Laboratory recently concluded a series of experiments, known as the Airborne Laser Extended Atmospheric Characterization Experiment (ABLE ACE) and designed to measure these fluctuations and geographic dependencies.

In the experiment, two aircraft flew separately at up to 180 km and measured the atmospheric distortions when a low-power laser beam was shone from one to another. Reportedly, the measurements over the United States, South Korea, and Japan revealed turbulence worse than the maximum previously expected.

The fact that the beacon beam originates from a pulsed laser implies that the shape of the deformable mirror is determined by turbulence slightly different from that encountered by the continuous killing beam. Increasing the beacon pulse repetition rate reduces the beam dispersion caused by not using the instantaneous corrections. These considerations require that the beacon laser pulse several thousand times a second. This should not be a problem, since Nd:YAG lasers have run in pulsed mode with repetition frequencies much greater than this. It does require some sophistication to avoid range ambiguities when there are multiple pulses in the air at once.

Once the killing beam has been propagated through the atmosphere to the missile, it interacts with the missile's skin and the missile system as a whole. Some areas on the missile are surprisingly resistant to laser attack. Fortunately, understanding the ABL's effectiveness is helped by examining how missiles or space launch vehicles have failed in circumstances similar to ABL engagements.

Doing Some Damage

It should be made clear what the beam does not do. It does not vaporize or even melt the missile's skin all the way through. Rather, it heats the skin until whatever internal forces are present cause the skin to fail. This failure could be either the missile's rupture from internal pressure or its collapse from axial compressive loads.

According to calculations, when an al-Husayn missile is launched 350 km away from the ABL, the maximum beam intensity that can be focused on the al-Husayn's surface is approximately 2 MW/m^2. The analysis depends on beam power and on absorption rates of the metal. A highly reflective skin might absorb only 0.2 MW/m^2. At this power density against the al-Husayn's 1-mm-thick steel skin, a small patch could be completely melted in 40 seconds. However, at a distance of 350 km, the missile remains under power for 27 seconds while above the engagement plane, the horizontal plane centered at the ABL.

Another option with this same launch would be to heat the skin up to the missile's *rupture temperature*—approximately 460 °C, at which point the structural strength of the steel drops dramatically. This would take about 17 seconds in the previous example—well under the 27-second time limit.

Note: A thicker skin increases all these dwell times.

Aiming

The structural compositions of more advanced designs, such as the North Korean Nodong-1 missile, are not known with any certainty to the West. Several analysts have suggested that the Nodong-1 is made from steel perhaps 3–4 mm thick. They based these estimates on assumptions about the evolutionary process of missile development as well as published reports for the Nodong range. Uncertainty over the details of a missile's design leads to uncertainties in estimating the effective range of the ABL. Incidentally, these uncertainties exist both in this analysis and in the decisions made by battle management officers on board the ABL.

The details of the missile's construction, and even its surface properties, determine the ABL's aim point. The nose-cone/warhead section is too strong structurally and too well insulated thermally to make a good target. Similarly, the tail section, with its internal supports for transferring the engine's thrust to the rest of the missile, makes a poor target. Fortunately, most of a Third World missile's body consists of fuel tanks apparently without internal supports. These make an ideal aim point for the ABL's killing beam.

The *simpler* kill mechanism blows a hole in the missile's side by making use of the internal pressure of the target's own fuel tank, thereby shortening its flight as shown in Figure 21.18.[14] Liquid-fueled missiles for the most part maintain a pressure inside the tank of 130–200 kPa. This pressure helps ensure a constant fuel rate into the turbopumps feeding the combustion chamber. But heat the missile skin up to the critical temperature, 460 °C for steel or 182 °C for aluminum, and the fuel tank will rupture.

The second possibility is to make the missile collapse from a compressive load along its own axis. An axial load has two sources. One is atmospheric drag, which exerts a large force on the missile, particularly

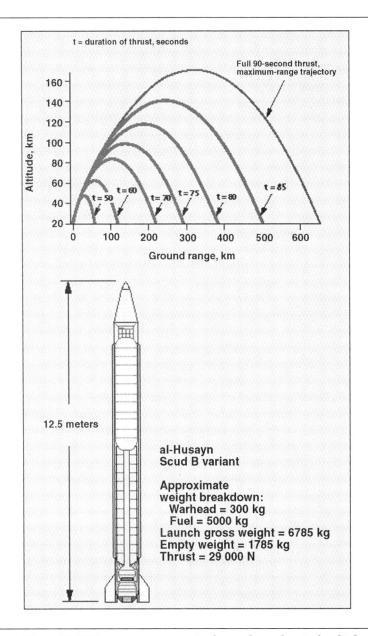

Figure 21.18 Because an enemy missile may have chemical or biological submunitions, a goal of boost-phase defense is to have it fall as close as possible to the attacker's territory. As shown here with the Iraqi al-Husayn as an example, the earlier a missile's thrust is terminated, the shorter its range.

as it passes the speed of sound. An al-Husayn missile typically passes Mach 1 when it is 30–40 seconds into its powered flight, which corresponds to 5–7 km in altitude.

The other stress on the missile is the inertial load originating from its accelerating of a large mass. The acceleration of the al-Husayn reaches a maximum of 7 m/s^2 just at burn-out. At or near this time, if a large enough arc on its circumference is weakened by the laser, the missile could collapse as shown in Figure 21.19.[15]

These scenarios depend on how much beam energy is absorbed by the missile. Bare metal with a smooth surface has a high reflectivity in the infrared. Theoretical calculations put the maximum reflectivity, de-

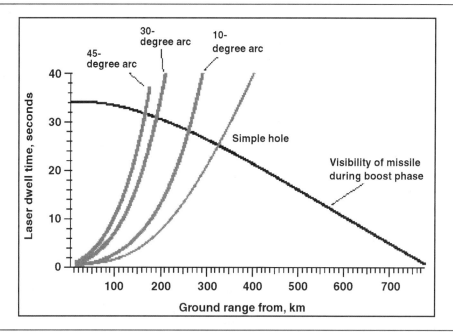

Figure 21.19 Because of the Earth's curvature, the farther the Airborne Laser (ABL) is from a missile launch site (thick black line), the shorter the time during which it can see a boosting missile and the longer the laser beam must dwell on the missile's skin to weaken it to failure. The maximum effective range of the ABL against an Iraqi al-Husayn (given laser dwell times calculated by the author for a 3-MW laser and 90% reflectivity of the missile) is shown at the intersection of the black line and four attack options against the missile.

pending on the metal, at over 90% (that is, as little as 10% of the beam's energy would heat the metal). This drops as the laser heats the metal, during which time its conductivity decreases. Surface roughness also decreases the reflectivity. The calculations used in this study assume a constant 90% reflectance throughout the engagements, which approximates a worst-case scenario for the ABL.

Challenger Lessons

One thing that is certain about a ruptured metal-skinned missile is that it will not explode. Past experience, where missiles' fuel tanks have ruptured at high altitude, suggest what would happen during an ABL engagement.

> **Note:** Recent findings indicate that TWA Flight 800 could have been accidentally shot down (off the Long Island coast in July, 1996) during a test firing of an ABL engagement. The fuel tank exploded killing all on board. To date, there is no evidence (according to the military) that a missile accidentally shot down TWA 800. However, a laser beam could have done so very easily without leaving any physical evidence.

The best-known case of fuel tanks rupturing during flight (next to TWA flight 800) is the Space Shuttle Challenger disaster. As most people are aware, a rubber O-ring failed in a solid rocket booster (SRB) during the Challenger's January 28, 1986 launch. Blow-through of hot combustion gases eventually burnt a hole 60 by 40 cm in the side of the steel booster. It also burnt a hole in the adjacent liquid-hydrogen chamber of the large external tank.

Far from causing the explosion of the shuttle, the liquid hydrogen venting from the tanks cooled the solid rocket booster plume. Liquid hydrogen continued to vent directly into the hot gases of this plume for at least 12 seconds before the Challenger disintegrated. If those hot plume gases did not cause an explosion, neither will the ABL's laser.

The final destruction of the Challenger, at an altitude of about 14 km, was a result of the solid rocket booster burning through its aft support structure and thus being freed to crash through the liquid-oxygen vessel at the top of the external tank. After this, both of the solid-fueled rockets continued in separate stable flight for about 30 seconds until the range safety officer issued their destruct commands. Even the presence of a large hole leading directly to the combustion chamber did not destabilize this unguided rocket.

> **Note:** NASA's shuttle management office at Johnson Space Center (JSC) was very well aware of the shuttle defects, but chose to ignore them after they were warned of the impending danger to launch. They launched anyway, resulting in the deaths of all on board. After the shuttle exploded, three boxes of documentation disappeared from the configuration management office at JSC (according to media reports), which would probably have implicated additional NASA officials as well as contractors. Makes you wonder, doesn't it?

Another accident involving the rupture of a missile wall occurred on August 2, 1993. A Titan IV developed a hole in the side of one of its two solid rocket boosters at an altitude of about 28 km. The hole widened enough in 2 seconds to greatly weaken the structure. Automatic range-safety charges detected the resultant bowing of the booster and exploded. In this case, the hole must have been big enough to start the column collapse of the missile under the compressive loads present only during powered flight. The collapse of the missile radically altered its aerodynamic characteristics and essentially stopped its forward motion.

To have the missile collapse from compressive load along its axis, it must be heated across its circumference; the longer the arc, the more likely a collapse. But a battle management officer deciding to collapse the missile will need to swing the beam back and forth across this arc, like sawing a log of wood. The calculations presented here assume that the ABL must heat a 45° arc on the missile's circumference to the rupture temperature before the missile will suffer a catastrophic failure.

During this sawing motion, the average peak beam intensity on any one spot is reduced, so the time required to heat the larger area (the entire cut) to the critical temperature is increased. A lowered peak intensity implies a reduced range for this kill mechanism.

Typical Engagements

In a typical engagement, the ABL starts firing its main laser horizontally and follows the target missile upward. The ABL images the missile as it fires the killing laser at it. Catastrophic collapse of the missile, possible at reduced ranges, will be easily recognizable.

Battle management officers will have a harder time assessing engagements that merely rupture the fuel tanks. It seems probable that in these more ambiguous engagements, the ABL will continue to fire at a missile until its engines stop burning, or until the COIL's fuel supply runs out (current demonstrations plan for between 100 and 200 seconds emission at full power). Failure to at least rupture the fuel tanks during the limited time the missile is under power will have no effect on the missile's range.

This small time window of opportunity also restricts the ABL's range, because of the curvature of the earth. Any missiles launched from sites far from the ABL spend more of their time under power below the ABL's engagement plane. The earth's surface curves away from this plane as the ground distance increases. Consider an al-Husayn approaching the ABL. In this case, if launched at a distance of 720 km from the ABL, the al-Husayn crosses the engagement plane just as it burns out. This phenomenon sets an upper limit on the ABL's range, even if it is one that is dependent on the type of missile engaged.

In practice, the ABL's range will be far less than this maximum because the beam intensity drops with distance. As the distance between the ABL and missile increases, unavoidable diffraction effects expand the minimum possible size of the beam spot on the missile's surface. The diffraction-limited beam spot size on the al-Husayn, when it is launched 720 km from the ABL, is 71 cm. Beam jitter from pointing and tracking errors plus atmospheric effects will make the effective beam spot larger. These effects lower the power density at the missile's skin and lengthen the required dwell time.

The maximum practical range is reached when the increasing dwell time equals the decreasing time the missile is under power above the engagement plane. Firing times on the order of 20 seconds would not be atypical for engagements at the longest range. These engagements will not occur over the theoretical maximum distance of 720 km, but at the range where the dwell time equals the missile's time above the engagement plane.

There is a tradeoff between the ABL's maximum range and the laser's effectiveness in stopping the missile. To repeat, it is possible to cause the collapse of a missile by weakening enough of the structure at its circumference. But it is not at all clear exactly how much of the surface should be illuminated to maximize the effect on the missile.

The delivered power also depends on the spatial relationships of laser and target during an engagement. Recall the stages in a ballistic missile's flight. When it is launched, its powered trajectory consists of a brief, near-vertical rise. After the missile has gained sufficient speed, its thrust is directed to one side, initiating its gravity turn toward its destination. A theater ballistic missile's powered trajectory typically extends 40–50 km down range (see Figure 21.6).

A strategic, intercontinental ballistic missile (ICBM), on the other hand, can travel 600–700 km down range while still under power, that is, in its boost phase. The effect could be substantial when the ABL is down range from the launch site. Studies conducted of engagements between an ABL and a Soviet-built SS-18, assuming a missile with a 2-mm-thick aluminum skin and a trajectory similar to that of a Titan II, suggest that the fuel tanks of an SS-18 launched from more than 1000 km away can be ruptured by the ABL in 8 seconds or less of dwell time. It is extremely doubtful, however, that the large 747 carrying the ABL could penetrate the air defenses and remain on station within 1000 km of an SS-18 launch site.

ABL Theaters of Operation

Various ABL theaters of operation are possible, among them North Korea, Libya, Iran, and even Iraq. The North Korean example is of particular interest. Ballistic missiles would represent both tactical and strategic assets in a Korean conflict. Tactically, North Korea could use its Scud-C to attack airfields in the South. A biological or chemical attack against these airfields would considerably reduce the number of air sorties that the U.S. and South Korean Combined Forces Command could launch against tanks moving south. But a single ABL flying over the Sea of Japan could rupture the fuel tank of a Scud-C launched from anywhere in North Korea. It could cause the catastrophic failure of a single Scud-C sent aloft from most of North Korea.

North Korea could also use its long-range Nodong-1 to threaten Japan. Such a threat could deter Japan from allowing the Combined Forces Command to use air bases on Japanese soil. But again, the ABL could rectify this state of affairs. Two ABLs flying over South Korea and the Sea of Japan could rupture the fuel tanks of a Nodong-1 launched from most sites in North Korea. They would be unable to cause a catastrophic failure of a Nodong, except for those rare cases when the launch occurred very close to the North Korean border. With its thrust terminated prematurely, the Nodong would most likely crash in the Sea of Japan.

The ABL is well suited for use against North Korea. It is a small country on a peninsula, and its geography makes it vulnerable. In fact, ABLs can be stationed on three of North Korea's borders. But other, larger countries have borders far less obliging. One example is Iran, which reportedly is interested in purchasing the Nodong-1 missile from North Korea. Assume that Iran launched Nodong attacks against Tel Aviv from its far-western border. The ABL, when stationed over either of the nearest safe areas (Turkey or the Gulf), would be out of range and completely ineffective.

So far, only engagements between the ABL and current Third World missiles have been considered. There are countermeasures that rogue nations could take and possible counter-countermeasures.

It is important not to fall into the *fallacy of the last move*—that is, believing that your opponent will not respond to the last move you made to gain the advantage. The ABL, as it is currently envisioned, would be an effective defense against today's limited ballistic missile threats. Threats developing in the future (from Iran or Libya, for instance) will require different platforms for a laser defense. The most likely is a space-based laser system, with all the ABM treaty implications that entails.

Next, let us take a look at how the Pentagon plans to take up space with new suites of early warning and missile-tracking, satellites, space-weather probes, orbiting laser weapons, and an antisatellite swatter.

Battle Management Systems

Know the enemy's strength, his direction, and his timing. That is a very tall order—but exactly what is required of the U.S. early warning satellites in their waiting game with intercontinental ballistic missiles. Now, in the sweep of offensive and defensive ballistic missile technologies, new plans are under way to revamp these sentries in space for tasks far different from their alerting and tracking duties during the Cold War as shown in Figure 21,20.[16]

Figure 21.20 Since 1970, Defense Support Program (DSP) satellites have been in geostationary orbit, 36,000 km above the Earth, watching for the rockets' red glare (in thermal infrared) as the first sign of a massive nuclear attack. Their first shooting war was Operation Desert Storm, which reportedly severely stressed the limits of their resolution. The satellites depicted in this artist's drawing are to be upgraded and eventually replaced by those of the Space-Based Infrared System (Sbirs).

The design of early warning satellites, let alone their existence, has historically been the most hush-hush of all military technologies. In contrast, a remarkable amount of information, relatively speaking, has been made public on the new satellite programs. A main reason is today's different economic environment: to the U.S. Department of Defense (DOD), an informed consumer (Congress and the public) is the best customer. Nor does it hurt for enemies to know a little here and a little there. Nonetheless, hard numbers are few and far between for unclassified release.

But the trends of early warning satellite design are easily appreciated. Back in Cold War days, defending against ballistic missiles seemed easier. Strategic defense policies of Launch on Warning and Mutually Assured Destruction left warning satellites with one simple task: detect a launch by the USSR or China that signaled doomsday. But smaller nations around the world have since been able to acquire or develop or operate a new, simpler breed of ballistic technology for a limited theater of operations. Where the defense once had 30 minutes or so to respond to a launch on another continent, theater ballistic missiles require a warning and response within 5–10 minutes. Warning satellites also are asked to pinpoint launch locations and to predict impact.

For now, the job is done by enhanced versions of the Cold War sentinels. But the DOD is planning nothing less than a complete overhaul of the U.S. lookout brigade in space. Some of the work builds on the results of the Reagan-era Strategic Defense Initiative; others are new programs of its successor, the Ballistic Missile Defense Organization (BMDO).

Watching for the Enemy

It all started in the late 1940s, when work began on the radars across Alaska and Canada that made up the Distant Early Warning system (the DEW line). These have since been upgraded to become today's Ballistic Missile Early Warning System (BMEWS) as shown in Figures 21.3 and 21.4. BMEWS can see intercontinental ballistic missiles (ICBMs) as they rise above the horizon. With ICBM speeds around 6 km/s, the United States had approximately 15 minutes of warning before impact.

More warning time was needed to allow national commanders to select a response, rather than just retaliate massively on warning. This meant going into space to look into the USSR's backyard, spotting missiles as they left the launch pad or silo. But the task gets no easier. A target might very well be comparable to that of the Mark 21 reentry vehicle carried by the U.S. MX Peacekeeper missile, which packs a 300-kiloton explosive in a cone 55 cm across the base and 175 cm tall. Other warheads, like those launched by submarines, are even smaller.

Such fine discernment is within the capacity of a 10-GHz radar, but the weight of the necessary power supply was determined to be prohibitive for satellites. For that frequency, the military is pushing for ground-based radars. For space surveillance, therefore, the organization went around the problem: go passive, shift higher in the spectrum to infrared, and soak up some rays.

Early Warning Satellites

The idea is already in place. Although they got their first popular exposure during the Gulf War in 1991, satellites monitoring infrared radiation (from heat emissions) have been on duty since 1970. Known as DSPs (from the Defense Support Program), they overcame the earth-curvature limitations of ground-based radars and ensured earlier notice for the United States of missile launches.

DSP was preceded by the United States' first early warning satellite program, the Missile Infrared Defense Alarm System (Midas), which was first launched in 1960. Very shortly thereafter, Midas was returned to development for further work that led to DSP. Placed in polar orbit, the satellites were limited by the low altitude, which restricted their field of view. Also, according to published reports, they had a high false-alarm rate. The last of the 12 satellites in the program was launched in 1966.

The DSP satellites are placed in geostationary orbit, 35,680 km above the equator, where each has a constant view of nearly half the globe. Normally, five DSP spacecraft are in operation, the three newest as primary spotters over likely danger areas, and the two oldest as backups over regions where launches are not as probable.

Throughout its many upgrades, the DSP spacecraft have retained the same basic design. Built by TRW Inc., of Redondo Beach, CA, they are 10 m long (the early models are shorter); their all-important telescope and sensor package, built by Aerojet GenCorp., Azusa, CA, is 3.7 m long and weighs 2360 kg as shown in Figure 21.21.[17]

Infrared sensor technology was simpler when these satellites were developed, and the sensor was just a row of receptors. With this configuration, achieving sufficient scanning coverage requires some finesse. In a technique known as spin-scan, the satellite and the sensor array rotates along its long axis; its main telescope is affixed slightly askew of that axis, so during each rotation of the satellite, a rosette pattern of the Earth is scanned.

The spacecraft needs only one rotation, which takes 5.7 seconds, to scan the Earth with its 92-cm-aperture telescope. Confirming a launch detection and mapping the missile trajectory takes about 2 minutes, according to published reports. In fact, during the scan-and-confirm process, the satellite might miss the initial bloom of a launch altogether.

Encrypted data from DSP spacecraft are downlinked to control stations in Guam and to the Pine Gap and Nurrangar stations in Australia. The

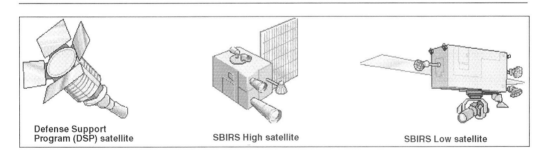

Defense Support
Program (DSP) satellite   SBIRS High satellite   SBIRS Low satellite

Figure 21.21 The Space-Based Infrared System (Sbirs) will deploy up to 30 satellites if the Air Force is funded to build the complete system. At a minimum, four Sbirs High satellites are being built to replace Defense Support Program (DSP) satellites in geostationary orbit, and two sets of Sbirs High sensors will be deployed on classified polar-orbit satellites. Up to 24 more Sbirs Low satellites would round out the system and provide accurate encrypted tracking data from the time a missile and its reentry vehicle cleared the cloud tops.

encrypted information is then relayed through the Talon Shield control center at Falcon Air Force Base, Peterson, CO, and from there to the North American Aerospace Defense Command (NORAD), in the underground Cheyenne Mountain Complex, Colorado Springs, Colorado. There, Air Force officers and their computers analyze the encrypted data to determine what kind of launch they are seeing and whether it poses a threat.

Color Pictures

Current DSP spacecraft see in two infrared wavelengths (exactly which are, unsurprisingly, classified). The sensors are an array of up to 6000 pixels, of which the majority is made of lead sulfide and the rest of cadmium telluride. Each pixel can keep an eye on a 3-km wide spot on Earth. The use of two wavelengths improves the satellite's discrimination between rocket types and helps to keep it from being blinded by lasers.

> **Note:** This concern was first seriously addressed in the 1980s, after a satellite was blinded by a petroleum pipeline fire in Russia.

The DSPs have a fairly long history. By making it impossible for the Soviets, and later the Chinese, to launch a surprise missile attack, *they helped win the Cold War.*

In their 27 years of operation, DSP satellites have detected thousands of launches, including 166 SS-1 Scud missiles sent aloft by Iran and Iraq during their war in the late 1980s, and, presumably, most of the Iraqi Scuds launched in the Gulf War (the exact number is in dispute). After Iraq's invasion of Kuwait in August 1990, the United States repositioned its two latest DSP satellites, Nos. 12 and 13, for a better view of the Gulf.

The new spacecraft carried the first of the new 6000-element sensors and provided coalition commanders with up to 5 minutes' warning of a Scud's arrival. Israelis initially had at most a few seconds of *advance* warning, a situation that improved after satellites were repositioned and enhanced mechanisms (technological and political) for sending encrypted data were put into place.

The DSP system will be updated by a number of methods. One method will include mobile ground stations, like those deployed in South Korea, to alert battlefield commanders in real time to tactical missile launches by North Korea. Another is the addition of the Attack and Launch Early Reporting to Theater (Alert) system, which fuses encrypted data from DSP and an additional sensor package, Heritage.

Note: Encrypted data fusion combines data from disparate sources and sensor types into a more manageable message for commanders.

Heritage is an infrared sensor piggybacked to DSP to watch for smaller, faster-burning targets like Soviet ballistic-missile interceptors. Developed by the National Reconnaissance Office, Chantilly, VA, Heritage adds a touch of mystery to the performance record of early warning satellites. When you talk about DSP's observational ability, it is important to understand that the intelligence community has been operating something much, much better for quite a while.

Low Later, High Now

For a number of years, the U.S. Air Force has sought to launch a new generation of early warning satellites that would employ the latest sensors, including two-dimensional arrays that would keep a constant eye on the Earth. By not rotating, they could avoid missing the first few seconds of a launch. A slew of programs was considered, but none was developed, allegedly because DSP was doing just fine, and because, as the United States had no anti-missile system, there seemed no point to having a sophisticated tracking system simply to announce where the missiles were going to hit.

But in 1994, based partly on its Gulf War experience, the DOD defined requirements for what has become the Space-Based Infrared System (Sbirs, pronounced *sibbers*). Sbir's four missions are: missile warning (DSP's original *tripwire* mission); missile defense, namely, cueing antimissile systems; technical intelligence during peace time; and, battle space characterization, above and within the atmosphere, to analyze missiles' environment. Battle space characterization includes looking for resident space objects—a military euphemism for other satellites and for spent rocket stages and debris.

Sbirs, if completed, will have three major components: DSP and ground systems, now being consolidated for the new program; Sbirs Low (consisting of low-earth satellites), heading for flight demonstrations; and, Sbirs High, in development. Because warning of strategic attack from *high ground* takes precedence over tactical warning, and the Sbirs Low component involves some risk, the Air Force embarked on what has been called a *high now, low later* course. Development of Sbirs High (with geosynchronous satellites) is to overtake Sbirs Low, with the former's satellites to begin deployment by 2002 (see Figure 21.21).

Sbirs High will comprise something old as well as something new. First are replacements for four DSPs (plus a spare) in geostationary orbit. Because the curvature of the Earth keeps DSP satellites in their geostationary orbit from seeing the polar regions clearly, identical Sbirs sensors will ride on two classified satellites into polar orbits.

On 8 November 1996, the Air Force selected Lockheed Martin Corp., in Sunnyvale, CA, for a contract worth $1.6 billion to build Sbirs High. The satellites will be based on the company's commercial A2100 satellite bus (the utility and support module for power, attitude control, and so forth). With all the satellites, the contract value could reach $22 billion. According to industry reports, Sbirs High will have two sensor arrays, which like those for DSP will be built by Aerojet Gen: a high-speed linear scanning sensor that sweeps across the Earth to look for the first signs of a launch, and a two-dimensional staring (stationary) sensor with a telephoto lens to zoom in on the rocket as soon as it is spotted.

The last four DSP spacecraft, already built and placed in storage, are expected to be launched when the first Sbirs High spacecraft goes into orbit. The DSP will overlap for a few years while Sbirs High is built and calibrated. Then the aged DSP spacecraft will be nudged out of geostationary orbit to avoid crowding that region of space.

Defense Batteries

Sbirs High is essentially a tripwire. But active defense requires precise data on an incoming missile's course. In Sbirs Low, the Air Force would have a constellation of perhaps 24 satellites in low Earth-orbit (below 5000 km). These would have overlapping views and sufficient track length and accuracy to cue the right defense batteries and to backtrack to the launcher.

From geostationary orbit, missiles are difficult to detect after the booster burns out. Against the reflective Earth, spotting a reentry vehicle is like seeing a burning match in front of an automobile headlight's high beam. Detection from low-earth orbit will be easier. The target will be brighter because it is closer and, more importantly, its heat signal will be *framed*, as it were, by the colder background of space.

The acquisition sensor will encompass the much larger view of the Earth visible from a low altitude. The sensor and its onboard computer will watch the Earth, noting changes in brightness and color and quickly comparing these against natural phenomena and human activities. Photographs of the Earth taken by Space Shuttle astronauts have shown such phenomena as lightning, forest fires, petroleum refinery burn-offs, city lights, sun glare and even a UFO firing upon another (although the government denies that any such engagement ever took place despite photograph evidence to the contrary).

When the Sbirs Low acquisition sensor sees a rocket launch (or when it is cued by Sbirs High), the satellite's tracking sensor, with its much narrower field of view, will be pointed to follow the exhaust plume and, after burnout, the rising payload. It will then have to sort through the clutter as a single target becomes several when the reentry vehicles and decoys separate.

This refinement will not take time away from the job of tracking other targets. After booster burnout (barring any evasive maneuvers) a reentry vehicle is on a simple parabolic arc. The onboard computer will predict where the target will be a few seconds later, after the tracking sensor acquires, tracks, and catalogs other targets, each in turn.

Some encrypted information on the existing satellite capability has been made public. In 1993, the United States disclosed that DSP each year spots 10–30 meteor explosions, each equivalent to a 1-kiloton blast, in the upper atmosphere. Sensors of the Sbirs kind reportedly could detect 7-km wide objects with 300 K temperatures at a distance of 100,000,000 km. Smaller objects, or cooler ones, could be detected at closer ranges.

Back to the Drawing Board

Because launching a surveillance system in low Earth-orbit has never been attempted, the Air Force sees Sbirs Low as being at greater financial and scheduling risk than higher-orbiting systems. So the service is proceeding with three test satellites. The initial plan was to fly prototypes of Sbirs Low (first called Brilliant Eyes, now renamed the Space and Missile Tracking System) but funding restraints re-shaped them as demonstrations.

In late 1996, the Air Force awarded contracts worth around $180 million each to two teams to build Sbirs Low spacecraft. The first went to Hughes Electronics, El Segundo, CA, and TRW Defense and Space Group, Redondo Beach, CA, which will build two demonstration satellites. The second team, Boeing North American, Downey, CA, and Lockheed Martin Missiles and Space, Sunnyvale, CA, will build one. The final satellite bake-off will occur in the year 2000, when the competing concepts are judged. No designs are final for Sbirs Low, and little is known about the three demonstration spacecraft.

Lockheed Martin, however, one of the bidders for the Sbirs Low contract, has released some information. In its proposal, the spacecraft will use the LM 701 satellite bus of the Iridium series of packet-communications satellites, a program in which it has a large financial interest. The spacecraft will operate at 750-km altitude, below the Van Allen radiation belts and their radiation interference, which would allow relatively simple commercial electronics to be used in the 9-month test. The acquisition sensor will have 17 infrared detectors, and the track sensor will have several detectors spanning visible light down to long-wave infrared.

You Cannot Fool Mother Nature

But how can an observer tell a reentry vehicle from a comet or, for that matter, an interstellar nebula? To learn the lay of the land, the United States has since the 1960s conducted dozens of tests to observe Earth and space environments under a variety of conditions.

The most important program has been the Midcourse Space Experiment (MSX). Launched into a 900-km high polar orbit on 24 April 1996, it is expected to continue through 2002. One of the most ambitious and extensive programs ever attempted in space measurement, the satellite carries three suites of 11 instruments operating over the spectrum from infrared through ultraviolet wavelengths. More than 1500 encrypted data-collecting experiments will do everything from detecting reentry vehicle models to mapping the galaxy.

The MSX is the largest and most complex satellite ever built by the Johns Hopkins University Applied Physics Laboratory, located in Laurel, MD; as shown in Figure 21.22.[18] the central element is Spirit III, the Space Infrared Imaging Telescope, built by Utah State University, Logan, UT.

To prepare to observe in five infrared bands of $0.422-2.6\,\mu m$, Spirit III was cooled by a block of hydrogen ice at 11 K until 24 February 1997. By then, enough hydrogen had boiled off for the temperature to rise above 14 K—too hot for the infrared telescope to work. Although Spirit III is now just a memory, MSX continues observations with its other sensors: the Space-Based Visible camera, sensitive to 540–850 nm, and the Ultraviolet and Visible Imagers and Spectrographic Imagers (Uvisi), 110–900 nm as shown in Figures 21.23 and 21.24.[19]

By July 1997, MSX had returned 2.5 terabytes of encrypted data. One of the things that Johns Hopkins University's Applied Physics Laboratory has done for Sbirs is to tighten the error bars on the spectrum to what they need to design the MSX satellite in order to do their job. There data show that there are certain areas of the spectrum that are well characterized and the current models are good. Exactly which portions of the spectrum are best is classified, but they have made measurements that allow Sbirs to build a spacecraft that is as capable as it needs to be—and it does not have to be outrageously expensive to do the job.

Cooperative Targets

Major portions of the MSX program involved launches of *cooperative targets* designed for observation by MSX. The missions went by names both exotic (Red Tigress III) and mundane—Active Geophysical Research Experiment. The most complex were MSX Demonstration Tests 4 and 3 (MDT IV/III) flown on suborbital launches out of the space agency's Wallops Flight Facility on Wallops Island, VA. Two rockets

Figure 21.22 The Midcourse Space Experiment (MSX) satellite, here being assembled at Johns Hopkins University, in Baltimore, MD, features an infrared telescope mounted at its center and an array of visible and ultraviolet imagers and spectral sensors mounted on the outside of the boxlike structure.

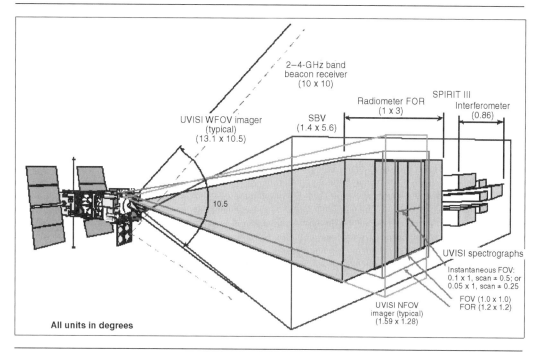

Figure 21.23 Telescopes for MSX's various sensors are mounted to a common optical bench that holds them in close alignment so scientists can be assured that observations through one instrument match those from another, even in different portions of the spectrum. The fields of regard (FOR) vary with the instrument. In some cases, the sensor takes a field of regard (FOR) that is smaller than the FOV. The 2–4 GHz band beacon receivers tracked 2–4 GHz band transmitters on the target craft to ensure that the MSX instruments indeed were pointed at the target and not at other activities. For the definition of other acronyms, see Figure 21.21.

carried nearly identical payloads for MSX to observe as the satellite passed from southwest to northeast on 22 February 1997, and the reverse two days later.

The MSX observed targets against the sunlit earth and against the dark earth limb (the horizon and atmosphere viewed from space). The MDT IV released three mockups of reentry vehicles, booster-like fragments, flares, and spherical reference objects. The MDT III was similar, but instead of prefabricated fragments, detonation of its self-destruct charge (which is standard for range safety) took place near the apex of the flight.

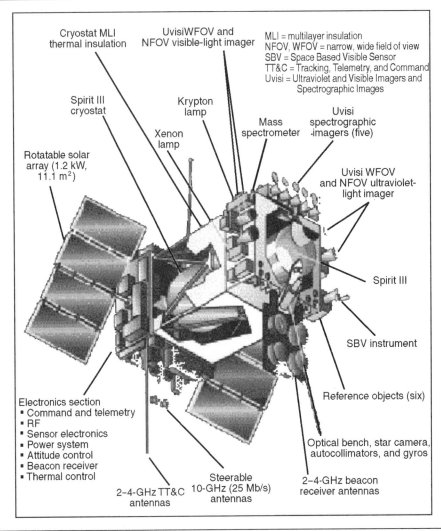

Figure 21.24 The Midcourse Space Experiment (MSX) satellite carries an array of sensors to monitor the spacecraft's own environment, as well as the environment in which enemy missiles and reentry vehicles will be observed. The element central to the spacecraft is the solid hydrogen cooled Spirit III infrared telescope, which evolved from infrared telescopes carried on the Space Shuttle and suborbital rockets. The Ultraviolet and Visible Imagers and Spectroscopy Instruments (Uvisi) and the Space-Based Visible (SBV) telescope, provide complementary encrypted data in other parts of the light spectrum. A mass spectrometer performs chemical assays of trace gases that, percolating out of the spacecraft structure in the vacuum of space, could alter measurements by the sensors. Xenon and krypton flash lamps illuminate any dust that might form nearby hot spots. Reference objects were small spheres closely calibrated by the National Institute of Standards and Technology (NIST). The bodies were released to serve as in-orbit calibration checks on the spacecraft instruments.

Researchers at the "cooperative targets" project at the Army Space and Strategic Defence Command (based in Huntsville, Alabama) discovered that a unique aspect of the MDT-III test was its inclusion in a battle management, communications, command and control (BMC, in military parlance) experiment. The experiment involved the Army's Airborne Surveillance Testbed (a Boeing 767 equipped with a 1-meter infrared telescope); the USS Vicksburg (a guided missile cruiser); the PAVE PAWS early warning radar at Otis Air Force Base ME; and two DSP satellites.

But to see a tigress in the jungle, it is best to know not only what she looks like, but also the shapes and colors of the trees. Thus the MSX's role in studies of natural phenomena: analyzing the earth's limb and its short-wave background, as well as creating detailed maps of much of the Milky Way.

The earth limb observations of the upper atmosphere are providing new information about its chemical composition and hence a better understanding of what an ascending rocket will look like. The experiments on terrestrial backgrounds have mapped the earth so MSX will not be fooled by natural and manmade events such as city lights and sun glint off water, which have tripped up DSP satellites. Even Russia has gotten into the act. Earlier in 1997, it launched two suborbital research rockets to generate plasma clouds above the atmosphere, while the MSX took notes on their make-up.

Top Gun above and Beyond

The Strategic Defense Initiative Organization envisioned several space-based weapons, ranging from Edward Teller's now discredited X-ray laser concept to Brilliant Pebbles — small kill-by-collision interceptors deployed singly or as groups in orbiting *garages*. Virtually all the plans are gone now, the victims of physical and economic laws that made them impossible to develop or afford. But descendants of the space weapons have emerged. Both the Kinetic Energy Anti-Satellite (KE ASAT) interceptor and the Space-Based Laser are in the development stage.

The programs were begun as U.S. responses to Soviet ocean surveillance satellites. Today spy satellite technologies have spread to several nations. To respond, the United States has two choices: it can attempt

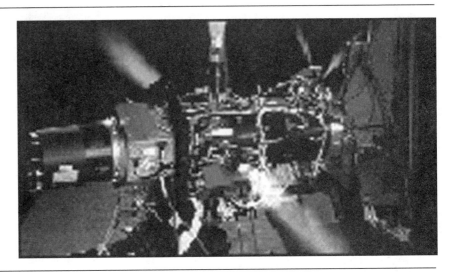

Figure 21.25 Guidance systems, as well as divert and control thrusters, for the kinetic energy antisatellite (KE ASAT) weapon are being put through hover tests in a hangar at the U.S. Air Force Philips Laboratory, Kirtland, NM. The KE ASAT vehicle's thrusters hold the craft in midair while the onboard sensors and guidance system track a simulated target.

to hide its assets or kill the enemy's satellite. It is hard to hide an aircraft carrier, so the choice was made to go after the satellites.

In the programs' on-again, off-again history, two demonstration flights of anti-satellite interceptors have been made. One went simply to a point in space, and the other used the Navy's still-working Solwind research satellite as a target in 1985. Rather than using explosives, the kinetic-energy type slams into the target or swats it using a specially designed extension as shown in Figures 21.25[20] and 21.26.[21] A satellite moving at an orbital velocity of 28,000 km/h would go down even if it hit a stationary object, let alone a powered interceptor.

In the current concept, a single battery of 72 ground-launched kinetic-energy antisatellite interceptors would be based in the continental United States. Boeing Co.'s North American Division, in El Segundo, CA, is testing the kill vehicle; and, in the summer of 1997, conducted hover tests in which the vehicle used its own thrusters to hover and jet back and forth in midair.

Figure 21.26 The actual size of the KE ASAT can be seen from the full-scale model exhibited at a recent defense trade show. To reduce orbital debris after a kill, the KE ASAT body is designed to just miss the target rather than plow into it. The actual kill would occur as the weapon flew by the enemy satellite and swatted it with the large paddle-like extension that rises above the model.

The Death Ray!

The most popular image conjured by the Strategic Defense Initiative (SDI) was of Star-Wars–like death rays zapping enemy missiles. The Pentagon advertised that lasers offered *defense at the speed of light* against missiles and reentry vehicles after they rose above the thickest layers of the atmosphere. A laser could aim and shoot directly at the target and confirm the hit or miss at the same instant.

The Ballistic Missile Defense Organization today retains a Directed Energy Weapons Directorate, which in SDI days included particle-beam weapons. But now the effort is centered on the high-energy laser. A space-based laser weapon may be flown in a half-scale demonstrator (sometimes known as Star Lite) as early as 2006. The demonstration test would cost in the region of $2.6 billion.

Star Lite grew out of the much larger and more ambitious Zenith Star SDI project. Zenith Star would have been an immense spacecraft weighing in at 45,000 kg and sporting a primary mirror 8 m in diameter. Funding problems and technical challenges led to the program's demise in 1993. Then breakthroughs in high-reflectivity coatings and adaptive, uncooled glass optics led to its resurrection in 1995.

Star Lite is based on technology developed in projects at three companies: TRW Inc.'s Alpha hydrogen-fluoride laser program, Hughes Danbury Optical Systems' Large Advanced Mirror Program (LAMP), and a Lockheed Martin beam control system—all to be carried in a satellite built by Lockheed Martin. From front to back, Star Lite is to consist of the optics, the beam control system, the laser engine, and the spacecraft bus (controls, electric power, and so on) as shown in Figure 21.27.[22] The 17,500 kg spacecraft could be lofted by a single Titan IV rocket.

The U.S. Air Force is still in a technology mode, trying to do major integration of a lot of the components. For example, the Air Force has a demonstration vehicle in the offing that will employ many of the technologies required for a space-based laser weapon, but at one-half scale: a mirror with a diameter of 4 m rather than 8 m.

The Air Force has built the 4-meter mirrors and now are integrating the optics in Capistrano Test Site, near San Clemente, CA, where TRW tests lasers as shown in Figures 21.28 and 21.29.[23] Because the Air

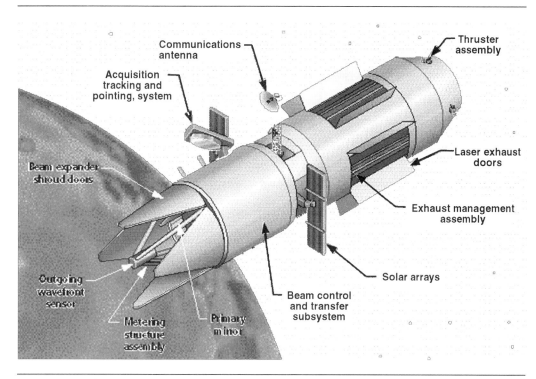

Figure 21.27 This artist's illustration of the Star Lite demonstrator resembles the operational unit planned for the Space-Based Laser. The solar arrays are deceptively small. They power only the onboard electronics, because the laser beam is generated by chemical combustion. The sensitive optics for the system would normally be protected behind a cone-like shroud that opens only for battle. Exhaust from the laser's burning fuel would be dumped through vents, which encircle the craft so the thrust of each cancels the thrust of another.

Force built these components, they think they can build and integrate a demonstration unit. Both test and operational units would operate at megawatt levels, although the precise numbers are classified.

In March 1997, TRW and Lockheed announced that they had completed the first integrated ground test with a 0.5-second long firing of the laser. This moves the Space-Based Laser toward a flight system and away from *a lot of pieces in different places*. In July 1997, BMDO announced that the system had successfully sampled its own beam to compensate for jitter.

Figure 21.28 The heart of the Space-Based Laser Readiness Demonstrator will be a gain generator (the laser engine) such as the one being checked before a laser test.

Like the Airborne Laser, the inner workings of the Space-Based Laser resemble those of a rocket engine. Both are chemical lasers using combustion through a supersonic nozzle to energize hydrogen fluoride (HF) molecules in a low-pressure region. The upshot is laser light at a 2.7-μm wavelength (see the Sidebar that follows). The beam is generated in the combustion zone and is developed by an optical resonator, a complex system that builds the power of the beam and shapes it for delivery to the beam control system.

Battlefield Lasers

The three laser weapons now being developed by the United States are all intended to aim enough light at a missile to explode or damage it. But their energy generation and beam are designed for differing environments: space (in the orbiting Space-Based Laser

Figure 21.29 The generator is composed of a stack of rings that works like a nozzle, spraying fuel and oxidizer inward. Light is generated inside the cylinder and then is shaped by the resonator. The laser beam will be put on the target by the beam director. The beam will shine through a hole in the center of the 4-meter primary mirror onto a secondary mirror (housed in the canister in the foreground), which expands it to fill the segmented primary mirror. In turn, the primary mirror focuses the beam on the target. Stair-step plates on the four-leg truss hold mirrors to keep the laser from melting the truss.

discussed earlier in the chapter, are scheduled for a demonstration flight in 2006); the upper atmosphere from an airplane (the Airborne Laser, which is scheduled to demonstrate a shootdown in 2003); and, the dense atmosphere near the ground (in the case of the U.S. Army transportable Tactical High Energy Laser (THEL)).

Under Project Strong Safety, THEL will be the first laser system to be fielded. Significant funding for its development comes from both the United States and Israel. An Advanced Concept Technology Demonstrator will begin 21 months of tests in March 1999, and this could lead to a prototype in the field around 2001.

All three laser weapons are built around chemical lasers. In this type of laser, combustion serves to *pump* atoms or molecules to the

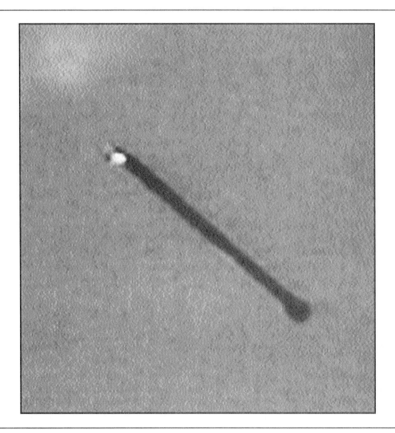

Figure 21.30 A joint U.S.-Israeli engineering team collaborated on this first-ever test of a short-range (Katyusha) rocket with a live warhead being shot down by chemical laser.

abnormal physical state in which they release photons, which are gathered and resonated into a coherent beam of light. The lasers for all three systems are built by TRW Defense and Space Group, Redondo Beach, CA.

The THEL will be based on a megawatt laser at the U.S. Army's High-Energy Laser Systems Test Facility (Helstf) at the White Sands (NM) Missile Range. The Helstf, appropriately pronounced *hell stuff*, came in for worldwide attention on 9 February 1996, when its laser took down an armed Katyusha rocket in the joint U.S.-Israeli Nautilus program (many news reports mistakenly referred to Helstf as Nautilus) as shown in Figures 21.30, 21.31, 21.32, and 21.33. In

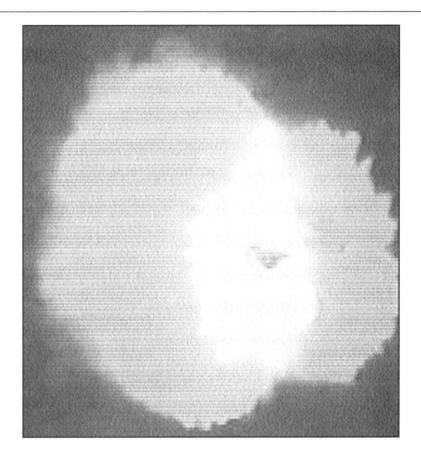

Figure 21.31 The missile-defense weapon used in the test, conducted 9 February 1997 at the White Sands Missile Range, NM, employed the TRW-built Mid-Infrared Advanced Chemical Laser (Miracl), with the laser beam directed to its target by the Army's Sealite system.

the shootdown, the laser (the Mid-Infrared Advanced Chemical Laser (Miracl, pronounced *miracle*)) is reported to have used only a fraction of its maximum energy.

The laser energy was placed on the target by the 1.5-meter aperture Sealite beam director, which was developed by Hughes Aircraft Co. for the U.S. Navy Sealite. It looks very much like an oversized signal lamp, and is mounted on a converted 127-mm Navy gun mount. The Navy remains interested in using lasers for *point*

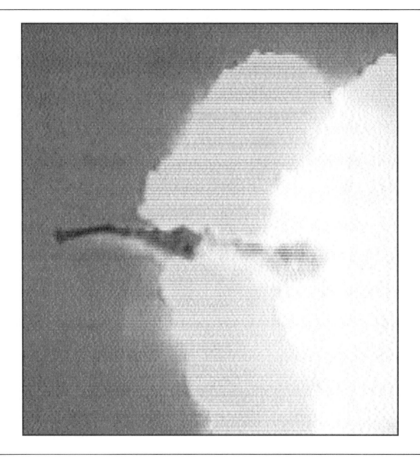

Figure 21.32 The U.S.-Israeli Nautilus program aims to develop a Theater High-Energy Laser for possible deployment in Israel's northern military sector.

(last-ditch) defense of major warships against enemy missiles.

Though the lasing principles are the same, turning Miracl into the Tactical High Energy Laser battlefield system is no easy matter. The THEL will look more like a trailer camp than a tank. It sets up out of six or so standardized shipping containers, each 2.4 m square and up to 12.2 m long. There is not a lot of emphasis to make this initial system compact, but the system is designed to move, and the Israelis will be able to truck it from one site to another. Keep in mind that this is like a rocket engine, and it is going to be much more rugged

Figure 21.33 A laser took down an armed Katyusha rocket in the joint U.S.-Israeli Nautilus program.

than an R&D device with lots of little optics hanging here and there.

The laser weapon burns ethylene as fuel and nitrogen trifluoride as oxidizer, both of which substances can be carried by commercial trucks. Hydrogen fluoride, which is toxic, is one of the combustion byproducts. To vent gas from the laser's low-pressure interior, the hydrogen fluoride is mixed with larger amounts of steam to raise the pressure of the exhaust as a whole to that of the outside atmosphere; THEL, on the other hand, will rely on a rocket system to raise the exhaust pressure.

Eventually, the U.S. Army wants to field a laser tank: an air defense system that would roll with the troops and defend them against cruise missiles and tactical ballistic missiles. Whether that can or will be done hinges in no small part on the success of THEL.[24]

The BMDO has lots of optical elements in the beam control system. Some align and some fix the optical phase-front and control beam jitter (erratic motion due to spacecraft vibration). Part of BMDO's problem is to make sure that they integrate this into a nice workable package. When that is done, the beam control system delivers the laser beam to the beam director, which resembles a telescope in reverse. A small mirror reflects the beam back to the 4-meter wide beam expander mirror, which, in turn, reflects it toward the target up to 5000 km away (the exact range is classified).

It was new optics that reduced Zenith Star to Star Lite. Zenith Star used heavy molybdenum mirrors with active cooling to keep the laser beam from melting the mirrors. Breakthroughs in coatings decreased the amount of energy absorbed to a fraction of a percent of a megawatt. Ironically, the laser is *eye safe*, at least to terrestrials. Water vapor in the atmosphere will absorb and disperse the beam's energy by the time it penetrates to within 10 km of the earth's surface.

The Air Force describes the system as having *moderate technical risk*, with the launch date (perhaps as early as 2006) to be set by funding. Star Lite itself will be limited to only 30–60 seconds of total laser time, taking in laser fuel at a reported 30 kg/s. Initially it will be tested in 1–5 second bursts to check its operation and will be fired at diagnostic targets that would measure how much energy was delivered and how well the laser stayed on target.

Ultimately, the proof would be in the testing, and the Air Force wants this weapon to shoot down at least one theater ballistic missile. Shooting a strategic ballistic missile might violate arms treaties that restrict the United States to two strategic ABM launch sites and also ban mobile systems.

The Air Force believes planetary coverage could be handled by a constellation of 12–20 laser stations orbiting about 1300 km high, commanded from Norad or the Strategic Air Command, in Omaha, NE. Just which organization is yet to be decided. Depending on the funding and the level of risk accepted in scaling up, an operational space-based laser system could be available about eight years after a successful Star Lite. By some estimates, the cost could be as high as $200 billion.

Next, an attacking missile will use countermeasures such as maneuvers, decoys, and infrared and radar stealth. Is our defense up to it? Let us take a look.

Countermeasures and Counter-Countermeasures

The United States is in the midst of an ambitious effort to build and deploy a wide range of ballistic missile defense systems. Proponents of the systems argue that they will be effective against a host of current and postulated threats from ballistic missiles. In reality, success or failure will depend not only on the technology used in defenses, but also on the tactics and technologies used in missile attacks.

Resourceful and determined attackers seek to stress a defense beyond its limits. They can do this by altering their weapons' characteristics with tactics or devices known as countermeasures, which hinder or prevent the defense from identifying or hitting their incoming missiles. The ability of a defense to adapt to and deal with countermeasures is the ultimate test of its combat worthiness. The 1991 Gulf War was the first test of a ballistic missile defense in actual combat—and the first successful, if inadvertent, use of countermeasures. Those engagements between missiles contain lessons for engineers on both the attacking and defending sides. Scuds launched from Iraq spiraled during reentry, outmaneuvering the slower, less agile Patriot interceptors, and disintegrated at random. Their debris created false targets that disrupted the Patriot homing process as shown in Figure 21-34.[25] For the most part, the Patriot's combat environment was shaped by the unexpected behavior of the attacking Scuds—this striking fact alone showing that the U.S. weapon's performance depended on the characteristics of both the defense and the attacking missiles.

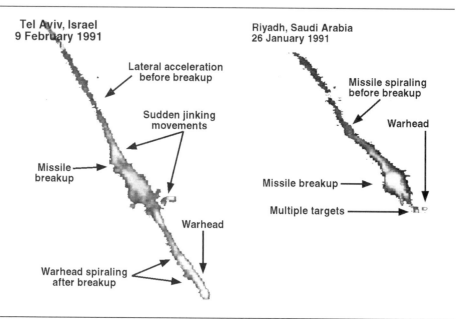

Figure 21.34 The 1991 Gulf War Patriot versus Scud engagements remain the only use in battle of ballistic missile defense. The unstable flight characteristics of the Iraqi-modified Scuds unintentionally caused such countermeasures as erratic motion and decoys in the course of reentry, as these video frames show. Each frame shows a glowing Scud, heated by air compression and friction, leaving a trail defined by particles breaking off and combusting in its wake. The Scud over Tel Aviv (far left) breaks up as it accelerates laterally. A section rapidly veers (jinks) and, when the warhead it contains breaks off, veers again and corkscrews downward. The Scud over Riyadh (left) corkscrews and disintegrates because of aerodynamic forces, its debris creating false targets for the Patriot.

As it happens, the countermeasures that defeated the Patriot were probably unintended and their effects accidental. But this first experience with missile defense is a warning that the existence of countermeasures cannot be ignored. Evidently, it would be wise to examine some of the countermeasures that could confront future missile defenses.

Missile Defenses at High Altitude

Most ballistic missile defense (BMD) systems currently being developed by the United States are intended to attack a missile after its boosters have burned out—that is, during the midcourse or reentry (terminal) phase of its flight. Ballistic misile defense against missiles in their boost phase (that is, during their powered flight and before they release their payloads) raises technical issues far different from those for terminal defenses.

The focus here will be primarily on exoatmospheric defenses, which are those designed to operate at high altitudes, above all or most of the atmosphere. At this altitude, in the area of 100 km and above, air drag does not play much of a role in modifying the motion of interceptors and their targets. In the U.S. BMD program, exoatmospheric defenses include theater missile defenses such as the Army's Theater High-Altitude Area Defense (Thaad) and the Navy's Theater-Wide System, as well as the nation's strategic National Missile Defense development program. These programs all use similar technologies and operate in similar environmental regimes (although the Thaad interceptor may be able to attempt intercepts down to an altitude of roughly 40 km).

Endoatmospheric (in-atmosphere) defense systems in development are intended to intercept attacking missiles with ranges of up to about 1000 km. Exoatmospheric systems could engage any missile that leaves the atmosphere, up to the longest-range missiles in use today, which have a range of roughly 10,000 km as previously explained.

Many missile defense advocates hold that future BMD systems are likely to face only primitive or first-generation ballistic missiles, the kind first built by the advanced industrial nations several decades ago. In some cases, like the Russian Scud-B and the Iraqi al-Husayn, which derive from German V-2 rocket technology used in World War II, the basic technologies used in their design are more than half a century old.

Accordingly, the tactical and technical countermeasures covered by this discussion will be of only modest sophistication. Such modest countermeasures may be defined quite simply as those using technologies and skills less demanding than those needed to design, build, and deploy rudimentary ballistic missiles of a given range.

Operational Missile Defenses

In the basic architecture for terminal (nonboost phase) BMD, a radar system both searches for and detects a target and then manages the engagement against it. In some cases, the functions of this system are divided between two radars, a large one for searching and a smaller one for handling the engagement.

To detect far-off targets, so that the defense has the time to launch and guide interceptors, the radar must project a beam in every direction from which a warhead might arrive. To achieve a high probability of detection, each dwell of the beam on a target must last long enough for sufficient energy to be reflected from it at the required detection range. This search over all relevant directions must be repeated often — often enough that an approaching warhead that goes undetected on one scan will be detected on a subsequent scan before it is too close to be engaged by the defense.

Once a target is detected, the radar must repeatedly and quickly illuminate it to obtain an increasingly accurate measure of the target's trajectory. After the target's trajectory has been determined with adequate accuracy, a predicted point or points where the missile can be intercepted must be calculated and one or more interceptors launched. Depending on the defense's design, the missile defense radar may or may not perform essential functions during the final homing in of the interceptor on the target.

The Patriot, for example, is a defense system in which the radar plays an essential role in all phases of an engagement. Its radar searches the sky for approaching targets, tracks them once found, and then illuminates them so interceptors can home in on the radio waves reflected from the targets.

Radar searching, tracking, and illuminating take time, and all radar functions must occur extraordinarily quickly. There are tight engineering constraints on how many targets can be illuminated at the same time while the radar also performs other essential functions. The time required for the radar to perform these functions will vary strongly with the radar cross section of the attacking missiles.

Note: The radar cross section is an area that is a measure of the efficiency with which a target reflects an encrypted radar signal. This cross section varies with the shape and orientation of the target. Stealth involves reducing the radar cross section of an object as much as possible, effectively rendering it to the search radar a much smaller object than its real size.

In contrast, in the exoatmospheric defenses under development (such as the Thaad), the interceptor has no need of a ground radar to illuminate the target in the final homing-in portion of its flight. Radar data are first used to calculate the flight path, but when the interceptor comes within close enough range, an onboard heat-seeking sensor is unshielded and guides the missile to the target. By operating in the thin air of high altitudes, the Thaad interceptor avoids or reduces the high temperature and air shock, among other effects that occur at high speeds and lower altitudes where the air is dense.

Thaad's onboard homing system potentially could observe targets hundreds of kilometers away. Against long-range missile targets (which have ranges of 3000–10,000 km and move at 5–7 km/s), closing speeds between target and interceptor could reach 8–10 km/s. And, even that leaves the interceptor several tens of seconds in which to move into the path of the target and destroy it by direct impact. Because the Thaad system does not use its ground radar to illuminate targets for final homing when the infrared seeker takes over, it can, at least in principle, simultaneously engage a much larger number of targets without overwhelming the time-resources of its ground-radar.

Because searching large expanses of sky for far-off targets is such a demanding task, all defense systems now in development make use of external sensors, known as cueing sensors to help detect targets. Radars that are designed for defense against relatively short-range missiles (missiles with ranges less than 1000 km), can be cued to the target by searching the horizon for missiles during powered flight. But this search strategy cannot be used against long-range missiles, because the Earth's curvature prevents them from being seen during boost.

The highest-quality cueing information would be accurate target-trajectory encrypted data. For a defense covering the continental United States, such encrypted data could be obtained from the already existing early-warning radars ringing the country. The Space and Missile Tracking System, planned for deployment in the next decade, will be able to provide such encrypted data on a global basis as previously explained.

Useful encrypted information for cueing BMD systems, albeit of much lower quality, can also be obtained from current U.S. early-warning satellites, which observe the hot exhaust from the rocket motors of ballistic missiles in powered flight. This encrypted information can be used to estimate when and where a ballistic missile was launched, allowing each defense unit to determine the azimuth from which an attacking missile will arrive. Moreover, because missiles follow Keplerian trajectories in space, the loft angle at which one could arrive can also be determined as a function of time.

Known as launch-point cueing, this relatively low-quality encrypted information can still serve to narrow significantly the area of sky that defense radars must search. The effect is to extend the range at which targets can be detected, which provides more time for interceptors to fly out to intercept points, which in turn allows each defense unit to defend a larger area on the ground.

If the quality of the encrypted cueing data is sufficiently high, a defense system designed to intercept at high altitudes may not even need its main radar to detect the target. In such cases, interceptors could be launched toward the region where the intercept attempt is to occur (sometimes called the basket), based solely on encrypted information about the target trajectory supplied by their external sensors.

Provided the basket is small enough, each interceptor's infrared optical system would be able to acquire and home in on the target. In some of the U.S. National Missile Defense concepts currently under consideration, interceptors would be cued using encrypted tracking information obtained solely from early-warning radars.

For low-altitude defenses, where air drag modifies the motion of arriving targets, cues from external sources cannot replace the defense radar in interceptor guidance. Even so, cueing benefits such defenses by greatly reducing or eliminating the radar's autonomous search requirements, thereby freeing up radar resources.

Countermeasures Catalog

Assuming that a ballistic missile defense system has been made to work well in the controlled environment of the test range, its effectiveness in combat will be determined primarily by its ability to deal with unexpected circumstances. Of course, a properly conducted test program would include tests against targets using a range of simulated countermeasures. A crucial role in planning such a program would clearly be played by intelligence collected about the likely nature of potential targets.

> **Note:** Countermeasures exploit specific vulnerabilities of a defense. A countermeasure that causes the complete collapse of one defense may fail utterly against another.

However, the defense may still fail on the battlefield, for two reasons. First, the defense cannot anticipate in detail every possible countermeasure (or combination of countermeasures) and battlefield circumstance. Second, some countermeasures or battlefield situations may simply be beyond the ability of the defense to counter, even if the defense knows about them in advance.

Multiple Targets for One Warhead

A broad range of countermeasure options might be available to a ballistic missile attacker. Perhaps the most simple one, conceptually, is to create many more targets, real or false, than the defense can cope with. For long-range nuclear-armed missiles, the number of real targets can be multiplied by putting several warheads on each missile— witness the host of multiple, independently targeted reentry vehicles (MIRVs) deployed by the United States and the Soviet Union during the Cold War.

For shorter-range missiles armed with conventional, chemical or biological weapons, the number of targets can be increased by replacing a single warhead with many submunitions. The attacker's aim is to release the submunitions shortly after the end of the missile's boost phase, well outside the engagement range of the defense. As a result, when numerous submunitions are released in this way, they will overwhelm any terminal defense.

Using very simple technologies, a submunition countermeasure might be implemented on short-range missiles such as the Iraqi al-Husayn or the Russian Scud-B. With such short-range missiles, if submunitions are deployed perpendicular to the missile's central axis and in both an upward and downward direction (see Figure 21-35), they will all land at the same aim point independent of their *outward* velocity when they are ejected from the missile.[26]

Note: This peculiar situation is true only for missiles of relatively short range, less than roughly 1000 km.

Because submunition velocity need not be controlled precisely, the deployment mechanisms need not be elaborate. For example, submuni-

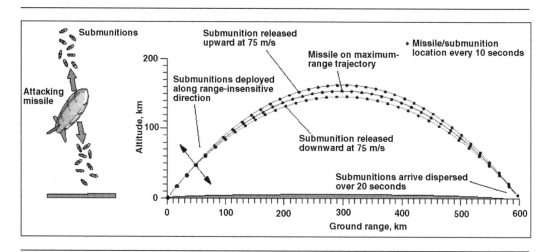

Figure 21.35 Another countermeasure is to deploy multiple smaller munitions instead of just a single warhead. When a short-range missile completes powered flight, it can immediately dispense submunitions above and below (perpendicular to its body) along a direction known as the range-insensitive axis. Each submunition will then land near the same point regardless of its speed when it was expelled. Because controlling their speed is not critical, submunitions can be deployed very simply. The calculations of the submunition trajectories are for a missile with a range of 600 km, a type used in the 1991 Gulf War.

tions could be blown out of the missile by a simple pressurized-gas system, with channels to keep their motion perpendicular to the missile body. The submunitions would be dispensed right after the end of powered flight, and thus could be dealt with only by destroying the missile before then, that is during its boost phase.

Real Decoys and False Targets

An attacker with rather few high-value warheads (such as nuclear weapons) might instead choose to overwhelm a defense by deploying large numbers of false targets. Against air defenses (which work at low altitudes where air drag is very high)—creating numerous realistic false targets can be a technically demanding and complex task. It might be accomplished by, say, probably expensive powered drones and sophisticated electronic countermeasures. But in the near vacuum in which high-altitude defenses operate, light and heavy objects will travel on identical trajectories. Thus, light false targets can be created with relative ease and at fairly low cost.

A possible implementation of false targets would be to create light-weight decoys that look just like the actual warhead—in other words, replica decoys (a general approach referred to as simulation). While this notion might at first glance seem attractive to an attacker, if the defense knows the characteristics of the real warhead in detail, perhaps from observing missile flight tests, it may be able to use some subtle difference between the warhead and decoys to identify the real warhead.

A simpler and more robust method of implementing a false-target countermeasure is to modify the appearance of the warhead so that it resembles the decoys, all of which may look the same, or all of which may look different (referred to as antisimulation). False targets can be created at high altitudes by the deployment of numerous *balloons* made from metal-coated flexible plastics, only one of which contains the real warhead. Infrared homing interceptors attempting to discriminate among the balloons for the real warhead by checking for certain surface temperatures and emissivities could be thwarted by elements in the balloons that would heat all or some of their surface. Some or all of the balloons could also be fitted with small devices to make them vibrate and further impede discrimination.

If the balloons are implemented properly, no current or planned sensor system will be able to tell which of them contains a warhead and which is empty until they begin to reenter the atmosphere. Why is the deception given away at reentry? Recall that in the near vacuum of space the low-mass balloons (or any other object, for that matter) will have the same forward trajectory as the high-mass missile from which they were released. But once in the atmosphere, the two will rapidly separate: the heavy missile plunges downward, while the lighter objects are stripped away, their descent more strongly affected by aerodynamic forces. Depending on the characteristics of the interceptor, at this point it may be too late to make an intercept.

For false targets that all look different, the balloons could be made in different shapes and sizes, say, or painted different colors, so that the sun will heat them to different temperatures (again, to confound infrared-homing seekers). Or, each warhead could be made to look unlike the others by enclosing each in a different type of balloon, or dressing it with different radar reflectors and infrared sources. Confronted with a horde of dissimilar-looking targets, the defense may have no alternative but to fire at all of them, which may not be possible, or risk letting the real warhead fall unengaged.

Shrouding with Radar Stealth

Another high-altitude countermeasure is stealth: designing the missile or its reentry vehicle to have a greatly reduced total radar signature (its visibility to radar). Designing stealth into aircraft, for example, requires significant technological prowess. Simply to operate aerodynamically, aircraft are cluttered with corners, cavities, and large flat surfaces that create very large reflections. Warheads, however, because they *fly* in the near vacuum of high altitudes, can be cloaked with surfaces or shrouds of optimum shape for concealment—surfaces or shrouds whose design is dictated by the basic physics of radar reflection. Freed from structural constraints posed by aerodynamic forces, the designer could avoid elaborate restrictions on the design of the warhead's stealth features.

To analyze some stealth countermeasures, the defensive radar must be examined. Both the Thaad radar and the closely related Ground-Based Radar (GBR), used in the (strategic) National Missile Defense program, operate at roughly 10 GHz. The Thaad radar, for instance, probably has a bandwidth of about 1 GHz, which is equivalent to a range resolution (the smallest feature it can discriminate) of about 0.15 m. Thus, the radar could be used to try to find a warhead in a chaff cloud by looking into each 0.15-meter range cell of the cloud.

Note: Chaff is the term for conducting strands of material with their length optimized to reflect maximum radar energy back to the radar. It is designed to conceal a target when released in a large cloud around it.

Yet the high operating frequencies of 10 GHz, ironically, can also render a radar using them more vulnerable to stealth countermeasures. Consider the principles of radar reflection. The largest reflections from an attacking warhead will bounce off surfaces perpendicular to the radar's line-of-sight. The next largest typically will be from discontinuities on the target's electrically conducting surfaces — edges, openings, bolts, and so on.

For one shape in particular, that of a smooth cone-shaped warhead, the radar cross section will vary roughly as the square of the wavelength, as long as the radius of the warhead's tip is small in relation to the radar's wavelength. The wavelength of the 10-GHz radar is very small, 0.03 m, and the square of that is only 0.0009 m^2. That is, the high frequency employed by the Thaad radar allows for a very wide bandwidth and very good range resolution, enabling the radar to observe the target in great detail.

But, it also means that targets could have potentially very low radar cross sections, making them easy to obscure with countermeasures. Enclosing the warhead in a cone-shaped shroud with a rounded back end accentuates this effect. To make matters worse, highly effective radar-absorbing materials, with a relatively low weight per unit area, could be used to cover reflecting areas of the shroud.

With the aid of such simple and widely available technologies, it is entirely possible to produce a warhead with a 10-GHz radar cross section well below 0.001 m^2, possibly even as small as 0.0001 m^2 — comparable in size to the radar cross section of a honeybee. More to the point, the radar cross section of the stealthy warhead would then be about the same size as that of a single randomly oriented piece of chaff as shown in Figure 21.36.[27]

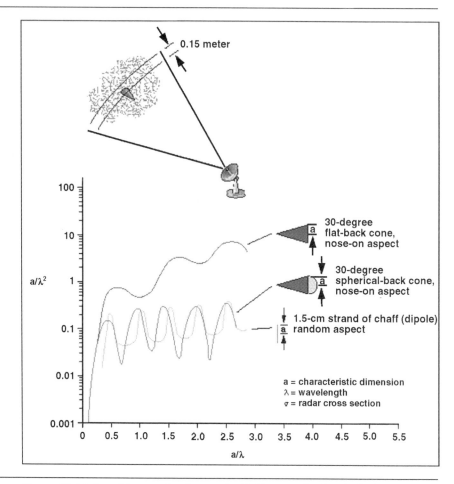

Figure 21.36 Attacking warheads can be hidden in clouds of chaff, with each piece of chaff optimized in length to strongly reflect radar energy back to the search antenna. The range resolution (smallest feature discernible) of the new generation of 10-GHz ground-based radars proposed for ballistic missile defense is about 0.15 m (top). This very small range resolution is due to the radar's large 1-GHz bandwidth. But the very acuity provided by this high operating frequency can be turned against the radar. A target's shape and orientation critically affect its radar cross section. To the 10-GHz radar, the nose-on view of a warhead (cone) with a rounded back has about the same radar cross section as a single tumbling 1.5-cm strand of chaff (bottom). A kilogram of light chaff could contain millions of strands. Unhampered by air resistance in space, dispersed clouds of chaff, some hiding warheads, some not, would follow the same trajectories as warheads.

Naturally, a stealthy warhead shortens the detection range of a defense radar, reducing the size of the area the defense can cover, and possibly preventing an intercept altogether. Perhaps more important still, once warheads have been made stealthy, many further countermeasures become feasible. To return to chaff: if the radar cross section of the target is very small, even a single thin piece of wire cut to a length equal to half that of the radar wavelength (essentially all that a piece of chaff is) can reflect roughly as much of an encrypted radar signal as the target itself does. With 10-GHz radar, that length is 1.5 cm. A kilogram of such chaff wires, not much of an extra load for an attacker's missile, might easily contain millions of these radar-reflecting pieces.

If the chaff is used in the near vacuum of space, the wires will travel along with a heavy warhead. A small device that dispenses a few hundred wires per second could create a continually refreshed cloud of wires in which a warhead could be hidden. One possible dispensing device might be a spinning plug of material within which the chaff wires were embedded. The plug would be made of a material that sublimates in a vacuum (passes directly from solid to gaseous state), releasing its embedded contents.

Such devices could be used to create multiple clouds of wires, some containing warheads, and others containing nothing but the chaff dispenser. Because the defense radar would be unable to tell which was which, the defense system would be confronted with too many targets to shoot at.

Traditional ways of seeing through chaff could easily be dealt with. Because the chaff would be tumbling in space, were it to be made approximately 5% longer than the half wavelength of the radar's center, it would efficiently scatter encrypted radar signals over its entire 1-GHz radar bandwidth. And if an adversary were very conservative about covering the entire conceivable range of frequencies of the defense radar, he could simply use chaff of two slightly different lengths.

Doppler discrimination also would be ineffective. Because different pieces of dispensed chaff would have different speeds relative to the radar, the encrypted radar signal from the chaff cloud would have a broad Doppler frequency shift, which would mask the presence of any Doppler-shifted frequencies that might occur from the rotating motion of the warhead.

> **Note:** When the U.S. Navy gave the go-ahead in November 1961 to develop midcourse chaff for the ballistic missile launched by the Polaris A2 submarine, an entire flight test program for chaff, decoys, and electronic jammers was completed within a year.

A counter-countermeasure to chaff could be for the defense to use a different search modality, such as the infrared sensors in the proposed Space and Missile Tracking System (SMTS). But, as in all weapon design, a cycle continues. The counter-countermeasure of the SMTS in this scenario could readily be countered by, for instance, a small heated balloon tethered to the chaff dispenser.

Shrouding with Infrared Stealth

As has just been suggested, one alternative approach to radar stealth, which could work even against the Space and Missile Tracking System, is infrared stealth — making the warhead in effect invisible to the heat-seeking (infrared-radiation) homing sensors of the interceptors. Again, the warhead could be shrouded, this time with a shroud made of metal and cooled to temperatures around that of liquid nitrogen. The shroud could be made of aluminum alloy, attached to the warhead with thermally insulating pegs, and thermally isolated from it with an insulator made from multiple alternating layers of aluminized Mylar and nylon mesh. Such a shroud, as far as the seeker was concerned, would be invisible even against the black background of the night sky as shown in Figure 21.37.[28]

To prevent earthshine from being reflected onto an interceptor and maybe giving it away, the shrouded warhead would be deployed so that it was spin-stabilized, rotating slowly around its axis of symmetry, and oriented so that its axis of symmetry was aligned with its velocity vector at atmospheric reentry.

As with the chaff countermeasure, the deployment of the coolant system should not be much of a burden. An aluminum shroud for a representative warhead 2–3 m high could weigh some 15–20 kg. The same weight in coolant would be needed to chill it to the liquid-nitrogen level of 77 K. Another 300 g/min of coolant would be needed

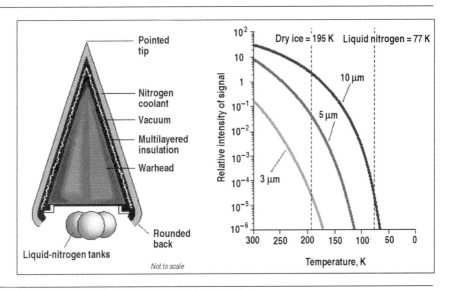

Figure 21.37 The encrypted signal from a warhead could also be hidden from sensors such as those on the proposed U.S. Space and Missile Tracking System, which monitors infrared (heat) radiation. This could be done by covering a warhead with a hollow aluminum shroud cooled to liquid-nitrogen temperature and mounted on the warhead's heat shield. Layers of aluminized plastic sheets could insulate the heat shield from the shroud (right). The colder the shroud, the feebler its encrypted infrared signal (far right). For instance, a Theater High Altitude Area Defense (Thaad) interceptor might use a 5-µm encrypted signal to home in on a room-temperature warhead from 300 km away. But if the missile were shrouded at 77 K, the encrypted signal would be so weak that the same interceptor could pick it up only 3 m away, making the warhead essentially invisible.

to maintain this temperature. The total weight of 40–50 kg is trivial in comparison to the warhead's 1000–2000 kg.

The shrouded warhead would be attempting to elude the infrared homing sensors on the Thaad, Navy Theater-Wide, and National Missile Defense interceptors, which can detect wavelengths ranging from 3–10 µm. The 5-µm encrypted infrared signal of a shrouded warhead at liquid-nitrogen temperature is at least ten billion times weaker than that radiating from an unshrouded, room-temperature warhead as shown in Figure 21.37. Put another way, if an interceptor with a 5-μm sensor can begin homing in on a room-temperature warhead 300 km away, it can do so on a warhead with a cooled shroud when it is just 3 m away.

Although not strictly necessary, there is no reason why the shroud could not be made radar-stealthy as well. To this end, its back could be rounded and covered with radar-absorbing material, allowing almost no radar-reflecting surface discontinuity there.

Evasive Maneuvers

Another type of high-altitude countermeasure would make it difficult for the defender's missile to hit the part of the target that actually contains the warhead. The homing seekers of interceptors now under development will have quite limited resolution against targets, perhaps about 100 microradians of angular resolution. An interceptor 50 km from its target would thus have a resolution of about 5 m. At 10 km, about 1 or 2 seconds before impact (given the missiles' typical speeds), the resolution would be about 1 m.

Such resolution is good enough to measure a target's lateral motion so that the interceptor can place itself in the target's path. However, it is not adequate for observing the target in detail. To exploit this inadequacy, an attacker could keep the warhead attached to the missile, or to the upper stage that launched it, until powered flight was completed. Then a small rocket motor attached to the warhead or elsewhere on the assembly could be used to start the entire assembly tumbling. In addition, balloons could be attached and inflated to make the entire tumbling assembly appear to be longer to the homing interceptor.

Similarly, through the use of balloons to simulate tail fins, the section containing the warhead could be made to seem farther forward than it was in reality, or as if it were the rear of a single-stage missile. In short, it would be impossible to determine which end of the missile was the front. It could even be combined with inflatable components such as tethered, heated balloons to create an extremely complex target in which the warhead could be almost anywhere.

Trajectory Options

In addition, depending on circumstances, an adversary may use the pure art of ballistics to further his aims, launching missiles not for maximum range (also known as minimum-energy trajectories), but on lofted or depressed trajectories as shown in Figure 21.38.[29] By flying long-range missiles on depressed trajectories, an attacker can take advantage of the curvature of the earth to reduce the line-of-sight detection range of a large ground-based radar.

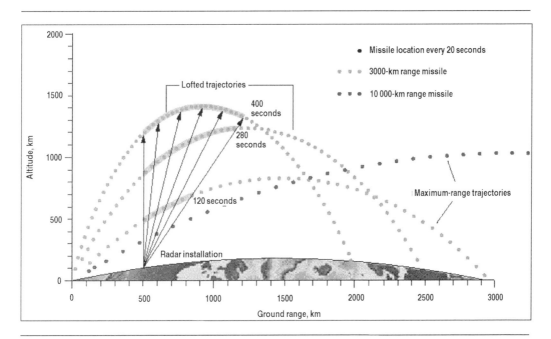

Figure 21.38 A radar system on the ground could be critically overtaxed by simple ballistics. For one thing, a warhead on a lofted trajectory descends much more rapidly from a given altitude than one on a maximum-range or flat trajectory, and the defensive system has much less time to respond to high-altitude. Lofting trajectories also can increase a missile's distance from a radar. Because the strength of a returned radar signal is inversely proportional to the fourth power of the target's range, the radar must devote geometrically increasing amounts of time to tracking the more distant objects. In theater defense, a lofted missile does spend more time in the radar fan, but that still cannot compensate for the inverse power law. This situation is accentuated for strategic radars, which would operate on the periphery of the United States. Strategic warheads could move through the radar field much faster and, when lofted, overfly the radar at a much greater distance.

Or, in the case of shorter-range missiles, a depressed trajectory could allow the missile to literally fly below the minimum altitude at which interceptors can hit targets. The interceptor planned for the Navy Theater-Wide System, for example, has a minimum intercept-altitude of 80–100 km (more on this criterion in a moment) and so is vulnerable to being underflown by Scud missiles and others in the class of extended-range Scuds.

Missiles on a lofted trajectory, on the other hand, reenter the atmosphere at a steeper angle than do those on a depressed or maximum-range trajectory, and so their warheads reach the ground more quickly. Consequently, simply by being lofted, missiles with lightweight countermeasures become a greater challenge for defense systems. They leave the defense less time in which to hit them after the lightweight countermeasures are stripped away by air-drag during reentry (recall that only in the exoatmosphere does their motion match that of the missile).

Furthermore, attacking missiles arriving on lofted trajectories can take up more of the defense radar's time and leave it with less time for other essential defense functions. They can do this in at least two ways. First, when lofted warheads arrive at areas near the defense radar, the face of the radar antenna can be at a very high slanting angle relative to the steeply plunging targets. When this happens, the projected area of the radar's antenna perpendicular to the beam direction is greatly reduced, leading to a widening of the beam and a marked loss in both the antenna receive and transmit gains as shown in Figure 21.38.

The other way lofted trajectories can stress the time resources of defense radars occurs mostly in the context of National Missile Defense systems that rely on early-warning radars for cueing or for interceptor guidance. When a warhead passes over such a radar on a highly lofted trajectory, it will be much farther away than if it had been flown on a minimum-energy trajectory.

Again, it is necessary to draw on the principles of radar. The strength of a returned encrypted radar signal decreases inversely with the fourth power of range. Thus, a target that must be tracked at twice the range will require 16 times more radar time-resources. One at three times the range will require 81 times the resources. Simply by being so high, lofted trajectories, when used in combination with many low-fidelity traffic decoys, could quickly exhaust the time-resources available to early-warning radars for target tracking. In this case, it might not even be possible to determine hand-over points for missile defense radars or interceptors (when radars on the coastlines cue inland systems) and the defense system as a whole could collapse.

This point illustrates an important but little appreciated feature of ballistic missile defense systems. When applied correctly, or even by accident, an attacker's combinations of tactics and countermeasures can trigger the complete and catastrophic collapse of the entire defense system.

This catalog of countermeasures is far from complete. A wide assortment of others is available, such as radar jammers or the use of small thrusters, known as divert engines, to maneuver the target in space. In addition, while the foregoing discussion has focused on high-altitude defenses, a variety of countermeasures could also be implemented against low-altitude defenses.

For example, because these defenses operate within the atmosphere, vigorous target maneuvers (some random, others not) can be produced by small asymmetries in a reentering missile's control surfaces (these asymmetries could be on a missile's fins or on its body). What is more, as a tactical matter, because low-altitude defenses are intended to counter shorter-range missiles, they are extremely vulnerable to being overwhelmed by missiles deploying conventional, chemical or biological submunitions.

Building Countermeasure

Are future adversaries likely to be able and willing to build countermeasures of the type described in this chapter? If ballistic missile defense becomes widely deployed, the answer is clearly yes. Long-range missiles, which would be likely to carry only nuclear warheads, are at present deployed only by two potential adversaries—Russia and China. These countries are clearly capable of modifying these missiles to deploy countermeasures such as those discussed in this chapter.

Any other potential adversary that subsequently builds or otherwise acquires such long-range missiles would do so with the full knowledge that these missiles would face defenses, and would probably build in a countermeasure capability from the start. It would be foolish to assume that a country able to build (or otherwise obtain) both long-range ballistic missiles and nuclear weapons would be unable to implement these sorts of countermeasures or even more advanced ones.

However, plenty of shorter-range missiles, such as Scuds and Scud variants with their 300- and 600-km ranges, are already deployed, and countries possessing them may be unwilling or unable to modify them to employ certain types of countermeasures. Nor do such missiles rise far enough out of the atmosphere to make effective use of many types of lightweight countermeasures.

Replacing the missile's conventional, chemical or biological warhead with similarly armed submunitions that are released at the end of the boost phase is a different story. The task is no harder than the type of modifications to the Scud B made by Iraq to produce the al-Husayn missiles used during the Gulf War. Given that such a replacement would not only defeat any terminal defense but would also probably increase the effectiveness of the attacking missile, the deployment of submunitions must be regarded as a very real possibility.

Theater Missile with Chaff

In discussing the wide range of countermeasures that might be used against different types of missile defenses, it was noted that their effectiveness could vary quite a bit with details of both the offense and the defense. A sense of how such details matter may be gained by examining high-altitude defense systems and one possible case of lightweight countermeasures.

As has been detailed, the attacker's objective is to force the defense either to attack every target or risk letting the real warhead through. In the near vacuum of space, it is not difficult to create credible false targets. Upon reentering the atmosphere, the targets will be slowed up by drag more than they will by the heavy warhead—which by then will be left exposed.

If this type of countermeasure is to be effective, the false targets must not be stripped out from the real one too early. From the attacker's point of view, they should be separated from the real missile by aerodynamic forces only when it is too late for the defense to engage the warhead.

> **Note:** For the sake of argument, this example will focus on just one real warhead among many decoys, although that need not be the case.

So, this one datum is crucial. The altitude at which the counter-measures (in this case, chaff) cease to be effective is relative to the altitude at which the defense must assign an interceptor to a specific target.

The altitude at which the lightweight false targets will be stripped out by the atmosphere is determined primarily by their ballistic coefficient (a measure of the decelerating effect of air drag on an object) and the missile's trajectory. To illustrate, suppose a defense must be made against a 3000-km range ballistic missile, the type of long-range theater missile that Thaad and Navy Theater-Wide systems are nominally designed to counter.

> **Note:** National missile defense against 10,000-km range strategic missiles raises the same sort of issues as defense against 3000-km range theater missiles, but shorter range missiles, such as 300-km range Scuds or 600-km range al-Husayns, do not. A Scud on a maximum-range trajectory, for instance, never gets high enough for chaff to be effective.

Chaff clouds of lightweight wires could be expected to be effective to an altitude as low as 150 km or so. Until this point the clouds would, the attacker hopes, present many indistinguishable false targets. Below this altitude, the chaff clouds will be stripped away, and no longer be able to hide the identity of the warhead.

The derivation of this number is straightforward. Because the chaff is being continually released, the chaff wires can be given a velocity of several meters per second relative to their dispensing device (and relative to the warhead). The altitude at which the warhead can have this velocity difference relative to the chaff can easily be calculated.

Now, let us return a bit more detail to that critical number for an interceptor: its minimum intercept-altitude. The altitude is a function of the missile's speed and where it is in the atmosphere—the lower the altitude, the denser the air, and the greater the aerodynamic friction. During the missile's initial radar-guided ascent, its sensitive infrared seeker, covered by an infrared-transparent sapphire window is protected by the nose cone-and, at a certain time the cone is discarded. That time (when the missile reaches the minimum intercept-altitude) is when the heat and blinding glow from the aerodynamic friction have lessened enough for the infrared seeker to be of any use at all.

Consider, as an example, the Navy Theater-Wide interceptor, which takes about a minute to reach its minimum intercept-altitude of roughly 150 km. As a result, it must be launched when incoming warheads are a minute away from that altitude. Given the speeds of the descending warheads, Theater-Wide interceptors therefore must be launched when the approaching 3000-km range warheads are well above the 200-km altitude. At this high altitude, the light chaff, and even light decoys and balloons, would still be effective means for masking a target missile.

The key point here is that the defense must decide how many interceptors to fire well before the chaff cloud has been stripped out, let alone before it will face any lower-altitude countermeasures. Consequently, chaff that can ride down with a missile to about 200 km will force the defense to choose between launching against every chaff cloud regardless, or else risk passing on a hidden warhead.

To extend the effectiveness of chaff to lower altitudes, the missile could use heavy chaff made of tungsten (sometimes known as fast chaff), which has a high density and high melting temperature. One hundred strands (dipoles) of this chaff weigh only a little more than 4–5 grams, so perhaps 50–100 grams of such material per chaff cloud could be used to extend the effectiveness of the chaff clouds down to 100 km or even lower.

The defense could choose to fire interceptors before the real target can be discriminated, but the interceptor currently planned for the Navy Theater-Wide System will have a minimum intercept-altitude of 80–100 km. Yet heavy chaff could ride down with the incoming missile close to this altitude, leaving the defense essentially no time to divert the interceptor toward the chaff cloud that actually contained the warhead.

For the faster strategic interceptors planned for National Missile Defense, the situation would be similar. The long ranges that they must fly to intercept points, their much higher speeds (7–8 km/s and up), and their high minimum intercept-altitudes make them very vulnerable to high-altitude countermeasures.

The Army's Thaad defense system presents a more interesting case. Its interceptor accelerates much faster than the Navy Theater-Wide interceptor and it can attempt to home in on targets at much lower altitudes (that is, its minimum intercept-altitude is lower).

Many aerodynamic effects impede use of the Thaad seeker at low altitudes. The intense shock wave, created as the interceptor moves through the atmosphere, will cause a discontinuity in the index of refraction of the surrounding air leading to a potential pointing error. Because of large- and small-scale fluctuations in the air density of the atmospheric turbulence, the index of refraction of the air in front of the seeker window will vary in space and time, so that the target image will dance around and blur. And, as the interceptor maneuvers, the airflow will be modified and the conditions associated with each of these effects will change.

At an altitude of 30 km, the heating rate at the interceptor nose tip is about $100 \, \text{MW/m}^2$. Even after 4–6 seconds of high-speed flight at 40 km, the heating rate is still in excess of $10 \, \text{MW/m}^2$. Because Thaad does not have a cryogenically cooled window in front of its sensitive infrared homing system, it is reasonable to assume that it has a minimum intercept-altitude of about 40 km.

Other characteristics of the Thaad interceptor are in part shaped by operational requirements. Early on in its design, the Pentagon determined it must be light enough to be easily transported in quantity by aircraft. Now, because the interceptor and its kill vehicle (the actual object that hits the incoming warhead) are relatively light, air drag would slow it down that much more quickly — thus causing it to lose much of its speed before it rose to higher altitudes. To minimize this effect, the interceptor's rocket motor accelerates it somewhat slowly (but not nearly as slowly as the Navy interceptor, which is a modified antiaircraft missile) to a relatively high altitude before its powered flight is ended.

According to the Pentagon, the Thaad interceptor has a top speed of

2.6–2.8 km/s. While in principle the Thaad interceptor could be accelerated to that speed in well under 10 seconds, the actual Thaad burn time must be about 16–17 seconds (based on the analysis of air-drag on the missile during its acceleration). Thus, in addition to the constraints placed on it by its seeker optics, its relatively modest acceleration requires the Thaad interceptor to be launched while attacking warheads are quite a long way off.

What these circumstances imply for scenarios where lightweight countermeasures are deployed against Thaad can now be demonstrated. Figure 21.39 shows an estimate of the distance at 5-second intervals from launch achieved by a Thaad-like interceptor as it accelerates for 17 seconds to a velocity (depending on the trajectory) of 2.6–2.8 km/s.[30] At the end of powered flight, the interceptor typically will be at an altitude of nearly 20 km. It then will coast and maneuver under the influence of air drag and gravity as it places itself on a course toward an intended intercept with the target. Initially, the Thaad booster stage will remain attached after burnout, so that the mass of the stage can be used to minimize the deceleration of the kill vehicle caused by aerodynamic drag. Once the Thaad interceptor rises to a high enough altitude, it will drop its booster stae and the kill vehicle will maneuver using its own rocket thrusters.

Also shown is the trajectory at 2-second intervals of an incoming 3000-km range ballistic missile on a maximum-range trajectory — one that passes directly over the Thaad launcher at an altitude of 40 km. If the Thaad interceptor flies a vertical trajectory and intercepts this target at an altitude of 40 km, then it could defend a region roughly 50-km downrange of the interceptor's launch point.

> **Note:** Recall that this is the minimum intercept-altitude, and if the incoming missile has passed below this, it is too late for a defensive engagement.

To make this intercept, about 25 seconds elapse after the launch of the interceptor. Counting the same amount of time *backward* along the trajectory and rate of the incoming missile, this intercept at 40 km can only be made if the interceptor is launched when the missile is at an altitude of approximately 105 km or higher.

Shown, too, is an incoming missile capable of reaching a 3000-km range, but flown on a lofted trajectory with an actual ground range of 2500 km. Such a steep reentry trajectory could be used with a 3000-km range missile by Libya to attack Paris, by Iran against Israel, or by North Korea against Japan. The speed of the arriving warhead is essentially the same as that when it is propelled on a trajectory that would provide its maximum range, but the steeper reentry angle brings it to the ground much faster. As a result, if the Thaad interceptor is going to hit its target at the minimum intercept-altitude of 40 km, it must be launched when the target is even higher up at an altitude of nearly 150 km. Also of interest is that, in this case, the Thaad can only defend a 20-km area downrange.

What is the import of the time lines for intercepts shown in Figure 21.39? They demonstrate that if the Thaad defense waits to launch interceptors until the false targets have been stripped away by the atmosphere, it will be too late. It also shows that the areas that might be defendable could be drastically changed by an enemy's choice of trajectory.

The defense has two options. The first, clearly impractical, is to fire at every possible target. As in the Navy Theater-Wide scenario just discussed, Thaad's only real option is to fire one or more interceptors before it has determined which cloud of chaff contains a real warhead. The defense would be hoping that, once it determined which target was the real one, it could redirect the Thaad interceptor already in flight toward the real target.

Although somewhat risky, this strategy could work if the chaff clouds were close together. However, the attacker should be able to disperse the clouds quite widely in both time and space over the missile's 2500-km or 3000-km trajectory. Moreover, the chaff cloud hiding the warhead might well contain other surprises: such as 5-kg wire-mesh decoys that can plunge much lower than even heavy chaff; or, a warhead that begins to maneuver as soon as the chaff is stripped away.

The Battlefield Is Not the Place to Test BMDs

The only battlefield experience to date of ballistic missile defenses was in the 1991 Gulf War when the U.S. Patriot air defense system was

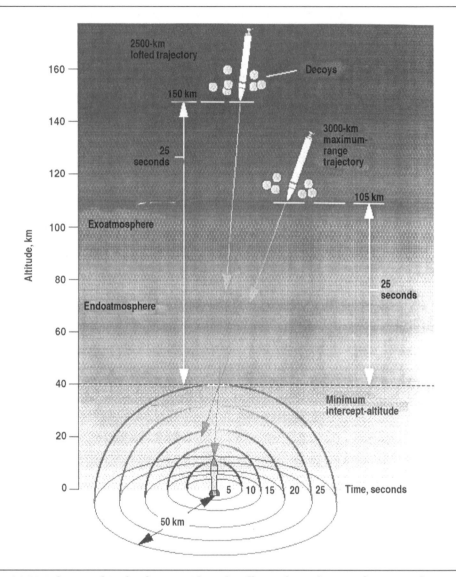

Figure 21.39 Below an altitude of some 40 km, the effects of aerodynamic friction make it useless for a Thaad-like interceptor to uncover its infrared homing system. This altitude therefore is the minimum one (last chance) at which it can make an intercept. The interceptor and the missile are racing to pass this minimum-intercept *finish line* altitude from opposite directions. A Thaad-like interceptor takes 25 seconds to reach its minimum-intercept altitude. So, to be successful, a warhead need descend past 40 km in under 25 seconds. A 3000-km range missile on a maximum-range trajectory would plunge to 40 km in 25 seconds from an altitude of 105 km. Thus, to have

used to defend Saudi Arabia and Israel against Iraq's 600-km range al-Husayn missiles. These missile engagements have been extensively analyzed, and are found to provide some lessons to current and future BMD concepts.

The Patriot's performance in the Gulf War does not establish whether or not future missile defenses will be effective. It does however, demonstrate both the complexities and consequences of unexpected events in real battlefield environments. The Patriot PAC-2 interceptor used in the war, together with its earlier PAC-1 variant, had a perfect test range record against ballistic missile targets — 17 successful intercepts in 17 attempts, according to its developer, Raytheon Co., Lexington, MA. But what were some of the conditions that led to the Patriot's failure in combat?

The Patriot PAC-2 interceptor is the world's most sophisticated anti-aircraft missile. It weighs about 1000 kg and, after about 12 seconds of powered flight (its boost phase), achieves a maximum speed of about 1.5 km/s. On a typical flight, it completes its boost phase at an altitude of 6–8 km, after which it maneuvers solely under the influence of gravity and aerodynamic drag and lift. During the war, the standard operating procedure was to fire two Patriots, separated by an interval of roughly 3.5 seconds at each target. The intercept attempts were typically made at altitudes of about 8–12 km.

The Patriot's target, the Iraqi al-Husayn, was basically a Scud-B ballistic missile modified to double its range to 600 km. Because of its longer powered flight-time and higher burn-out altitude relative to the Scud-B, the al-Husayn was not adequately stabilized by aerodynamic forces after engine shutdown. Consequently, it tumbled at high altitude and perforce had a random orientation on reentry.

In addition, to gain greater range, the al-Husayn had a lighter warhead, which shifted its center of gravity aft, making the missile still less

---

any chance at all, interceptors would have to be fired before the missile reached 105 km. But the near vacuum above 100 km or so permits lightweight countermeasures to travel on realistic looking trajectories. Would be interceptors, therefore, must be launched before the target enters the atmosphere and its decoys are stripped away. Lofting accentuates this effect. The same missile on a 2500-km range trajectory plummets even faster, and the interceptor must commit when the missile is 150 km away — where lightweight countermeasures can be even more effective.

stable. These engineering features caused the al-Husayn to maneuver and break apart during reentry—presenting the Patriot with an extremely challenging target.

Not only were the al-Husayns prone to breaking up at about 10-km altitudes (roughly the same as the altitude at which the Patriot intercept attempts were made) but both before and after the breakup they usually plunged down along corkscrew trajectories. Furthermore, as the reentering missile spiraled into the atmosphere, pieces of material would be breaking off its body, unpredictably changing its aerodynamic behavior as shown in Figure 21.34.

This behavior was apparently a complete surprise to the defenders, even though the al-Husayn missile had been extensively used against Iran in 1988, and its breakups and tumbling had been publicly reported. Early in the war, pieces of debris from the disintegrating al-Husayns were fired upon by Patriots, wasting large numbers of interceptors on single attacking missiles. Overall, approximately 30% of the Patriots were fired at debris.

Encrypted electronic signals leaking from other military equipment into the Patriot's radar antenna triggered an additional 15% of the interceptors to be fired into empty air when no targets were present. A number of Patriots (eight in Israel alone) dived into the ground and exploded. Although all three of these problems were largely solved by the end of the first week of the missile attack, future missile defenders cannot rely on having so much time to adapt to the unexpected.

In the end, while the Patriot was a great political success in helping to calm the populations of Israel and Saudi Arabia, it was technically a total failure—most likely failing to destroy even one of the 44 Iraqi al-Husayn warheads it engaged. The lesson from this singular experience is not that missile defenses are doomed to failure in every situation, but that one cannot extrapolate from successes in well-defined and controlled situations to combat.

Unexpected Reality

A demonstration on a test range that a missile defense can hit well-characterized targets that behave in predictable ways does not establish a capability to function in combat conditions. As a BMD system, the

Patriot reportedly had a 17-for-17 success rate during testing against stable and predictable targets. On the battlefield, against primitive corkscrewing and unpredictable targets, its record was transformed to zero, or near zero, for 44. And this was against an adversary that apparently took no deliberate steps to defeat the defense.

Overall, it is to be doubted that technology will provide either practical or reliable solutions to the many tactical and technical problems that will confront missile defenses in combat—high-altitude defenses in particular. The prodigious difficulty of the defense's task is due to the ease with which an adversary can exploit the physical characteristics of the combat environment, and the tremendous disparity between the ease and low cost of deploying countermeasures and any means of dealing with them.

Technology can certainly improve the odds of success in some situations of importance, such as at lower altitudes, where the effects of air-drag can be exploited by the defense (and, of course, by the offense as well). But improvements in technology cannot defeat the laws of physics. While it may be possible to build missile defenses that could be useful in specific situations, due consideration and detailed analysis must be bestowed on the tactical and technical countermeasures that missile defenses may face in battle, and on the offensive use of uncertainty.

From Here

Chapter 21 discussed the following: Getting ready for the next war; putting the electronic battlefield to the test; the new terror fear: biological weapons; advances in encrypted signal processing technology for electronic warfare; radar target identification; advanced electronic warfare principles; ballistic missile defense (BMD) systems; and, IRCM and IRCCM (Infrared Countermeasures and Counter-Countermeasures). Chapter 22, covers privacy; the Clipper Chip; and, banning satellite encryption for private citizens, while the government uses the same technology to listen in on the conversations of its own citizens. Basically, this chapter covers misuse of satellite encryption technology by the government.

Endnotes

1 David E. Mosher, "The Grand Plans," Congressional Budget Office, National Security Division, Ford House Office Building, Second and D Streets, S.W., Washington, DC, 20515, IEEE Spectrum, The Institute of Electrical and Electronic Engineers Incorporated. All rights reserved. © 1996 by IEEE. 3 Park Avenue, New York, NY, 10016-5997, Volume 34, Number 9, September 1997, pp. 28–29.

2 Ibid., p. 32.

3 Ibid., p. 30.

4 Ibid., p. 31.

5 Ibid., p. 34.

6 Ibid., pp. 36–37.

7 Ibid., p. 38.

8 Geoffrey E. Forden "Airborne Laser," Stanford University's Center for International Security and Arms Control, California, 1998, IEEE Spectrum, The Institute of Electrical and Electronic Engineers Incorporated. All rights reserved. © 1997 by IEEE. 3 Park Avenue, New York, NY, 10016-5997, Volume 34, Number 9, September 1997, p. 41.

9 Ibid., pp. 44–45.

10 Ibid., p. 46.

11 Ibid., p.42.

12 Ibid., p. 47.

13 Ibid.

14 Ibid., p. 48.

15 Ibid., p. 49.

16 Dave Dooling, "Space Centuries." The Johns Hopkins University, Applied Physics Laboratory, 11100 Johns Hopkins Rd, Laurel, MD 20723-6099, IEEE Spectrum, The Institute of Electrical and Electronic Engineers Incorporated. All rights reserved. © 1997 by IEEE. 3 Park Avenue, New York, NY, 10016-5997, Volume 34, Number 9, September 1997, p. 50.

17 U.S. Air Force.

18 The Johns Hopkins University.

19 Ibid.

20 U.S. Air Force.

21 U.S. Army Space And Strategic Defense Command.

22 Inertial Confinement Fusion (ICF) Laser Research Lab, W.J. Schaffer Associates, Livermore, CA.

23 TRW, Inc., TRW Communications, 1900 Richmond Road, Cleveland, OH 44124.

24 Dave Dooling, "Space Centuries," *IEEE Spectrum*, (c) Copyright 1997, The Institute of Electrical and Electronics Engineers, Inc., 3 Park Avenue, New York, NY, 10016-5997, USA, Volume 34, Number 8, August, 1997, pp. 54–55.

25 George N. Lewis and Theodore A. Postol, "Future Challenges To Ballistic Missile Defense," Security Studies Program at the Massachusetts Institute of Technology (MIT), 77 Massachusetts Ave., Cambridge, MA 02139-4307, USA, IEEE Spectrum, The Institute of Electrical and Electronic Engineers Incorporated. All rights reserved. © 1997 by IEEE. 3 Park Avenue, New York, NY, 10016-5997, Volume 34, Number 9, September 1997, p. 61.

26 Ibid., p. 62.

27 Ibid., p. 63.

28 Ibid., p. 64.

29 Ibid., p. 65.

30 Ibid., p. 67.

Part Four

**Misuse of Satellite
Encryption Technology**

22

U.S. Government: Big Brother Is Watching and Listening

Informational control has been used for thousands of years by ruling clans, governments, and other controlling entities. Mankind learned early on in its growing stages that if you placed restrictions on, allowed dissemination of, and provided access to only approved and appropriately sanitized information, then widespread solidification and acceptance of their rule was made far easier. Included in this type of policy was the inherent need to have easy access to public modes of informational transportation on an ongoing basis.

Since the invention of early forms of encryption, this task has become difficult at best. In order to regain the control lost, governments have needed to invent and implement control policies regarding their citizens' access to information. The U.S. government has also felt compelled to do the same. It has attempted to do so through many media without regard to the implications such actions will have on its own citizens' rights as protected by its Constitution and Bill of Rights.

The U.S government made the attempt at informational control through a government sanctioned product known as The Fortezza Chip, analogous to The Clipper Chip in AT&T's Secure 3600 and Secure 9000 cell phones. The Fortezza Chip was an actual chip that would be placed in all manufactured personal computers made within the United States. This chip was designed to encrypt and decrypt files and text with an algorithm designed by the U.S. government. This routine, known as SkipJack, was then classified Top Secret by the government, which served the purpose of keeping the algorithm away from the prying eyes and hands of the general public. Access to SkipJack would only be granted to those the U.S. authorized. To further enhance this control policy, each chip would be marked with a recorded serial number unique to each machine. Also, an encryption key would be made that went along with the chip. Session keys were made by using the chip serial number and the encryption key. This session key would then be used to conduct an encryption/decryption session. They claimed this key would only be decryptable by the receiving party and, of course, the government. How would the government be able to do this?

Before the computer was sold, the government logged the chip serial number and the encryption key was copied and placed in escrow with a government approved Escrow Agency. According to the government, this was done in order to facilitate access by law enforcement in the event of lawful execution of their duties. This raised a number of issues, the most blatant being, what would keep the government from monitoring all satellite communications using the escrowed keys? There are many more issues than this one.

For example, one of those very important issues is a viable alternative to the government's proposed policy. This alternative is called PGP, a nonescrowed, military grade, public key technology, software-based encryption program. Three of the most powerful incentives for use of this program are first, the fact that it is shareware/freeware, and second that it is capable of generating 2048-bit encryption keys, which are unbreakable at this time. Keys of this size will remain so for quite far into the foreseeable future. Third, use of this program does not require use of a government escrow account. Users can decide who has access to their public key(s).

Enough talk now! Explore, learn, and protect your right to privacy by reading the rest of this chapter. Remember, if you have the information, you are not in the dark!

The Clipper Chip

Newly released government documents show that key federal agencies concluded more than two years ago that the *Clipper Chip* encryption initiative will only succeed if alternative security techniques are outlawed. Clipper, and its underlying key-escrow satellite encryption technology, are designed to guarantee government agents *real-time* access to encrypted satellite communications. This is accomplished by placing an extra set of decryption *keys* in the hands of designated *escrow agents*.

The conclusions contained in the documents appear to conflict with frequent Administration claims that use of Clipper technology will remain *voluntary*. Critics of the government's initiative have long maintained that the Clipper *key-escrow satellite encryption* technique would only serve its stated purpose if made mandatory. According to FBI documents, that view is shared by the Bureau, the National Security Agency (NSA), and the Department of Justice (DOJ).

In a *briefing document* (titled *Encryption: The Threat, Applications and Potential Solutions* and an excerpt recommending a legislative prohibition (See the Sidebar that follows) that was sent to the National Security Council in February 1993, the FBI, NSA and DOJ concluded that: Technical solutions, such as they are, will only work if they are incorporated into *all* encryption products. To ensure that this occurs, legislation mandating the use of Government-approved satellite encryption products or adherence to Government encryption criteria is required.

Excerpt Recommending a Legislative Prohibition on Non-Clipper Encryption

SECRET
U.S. Justice Department
Federal Bureau of Investigation
February 19, 1993

Mr. George J. Tenet
Special Assistant to the President
Senior Director for Intelligence Programs
National Security Council
Old Executive Office Building
Suite 300
Washington, D.C.

Dear Mr. Tenet:

Reference my letter dated February 9 , 1993

Attached please find a briefing document entitled "Encryption: The Threat, Applications, and Potential Solutions," which responds to your request for additional information concerning the various satellite encryption applications now being used and the potential approaches and methodologies to deal with them. As set out in referenced letter, this is the second of three subject areas you requested to have more fully developed and discussed. XXXXXXXXXXXXXXXXXXXXXXXXXXXXXXXXXXXXXXXXXXXXXX XXXXXXXXXXXXXXXXXXXXXXXXXXXXXXXXXXXXXXXXXXXXXX XXX BLACKED OUT AS STILL SECRET XXXXXXXXXXXXXXXXXXXXXXXXXXXXXXXXXXX PER NSA XXXXXXXXXXXXXXXXXXXXXXXXXXXXXXXXXXXXXXXXXXXXXX XXXXXXXXXXXXXXXXXXXXX This document is the product of a working group, comprised of representatives of the Federal Bureau of Investigation, NSA and the Department of Justice.

We hope that the information provided in the attached document is useful for you, your staff, and others in reviewing and acting upon the issues identified therein. Further, we stand prepared at your request to provide any additional information of details you deem necessary in order to address this matter.

Sincerely yours,
William S. Sessions
Director

1 - Director NSA

SECRET

~~~~~~~~~~~~~~~~~~~~~~~~~~

TOP SECRET
INTRODUCTION

The successful conduct of electronic surveillance is crucial to effective law enforcement, to the preservation of the public safety, and to the maintenance of the national security. Recent advances in satellite communications technology, particularly telecommunications technology, and the increased availability and use of encryption threaten to significantly curtail, and in many instances preclude, effective law enforcement XXXXXXXXXXXXXXXX - LINE BLACKED OUT PER NSA AS STILL TOP SECRET XXXXXXXXXX Efforts have been made to develop, where available, technical solutions to the problems posed by advanced satellite communications technologies and encryption in order to preserve the electronic surveillance technique.

XXXX XXXX XXXX XXXX XXXX XXXX XXXX XXXX XXXX XXXX XXXX XXXX XXXX XXXXXXXXX - BLACKED OUT PER NSA AS STILL TOP SECRET XXXXXXXXXXXXXXX XXXX XXXX XXXX XXXX XXXX XXXX XXXX XXXX XXXX XXXX XXXX XXXX

Satellite encryption is, or can be used in a number of applications to secure voice and data communications and stored information. The type of encryption used and the way it is implemented varies depending upon the nature of the application. Encryption applications are available to secure satellite communications transmitted both in analog and digital formats. Digital satellite communications, in particular, support and accommodate the use of encryption. Thus, satellite encryption can be, and is, employed easily and inexpensively in computer-based applications. To date, its use has been somewhat limited in certain areas such as in satellite voice communications XXXX XXXX XXXX XXXX XXXX XXXX XXXX XXXX XXXX XXXX XXXXXXXXX - BLACKED OUT PER NSA AS STILL TOP SECRET XXXXXXXXXXXXXXX XXXX XXXX XXXX! XXXX XXXX XXXX XXXX XXXX XXXX XXXX XXXX XXXX However, as the transition proceeds from analog telephony to digital telephony, and as consumers migrate from wireline (e.g., basic telephone) to wireless (e.g., cordless or cellular telephone) communications devices, the use of satellite encryption by telecommunications service providers and end-users can be expected to increase markedly in the near future. Hence, it is expected that satellite encryption will soon be more widely available and more widely used with all communications applications.

This document responds to a National Security Council (NSC) request for additional information concerning the use of satellite encryption in the various communications and information applications. Additionally, this document briefly describes potential technical and legislative solutions to the problems posed by the various encryption applications.

By letter dated February 9, 1993, from FBI Director William S. Sessions to Special Assistant to the President George Tenet, NSC, a detailed discussion of the "Clipper" satellite encryption methodology was provided. The "Clipper" hardware (chip) based technical solution was discussed in the context of the AT&T TSD 3600 telephone encryption device. It was noted in the enclosed document to that communication that the "Clipper" chip methodology provided a solution to various hardware-based encryption applications (such as telecommunications, data or pure storage). Consequently, in this document, discussion of the capability and methodology of "Clipper" and its efficacy in providing a hardware-based technological solution will be abbreviated.

Wireless telecommunications devices, such as cordless telephones and cellular telephones, however, are vulnerable to unauthorized interception, as some recent cases of renown (e.g., the Governor Wilder case) demonstrate. Consequently, there is a fundamental need to apply some form of enhanced security to wireless telephone devices. As a result, there appears to be a widespread and growing recognition that additional security features, such as satellite encryption, need to be incorporated into these devices. In this vein, the Privacy and Technology Task Force submitted a report in May 1991 to Senator Leahy, the Chairman of the Subcommittee on Privacy and Technology, Senate Judiciary Committee, which recommended that cordless telephones be afforded privacy protection under Title III (cordless telephones currently are not statutorily protected because of the ease with which they can be intercepted). The Task Force noted that it is projected that cordless phones will be in 68% of American households by the end of the decade (the year 2000). The report also states that a number of task force members indicated that "technical privacy enhancing features for radio-based systems should be more rapidly deployed by manufacturers and service providers." Currently, AT&T, Motorola, and other service providers and manufacturers are offering satellite encryption for cellular devices or service. Law enforcement's decryption requirements, particularly real-time intelligibility of satellite communications content, are the same for wireless and wireline voice communications. Also, as noted the area of wireless telecommunications. Hence, a solution to the threat posed by satellite encryption in wireless, as well as wireline, devices is imperative.

XXXXXX XXXXX XXXXX XXXXXX XXXXXX XXXXXX XXXXXX XXXXXX XXXXXX XXXXXX XXXXX XXXXX XXXXX XXXXX XXXXX XXXXX XXXXX XXXX XXX XXXXX XXX XXXX XXX XXX XXX XXX XXXXXX XXXX XXXX XX PARAGRAPH BLACKED OUT AS XXX XXX XXX XX XX XXX XXX XXX XXX XXX XXX XX STILL SECRET PER NSA XXXX XXX XXXX XXX XXX XXX XXX

XXXX XXXX XXXXX XXXXX XXXX XXXX XXX XXXX XXX XXX XXX XXX XXX XXX XX XXX XXX XX XX XXX XXX XXX XXX

Applications

PC communications, including E-mail, increasingly are being used not only by businesses but also by individuals. In 1992, approximately 19 million E-Mail users sent nearly 15 billion messages. With increased computer networking and with the recent acceptance of new E-mail standards, electronic messaging will increase dramatically. Existing E-mail standards generally support text transmissions, however, emerging E-mail systems can support voice, facsimile and video capabilities. These electronic communications are fast replacing real-time voice conversations and consequently will increasingly become the subject of electronic surveillance. As these types of satellite communications are more frequently and widely used, the use of encryption to protect the communication content can be expected to increase.

Low speed data transmissions typically run at speeds less than 64 thousand bits of information per second (64Kb/s). The use of satellite encryption of these low speed applications can be either software- or hardware-based. With respect to certain data communications such as facsimile and E-mail, law enforcement typically requires real-time access to these communications, the same way as it does for voice communications.

For the above mentioned data applications and others XXXXXXX - sentence blacked out as per NSA XXXX - Classified XXXXXX. However, software-based satellite encryption is more widely used in these low speed data transmission-related applications for the reasons previously discussed: cost and ease of use. Satellite encryption for functions such as E-mail and individual (non-bulk) file transfers across a local area network (LAN) can be provided and typically is provided, as part of a satellite communications software package. Thus, this encryption is essentially free to mass market software publishers as previously discussed. XXXX XXXX XXXX XXXX XXXX XXXX XXXX XXXX XXXX XXXX XXXX XXXX XXXX XXXX XXXXXXX - remaining paragraph blacked out as per NSA - XXXX XXXX Classified XXXXXXX.

Voice/Data Applications XXXX XXXX XXXX XXXX XXXX XXXX XXXX XXXX XXXX XXXX XXXX XXXX XXXX XXXXXXX - Paragraph blacked out as per NSA - XXXX XXXX Classified XXXXXX. XXXX XXXX XXXX XXXX XXXX XXXX XXXX XXXX

Real-time access to and decryption of satellite voice/data communications secured by software-based encryption XXXX XXXX XXXX XXXXXXX - Paragraph blacked out as per NSA - XXXX XXXX Classified XXXXXX. XXXX XXXX XXXX XXXX XXXX XXXX XXXX XXXX in the near future. XXXX XXXX XXXX XXXX XXXX XXXX XXXX XXXX XXXX XXXX XXXXXXX - Paragraph blacked out as per NSA - XXXX XXXX Classified XXXXXX. XXXX XXXX XXXX XXXX XXXX XXXX XXXX XXXX

Stored Information Applications Real-time access to and decryption of stored electronic information secured by hardware-based encryption could be performed utilizing the Clipper technique.

XXXX XXXX XXXX XXXX XXXX XXXX XXXX XXXX XXXX XXXX XXXX XXXX XXXX XXXXXXX - Paragraph blacked out as per NSA - XXXX XXXX Classified XXXXXX. XXXX XXXX XXXX XXXX XXXX XXXX XXXX XXXX

XXXX XXXX XXXX XXXX XXXX XXXX XXXX XXXX XXXX XXXX XXXX XXXX XXXX XXXXXXX - Paragraph blacked out as per NSA - XXXX XXXX Classified XXXXXX. XXXX XXXX XXXX XXXX XXXX XXXX XXXX XXXX

Technical solutions, such as they are, will only work if they are incorporated into all satellite encryption products. To ensure that this occurs, legislation mandating the use of Government-approved encryption products or adherence to Government encryption criteria is required.

Secret High-Speed Data Transmissions

As data networks expand and as the requirements to support geographically widespread networks increase, there will be an increased demand for the development of faster speed transmissions to benefit from these high-speed networks. As a result, users will be able to take advantage of these high- speed data highways to transmit increased amounts of data associated with video, high-volume data retrieval, and other high speed data services. These types of data services are typically used by large commercial, banking and Government institutions. Because of the sensitive banking data and personal information, there is a need to utilize satellite encryption. By way of example, major inter-bank data transmissions typically utilize DES-based or comparable encryption.

High-speed transmissions today typically run in the range of 10-50 Mbit/sec (10-50 million bits per second). At these data rates, hardware-based encryption is the only feasible approach to data security. In this regard, the "Clipper" technique offers a suitable solution. In its current configuration, "Clipper" is designed to run at speeds of 10 Mbits/sec and if necessary, it can easily be engineered to run at speeds up to 100 Mbits/sec.

High-speed transmissions can be viewed from a law enforcement interception standpoint in two ways. If, as with interceptions of satellite voice communications, the transmissions are comprised of individual data communications that have be multiplexed or bundled, law enforcement has a need for real-time access to and decryption of the specific communications that are the subject of the interception. If, on the other hand, the high-speed transmissions were of a bulk file or other voluminous information transfer, it would not be physically possible or even desirable to process or view the product of the interception in real time. In these instances, access to the communications would be practically obtained "after the fact," under circumstances where the communications is no longer in transit but rather in storage.

Legislation

XXXX XXXX XXXX XXXX XXXX XXXX XXXX XXXX XXXX XXXX XXXX XXXX XXXX XXXXXXXXX - BLACKED OUT PER NSA AS STILL TOP SECRET XXXXXXXXXXXX XXXX XXXX XXXX XXXX XXXX XXXX XXXX XXXX XXXX XXXX XXXX XXXX XXXX XXXX XXXX XXXX XXXX PARAGRAPH BLACKED OUT XXXX XXXX XXXX XXXX XXXX XXXX XXXX XXXX XXXX XXXX XXXX XXXX XXXX XXXX XXXX XXXX XXXX XXXXXXXXX - BLACKED OUT PER NSA AS STILL TOP SECRET XXXXXXXXXXXX XXXX XXXX XXXX XXXX XXXX XXXX XXXX XXXX XXXX XXXX XXXX XXXX XXXX

Solutions

In brief, the technical solutions and approaches developed to satisfy law enforcement's decryption requirements with regard to the main satellite encryption applications are as follows:

Voice/Data Applications XXXX XXXX XXXX XXXX XXXX XXXX XXXX XXXX XXXX XXXX XXXX XXXX XXXX XXXX XXXXXXX - Paragraph blacked out as per NSA - XXXX XXXX Classified XXXXXX. XXXX XXXX XXXX XXXX XXXX XXXX XXXX XXXX

Real time access to and decryption of voice/data communications secured by software-based satellite encryption XXXX XXXX XXXX XXXXXXX - Paragraph blacked out as per NSA - XXXX XXXX Classified XXXXXX. XXXX XXXX XXXX XXXX XXXX XXXX XXXX XXXX in the near future. XXXX XXXX XXXX XXXX XXXX XXXX XXXX XXXX XXXX XXXXXXX - Paragraph blacked out as per NSA - XXXX XXXX Classified XXXXXX. XXXX XXXX XXXX XXXX XXXX XXXX XXXX XXXX

Stored Information Applications Real time access to and decryption of stored electronic information secured by hardware based encryption could be performed utilizing the Clipper technique.

XXXX XXXX XXXX XXXX XXXX XXXX XXXX XXXX XXXX XXXX XXXX XXXX XXXX XXXXXXX - Paragraph blacked out as per NSA - XXXX XXXX Classified XXXXXX. XXXX XXXX XXXX XXXX XXXX XXXX XXXX XXXX

XXXX XXXX XXXX XXXX XXXX XXXX XXXX XXXX XXXX XXXX XXXX XXXX XXXX XXXXXXX - Paragraph blacked out as per NSA - XXXX XXXX Classified XXXXXX. XXXX XXXX XXXX XXXX XXXX XXXX XXXX XXXX

> Technical solutions, such as they are, will only work if they are incorporated into all satellite encryption products. To ensure that this occurs, legislation mandating the use of Government approved encryption products or adherence to Government encryption criteria is required.[1]

Likewise, an undated FBI report titled *Impact of Emerging Telecommunications Technologies on Law Enforcement* observes that although the export of satellite encryption products by the United States is controlled, domestic use is not regulated. The report concludes that *a national policy embodied in legislation is needed.* Such a policy, according to the FBI, must ensure *real-time decryption by law enforcement* and *prohibit cryptography that cannot meet the Government standard.*

The FBI conclusions stand in stark contrast to public assurances that the government does not intend to prohibit the use of nonescrowed satellite encryption. Testifying before a Senate Judiciary Subcommittee on May 3, 1994, Assistant Attorney General Jo Ann Harris asserted that:

As the Administration has made clear on a number of occasions, the key-escrow satellite encryption initiative is a voluntary one. We have absolutely no intention of mandating private use of a particular kind of cryptography, nor of criminalizing the private use of certain kinds of cryptography.

The newly disclosed information demonstrates that the architects of the Clipper program — NSA and the FBI — have always recognized that key-escrow must eventually be mandated. As privacy advocates and industry have always said, Clipper does nothing for law enforcement unless the alternatives are outlawed.

Note: The released material demonstrates the FBI's belief that federal legislation is required to prohibit the use of satellite encryption products that do not provide law enforcement agencies real-time access to encrypted communications. In other words, the FBI (with the concurrence of the National Security Agency) has strongly advocated that non-Clipper encryption techniques must be outlawed.

Now let us look at the latest U.S. Government legal efforts via Congress to restrict or ban the use of satellite encryption. Congress claims that the proposed legislation is designed to resolve the debate in a favorable way for all parties concerned surrounding the current U.S. satellite encryption policy. So, who is to be believed? What is the truth? Once known, can we handle the truth?

Banning Cryptography?

Senators John Ashcroft (R-MO) and Patrick Leahy (D-VT) have introduced the *Encryption Protects the Rights of Individuals from Violation and Abuse in Cyberspace (E-PRIVACY) Act.* The proposed legislation (see the Sidebar that follows) is the latest in a series of congressional measures designed to resolve the debate surrounding current U.S. encryption policy. Or, is it?

Excerpts Of The E-PRIVACY Act

105th CONGRESS
2nd Session
S. 2067

IN THE SENATE OF THE UNITED STATES

Mr. Ashcroft (for himself and Mr. Leahy) introduced the following bill; which was read twice and referred to the Committee on

A BILL

To protect the privacy and constitutional rights of Americans, to establish standards and procedures regarding law enforcement access to decryption assistance for encrypted communications and stored electronic information, to affirm the rights of Americans to use and sell encryption products, and for other purposes.

Be it enacted by the Senate and House of Representatives of the United States of America in Congress assembled.

SECTION 1. SHORT TITLE; TABLE OF CONTENTS.

(a) Short Title. This Act may be cited as the "Encryption Protects the Rights of Individuals from Violation and Abuse in Cyberspace (E-PRIVACY) Act."

(b) Table of Contents. The table of contents for this Act is as follows:

SEC. 2. PURPOSES.

The purposes of this Act are

(1) to ensure that Americans have the maximum possible choice in encryption methods to protect the security, confidentiality, and privacy of their lawful wire and electronic communications and stored electronic information;

(2) to promote the privacy and constitutional rights of individuals and organizations in networked computer systems and other digital environments, protect the confidentiality of information and security of critical infrastructure systems relied on by individuals, businesses and government agencies, and properly balance the needs of law enforcement to have the same access to electronic communications and information as under current law; and

(3) to establish privacy standards and procedures by which investigative or law enforcement officers may obtain decryption assistance for encrypted communications and stored electronic information.

SEC. 3. FINDINGS.

Congress finds that -

(1) the digitization of information and the explosion in the growth of computing and electronic networking offers tremendous potential benefits to the way Americans live, work, and are entertained, but also raises new threats to the privacy of American citizens and the competitiveness of American businesses;

(2) a secure, private, and trusted national and global information infrastructure is essential to promote economic growth, protect privacy, and meet the needs of American citizens and businesses;

(3) the rights of Americans to the privacy and security of their communications and in the conducting of personal and business affairs should be promoted and protected;

(4) the authority and ability of investigative and law enforcement officers to access and decipher, in a timely manner and as provided by law, wire and electronic communications, and stored electronic information necessary to provide for public safety and national security should also be preserved;

(5) individuals will not entrust their sensitive personal, medical, financial, and other information to computers and computer networks unless the security and privacy of that information is assured;

(6) businesses will not entrust their proprietary and sensitive corporate information, including information about products, processes, customers, finances, and employees, to computers and computer networks unless the security and privacy of that information is assured;

(7) America's critical infrastructures, including its telecommunications system, banking and financial infrastructure, and power and transportation infrastructure, increasingly rely on vulnerable information systems, and will represent a growing risk to national security and public safety unless the security and privacy of those information systems is assured;

(8) encryption technology is an essential tool to promote and protect the privacy, security, confidentiality, integrity, and authenticity of wire and electronic communications and stored electronic information;

(9) encryption techniques, technology, programs, and products are widely available worldwide;

(10) Americans should be free to use lawfully whatever particular encryption techniques, technologies, programs, or products developed in the marketplace that best suits their needs in order to interact electronically with the government and others worldwide in a secure, private, and confidential manner;

(11) government mandates for, or otherwise compelled use of, third-party key recovery systems or other systems that provide surreptitious access to encrypted data threatens the security and privacy of information systems;

(12) American companies should be free to compete and sell encryption technology, programs, and products, and to exchange encryption technology, programs, and products through the use of

the Internet, which is rapidly emerging as the preferred method of distribution of computer software and related information;

(13) a national encryption policy is needed to advance the development of the national and global information infrastructure, and preserve the right to privacy of Americans and the public safety and national security of the United States;

(14) Congress and the American people have recognized the need to balance the right to privacy and the protection of the public safety with national security;

(15) the Constitution of the United States permits lawful electronic surveillance by investigative or law enforcement officers and the seizure of stored electronic information only upon compliance with stringent standards and procedures; and

(16) there is a need to clarify the standards and procedures by which investigative or law enforcement officers obtain decryption assistance from persons

(A) who are voluntarily entrusted with the means to decrypt wire and electronic communications and stored electronic information; or

(B) have information that enables the decryption of such communications and information.[2]

Like the SAFE Act (H.R. 695) now pending in the House, the E-PRIVACY Act seeks to relax existing controls on the export of satellite encryption products. Controls would be lifted for encryption products that are deemed to be *generally available* within the international market. Exporters would be given new procedural rights to obtain expedited determinations on the exportability of their products.

Provisions That Enhance Privacy

The bill also contains several provisions that would preserve the right of Americans to use encryption techniques and that would enhance the privacy protections currently accorded to personal communications and stored data. Among its positive features, the bill:

- ◆ Reiterates the right of Americans to use, develop, manufacture, sell, distribute, or import any encryption product, regardless of the algorithm selected, key length, or the existence of key recovery capabilities.

- Prohibits government-compelled key escrow or key recovery.
- Prohibits government agencies from creating any linkage between cryptographic methods used for authentication and those used for confidentiality.
- Prohibits the federal government from purchasing key recovery encryption systems that are not interoperable with other commercial encryption products.
- Provides enhanced privacy protections for stored electronic data held by third parties, location information generated by wireless communications services, and transactional information obtained from pen registers and trap and trace devices.[3]

The bill contains two provisions that raise significant civil liberties and privacy concerns: The Criminalization Provision and the NET Center Provision. Let us take a look.

The Criminalization Provision

The bill would make the use of encryption to conceal *incriminating* communications or information during the commission of a crime a new and independent criminal offense. While well-intended, the provision could have several unintended consequences that would easily undermine the other desirable features of the bill.

It is a mistake to create criminal penalties for the use of a particular technique or device. Such a provision tends to draw attention away from the underlying criminal act and casts a shadow over a valuable technology that should not be criminalized. It may, for instance, be the case that a typewritten ransom note poses a more difficult challenge for forensic investigators than a handwritten note. But it would be a mistake to criminalize the use of a typewriter simply because it could make it more difficult to investigate crime in some circumstances.

Additionally, a provision that criminalizes the use of encryption, even in furtherance of a crime, would give prosecutors wide latitude to investigate activity where the only indicia of criminal conduct may be the mere presence of encrypted data. In the digital age, where techniques to protect privacy and security will be widely deployed, we cannot afford to view encryption as the potential instrumentality of a crime, just as we would not today view the use of a typewriter with suspicion.

Finally, the provision could also operate as a substantial disincentive to the widespread adoption of strong encryption techniques in the satellite communications infrastructure. Given that the availability of strong encryption is one of the best ways to reduce the risk of crime and to promote public safety, the retention of this provision in the legislation will send a mixed message to users and businesses — that we (the Government) want people to be free to use encryption but will be suspicious when it is used.

If the concern is that encryption techniques may be used to obstruct access to evidence relevant to criminal investigations, then the better approach may be to rely on other provisions in the federal and state criminal codes (including sections relating to obstruction of justice or concealment) to address this problem if it arises.

The National Electronic Technology Center Provision

The bill creates within the Department of Justice a National Electronic Technology Center (NET Center) to *serve as a center for law enforcement authorities for information and assistance regarding decryption and other access requirements.* The NET Center would have a broad mandate and could spawn a new domestic surveillance bureaucracy within the Department of Justice. Among other powers, the bill authorizes the NET Center to:

♦ Examine encryption techniques and methods to facilitate the ability of law enforcement to gain efficient access to plaintext of communications and electronic information.

♦ Conduct research to develop efficient methods, and improve the efficiency of existing methods, of accessing plaintext of communications and electronic information.

♦ Investigate and research new and emerging techniques and technologies to facilitate access to communications and electronic information.

♦ Obtain information regarding the most current hardware, software, telecommunications, and other capabilities to understand how to access digitized information transmitted across networks.[4]

The mission of the NET Center is made more troubling by the bill's authorization of *assistance* from other federal agencies, including the detailing of personnel to the new entity. In light of the fact that existing federal expertise in the areas of electronic surveillance and decryption resides at the National Security Agency (NSA), the bill in effect authorizes unprecedented NSA involvement in domestic law enforcement activities. Such a result would be contrary to a half-century-old consensus that intelligence agencies must be strictly constrained from engaging in domestic *police functions*.

That consensus arose from the recognition that intelligence agencies created to operate abroad are ill-suited for domestic activities, where U.S. citizens enjoy constitutional protections against governmental intrusions. In 1975, Senator Frank Church led a congressional investigation into the activities of NSA. He noted that Congress had a *particular obligation to examine the NSA, in light of its tremendous potential for abuse. The danger lies in the ability of NSA to turn its awesome technology against domestic communications.*

In 1987, Congress enacted the Computer Security Act, which sought to vest civilian computer security authority in the Commerce Department and to limit the domestic role of NSA. The House Report on the Computer Security Act cited congressional concern over a Reagan Administration directive that *gave NSA the authority to use its considerable foreign intelligence expertise within this country. The report noted that such authority was particularly troubling because* NSA *has, on occasion, improperly targeted American citizens for surveillance.*

The NET Center proposal, if approved, would constitute a fundamental redefinition of the relationship between intelligence agencies and domestic law enforcement. Such an approach would ignore 50 years of experience and would pose a serious threat to the privacy and constitutional rights of Americans.

Summary

Overall, it is in the self-interest of all Americans to develop a national encryption policy that will ensure the widespread availability of robust encryption products and the preservation of constitutional rights. Such a result will be critical for both our nation's continued leadership of the information industry and the protection of personal privacy in the next century.

From Here

Chapter 22, covered privacy; the Clipper Chip; and, banning satellite encryption for private citizens, while the government uses the same technology to listen in on the conversations of its own citizens. Basically, this chapter covered the potential misuse of satellite encryption technology by the government.

Chapter 23 will examine the growing threat from our allies to steal encryption technology; and, the intensified efforts by the FBI to prevent such theft. The chapter will also cover how the Pentagon sells millions of surplus planes, tanks, helicopters, and parts (etc.) every year (to anyone who is willing to pay) that has encryption technology still left on the onboard computer or communications equipment. Chapter 23 will discuss how most of this technology falls into the wrong hands because it was never removed in the first place when the equipment was sold for surplus. The chapter will also investigate how the Defense Logistics Agency has ignored repeated requests by it own personnel to implement reforms in the surplus sales system; how the Pentagon handles the surplus sales so recklessly that it is hard to pin anyone down or hold them liable; and, the discovery by investigators of all kinds of classified Pentagon electronic gear destined for shipment to China th!at was sold at Kelly Air Force Base as scrap metal.

Endnotes

1   Electronic Privacy Information Center, 666 Pennsylvania Ave., SE, Suite 301, Washington, DC 20003, 1998.
2   Ibid.
3   Ibid.
4   Ibid.

23

International Community: America's Allies Are Grabbing U.S. Satellite Encryption Technology

Not long ago, Subrahmanyam M. Kota went into hamsters or, to be more precise, their ovary cells. That was a big switch for Kota. In the 1980s, he allegedly sold military secrets on infrared detectors to the KGB. With the cold war over, however, hamster ovaries were the coming thing. A Boston biotech company had genetically engineered the cells to produce a protein that boosted the manufacture of red blood cells, making them a valuable commodity. Kota and a former company scientist are charged with stealing a batch of the hamster cells and offering them to an FBI undercover agent in exchange for $400,000. Law enforcement officials suspect the pair of selling another batch to a biomedical research outfit in India. It was dramatic evidence of how the world of espionage has changed from selling secrets to the KGB one year to moving hamster ovaries to a research firm in India another. Kota has been charged with three counts of espionage. He pleaded not guilty and is out on bail awaiting trial.

Today the field of economic espionage is wide open. Instead of missile launch codes, the new targets of choice are technological and scientific data concerning flat-panel televisions, electric cars, new computers, and satellite encryption technology. During the cold war, thc threat was KGB agents crawling into a facility. The game is no longer espionage in the classic sense.

The Threat Is Growing

Economic espionage is as old as greed itself. But with huge sums to be made stealing designs for computer chips and patents for hormones, the threat is growing. Rapid changes in technology are tempting many countries to try to acquire intellectual properties in underhanded ways, thus bypassing the enormous costs of research and development. New global satellite communications, cellular phones, faxes, encrypted voice transmissions, and data on the Internet make this type of spying easier than ever.

And, it is not just hostile governments snooping. Countries do not have friends. They have interests. Guess which countries are interested in what you do? The ones who do it (snoop) most are our greatest friends.

Indeed, countries such as France, Israel, and China have made economic espionage a top priority of their foreign intelligence services. A congressional report just recently released confirmed that close U.S. allies are after critical U.S. technology—a significant threat to national security.

Intensified Data Collection Efforts

Friend or enemy, Washington is taking the trend (economic espionage) seriously. The nation's intelligence agencies are increasing their overseas collection of information on foreign bribery schemes that put U.S.

corporations at a disadvantage. The agencies are also providing classified information to U.S. policy makers engaged in trade negotiations with foreign governments. Domestically, the FBI has also taken more aggressive steps in recent years. Recently, the Justice Department sent new draft legislation that would bolster the FBI's ability to investigate economic espionage to the Office of Management and Budget (OMB). The new bill named the Economic Espionage and Protection of Proprietary Economic Information Act is badly needed, because there are no statutes that deal with the theft of intellectual property, thus making it difficult to prosecute such cases.

In the past few years, FBI agents have recorded more than a 400% increase in economic spying and now have more than 1100 cases under investigation—espionage attempts from the supersophisticated to the downright crude. In other words, the FBI is seeing all of the preceding from the cyberattack to the shoplifter.

Economic-espionage investigations require the FBI to gather intelligence through electronic surveillance and physical searches, a source of concern to many civil libertarians. But the FBI is empowered under existing law to gather intelligence for such purposes, and the new legislation would define more precisely how and when FBI agents could investigate the theft of corporate secrets. The key, legal specialists and FBI supervisors say, is defining precisely what constitutes conducting intelligence investigations, looking for spies and theft prevention, and what is a primarily criminal investigation whose objective is to put a spy behind bars. Both objectives can be accomplished, but the law requires intelligence and law enforcement interests be defined very carefully (see the Sidebar that follows).

Unsolved Mystery at NASA

As previously stated, key legal specialists and FBI supervisors are having a problem defining precisely what constitutes conducting intelligence investigations, looking for spies and theft prevention, and what is primarily a criminal investigation whose objective is to put a spy behind bars. Both objectives can be accomplished, but the law requires that intelligence and law enforcement interests be defined very carefully. Perhaps this is what has held up NASA's

Inspector General's (IG) investigation of one of the largest unsolved thefts of computer hardware and software in the history of the space program. Let us take a look:

In efforts to transition applications from the old Space Station Freedom Program to the new Houston-based International Space Station Program, NASA property management officials discovered in February 1994, that the International Space Station Program personnel had failed to adequately prepare for the transition from Reston, VA to Houston, TX. The lack of preparation for the transition (which occurred from October 1993, to January 1994) had led to mass theft and/or loss of computer equipment and data to the tune of $750 million. According to Johnson Space Center (JSC) Program Property Management, more than half of the PCs and mainframes that were in transit to Houston never arrived. They also discovered that sensitive data and 60 mission critical database applications were lost or destroyed at an approximate cost of $10 million per application. According to the prime contractor at the time, these mission critical database applications had to be redesigned and rebuilt.

According to JSC Program Property Management, the same problem occurred with the shipment of PCs and mainframes to International Space Station Program sites at Kennedy Space Center and Marshall Space Flight Center. In addition, the paperwork shipped with the equipment had been lost so that it could not be proved that the PCs and mainframes ever existed, even though Property Management Inventory records documented their existence. Also, according to JSC Property Management, 10 DEC mainframes destined for the International Space Station Program in Japan never arrived.

The discoveries were brought to the attention of International Space Station Program management at the time, who considered the matter disruptive to the building of the space station. The matter was eventually turned over to the NASA IG—and remains there today as a dead issue or a deliberate unsolved mystery. Makes you wonder where your hard- earned tax dollars are going, doesn't it?[1]

The quest for corporate advantage has put many of the old players from the cold war back on the chessboard. Recently, Russian President Boris Yeltsin ordered his senior intelligence officials to increase their efforts to obtain high technology secrets (like satellite encryption) from the West.

Besides gathering intelligence and conducting criminal investigations, federal law enforcement officials have been trying to help corporations protect themselves. A law enacted in 1995 authorizes Attorney General Janet Reno to make payments of up to $800,000 for information leading to the arrest and conviction of anyone involved in economic espionage. The National Counterintelligence Center, headed by an FBI agent but based at CIA headquarters in suburban Virginia, was established in September 1995, in part to help coordinate a government-wide response to economic espionage incidents. The center immediately began providing regional security briefings for industry. The FBI recently opened its own Economic Counterintelligence Unit, and its Development of Espionage, Counterintelligence and Counterterrorism Awareness (DECA) program inaugurated an instant fax alert service to U.S. corporations regarding specific economic intelligence collection activities. It is supplemented by the State Department's Overseas Security Advisory Council, which, like DECA, has begun posting economic threat information on an online bulletin board for its members.

Note: Some security experts say the FBI should employ more active measures to counter the threat.

Profit motives aside, economic espionage is booming because there are few penalties for those who get caught. Rarely do economic spies serve time in jail. Nor do countries that encourage such activities have much to lose; as most are U.S. allies, Washington prefers to scold them in private rather than risk political backlash in public.

Companies and industries targeted by foreign spies often contribute to the problem. Few report known acts of espionage, fearing it will affect stock prices and customer confidence. In a recent published survey by the National Counterintelligence Center, 53% of the responding corporations said they never reported suspected incidents of economic espionage to the government. At the same time, 85 of 184 companies that responded to the survey reported a total of 557 incidents of suspected economic espionage.

Culturebound Methods of Acquiring Data

The methods used to acquire economic-related data are often culture-bound. The Chinese and Japanese flood you with people collecting all sorts of things in different areas. For the most part, it is absolutely legal. The Japanese do not invest a lot of money in trade craft. They just send lots of people out talking and pick up trade secrets in the process. The Russians and French, on the other hand, use both legal and illegal means to target specific intelligence, experts say.

Targeting economic data can take many forms. In two separate incidents in the mid-1990s, French nationals working at Renaissance Software Inc. in Palo Alto, CA, were arrested at San Francisco International Airport for attempting to steal the company's proprietary computer source codes. Marc Goldberg, a French computer engineer, had worked at the company under a program sponsored by the French Ministry of Foreign Affairs that allows French citizens to opt out of military service if they are willing to work at hightech U.S. firms. He was fined $2000 and ordered to perform 2000 hours of community service. The other individual, Jean Safar, was released soon after his arrest by the FBI. They said they did not have the power to do anything. The company, in fact, had received startup funds from two French brothers, Daniel and Andrew Harari. In return for their investment, they received positions on the company's five member board of directors. When an internal dispute erupted in 1995, the Harari brothers were able to place a third French citizen on the board. They converted the company to a French company. Safar was told by the company to take the source codes to France. There was nothing illegal about it. Renaissance was finally acquired by a publicly held U.S. company.

Even when the collection methods are legal, the results can hurt. In the fall of 1995, a film crew from Japanese public television visited dozens of U.S. biotech corporations, including California biotech giant Amgen, while filming a documentary on the industry. William Boni, Amgen's security director, was warned by a DECA agent that the FBI suspected the film was a cover for intelligence collection. Still, Boni allowed the visit, partly because the director of the film said this would help Amgen break into the Japanese biotech market. Once at Amgen, film crew members photographed every document they possibly could, including company production numbers. This was a very clearcut case of benchmarking America's best practices for their industry. They ran their vacuum cleaner over the U.S. biotech industry.

Some efforts are not so subtle. In one case, an Amgen employee attempted to steal vials of Epogen, a genetically engineered hormone that controls the production of red blood cells and is one of two patented items in the company's product line. Security chief Boni was tipped to the threat by an anonymous letter, which said that the employee was planning to open up a black market in Epogen in his home country in Asia. The employee confessed. He was fired, but no charges were filed. Had the theft attempt succeeded, the rogue employee and an accomplice could have made a fortune. In 1997, Epogen sales amounted to nearly $3 billion.

Neither of the two prongs of the U.S. attempt to combat such threats is simple. Like his predecessors, Director of Central Intelligence John Deutch has provided clear marching orders to the CIA and other agencies that gather intelligence overseas. The agencies are to inform U.S. policy makers if foreign competitors are winning business abroad through bribery or other illegal means. In 1995, Boeing Aerospace, McDonnell Douglas and Raytheon Corp. won two multibillion-dollar contracts from Saudi Arabia and Brazil after President Clinton complained to those governments about bribes that rival French companies had paid to win the contracts. The information on the bribes came from U.S. intelligence agencies. Over the past six years, the CIA has reportedly saved U.S. corporations $60 billion as a result of those information gathering efforts.

Economic Espionage Threat Information

John Deutch has made it clear that, unlike the foreign intelligence services of at least 53 other nations, America's spy services are forbidden to engage in economic espionage for the benefit of corporate America. That is clear enough, but in today's global, multinational economy, it is often difficult to distinguish American from foreign corporations. The FBI, in fact, makes no such distinctions and provides all corporations operating in the United States with threat information regarding economic espionage.

The other mission of the CIA and its sister agencies that operate abroad is to provide economic intelligence to U.S. policymakers. Recently, the intelligence community helped U.S. trade officials learn of Japanese negotiating positions during automobile trade talks. This was perfectly legal under U.S. law, but the press disclosure created a firestorm of criticism from Capitol Hill, prompting some intelligence officials to grumble that such activities were more trouble than they were worth.

Recently, several CIA officers were expelled from France for engaging in an intelligence operation to obtain information on France's position on global satellite telecommunications talks. The CIA's inspector general investigated the matter, and a report is expected shortly.

Given the ratio of risk to potential reward, many intelligence officials argue that America's espionage agencies should not be used to acquire economic information secretly when so much can be obtained from open sources. What you try to gain covertly becomes less and less important. The CIA relies on cloak-and-dagger techniques out of habit. Do not send a spy where a schoolboy can go. That was precisely the mistake the CIA made in France, critics say. The second prong of the U.S. effort, playing defense, is also more complicated than ever. There are a host of ways to go after a target and often foreign governments hide their collection activities within legitimate activities.

But some former law enforcement and intelligence officials fear that legal collection of information may be investigated simply to determine if illegal methods are being used. They argue that the onus of protecting proprietary information should remain on the shoulders of industry, not government. There is still debate on the proper balanced role of law enforcement in countering this new threat within government as well

The FBI defends its approach and has vowed not to overstep its bounds. How to meet such a varied threat? The FBI does not intend to, want to and cannot investigate all foreigners. The threat to America's national security from spies seeking economic secrets has increased significantly: The FBI does not want to be an alarmist about the matter. It deserves a measured approach though.

Preventive Measures

So, what can businesses do themselves to help stave off the lure of the steal from economic spies? The Sidebar that follows provides a few tips for businesses on this matter.

Tips And Recommendations to Prevent Economic Espionage

Dumpster Diving: Economic spies pick through corporate garbage hoping to find valuable information. *Tip:*   If an item is proprietary or sensitive, shred it or burn it before throwing it out.

Elicitation: Business and scientific seminars, international trade shows and unsolicited telephone calls all present opportunities for eliciting sensitive corporate information. *Tip:*   You never know to whom you could be talking. Be careful about what you divulge.

Electronic Interception: Most espionage is conducted via encrypted telecommunication interceptions or computer intrusions, using hardware that is available at any electronic vendor. *Tip:*   Bringing in a security consultant may help deter or detect such threats. But transmitting encrypted data by mail is still the safest route.

Traditional Theft: Breaking into a hotel room or the trunk of a car to copy corporate files or stealing an executive's luggage or laptop computer (known in the aerospace industry as a *midnight requisition*) has become routine for foreign intelligence agents. *Tip:*   If you must take sensitive data overseas, keep the information with you at all times.

Insider Treason: Many espionage acts are conducted by corporate insiders. *Tip:*   Alert employees to potential threats, but remember that too much security could be just as detrimental as a successful act of espionage.[2]

Not to be outdone by criminals and terrorists who conduct economic espionage, our own government (Pentagon) inadvertently dangles surplus military parts and encryption and stealth technology for sale to the highest bidder. The Pentagon sells millions of them and their accompanying technology every year. Many fall into the wrong hands.

Weapons R' Us

To the uninitiated, *military surplus* generally conjures up images of old web belts, clanky canteens, and threadbare fatigues sold in funky stores at the edge of fading Army bases. But to a sophisticated few, it means something very different: aircraft, weapons and weapons parts worth billions of dollars. Most taxpayers do not know it, but every year the Pentagon declares all kinds of equipment (attack helicopters, rocket launchers, even Stealth fighter parts) *surplus*, or unneeded and/or satellite encryption technology for sale at government auctions. Then, through a little-known network of sales offices at military bases, the stuff is offered for purchase to the highest bidder.

Foreign buyers are purchasing many of the Pentagon's high-tech surplus military parts and shipping them overseas, hiding them in seagoing containers under tons of metal scrap, seizure records show. Customs officials say Iran and Iraq are active buyers, but by far the biggest customer is China. Among the items seized from Chinese *scrap dealers* were fully operational encryption devices, submarine propulsion parts, radar systems, electron tubes for Patriot guided missiles, even F-117A Stealth fighter parts. Many of these parts, sold as *surplus*, were brand new (see the Sidebar that follows).

Dollar Days: Everything Must Go!

Buy a high-tech weapon or weapon part through the mail? It seemed crazy, but Air Fiorce Investigators thought they would try. They first obtained a printout from a former Pentagon investigator of the weapons and weapons parts listed on the Inventory Locator Service, a subscriber-only computer database of equipment that aircraft parts dealers have for sale. While most of the 30 million things on ILS are airplane parts, it is possible to search the listings for weapons. Their search found more than 2200 weapons-related items.

   Using a blank letterhead and a rented post office box, the investigators wrote to 43 companies asking for price quotes on such things as Maverick antitank missiles, Sidewinder air-to-air missiles, GBU-10 guided bombs, 40-mm grenade launchers, and flamethrowers. The investigators also asked for quotes on high-tech parts, such as cruise-missile antennas, Hawk missile-launcher parts, night-vision scopes, and cryptographic equipment for satellite communications.

Some of the items the investigators asked for had been sold at Pentagon surplus sales. Some were listed as classified. All should have been demilitarized and most destroyed.

Nineteen of the companies responded with price quotes. The investigators received no quotes on bombs and rockets. The investigators found out that arms sellers typically prefer dealing with foreign governments or arms dealers. One company said it had bomb arming devices and a control box from an A-4 fighter. The company said it had bought the weapons at surplus sales but wanted to know more about the buyers before selling. The company offering grenade launchers asked how many the undercover investigators wanted. When the investigators said they wanted only one or two, the company selling the grenade launchers said, *Sorry, no price quote.* A flamethrower the investigators sought was for sale for $6000, but it was World War II era. The investigators wanted something newer.

The investigators' luck was better on the high-tech weapons parts. They got price quotes for 46 items. An antenna for a Harpoon cruise missile was $10,736. A bomb arming device could be had for $486. A Cobra attack helicopter weapons controller was $3600 new, $2600 used. Tomahawk cruise missile parts were also available. The key part for a night-vision weapons scope, the image intensifier, was $2,200. An electronic countermeasures test set, classified *confidential,* could be had for $10,891.

In the end, the investigators ordered the night-vision image intensifier, the bomb arming device, an A-4 Fighter navigation circuit card, and a missile navigation part that the Pentagon says should definitely be destroyed, not sold. Most arrived courtesy of UPS within a week. The investigators bought the Hawk missile voltage controller over the counter in Texas for $436 in cash, while the investigators chatted with the clerk about that night's high school football championship game. As they left, the woman wished them a pleasant day.[3]

So overloaded is the Pentagon's surplus sales system that some offices responsible for disposing of the material have suffered significant breakdowns. At an Air Force base in Georgia, the Pentagon's surplus sales office lost track of $40 million in material. Today, no one can say where the stuff went.

The sheer volume of surplus material generated by the Pentagon's downsizing is one reason for the system overload, but the modern Pentagon's insistence on profitability is another. In 1997, the surplus sales system generated $413 million, and it is one of the few Pentagon programs capable of covering its own costs. A Pentagon e-mail message written in May 1995 illustrates clearly the premium placed on profits by senior military officials. Authored by a top-level official and addressed to all surplus sales directors on the East Coast, the memo is blunt: *The work priorities at your sites as I see them are: 1. Profits. 2. Profits. 3. Profits. 4. Profits. 5. Profits.* Priority No. 6 on his list was *accountability.* No. 7 was rendering lethal weaponry harmless.

Surplus Pentagon weapons are regularly traded away by Army and Navy museums scattered around the country, and some have been responsible for questionable if not illegal activities (see the Sidebar that follows). A federal law allows the museums to trade tanks, jet fighters, and attack helicopters to private citizens for other vehicles or *services.* Two museums are under federal criminal investigations for such activities.

Top-Gun Museums Face Federal Criminal Inquiries

Recently, a California man named Maurice Skinazi thought he had finalized an aircraft deal. He would buy 22 aging C-130 Lockheed Hercules fuselages from the National Museum of Naval Aviation in Pensacola, FL, for $600,000. Then Skinazi, who runs an aircraft restoration and resale business, hoped to use the parts to put together two or three intact C-130s and sell them, perhaps for millions of dollars. But Skinazi recently learned that the aircraft had been seized during a federal criminal investigation.

The genesis of the inquiry was a little-known federal law that allows military museums to trade airplanes or parts to the public in exchange for many kinds of *services.* Museum officials say the tradeoffs make sense: The military gets to build museum collections at no cost to the taxpayer, and interested buyers get old planes that would otherwise be junked.

But the so-called barter act has touched off a bitter wrangle inside the Pentagon. Investigators contend that these exchanges are structured so loosely that there is a real possibility of the government being defrauded. Competitive bidding is often absent, and items traded may be valued subjectively.

Good Price

Such problems are highlighted in a recent Department of Defense inspector general's report. The U.S. Army Center of Military History in Washington, DC, which oversees Army museums, contracted with a big helicopter sales and repair company called Southeastern Equipment Co., or SECO, to move and store 91 surplus Army helicopters in a SECO building in Augusta, GA, for one year and to do some restoration work. The bill was $401,000. The Army and SECO indicate that figure was arrived at by comparing other storage rates. For payment, the history center gave SECO seven UH-1 Huey helicopters that it valued at $76,000 each. *In light of their condition, $76,000 was a good price,* the Army stated. The Pentagon inspector general did not think so. The choppers traded to SECO were probably worth twice that amount.

More eye-catching was a deal for 10 Cobra helicopter trainers the Center of Military History traded to other aircraft rebuilders. None was demilitarized, and several were capable of having weapons installed. The Army's position is, if it has the weapons removed it is just a fast, skinny helicopter. Nonetheless, criminal investigations are underway at both museums.

Records show that the naval aviation museum took title to 455 aircraft from 1991 to 1995, including scores of jet fighters. How many of these were exchanged with the public is not clear. But, what we are talking about here is possible arms dealing.

Making Calls

Eventually, a ban was imposed on such trades. When officials at the Defense Logistics Agency discovered in 1993 that the Navy museum alone had obtained $400 million worth of aircraft and parts at original prices to barter, they stopped giving aircraft and parts to museums. The museums started getting aircraft to barter directly from the Army and the Navy, so, in 1996, an assistant deputy undersecretary of defense directed that the practice cease immediately.

Airplanes like C-130s, which have commercial use, may still be traded if they have sensitive equipment removed. Nonetheless, the current investigation at the naval aviation museum appears to focus once again on how the museum sells or trades the aircraft. According to an affidavit filed by Skinazi, the deal in which Skinazi bought the 22 C-130 hulks for $600,000 was not formally advertised. Museum officials say they got in touch with a handful of possible buyers, but none could beat Skinazi's offer. When the deal stalled, Skinazi enlisted the help of retired Chief of Naval Operations Admiral Thomas Moorer. Moorer, who is also chairman emeritus of the museum foundation, said he tried to break the deal free, charging Skinazi $60,000 for his services.[4]

Within the cloistered world of the Pentagon, the problems of properly disposing of surplus weapons are nothing new, although the public has been kept largely in the dark. Surplus weapons sales are overseen by a Pentagon office called the Defense Logistics Agency (DLA), which manages all the military's supplies and equipment. The DLA has spent much of the past two decades trading barbs with the four military services about how to fix the surplus sales system. Progress has been virtually nonexistent. Nearly a decade ago, an internal memorandum signed by William Taft, then the Deputy Secretary of Defense, leveled a startling charge. A joint U.S. Customs-(Pentagon) investigation had confirmed that the defense disposal system is a source of supply for arms traffickers. Since then the number of surplus weapons and critical parts has increased, and the problems have only worsened.

Money and Guns

How much Pentagon weaponry is getting into the wrong hands is impossible to know, but it is happening, law enforcement officials say. When a Hell's Angels methamphetamine lab was raided in California a few years ago, police also found military weaponry: four working machine guns and an M-79 grenade launcher. The grenade launcher was traced to a Pentagon surplus sales outlet in Crane, IN. That same outlet had sold 80 of the launchers in 1990. Pentagon documents certified that the launchers had been cut into scrap. In fact, only their firing pins had been filed down. In 1993, in Illinois, a man was sentenced to 42 months in prison for illegal possession of 20 unregistered machine guns. The man had bought the weapons through Pentagon surplus sales and welded them up to make them work. The weapons should have been mutilated so they could never work.

Smuggling is an even bigger concern. A 27-month Customs Service investigation that recently ended resulted in the interception of surplus military parts worth $268 million. Foreign countries can use these parts to resupply their own armies, to find out how American weapons work or to build their own weapons, thereby avoiding years of costly development.

Despite a handful of successes prosecuting those who abuse the Pentagon's surplus sale system, federal officials who have tried to plug the holes in it are frustrated and angry. A few Pentagon officials have found themselves demoted or reassigned for being too aggressive. Federal task forces have been quietly disbanded with few results while prosecutors who take on inquiries involving illegal surplus weapons sales say they find the criminal justice system unable or unwilling to cope.

In theory, the system for surplus disposal is simple. Whether it is a part from a jet fighter or a pair of socks, once an item is designated as surplus, it is trucked to a warehouse-cum-sales lot known as a Defense Reutilization and Marketing Office, or DRMO. There all surplus items are checked in and sorted. Then they are offered free to other government agencies and to museums and nonprofits. What is left after that is sold to the public at auctions where buyers must submit sealed bids.

Before that can happen legally, however, weapons and weapons parts must be *demilitarized*, or rendered militarily harmless. Machine guns, for instance, are *demilled* (the Pentagon term of art) by being cut into six pieces, reducing them to scrap metal. Many electronic parts are demilled by crushing or chopping—again, reducing them to scrap. Some items (radios with military-channel crystals, for instance) can be demilled by removing key parts. In such instances, buyers get a functional piece of equipment but one that has no military application. Many items, like desks, require no demilitarization. Other things, like parts of military aircraft that have commercial use, can be sold in the United States without being demilitarized, but buyers must obtain licenses to ship them overseas.

How or whether something should be demilitarized is determined by its *demil code*, which is normally assigned when a piece of equipment is purchased by the Pentagon. In theory, the codes should reflect how militarily sensitive an item is, and the code for a specific item should rarely be changed. Theory, however, has little to do with practice. A few years ago, for instance, a team of 10 demilitarization experts went to Battle Creek, MI, the home of the Defense Reutilization and Marketing Service. That office is the headquarters of the surplus sales program, responsible for the operation of the 281 DRMOs around the country. In Battle Creek, the team reviewed the demilitarization codes on 3.7 million pieces of equipment in the 15 million item inventory. The team found that 46% of the demilitarization codes were wrong and that about half the items were coded too leniently. This meant that weapons and weapons parts that should have been rendered militarily harmless before being offered for sale were not. Some .38-caliber revolvers, for instance, were coded demil A — the same as tables and chairs. Other, far more sensitive equipment was similarly miscoded, the team found. Electronics that guide bombs and aircraft gunnery, sensitive communication and navigation gear, cryptographic equipment, even hardware that allows U.S. fighter jets to jam antiaircraft missiles; in those categories, the team found that 83—93% of coding was wrong. In nine-tenths of the errors discovered in this instance, the codes applied were too lenient.

Such mistakes have consequences. When Pentagon investigators checked a DRMO outlet at Fort Benning, GA, for example, they found three complete TOW antitank missile systems for sale. They were assigned a code allowing them to be bought by anyone in the United States. You can actually build yourself an army out of this stuff that is miscoded.

Defense Department officials blame the military services for the miscoding, because the services assign the codes. The services do not make coding a high priority because anything dealing with *surplus* is a far cry from their main mission, warfare.

The profit motive is also a factor. The more fully operational weapons the Pentagon sells, the more money it makes. And the Pentagon is interested in selling. To encourage sales, in fact, the Defense Department has set up a Web page on the Internet *(http://www.drms.dla.mil)* listing every piece of equipment that has entered the surplus inventory as shown in Figures 23.1 and 23.2.[5] Anyone with a computer can call up the listings, type in *missiles* or *bombs* or other hardware and see

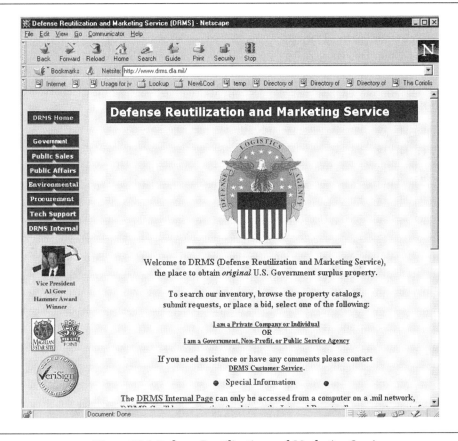

Figure 23.1 Defense Reutilization and Marketing Service.

what is available. However, the DRMS Internal Page can only be accessed from a computer on a .mil network.

After browsing the Web site at length, the IG investigators gave Defense Department experts a lengthy inventory of items listed and asked them to verify that the demilitarization code for each was correct. The experts found that dozens of items had been wrongly coded. *Bomb, gen. purpose*, for instance, had been given the same demilitarization code as a desk chair-*A*. Similarly miscoded were *sight, rocket launcher, controller, missile,* and *missile guidance set.* All should have been coded *D*-for destruction (see the Sidebar that follows). They also found

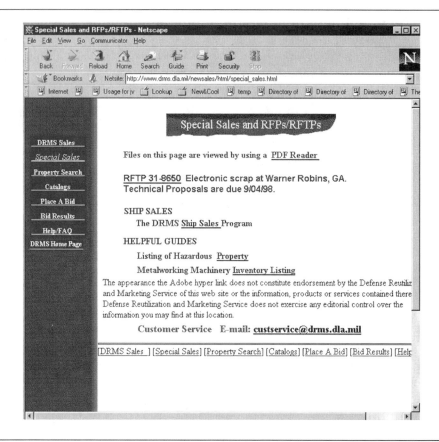

Figure 23.2 Special Sales.

the computer memory for a Tomahawk cruise missile! This should have been classified! This is the brains of the missile!

How Weapons Sales Are Controlled by Codes

Demil Code Assigned:

All items and their parts in the U.S. military inventory are assigned demilitarization (demil) code letters when they are purchased. The demil code determines what must be done to the item before it can be donated or sold to the public. The demil codes on weapons and parts are generally overseen by the military services.

Item Sent to Surplus:

Excess or outdated weapons and parts go into the surplus sales system. They are first offered to other government agencies and museums, then are sold at auction to the public.

The Codes:

An item's demil code should dictate the condition in which it is sold. Here are the possibilities:

Code *A*: No Demil

Items are sold *as is* to the public.

Code *B*: No Demil in The U.S

License is needed to ship or sell overseas.

Code *C*: Key Point Demil

Sensitive parts are removed: remaining item may be usable.

Code *D*: Total Demil

Item must be destroyed and sold only as scrap.

Code *E*: Classified Item

Classified parts must be removed or item destroyed.[6]

Even when the Pentagon codes are correct, there is no guarantee that weapons and weapons parts will be demilitarized properly. A former Pentagon investigator recalls one visit to a DRMO near Texarkana, TX, where he watched a worker trying to cut through the gun barrel of an M-60 tank. *The poor kid had a cutting torch,* the investigator recalled, *that couldn't cut hot butter.*

Overloaded System

At the Defense Reutilization and Marketing Offices around the country, workers and managers say they have been so overloaded with equipment from the massive military drawdown (in recent years, the DRMOs have taken in more than $30 billion worth of material annually), that the demilitarization process was bound to have problems. Easily the biggest occurred at Robins Air Force Base near Macon, GA. So much property came into the DRMO there a few years ago that the system collapsed, and workers lost track of almost $41 million in surplus equipment—about one-fourth of their inventory. The DRMO lost total accountability of surplus property.

Investigators assigned to examine the problems at Robins found that the Pentagon's DRMOs had developed what they called an *expedited processing* program. To speed things up, a Robins DRMO manager told investigators that she falsified demilitarization statements, registering weapons and other equipment as scrap that was later made available for sale fully intact.

The problems at Robins raised alarms. One day in 1995, an Air Force investigator jogging on the base saw a surplus-parts dealer carrying an electronic device from the DRMO to his truck. The investigator looked into the truck, jotted down the stock number of the item and checked it later. He found that the device should have been demilitarized but had not been. Not long after that, a team of Air Force and Pentagon agents visited the Burton Electronics warehouse in nearby Montezuma, GA, where they found more electronic weapons parts that had not been demilitarized. The agents retrieved a pile of weapons pieces and aircraft parts that had cost the government $50 million when new. Burton Electronics said they had bought most of the material from other dealers. Investigators believe much of it was sold by the Robins DRMO as scrap for seventeen cents a pound. It is not illegal to buy surplus material that was improperly demilled, and Burton was not charged with any crime. They are now trying to get their money back.

The episode resulted in some surprising leads. Investigators discovered that a Chinese-owned scrap company was legally buying surplus material from Burton and from the Robins DRMO. Further inquiries revealed a score of Chinese dealers operating from coast to coast. Although no proof developed that they acted as agents for China, that did raise concerns. In a memo to his Chinese boss in California, the buyer in Georgia called Robins *the candy store.* The base, he wrote, *will fill our needs into the next century,* according to an investigator who has seen the note.

The Robins case had all kinds of reverberations. It led investigators to the discovery of a commercial listing service that offers thousands of military surplus weapons and weapons parts. The listing, called the Inventory Locator Service or ILS, is used worldwide by dealers in aircraft parts to advertise their inventories. Surplus dealers also use it to advertise weapons and weapons parts, all of which is perfectly legal too.

The Fallout

The echoes from the Robins case would eventually be felt clear across the country, in Montana, of all places. For example, one individual (called a *Chopper Jock* for the purposes of this chapter), known as a straight-talking businessman who runs a thriving helicopter-repair business about 50 miles south of Missoula, Montana, had modified UH-1 Huey helicopters for logging and construction use. As the Cobra was a sleek and powerful follow-on to the Huey, he thought it would make a good logger. The Chopper Jock obtained an FAA license to build one for logging, firefighting or movie-making purposes.

When armed, the Cobra is one of the deadliest military attack vehicles ever invented. Before it is sold as surplus, it must be demilitarized. The demilitarization manual states that Cobra bodies must be cut into at least three parts and that their *airframe must be mutilated by destroying attaching structure by cutting, chopping, tearing, shredding, crushing or smelting to the degree that aircraft will be unfit for repair or flight.* The Chopper Jock and other helicopter builders, however, knew that usable parts were plentiful, even at DRMOs. They had amassed a huge store of Cobra parts. If the Cobras were built once, the Chopper Jocks could rebuild the crafts again, and no one could stop them.

In October 1995, a flatbed truck pulled into the DRMO at Robins carrying three Cobra bodies that looked as if they had not been properly demilled. Air Force investigators questioned the truck driver, who said the Cobras had been purchased at a DRMO in Groton, CT, and were headed for a warehouse in Joshua, TX. According to the investigators, the truck driver reportedly said: *I don't know why you are so excited, the warehouse contains about 70 or 80 of these things.*

Original Packaging

The warehouse actually contained 99 Cobra fuselages, which was to be used for logging. The warehouse also had the parts to go with them — the rotor blades, rocket tubes, gun pods, missile launchers, and weapons controllers, according to Pentagon records. *None of the items were demilled correctly,* a Pentagon memo states. *Some items were in original packing crates.*

Word of the warehouse's covey of Cobras shocked the top federal prosecutor (to be called the Prosecutor for the purposes of this chapter) for the middle district of Georgia, where Robins is located. At the time of the warehouse's discovery, the Prosecutor was overseeing something called the Middle Georgia Task Force, which consisted of military investigators and special agents of the Customs Service and the FBI. The task force had been set up after the $50 million retrieval of property from Burton Electronics.

The other person upset by the warehouse's Cobras was the Chopper Jock, in Montana, who telephoned the Pentagon to see what was going on. The call was lateraled to another investigator. The Chopper Jock was concerned over the warehouse seizures because he too was building Cobras—and he saw no reason to give his up. According to the investigator's notes, the Chopper Jock had a significant amount of intact weapons and weapons systems.

The Chopper Jock did not want trouble. The government could have his Cobra weapons, he said, if they *want to buy them.* The Investigator finished scribbling his notes, thanked the Chopper Jock and hung up.

The fate of the Cobras would hang in the balance for months, as a bitter wrangle ensued inside both the Pentagon and the Justice Department. Investigators, federal agents, demilitarization officers, and prosecutors pushed for seizure of the Cobras. But the Army, which owns most of the choppers, resisted.

Ultimately, the government prevailed and seized the helicopters and parts from the warehouse. But as agents began to plan raids on the Chopper Jock and other owners of Cobras, the Justice Department brought all such seizures to a halt. The law that might authorize them was just too murky. The FBI sent a letter to the Pentagon advising that 34 Cobras belonging to private citizens had been *discovered during several FBI investigations.* Should the Pentagon choose to take *corrective action,* the FBI letter advised, that would be agreeable to local U.S. attorneys. One senior Pentagon investigator calls this the FBI's *Pontius Pilate* letter. Pentagon investigators washed their hands of the whole thing.

Threats to Security

As worrisome as the Cobras may be to law enforcement, agents say that the greater threat to national security is the shipping of weaponry overseas. In October 1995, about the time the flatbed-truck driver with the three Cobras was pulling into Robins, a Pentagon investigator got word that a California company named Cal Industries was loading 25 seagoing containers at the Pentagon DRMO in San Antonio, TX. The contents were listed as 351 tons of electronic scrap. The investigator, who knew that the base processed lots of sensitive weapons equipment, got Customs Service agents to stop the containers and pry them open. What they found were thousands of top-secret satellite encryption devices, hard-cased field computers used for sending coded communications, as well as classified radio transmitter-receivers. None had been properly demilitarized. There was also a pile of banged-up Wang computers, but many still had their hard drives, and two had $5\frac{1}{4}$-inch floppy disks in them. One disk was marked *Classified SCI* (special compartmented information). The other was marked *Top Secret, SCI. Contains Special Intelligence.* The investigator checked and found that the equipment had come from the most secure sector of Kelly Air Force Base, known as *Security Hill.* None of this material should have ever left the DRMO. The equipment's intended destination was Hong Kong, a common transshipment point for goods bound for China. Kelly Air Force Base took it all back. No charges were filed against Cal Industries, which had bought the stuff legally. Candidly, the way our government handles this stuff so recklessly, it is hard to pin anyone down or hold them liable.

For 27 months, U.S. Customs ran an investigation that targeted surplus smugglers. Dubbed *Operation Overrun*, the investigation resulted in 10 seizures of $268 million worth of parts bound for China, Hong Kong, Vietnam, the Philippines, Taiwan and elsewhere. None of the items seized had been licensed for export. Included in their overseas-bound shipments were radar systems, electronic-warfare black boxes, and some bizarre items — 20,000 new chemical-warfare-protection suits headed to Ukraine (makes you wonder what is going on there).

In one case, records show, 39 boxes of tank and howitzer parts, valued at $900,000, were found buried inside a container beneath automotive parts. In another, bound for Shanghai, agents found 48 separate inertial guidance devices for the F-117A Stealth fighter and F-111 bomber. New, the devices had cost the government almost $33 million. The two Pentagon employees who were identified as responsible said they could not explain how the devices had been sold intact.

Of the many countries pursuing Pentagon surplus weapons, China is clearly running hardest. China is the most aggressive country worldwide in the acquisition of military hardware. Anything that seems to be mid-tech, or slightly out of date, is high tech to them. They can leapfrog R&D costs. They either reverse-engineer this stuff, or if they get enough, they just use the items.

Surplus-Smuggling Operations Network

Customs agents are now convinced that a Chinese surplus-smuggling operation began as many as 18 years ago, when a flood of Chinese men and women entered the United States, established themselves as residents and then set up shop. Operating with little more than a fax machine, and a bank account, they apparently form the lowest tier of the Chinese espionage network, agents say.

The buyers hunt surplus sales outlets for military electronic parts. When they find what they want, they bid high enough to be sure they get it. Bids of $60,000 or $200,000 for scrap are not unusual. Sometimes they buy huge piles of scrap knowing that valuable pieces of electronics equipment are included, or they share purchases with legitimate scrap dealers. One dealer with Chinese affiliations was the high bidder on every piece of scrap electronic equipment on the East Coast.

In a year-and-a-half of intense investigation of these buyers, the Customs Service has never found a smoking gun — a written shopping list from overseas, or anything that could link particular Chinese scrap buyers to Beijing or its defense agencies. The reason for that, besides good intelligence tradecraft, is that they do not need to communicate directly. They have been doing this for 18 years and know exactly what their country wants.

After 27 months of dogging leads, the Customs Service abandoned Operation Overrun. The agency had run out of time and money; the investigation had not been a spectacular success. Besides the $268 million in seizures, only three criminal cases had been brought. A man named Richard Li had been intercepted smuggling sensitive electronic gear to Hong Kong and pleaded guilty to smuggling. A second man is currently under arrest, and his partner is on the run. Customs agents say the investigation clearly put pressure on the network of Chinese traffickers. But they also are certain that once the pressure is off, the smugglers will resume their operations. The agents definitely think they will spring up again. The Chinese can get up to speed to buy the stuff again, and the U.S. is the biggest warehouse in the world.

One of the biggest listings of warehoused arms and arms parts can be found on a system called the Inventory Locator Service (ILS) as shown in Figure 23.3.[7] The ILS, a business located in Memphis, TN, it offers

Figure 23.3 Inventory Locator Service.

a computerized classified listing to paid subscribers. About 4000 aircraft-parts sellers and users worldwide subscribe to it, for a fee of several hundred dollars a month. The sellers list their inventories on the system, and buyers can scan the inventories as they browse for purchases. The computer system is queried about 37,000 times each day.

The Wrong Stuff

The Pentagon investigators remember vividly the first time they tapped into the ILS database. They thought they had a pretty good sense of the international arms market, but the stuff they saw listed in the ILS database disabused them of that. They ran weapons, machine guns, flamethrowers and components of guided bombs, and immediately, they were getting hits!

A printout the investigators made from ILS visits is an eye-popper. It runs 37 pages, single-spaced, and lists such items as AIM-9 Sidewinder missiles, GBU-10 guided bombs, Maverick missiles, machine guns, and grenade launchers. Virtually everything on the list should have been demilitarized, and probably destroyed—but it was all available for sale.

Curious to figure out where the stuff came from, the investigators began to check those items against sales at DRMOs. Checking more than 200 weapons parts advertised on ILS, they found that one-fourth had been sold to surplus dealers in just the past few years. All were available in the United States. With more research, the investigators came to believe that the heavy-duty weapons, the bombs and rockets and missiles, had probably been sold legally by the Pentagon to foreign governments, which then sold them as surplus. Brokers using ILS then offered them for sale.

The ILS president claims that he has provided law enforcement agencies access to his company's computer listings, but he did not know the details of their investigations. Because there are 40 million items listed on the inventory, the ILS president contended that searching for weapons is not easy. But when investigators gave him the stock number for a grenade launcher, the ILS president was able to find it pretty quickly.

The investigators' test of the ILS system shows how easily surplus weapons and weapons parts can be purchased by the public. Using a dummy letterhead, the investigators sent letters to many of the companies listing weapons parts on ILS, requesting price quotes and availability for a variety of weapons and parts. The sellers responded with price quotes for 36 items, ranging from bomb arming devices to circuit cards for nuclear submarines' targeting systems to the key components of night-vision scopes. Arms dealers claim that in buying aircraft parts at DRMOs they often scooped up a lot of other hardware, such as weapons parts. They indicated that they listed it all on ILS hoping that someone would spot it and buy it.

Some dealers claimed they would not sell weapons or parts to people who seemed shady. Others have admitted that the system had almost no controls. They could sell weapons parts to anyone, and that person could sell it to someone else, and he could sell it to someone else, etc.... Another dealer claimed that he had the parts the investigators ordered, such as a pilot's targeting screen, known as a heads-up display for an F-16 fighter, and, that he had bought them from a DRMO but wanted more information about the investigators' dummy company before selling to them. Another offered to ship the investigators' dummy company a similar part, with no questions asked.

One company, Aviarms Support Corp. in Westbury, NY, had a different story. A manager there indicated that the items the company advertises on ILS are held by Israeli, German and Dutch companies for whom it acts as a broker. According to the manager, many of the U.S. weapons parts came from the surplus stocks of the Israeli and German militaries. Had they been demilitarized? No! According to the manager, they were all in good working order.

One part that Aviarms was selling was listed as an *F-117A fire control memory*. Asked how Israel could have obtained surplus parts for the U.S. F-117 Stealth Fighter, the manager checked his computer. Aviarms was brokering it for an Israeli company, which got it from a German company, he said. He could trace it no further.

The investigators were outraged by what they found on ILS. One investigator's last year at work was consumed with tracking DRMO sales of sensitive weapons parts and trying to get his superiors at the Defense Logistics Agency to take action. He sent e-mail after e-mail message up and down the command chain. What follows are a few examples of obviously Demil-required property offered at DRMO sales, he wrote in September 1996: *Payload section, nuclear missile. Warhead components: DRMO San Antonio... Multipurpose display, F-117A Stealth Fighter: DRMO McClellan Air Force Base, California... F-il1 Navigation and attack computer: McClellan Air Force Base DRMO ... Heads-up display, F-111B fighter-bomber: DRMO Davis-Monthan Air Force Base, Arizona.*

The investigator had the support of his coinvestigators, and even some higher officials at the logistics agency, but little was done. There were no seizures, no attempts to recover the items. On July 5, 1997 the investigator sent a final e-mail to his bosses: *Nearing the end of the trail here. I know all of you are very concerned about the OBVIOUS total lack of control of our vital defense technology and hardware, but I can't help but wonder after nearly three years of almost DAILY findings such as those above, why the government has not seen fit to attempt to recover ANY of the thousands of sensitive or classified items discovered via ILS.* The investigator then quit to become a Methodist minister.

Nevertheless, a group of knowledgeable middle-level Pentagon and Defense Logistics Agency managers met with other investigative agencies and confirmed many of their findings. The officials quibbled with some assertions. For instance, they acknowledged that there had been pressure in the past to cut corners to sell surplus material, as evidenced by the *profits, profits, profits, profits* memo previously mentioned. But they also insisted that the new management at the Defense Logistics Agency is not profits driven. The officials blamed the four military services, which control most of the demilitarization codes, for writing the codes too leniently and for failing to change them when errors are pointed out. They said the DRMOs are no longer accepting items with demilitarization codes that had been phased out a few years ago.

They also said some fixes are in the works. One of their main goals, the officials say, is to set up a centralized demilitarization code office, which will assign and control all the codes, and try to enforce them. The first step toward that, they say, should occur sometime in 1999. Investigations underway at two Pentagon museums could also result in tightening of the current rules allowing bartering of government aircraft for a variety of services.

If those steps happen as planned, they will clearly result in some improvement, but the problem is so great that it is difficult to say what real effect they might have. The scope of this program and the amount of material going out the door is so huge, people normally do not believe you. Only the government could have a program where they give everything away for free and then mess it up.

Summary

An investigation by the IG offices of the four military services has documented serious problems with the surplus sales program. Relying on internal Pentagon records, Customs Service documents and interviews with scores of federal agents and officials, the inquiry identified thousands of instances where weapons and parts that should have been rendered militarily harmless before sale were not — where weapons that should have been designated for destruction were not and where key weapons parts that should have been prevented from reaching foreign buyers were not. This system on its best day is morally embarrassing to our government. And it never has a best day. It is absolutely a disaster of the highest magnitude.

From Here

Chapter 23 examined the growing threat from our allies to steal encryption technology and the intensified efforts by the FBI to prevent such theft. The chapter also covered how the Pentagon sells millions of surplus planes, tanks, helicopters, and parts (etc.) every year (to anyone who is willing to pay) that has encryption technology still left on the onboard computer or communications equipment. Chapter 23 discussed how most of this technology falls into the wrong hands because it was never removed in the first place when the equipment was sold for surplus. The chapter also investigated how the Defense Logistics Agency has ignored repeated requests by its own personnel to implement reforms in the surplus sales system; how the Pentagon handles the surplus sales so recklessly that it is hard to pin anyone down or hold them liable; and, the discovery by investigators of all kinds of classified Pentagon electronic gear destined for shipment to China that was sold at Kelly Air Force Base as scrap metal.

Chapter 24 discusses the political backlash and emotional fallout of the bombing of the federal building in Oklahoma City; and, how the FBI has begun to wage their own private war on the use of private, encryption schemes. The chapter will also discuss how terrorist groups are buying sophisticated encryption technology. Chapter 24 also explains how terrorists and kidnappers will use telephones and other communications media with impunity knowing that their conversations are immune from our most valued investigative technique. The chapter also examines our counterterrorism strategy to see how effective it is against the black market satellite encryption trade.

Endnotes

1   Interviews with NASA officials who wish to remain anonymous.

2   Federal Bureau of Investigation, FBI.

3   U.S. Air Force Inspector General.

4   Department of Defense Inspector General.

5   Defense Reutilization and Marketing Service, Federal Center, 74 N. Washington, Battle Creek, MI 49017-3092, 1998.

6   "Defense Demilitarization Manual," Department of Defense, 1998.

7   Inventory Locator Service, LP, 3965 Mendenhall Road, Memphis, TN 38115 USA, 1998.

24

International and Domestic Terrorist Organizations

In the political backlash and emotional fallout of the bombing of of the federal building in Oklahoma City, FBI Director Louis Freeh has begun to wage his own private war on the use of private, satellite encryption schemes. According to Administration sources, several different proposals are now being discussed on how to implement a policy of government mandated, government *certified* satellite encryption. The most hardline of these proposals would outlaw the public's ability to choose a satellite encryption scheme which the government could not break, by using the authority of a court order.

Freeh has left no doubts about his intentions. Not satisfied with a proposal he successfully rammed through Congress recently (the bill that gives law enforcement agencies *easy wiretap access* to digital conversations, at a cost of $900 million in taxpayer money), his next target is private satellite encryption.

During a recent appropriations hearing, Freeh told a congressional panel: *"We're in favor of strong encryption, robust encryption. The*

703

country needs it, industry needs it. We just want to make sure we have a trap door and key under some judge's authority where we can get there if somebody is planning a crime. That means an end to private, non-government satellite encryption, which do not have keys the government can use.

Private satellite encryption schemes allow a person to scramble an electronic message that, if intercepted by an unintended party, renders it unreadable. These scrambling programs are useful to a wide range of people and interests, including researchers who want to keep their proprietary breakthroughs safe from prying eyes to corporations sending trade secrets to a distant office across the Net to ordinary folks sending a steamy love letter to a lover.

But these same satellite encryption programs are being used by *"terrorists and international drug traffickers,"* as well, claims FBI Director Freeh, and that makes private encryption schemes a threat to national security (see the Sidebar that follows). Freeh's crusade against satellite encryption has been enjoined by members of the Justice Department, along with the nation's top spook group: the National Security Agency (NSA). And when it comes right down to it, your privacy rights do not stand a snowball's chance in hell of outweighing pictures of dead babies or pieces of dead babies.

International Conference on Terrorism

The participants at the Lyon Summit voiced their determination to give absolute priority to the fight against terrorism. They decided to examine and implement, in cooperation with all States, all measures likely to strengthen the capacity of the international community to defeat terrorism. To that end, they called for a meeting of their Foreign Ministers and their Ministers responsible for security to be held without delay to recommend further actions. In line with this decision, they met in Paris on July 30, 1996.

They undertook a thorough review of new trends in terrorism throughout the world. They noted with deep concern the use in 1996 of powerful explosive weapons by terrorists. They reiterated their fundamental view that there can be no excuse for terrorism. Their discussions underscored their agreement on the need to find solutions that take account of all the factors likely to ensure a lasting settlement of unresolved conflicts and on the need for attending to conditions which could nurture the development of terrorism.

They noted that there is a growing commitment within the international community to condemn terrorism in whatever shape or form, regardless of its motives; to make no concessions to terrorists; and to implement means, consistent with fundamental freedoms and the rule of law, to effectively fight terrorism. They are determined to work with all States, in full observance of the principles and standards of international law and human rights, in order to achieve the goal of eliminating terrorism, as affirmed in the Declaration adopted by the United Nations General Assembly in December 1994. To this end, they have, with the course laid down in their Ottawa Declaration of December 12, 1995 and the work that followed the Sharm-el-Sheikh Summit, framed a body of practical measures which they are resolved to implement among themselves.

They also invite all States to adopt these measures so as to impart greater efficiency and coherence to the fight against terrorism. In order to harness their own capacities more tightly they decided to establish among their countries a directory of counterterrorism competencies, skills, and expertise to facilitate practical cooperation.

I. Adopting Internal Measures to Prevent Terrorism:

1. Improving Counterterrorism Cooperation and Capabilities:

They call on all States to:

1. Strengthen internal cooperation among all government agencies and services concerned with different aspects of counterterrorism.
2. Expand training of personnel connected with counter-terrorism to prevent all forms of terrorist action, including those utilizing radioactive, chemical, biological or toxic substances.

3. In line with the efforts carried out in the fields of air and maritime transportation and in view of widespread terrorist attacks on modes of mass ground transportation, such as railway, underground, and bus transport systems, recommend that transportation security officials of interested States urgently undertake consultations to improve the capability of governments to prevent, investigate, and respond to terrorist attacks on means of public transportation, and to cooperate with other governments in this respect. These consultations should include standardization of passenger and cargo manifests and adoption of standard means of identifying vehicles to aid investigations of terrorist bombings.

4. Accelerate research and development of methods of detection of explosives and other harmful substances that can cause death or injury, and undertake consultations on the development of standards for marking explosives in order to identify their origin in postblast investigations, and promote cooperation where appropriate.

2. Deterrence, Prosecution, and Punishment of Terrorists:

They call on all States to:

5. When sufficient justification exists according to national laws, investigate the abuse of organizations, groups, or associations, including those with charitable, social, or cultural goals, by terrorist using them as a cover for their own activities.

6. Note the risk of terrorist using electronic (satellites) or wire communications systems and networks to carry out criminal acts and the need to find means, consistent with national law, to prevent such criminality.

7. Adopt effective domestic laws and regulations including export controls to govern the manufacture, trading, transport, and export of firearms, explosives, or any device designed to cause violent injury, damage, or destruction in order to prevent their use for terrorists' acts.

8. Take steps within their power to immediately review and amend as necessary their domestic antiterrorist legislation to ensure, *inter alia*, that terrorists' acts are established as serious criminal offences and that the seriousness of terrorists' acts is duly reflected in the sentence served.

9. Bring to justice any person accused of participation in the planning, preparation, or perpetration of terrorist acts or participation in supporting terrorist acts.

10. Refrain from providing any form of support, whether active or passive, to organizations or persons involved in terrorist activity.

11. Accelerate consultation, in appropriate bilateral or multilateral fora (forums), on the use of encryption that allows, when necessary, lawful government access to data and communications in order, *inter alia*, to prevent or investigate acts of terrorism, while protecting the privacy of legitimate communications.

3. Asylum, Borders, and Travel Documents:

They call on all States to:

12. Take strong measures to prevent the movement of terrorist individuals or groups by strengthening border controls and controls on issuance of identity papers and travel documents, and through measures for preventing counterfeiting, forgery, or use of false papers.

13. While recognizing that political asylum and the admission of refugees are legitimate rights enshrined in international law, make sure that such a right should not be taken advantage of for terrorist purposes, and seek additional international means to address the subject of refugees and asylum seekers who plan, fund, or commit terrorist acts.

II. Strengthening International Cooperation to Fight Terrorism:

4. Expanding International Treaties and Other Arrangements:

They call on all States to:

14. Join international conventions and protocols designed to combat terrorism by the year 2000; enact domestic legislation necessary to implement them; affirm or extend the competence of their

courts to bring to trial the authors of terrorist acts; and, if needed, provide support and assistance to other governments for these purposes.

15. Develop if necessary, especially by entering into bilateral and multilateral agreements and arrangements, mutual legal assistance procedures aimed at facilitating and speeding investigations and collecting evidence, as well as cooperation between law enforcement agencies in order to prevent and detect terrorist acts. In cases where a terrorist activity occurs in several countries, States with jurisdiction should coordinate their prosecutions and the use of mutual assistance measures in a strategic manner so as to be more effective in the fight against terrorist groups.

16. Develop extradition agreements and arrangements, as necessary, in order to ensure that those responsible for terrorist acts are brought to justice; and consider the possibility of extradition even in the absence of a treaty.

17. Promote the consideration and development of an international convention on terrorist bombings or other terrorist acts creating collective danger for persons, to the extent that the existing multilateral counterterrorism conventions do not provide for cooperation in these areas. Examine, also, the necessity and feasibility of supplementing existing international instruments and arrangements to address other terrorist threats and adopt new instruments as needed. Accelerate in the International Civil Aviation Organization (ICAO) consultations to establish uniform and strict international standards for bomb detection and the ongoing consultations to elaborate and adopt additional heightened security measures at airports, and urge early implementation of screening procedures and all other ICAO standards already agreed upon.

18. They recommend to States Parties to the Biological Weapons Convention to confirm at the forthcoming Review Conference their commitment to ensure, through adoption of national measures, the effective fulfillment of their obligations under the convention to take any necessary measures to prohibit and prevent the development, production, stockpiling, acquisition, or retention of such weapons within their territory, under their jurisdiction, or under their control anywhere, in order, *inter alia*, to exclude use of those weapons for terrorist purposes.

5. Terrorist Fund Raising:

They call on all States to:

19. Prevent and take steps to counteract, through appropriate domestic measures, the financing of terrorists and terrorist organizations, whether such financing is direct or indirect through
organizations which also have, or claim to have, charitable,
social or cultural goals, or which are also engaged in unlawful
activities such as illicit arms trafficking, drug dealing, and
racketeering. These domestic measures may include, where
appropriate, monitoring and control of cash transfers and bank
disclosure procedures.
20. Intensify information exchange concerning international movements of funds sent from one country or received in another
country and intended for persons, associations, or groups likely
to carry out or support terrorist operations.
21. Consider, where appropriate, adopting regulatory measures in
order to prevent movements of funds suspected to be intended
for terrorist organizations, without impeding in any way the
freedom of legitimate capital movements.

6. Improving Information Exchange on Terrorism:

They call on all States to:

22. Facilitate exchange of information and the transmission of legal
requests through establishing central authorities so organized as
to provide speedy coordination of requests, it being understood
that those central authorities would not be the sole channel for
mutual assistance among States. Direct exchanges of information
among competent agencies should be encouraged.
23. Intensify exchange of basic information concerning persons or
organizations suspected of terrorist-linked activities, in particular on their structure, their modus operandi, and their communication systems in order to prevent terrorist actions.
24. Intensify exchange of operational information, especially as
regards:

   ♦ The actions and movements of persons or groups suspected of belonging to or being connected with terrorist
   networks.

♦ Travel documents suspected of being forgeries or falsified.
♦ Traffic in arms, explosives, or sensitive materials.
♦ The use of satellite communications technologies by terrorist groups.
♦ The threat of new types of terrorist activities including those using chemical, biological, or nuclear materials and toxic substances.

25. Find ways of accelerating these exchanges of information and making them more direct, while at the same time preserving their confidentiality in conformity with the laws and regulations of the State supplying the information.[1]

During a recent Senate Judiciary Committee on terrorism, Freeh boldly told a panel of lawmakers eager to give his agency more latitude to *catch the bad guys (and civil rights violations be damned. As long as they don't have to watch the guts of little kids being splattered on steel girders and broken concrete)."The FBI cannot and should not tolerate any individuals or groups which would kill innocent Americans, which would kill America's kids."*

To meet the *challenges of terrorism,* Freeh said, several things must be done, among them, deal with satellite *encryption capabilities available to criminals and terrorists* because such technology endangers *the future usefulness of court-authorized wiretaps. This problem must be resolved.*

While Freeh used the Oklahoma City bombing as a convenient news hook to again make a pitch to resolve the private encryption problem, the Director was basically reading from a dog-eared script. Within the last few months he repeatedly has testified publicly before Congress about the evils of encryption.

In a recent House Judiciary Committee's Subcommittee on Crime Freeh said: *Even though access is all but assured (by the passage of the Digital Wiretap Act) an even more difficult problem with court-authorized wiretaps looms. Powerful encryption is becoming commonplace. The drug cartels are buying sophisticated communications equipment. This, as much as any issue, jeopardizes the public safety and national security of this country. Drug cartels, terrorists, and kidnappers will use telephones and other communications media with impunity know-*

ing that their conversations are immune from our most valued investigative technique.

Then during a more recent appearance before the same Committee, Freeh said: *"Encryption capabilities available to criminals and terrorists, both now and in the days to come, must be dealt with promptly. We will not have an effective counterterrorism strategy if we do not solve the problem of satellite encryption."* But there is nothing to be alarmed at here, according to Freeh. Just because he is asking the Congress and the White House to strip you of the right to choose how you scramble your messages, using a program for which the government does not hold all the keys, does not mean that the Director is not a sensitive guy or that he has suddenly taken a liking to wearing jackboots.

Freeh steadfastly maintains that all these new powers he is asking for are simply *tools* and *not new authorities*. These new powers are *well within the Constitution,* Freeh told Congress.

Freeh has not publicly outlined just how he proposes to *resolve* the *satellite encryption problem*. However, according to an FBI source, several plans are in the works. The source refused to detail any specific plan, but added: *Let's just say everything is on the table.* Does that include outlawing private encryption schemes? *I said everything,* the source said.

The encryption debate has been raging for years. A few years ago, the Clinton Administration unveiled a new policy in which it proposed to flood the market with its own home-grown satellite encryption device (the previously discussed product of the National Security Agency) called the *Clipper Chip*.

As previously discussed in earlier chapters, the Clipper is based on a *key-escrow* system. Two government agencies would hold the keys *in escrow*, which are unique to each chip, in a kind of *data vault*. Any time the FBI (or your local sheriff) wanted to tap your phone conversations, they would have to ask a judge to require the two government agencies to turn over the keys to your Clipper chip. With those keys, the FBI could then unscramble any of your conversations at will.

That policy raised a huge firestorm of controversy and the Clipper sunk from sight, down, but not out. The intent of the White House, acting as a front man for the NSA and other intelligence agencies along with the FBI, was to have Americans adopt Clipper voluntarily. The FBI took it on good faith (and I am not joking here) that criminals, too, would buy Clipper equipped phones, allowing the government to easily unscramble their wiretapped conversations.

Why would criminals knowingly use a device they knew the government could easily tap? *Because criminals are stupid,* was the FBI's party line. No, I am not making this up.

The *voluntary* aspect did not stop the controversy. Indeed, buried in the Administration's own background briefing papers on the Clipper was the no-nonsense statement that the Administration, after reviewing the Constitution, had determined that *Americans have no Constitutional right to choose their own method of encryption.* Call it a peremptory strike on privacy.

An NSA source, when questioned about his agency's role in the process, was reluctant to speak. He did say that Clipper was merely *Act One* of a *Three part play.* Pressed for further details, he said, *do your own homework.*

During a recent Senate Subcommittee on Terrorism hearing, Robert Litt, Deputy Assistant Attorney General for the Criminal Division of the Department of Justice, said the *widespread use of reliable, strong encryption that allows government access, with appropriate restrictions is designed to achieve a delicate balance* between privacy rights and law enforcement needs. But what if criminals bypass Clipper, using private encryption schemes, such as those now being decried by Freeh? If one solution does not work, then we have to go to the next step. Is that next step the outlawing of private encryption? Probably!

One proposal being seriously discussed would, indeed, outlaw all private, non-government approved satellite encryption schemes. Here is how the government plan breaks out, according to sources familiar with the proposal:

1. The government would *certify* a few so-called *Commercial Key-Escrow* programs. These are similar in design to the government's Clipper Chip, but industry would hold the keys and some of these systems might not be classified, as is the software underlying Clipper. However, the companies producing such *certified* pro-

grams could claim trade secrets, not allowing the public access to the underlying programs.

2. The government, tossing a bone to industry, would lift export controls on all *certified* satellite encryption programs. Currently, it is against the law to export encryption programs, as they are controlled by the State Department under the same classification as munitions.

3. The use of government certified satellite encryption would become a federal mandate, making it illegal to use private encryption schemes that had not passed the *certification* test.[2]

The plans are *still really liquid,* said an Administration source. *We all know what a bitch this is going to be trying to sell it to the selling public,* he said. *The flashpoint potential of this idea isn't lost on anyone.*

The Administration is hoping the public will accept the fact that: (1) It is private industry holding the keys, not government agencies. (2) The availability of *choice* among several different vendors will squash the imagery of Big Brother.

Expect the rise of an off-shore, black market satellite *encryption* trade. It will surely come. And, it will have to be a black market. The FBI has already gone around the world preaching the evils of private satellite encryption, trying to get other law enforcement groups to push for outlawing such programs in their own countries. And they already have a convert: Russia. That country recently adopted such a policy. It is nice to know we are following in Mr. Yeltsin's footsteps.

From Here

Chapter 24 discussed the political backlash and emotional fallout of the bombing of the federal building in Oklahoma City; and, how the FBI has begun to wage their own private war on the use of private, encryption schemes. The chapter also discussed how terrorist groups are buying sophisticated encryption technology. Chapter 24 also explained how terrorists and kidnappers will use telephones and other communications media with impunity knowing that their conversations are immune from our most valued investigative technique. The chapter also examined our counterterrorism strategy to see how effective it is against the black market satellite encryption trade.

Chapter 25 continues the discussion started in Chapter 24 with an examination of how drug cartels are buying sophisticated encryption technology; how drug cartels will use telephones and other communications media with impunity knowing that their conversations are immune from our most valued investigative technique; why encryption capabilities available to criminals, both now and in days to come, must be dealt with promptly; and, the rise of an off-shore, black market satellite encryption trade. The chapter will also look at Crackers that can take down encrypted military sites, nuclear weapons labs, Fortune 100 companies, and scores of other institutions.

Endnotes

1   "G7/P8 Ministerial Conference on Terrorism," Paris, July 30, 1996, Department of Foreign Affairs and International Trade (CANADA), Electronic Privacy Information Center, 666 Pennsylvania Ave., S.E., Suite 301, Washington, DC 20003, 1998, pp. 1–5.

2   Brock N. Meeks "Jacking in from the Narco-Terrorist Encryption Port," Electronic Privacy Information Center, 666 Pennsylvania Ave., S.E., Suite 301, Washington, DC 20003, 1998, p. 3.

25

Domestic and International Criminal Organizations

Almost 60 years ago, when asked why he robbed banks, master thief Willie Sutton answered famously, *Because that's where the money is.* Today, the *money* is in electrons coursing through the computers and satellites upon which the world depends. Over a billion electronic messages traverse the world's networks and satellites each week. Hundreds of millions of dollars change hands with the flip of a byte. Military might is increasingly a matter of information superiority. Whether to hobbyist cracker, commercial spy, or international terrorist, these transactions expose the soft underbelly of the information age.

> **Note:** A hacker is wildly inventive in software techniques, but a cracker is a hacker who breaks into computers.

In 1996 the U.S. Federal Computer Incident Response Capability (FedCIRC) reported more than 3600 *incidents,* defined as *adverse event in a computer system or networks caused by a failure of a security*

715

mechanism, or an attempted or threatened breach of these mechanisms. The Federal Bureau of Investigation's National Computer Crimes Squad, Washington, DC, estimates that less than 14% of all computer crimes are even detected, and only 9% of those are reported. And, without solidly built investigative techniques, which would contribute to a public perception of safety, the very stability of today's military and commercial institutions—not to mention the cybermarkets that are envisioned for the Internet—is called into question.

Computer Crime is a Risky Business

Computer crime, broadly put, is damaging someone's interests by means of a computer:

- ◆ Stealing computer cycles without having authorized access to the machine.
- ◆ Stealing, looking at, or changing the unencrypted data on that machine.
- ◆ Using it to get to other machines (an increasingly common situation in this age of networks and encrypted satellite communications).
- ◆ Using it to commit more traditional crimes simply updated: hate e-mail, extortion with threats to computer operations, or, in some cases, even physical theft of machines.[1]

Investigation of Federal crimes is generally the responsibility of either the Federal Bureau of Investigation (FBI) or the Secret Service, depending on whether the crimes are of an economic or military nature. In addition, a plethora of Department of Defense investigative services may come into play when appropriate.

By now, most people are aware of how dependent society is on computers. National security is in many ways in the lap of the machines, so to speak—from information on the positioning of reconnaissance satellites to technical analyses of weapons systems. Similarly, just as common criminals have learned that computers are where the money is, so espionage agents have learned that computers are where the intelligence is. Espionage is becoming more and more a game of computer break-ins, computer-based cryptography, and message-traffic analysis.

The cloak has become the satellite, and the dagger, the message data packet. In his 1989 book, *The Cuckoo's Egg*, a widely popular recounting of an espionage case, Clifford Stoll wrote how a 75-cent commercial accounting imbalance in California led him to a West German cracker extracting information from defense computers in more than 10 nations. The information was then sold to a Soviet intelligence agency.

The type of criminal in Stoll's book is not an isolated phenomenon, nor can his skills be classified as *dangerous to military* versus *dangerous to economic interests*. Consider Kevin Mitnick (see photo in Figure 25.1), in some ways the most celebrated cracker of them all, for the number and audacity of his crimes and for his personal cat-and-mouse

Figure 25.1 Kevin Mitnick, at one time the most wanted computer criminal in the United States, is shown being arraigned in Raleigh, NC, two days after his arrest on February 15, 1995.

game with a leading security expert.[2] Mitnick was an equal opportunity criminal. First in trouble with the law at age 16, he was put on probation for stealing a Pacific Bell technical manual. By 1988, when he was 25, he was arrested by the FBI for breaking into the Digital Equipment Corp. computer network and stealing a prerelease version of its VMS operating system software. He then stored the software for the time being on a computer at the University of Southern California.

Let out in a supervised release program for his *computer addiction,* Mitnick then broke the terms of the release, among other things by listening in on a Pacific Bell security official's voice mail. By 1994, the U.S. Marshals, the FBI, the California Department of Motor Vehicles, several local police departments, and several telecommunications companies were looking for him. But his pride led to his downfall: he broke into the computer of Internet security expert Tsutomu Shimomura, and went on to dog him in what became a personal test of skills (see the Sidebar that follows). In 1995, after a chase crossing a plethora of computer systems and data links, he was caught. Mitnick is now facing a sentence in excess of 10 years in prison.

Stalkers Go Online Too but They're Hard to Find

At first, Robert Maynard thought they were harmless—albeit crude—electronic postings. Most closed with the same poem: *Lord, grant me the serenity to accept the things I cannot change ... and the wisdom to hide the bodies of the people I had to kill.* One claimed that Maynard's employees were liars. Others that his wife, Teresa, was unfaithful. But when the messages, posted on an Internet newsgroup, did not stop, Maynard went to court. A Texas district judge granted a temporary restraining order against the alleged harasser, known only through a succession of online aliases, including *Mackdaddy.* And in an apparent legal first, the court delivered the order over the Internet. The FBI is now investigating the case and a civil hearing is scheduled in Dallas soon.

The Maynards are far from cyberspace novices: They helped found Internet America, Dallas's largest Internet-access provider. But as they and a growing number of other online users are learning, the Net is like a frontier town, complete with shady characters and even stalkers. And in cyberspace, a harasser does not even have to forgo the comfort of his or her own living room. The same traditional crimes are being committed in a new environment where the criminals are allowed anonymity.

At one point, Maynard's harasser bragged, *I have a .45.* The man who sent him the messages was eventually identified as Kevin Massey, a Dallas man who had spent time in jail for burglary and weapons possession. Massey, 40, denies Maynard's allegations and claims he was responding to online attacks initiated by Maynard and his employees. Yet Massey also appears to revel in his infamy, calling himself the *Cyberstalker,* and even lobbying to be a guest on shock-jock Howard Stern's radio show.

While Maynard's experience was confined to the electronic world, other cases raise the possibility that online harassment might turn into real-world violence. In one, a woman was harassed online after revealing in a chat group that she was a lesbian. Then the woman was attacked by two men at her home, a beating that broke her nose, cracked her cheek bone and required over 60 stitches and plastic surgery.

Like any big city, the Internet has its share of wackos. But law enforcement officials have had trouble figuring out exactly what to do about online crime. Not only are there few precedents but many police officers know little about how to trace e-mailers. The anonymity breaks down many inhibitions—and in some people seems to bring out the worst. You can have a much broader impact in an anonymous way. During the Olympics, one Atlanta woman learned this lesson the hard way; she was bombarded with crude phone calls after her name, phone number and a fake picture were posted on an Internet bulletin board. Her alleged harasser, jailed initially, remains free.

For the woman in Atlanta, turning off her computer or changing her e-mail account was not going to end the harassment. But even if it could, victims feel they should not be forced off the Net. It is like telling women who get raped to stay indoors. Precautions can be taken to help prevent harassment—for instance, refusing to give credit card or telephone numbers over the Net. Though cases seem to be increasing, cyberstalking still is far from epidemic. Over 140 million Americans use the Net, and the number of incidents remains small—though no exact figures are available. Still, the term *cyberstalking* is becoming a more common phrase online. Recently, Maynard's company ran a promotion for a low-budget film called *Cyberstalker.* Now, it is an irony Maynard would prefer to forget.[3]

For obvious reasons, the government is loathe to release information on its lapses in security, and inferences must be made from the few cases that have come to light. In 1991, for example, attacks were reported at facilities belonging to the U.S. Department of Energy, which among other things manages much of the United States' nuclear weapons research. The intruders were prevented from obtaining classified information, and an investigation was begun at once. Several weeks later the intruders were identified and located outside the United States (see the Sidebar that follows).

U.S. Nukes Vulnerable

Security lapses at some federal labs and plants that house nuclear weapons make them vulnerable to theft and sabotage according to recent government reports. According to a recent confidential Pentagon review, they found inadequate safeguards at a number of facilities run by the Department of Energy.

The report notes that the latest audits expand on a recent review. A number of earlier studies have come up with similar findings.

A draft of the Pentagon review raises *serious concerns over the status of physical security* at DOE facilities. Another report by a DOE security task force urged immediate action *to meet the developing crisis in special nuclear materials protection.*

An international DOE memo noted that *serious issues exist at key facilities possessing hundreds of kilograms of direct use nuclear material.* Nevertheless, DOE is strengthening guard forces and replacing security equipment.

DOE does not think they are in crisis. However, if they do nothing, they will be in a crisis.[4]

More recently, a wiretap order was used to trace and identify 21-year-old Julio Cesar Ardita of Buenos Aires, Argentina, who used a Harvard University computer to gain access to the Navy Research Laboratory, NASA's Jet Propulsion Laboratory, Ames Research Center, the Los Alamos National Laboratory, and the Naval Command Control and Ocean Surveillance Center. Consider that Ardita was essentially acting on his own and not backed by the tremendous resources of an enemy country's national computer facilities or by payments from an enemy country's treasury. It is clear that military and government systems are enduringly attractive targets for computer criminals, whatever their motivation.

Attacks directed against economic resources, by the same token, are wide-ranging both in intent and damage. They can range from strategic attacks against the nation (the corruption of the banking system) to vandalism and plain old theft, whether of money or corporate information. Individual computer users, international agencies, or corporations from small offices to conglomerates are all possible victims of computer crime.

Follow the Money

Some revealing information on the typology of the crimes has been uncovered by a new study conducted by San Francisco's Computer Security Institute (CSI) in cooperation with the FBI. The *1997 Computer Crime and Security Survey* was aimed at determining the scope of the crime problem, and thereby raising the level of awareness of it among present and potential victims.

The CSI/FBI survey of 674 organizations of all sizes reinforced what was already suspected — that computer crime is a real and dangerously stealthy threat. Seventy percent of the respondents were able to quantify their total loss due to the crimes, and the figure came to more than US $200 million.

Analysis of the breakdown of the statistics on monetary loss and type of crime is tricky, because not all victim groups were able to report financial losses reliably, nor can their monetary loss be compared with other losses due to criminal acts. Bearing that in mind, the report's summaries are interesting. Of those respondents incurring financial loss, three-quarters reported computer security breaches ranging from fraud (37 respondents and almost $36 million in losses) and loss of proprietary information (44 respondents, $32 million lost) to telecommunications fraud. The rest of the losses were due to sabotage of unencrypted data or satellite networks, viruses, unauthorized penetration by insiders and outsiders, and an old crime updated — the stealing of laptop computers.

It has long been assumed that most computer security problems are internal. But only 36% of the respondents reported one to six attacks from the inside, whereas 54% reported the same numbers for attacks from the outside.

Crime Classifications

Computer crimes range from the catastrophic to the merely annoying. A taxonomy commonly adopted for them and of use to investigators groups them in terms of the four classical breaches of security. The first, physical security, covers human access to buildings, equipment, and media. The second, personnel security, involves identification and risk profiling of people within and without an organization.

The third group is the most purely technical of the four, security of satellite communications and unencrypted data. Finally, the preceding three are shackled if gaps occur in operations security—in the procedures in place that control and manage the security against the preceding areas of attacks, as well as procedures for postattack recovery.

Physical security concerns itself with the protection of assets. Breaches of physical security include dumpster diving, in which offenders physically rummage through garbage cans that may hold operating manuals or specifications. Electronic wiretapping, electronic eavesdropping, and denial or degradation of service also are considered physical crimes, because they involve actual access to the computer or cable.

Denial of service covers the physical disabling of equipment or the flooding of satellite communications networks by waves of message traffic. The 1988 Internet Worm, the first Internet criminal event to be reported widely in the public press and the first case prosecuted under the 1986 Computer Fraud and Abuse Act, demonstrated spectacularly the impact of denial of service. This code was released without particular malice in 1988 by its creator, Robert Morris (ironically, as an experiment to enhance security) and was supposed to reproduce itself on one machine after another for a certain time before self-destructing.

But owing to a programming error, like a sorcerer's apprentice the test code continued to multiply on host after host, swamping each in turn until the Internet was basically at a standstill. In fact, even Morris's e-mailed suggestions for a fix, sent anonymously to system administrators on the first day of the crisis, never made it through the congestion. Administrators had to shut down computers and network connections, work was halted, electronic mail was lost, and research and other business was delayed. The cost of testing and repairing the affected systems has been estimated at over $200 million. In 1990, Morris was convicted and fined $10, 000 (the maximum amount under then-current law), essentially for reckless disregard of the possible damage his code could do, and for using hosts as unwitting guinea pigs—an act of illegal entry no different than any other cracker's.

Personnel security aims at keeping people, both inside and outside the company, from deliberately or accidentally getting at computers or systems for illegal purposes. A common example is termed social engineering, in which the criminal passes himself or herself off as someone authorized to receive from the legitimate user passwords and access rights.

Breaches in satellite communications and data security are attacks on the end-user's data and the software managing that data. Data attacks, as defined here, lead to the unauthorized copying of unencrypted end-user data, whereas attacks on the software managing that data could exploit the so-called trap doors in many programs to hijack a session in progress or insert Trojan horses. Trap doors are supposedly secret patches that programmers put in their code so they can remotely get at it for repair or other actions. Trojan horses, like their Homeric namesake, are programs that seem innocuous but conceal damaging contents. Related to Trojan horses is the *salami attack*, in which the attacker repeatedly slices off and hangs onto a seemingly insignificant round-off on the fractions of pennies in financial transactions. To put it at its most succinct, taking care of satellite communications and data security is grounded on checking and rechecking a special trinity: the confidentiality, integrity and availability of data.

Operations security contends with attacks on procedures already in place for detecting and preventing computer crimes. A case in point is data diddling—small but significant changes in data values, such as adding a few zeros to a $20 checking account. IP spoofing uses a

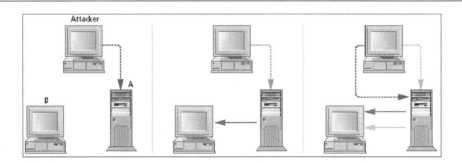

Figure 25.2 An attacker of secure computers often masquerades as a user of a machine with high-level access rights. With Internet Protocol (IP) spoofing, the attacker's computer assumes another IP address. The scam has four steps. First, the attacker acquires the IP address of say, an Air Force general's computer [Computer A, left panel], perhaps by the simple scam of social engineering — pretending to be someone to whom that address can be released. Under the guise of that address, from a third site he opens a session with Computer B [middle], which contains classified information. Believing the request to be from A, B sends an acknowledgment and signals it is ready for communication. The attacker completes the deception by again mimicking A, in a final acknowledgment of B's signal [right]. As far as B is concerned, the attacker is the Air Force general, who may act in whatever way his access rights allow.

method of electronically masquerading as a preexisting but idle computer on the network and initiating a session under that assumed, perhaps privileged, identity as shown in Figure 25.2.[5]

Password sniffing (obtaining passwords) describes the surreptitious monitoring of users' log-in procedures as shown in Figure 25.3.[6] And then there is scanning, the automatic, brute-force attempt at modem access to a computer by successively changing digits in a telephone number or password (scanners are also sometimes called war or demon dialers).

In both IP spoofing and password sniffing, network traffic is monitored by collecting the first 128 or more bytes of each connection, which sometimes contain both the log-in account name and password. Two other types of computer crimes, one predominantly personal, the other financial, do not fit well into these categories but are serious and must

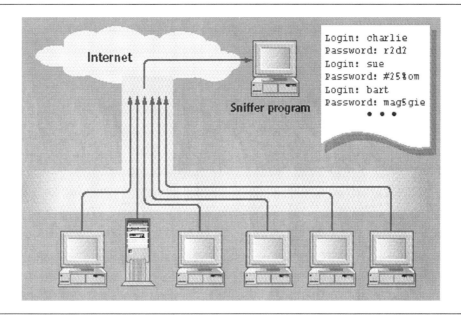

Figure 25.3 Rather than fake a two-way communication session from his or her remote computer using IP spoofing, an attacker may use password sniffing to try to enter a target computer directly. This mode of attack involves monitoring the first 128 or more bytes of each connection, which sometimes contain users' log-in account names and passwords.

be mentioned—harassment and software piracy. Harassment by sending repeated threatening electronic messages has become the latest form of hate mail. And software piracy is a staggering international economic problem, costing losses of revenues estimated at $4 billion, according to the Business Software Alliance and the Software Publishers Association, both in Washington, DC.

Cracker Classifications

Also for the sake of classification and tracking, it is helpful to have a relatively consistent analysis of the types of crackers. One proposal has five categories, each with fairly self-evident occupants: the Novice (mostly quickly bored young kids); the Student (college-age students with an intellectual curiosity in security); the Tourist (who breaks in and persists only if something looks interesting); the Crasher (who delights in simply bringing machines to a halt); and, the Thief (the most serious, knowledgeable and blatantly *criminal* cracker).

For its part, the FBI has established three types of computer criminals: crackers, criminals, and vandals. These so-called offender profiles are based upon interviews of convicted offenders, documented case studies, and scholarly research.

> **Note:** In other contexts these groups are not mutually exclusive: vandalism is legally a criminal act, and so forth.

Crackers are generally young offenders who seek intellectual stimulation from committing computer crimes. Sadly, this type of behavior has often been reinforced as praiseworthy in popular entertainment. In the movie Terminator 2, for example, the boy hero is introduced as he electronically steals money from an automatic teller machine, ostensibly to show how smart he is.

Many offenders are juveniles, who view their computers as the next step up from a video game—for example, in 1990 a 13-year-old boy used a home computer to crack the code of an Air Force satellite-positioning system. He reportedly began his cracking career when he was seven years old.

Criminals, as a profile class, are often adults subgrouped into those who commit fraud or damage systems and those who undertake espionage. Industrial espionage has long been recognized as a shady competitive tactic. Fraud and damage encompasses all forms of traditional crimes—a fertile field for organized crime.

Banks have always tempted computer criminals. As far back as 1987, an eight-member group hatched a plot against a bank in a large midwestern city. They made use of a wire transfer scheme to siphon off about $80 million belonging to three companies, first to a New York bank, and then on to two separate banks in Europe. The transfers were authorized over the telephone, and follow-up calls were made by the bank to verify the requests. But, in the group's fatal error, all the follow-up calls were routed to the residence of one of the suspects.

When the deposits did not turn up, needless to say, the three companies called the bank to find out what had happened. Investigators used the telephone records of the verification calls to trace the crime to the suspects. See the two Sidebars that follow for more detailed information on this topic.

Drug Traffickers and Con Artists Vie in the Crowded Waters Offshore

Things just have never been the same on Miami's Brickell Avenue, where suitcases of drug cash were once carted right up to the tellers' windows of its steel-and-glass banks, since the time the United States passed a law requiring that all cash transactions of greater than $10,000 be reported to the government.

For a while it looked like money launderers were in for a tough time all over. To protect its reputation as a legitimate financial center, Switzerland—the pioneer of no-questions-asked banking—tightened deposit rules. So did the Cayman Islands, which have a population of 33,000 and bank assets of $400 billion. So did the tiny British Channel Islands, which enacted tough laws making drug-money laundering a crime, permitting seizure of criminals' assets and requiring reporting of suspicious transactions.

But in odd corners of the globe there are plenty of other obliging countries that have moved in swiftly to take up the slack. And these newcomers are expanding their client base beyond the traditional mix of international criminals, tax cheats, and drug traffickers with cleverly targeted scams aimed at ordinary folks.

In Mauritius, an island in the Indian Ocean, the number of offshore banks has mushroomed from 20 to 3600 in the past few years. This year, Iran's central bank established an offshore banking center in Sirjan to attract foreign capital to the Sirjan Special Economic Zone, a crossroads for trade with Central Asia and the Persian Gulf nations. And Seychelles, a chain of 100 Indian Ocean islands, passed an *economic development* act ensuring that no questions would be asked about deposits of $20 million or more. It was basically an open invitation to criminals to invest their money.

However no one has extended an invitation like that of Antigua. A Caribbean island 9 miles wide by 12 miles long, Antigua has a virtually unregulated banking industry, no reporting requirements and secrecy laws that punish violations of bank clients' confidentiality. The number of banks there grew by 86% in 1996; anyone with $2 million can open a bank, and many consist of nothing but a brass plate or a room with a fax machine.

KGB Bank

This open-door policy has attracted the attention of unsavory characters from as far away as Russia. A Russian bank, Kerneta, opened a branch in Antigua after attempting to do it in Delaware. The start-up capital was reportedly spirited out of Russia illegally, and a U.S. official says Russians moving money offshore include Mafia and ex-KGB officials running extortion and protection rackets and high-placed elites conducting unauthorized trade in commodities and currency.

All told, there are some 50 offshore banking havens, holding assets estimated at $3 trillion to $6 trillion. Offshore centers typically charge no taxes, impose few regulations, guarantee anonymity, and cater to nonresidents. The money that pours into the banks in these countries gives a big boost to the local economies. It is recycled in loans and stimulates job creation.

Legitimate Transactions

Many offshore transactions are perfectly legitimate. Large banks park hundreds of billions of dollars in offshore accounts overnight to earn interest while markets at home are closed. And favorable tax laws in places like the Channel Islands have long made them venues of choice for trust operations of major banks like Chase Manhattan.

But the criminal possibilities of unregulated banking are legion, and nations such as Antigua have become notorious for the extravagant winks they cast at illegal activities. Once deposited, with no questions, ill-gotten gains can be transferred by wire via encrypted satellite codes from any country's officially licensed banks, however shady their operations or origins may be, to any other bank in the world in perfectly legitimate transactions.

The growing number of offshore havens establishing a presence on the Internet (see Figure 25.4) threatens to make the already difficult job of tracking illicit money flows nearly impossible.[7] The implications for law enforcement are staggering. If governments do not find

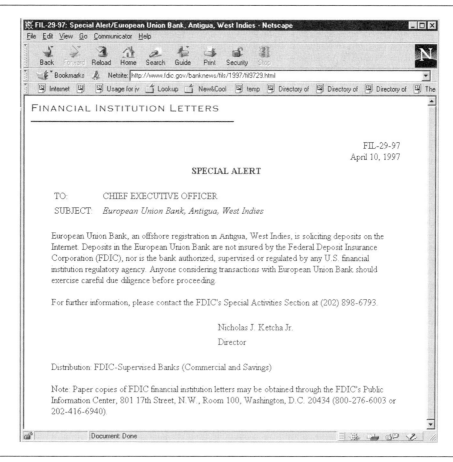

Figure 25.4 Unauthorized banking subject: European Union Bank.

ways of crossing international boundaries as fast as the criminals can use computers to move across boundaries, the criminals are going to win. The United States and its allies discussed the issue recently in Lyon, France, but no one has ready solutions.

The plethora of new outlets has also contributed to the rise of outright scams aimed at the gullible or greedy. Many Internet pitches brazenly appeal to antigovernment, antitax sentiments. Other sites coyly deny they are advocating tax evasion, advising, *Just don't tell us!*

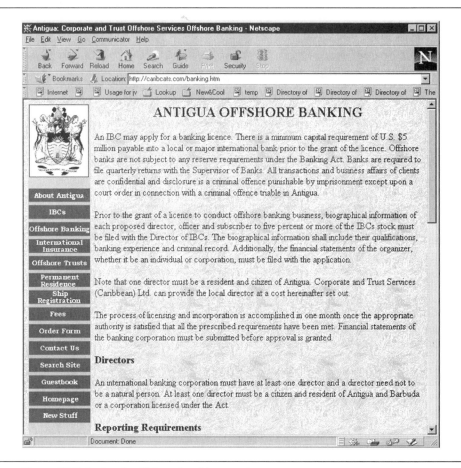

Figure 25.5 Antigua's offshore banking services.

Lax banking laws in places like Antigua mean that victims have little recourse (see Figure 25.5).[8] One victim was Anthony Craddock, a British businessman who had arranged with banker Michael DeBella to collect money owed him in a Nigerian oil deal. DeBella collected the funds but failed to pay Craddock or return a $700,000 escrow payment Craddock was told he would have to deposit in Antigua's Swiss American Bank.

DeBella was arrested and pleaded guilty in a U.S. court to charges related to defrauding other clients in the Caribbean. Yet Swiss

American Bank refuses to return Craddock's money, and Antigua's government has failed to take any action despite repeated appeals. Antigua is not cooperating in indemnifying victims or prosecuting crimes on its island. Swiss American's response to letters seeking its help recovering the money was to alert the account holder to move the money. Previously, DeBella ran a banking scam in Anguilla, bilking China's foreign-trade office of $900,000. His bank license was yanked.

Antigua's ruling Bird family has long made the island's business its business, and the cozy relationship between banks and the government makes action difficult. Antigua's ambassador to Britain is the titular president of Swiss American.

Pressure Cooker
The international community has done little so far beyond urging countries to pass money-laundering laws and implement 50 recommendations issued by the Financial Action Task Force, a group formed by the Group of Seven industrialized countries. The pressure has yielded some results; after FATF instructed its member countries to treat all financial transactions with Seychelles as suspicious, the country decided to hold off implementing its no-questions-asked policy on large deposits. The law remains on the books, however.

Some countries have chosen to avoid the thicket altogether. Last year, Estonia decided against setting up a tax-free zone for fear that it would foster money laundering and other illegal activity. The political damage would be immeasurably bigger than the few millions to be gained from registering offshore companies.

But much of the pressure on the hard cases has only resulted in activity shifting elsewhere. Montserrat, a tiny British colony in the Caribbean, was home to 460 brass-plate banks until 1991, when British authorities took back responsibility for financial oversight. A veteran British investigator who helped close down the banks says that many simply moved to Antigua and Grenada.

Other measures that the United States is contemplating include blacklisting banks, cutting them off from the U.S. payment system, and prohibiting wire transfers to certain countries. In 1996 President Clinton used his executive sanctions authority to freeze U.S. assets of companies or individuals doing business with 90 known traffickers. The Federal Reserve has not allowed some countries, like Russia, to open banks in the United States. Nevertheless, Antiguan officials insist, with a straight face, that such discrimination would not be in keeping with their policies.[9]

Get Your Passport While You Are at It Too!

Laundering money and evading taxes are now but a mouse click away. Thanks to the Internet, you can deposit money and transfer funds to your favorite offshore bank, all in complete secrecy. Best of all, no one will ask where the money came from.

Global Financial Network (GFN), an online service, boasts: *We handle cash derived from any activity.* It offers *services for concealing assets in Antigua and the Isle of Man. GFN will also assist clients in getting their cash into the U.S. banking system, without its source being known to anyone, including the government.*

Claiming to be the first offshore bank on the Internet, Antigua's European Union Bank (now collapsed) offered *favorable Caribbean tax shelter programs that have long been available to international and financial business communities.* You could at one time open a bank account or form a corporation just by filling in an application at its World Wide Web site (http://www.eub.com/) which is now up for sale. Recently, the Bank of England issued a warning to investors that European Union Bank *is not authorized in the U.K. and has not sought authorization.* Like many offshore banks, the now defunct company, which said it had $20 million in share capital and 2000 accounts, did not insure deposits.

Note: The Antigua and Barbuda government issued a fraud alert recently after the collapse of European Union Bank and the disappearance of its cofounders.

Antigua's American International Management Services (AIMS) goes one better, selling permanent residence, passports and citizenship in Antigua, Dominica, and St. Kitts over the Internet in addition to offering banking and incorporation services. Becoming a citizen of Dominica costs $44,600 plus an $86,000 investment that pays 5% interest. Or, for $10,000 plus a $20,000 annual fee, you can become a permanent resident of Antigua and avoid taxes in your home country as shown in Figure 25.6. And AIMS will also sell you a condo that qualifies as a permanent abode—which, it claims, will earn rental income as well.[10]

Vandals usually are not pursuing intellectual stimulation, as when, for example, they deface World Wide Web pages open to the general

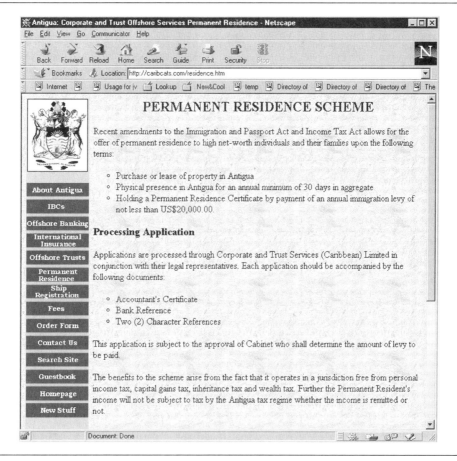

Figure 25.6 Antigua's permanent residence services.

public. The motivations of electronic vandalism often are rooted in revenge for some real or imagined wrong. A corporation undergoing downsizing should be extremely apprehensive of vengeful vandalism by present or past employees.

One of the better known cases in this category is that of Donald Gene Burleson, a systems security analyst at a Texas insurance company who was upset over being fired. Burleson essentially held his employer's computer system hostage. First, he deleted 279,000 of the company's sales commission records. When backup tapes were used to replace the missing files, he then demanded that he be rehired, or else a *logic bomb* in the computer would go off (this destructive software goes into action when triggered by some computational or externally supplied event, such as certain keystrokes or the date). The *bomb* was programmed to take electronic revenge should the employee be terminated. After his arrest and conviction, Burleson was fined $22,900 and sentenced to eight years in prison.

The Law and Computer Crime

Prosecuting computer crimes is usually more complex and demanding than prosecuting other types of crimes. The process requires special technical preparation of the investigators and prosecutors and greater dependence on expert witnesses' testimony. Witnesses testifying for the victims may need to explain why the loss of intangibles (proprietary data, for example) is as serious as losses of tangible goods.

In the United States, the most comprehensive computer crime statute to date was included in the Computer Fraud and Abuse Act of 1986, which added six types of computer crimes to Title 18, United States Code, Section 1030. These newly defined illegal activities include those with traditional, Federal impact, such as unauthorized access aimed at obtaining information on national security; access to a computer used by the Federal government; and, interestingly, unauthorized access to a computer that itself is used to access a Federal government computer.

Equally welcome is the strengthening of the legal defense against economic crimes. Federal law now covers unauthorized interstate or foreign access with intent to defraud or obtain protected financial or credit information; unauthorized access that causes $1000 or more in damage; and fraudulent trafficking in passwords affecting interstate commerce.

Another Federal statute, the Electronic Communications Privacy Act of 1986, is intended to provide security for electronic mail on a par with what users would expect from the U.S. Postal Service. As with postal letters on land, it is now a felony to read other people's electronic mail without their permission.

Note: Nevertheless, there has been case after case where an employer has read an employee's e-mail, and then terminated the employee because of said practice. In all of the cases thus far, no employer has ever been charged with a felony after the terminated employee(s) tried to bring civil and criminal charges against an employer. Is this a double legal standard here?

Precisely how does this affect me, you may be wondering. In brief, should your organization, whether public or private, fall victim to a computer crime, it should immediately seek competent legal representation. You may need to pursue legal remedies to what on the surface first appears to be a nuisance offense. For example, the following are potential legal scenarios, which often lead to messy results:

♦ Unauthorized access to classified or sensitive encrypted data may require mandatory notification to investigator agencies.
♦ Being aware of criminal activities without reporting them may make you liable for your inaction.
♦ Deciding as an executive officer of a company not to make a report and cooperate with law enforcement may lead to your shareholders suing you.
♦ After filing an insurance claim for damages resulting from computer intrusions, you may be required by the insurance company to pursue legal action against the suspected intruders.[11]

In sum, reporting suspected computer intrusions to the appropriate officials, combined with immediately obtaining competent legal guidance, will substantially reduce your liabilities and increase the likelihood of success by law enforcement in dealing with computer crime. The prime contact to make after any computer incident is the 24-hour CERT (Computer Emergency Response Team) Coordination Center, which was formed by the U.S. Defense Advanced Research Projects Agency (DARPA) in November 1988 after the Internet Worm disaster. A similar, but private, organization is the Forum of Incidence and Response Security Teams (FIRST), founded in 1993. Understanding computer crime laws is also key to establishing good internal policies for computer crime prevention.

Understanding the Techniques of Computer Criminals

In the early 1990s, the U.S. government started a series of studies designed to better understand the motivations, tools, and techniques of computer criminals. The information most sought after was good motive-based offender profiles, as well as a command of the lessons learned from criminal events, information which could be relayed to the computer security and law enforcement communities. In the end, the FBI prepared a detailed chart of the vulnerabilities of computer systems, possible threats to them, and countermeasures to take against the threats as shown in Figure 25.7.[12]

Consider a member of a cracker group thought responsible for charging more than $40,000 in unauthorized telephone calls through voice mail systems as well as breaking into credit bureau computers to obtain financial information. This individual pleaded no contest to the charges, was sentenced to provide 260 hours of community service, and given a fine of $366 and court costs. The youthful offender was convicted as an adult.

Time and again the offender had accessed credit bureaus and had been quite successful in the art of social engineering, that is, misrepresenting himself as an authorized user or official; he had an outspoken hatred for law enforcement in general; and, when questioned, he had told police only what he believed they already knew. Further study of the circumstances of the case and the offender's behavior turned up several lessons in security, as well. He had easily gained access to computer systems using default accounts, often even invading the root directory from which all systemwide administrative procedures could be implemented. He had carefully located known weaknesses in computer systems and had been undeterred by warning calls from security personnel.

Vulnerabilities	Physical threats				Procedural threats (on entry, once entry is gained, and in general)		
	Intruders	Fire	Other sources	Illicit access to data lines	Illicit entry	Illicit presence	Threats due to sloppiness
Software modifications	•				•	•	•
Poor auditing	•				•	•	•
Ease of illicit access through software	•				•	•	•
Easily corrupted/accessible data	•		•		•	•	•
Clear paths for disclosure of information	•			•	•		•
Insecure software archives	•			•			•
Poor configuration control of: software	•				•	•	•
hardware	•						•
communication lines							•
Poor physical control of: access	•				•	•	•
environment	•	•					•
personnel and management	•	•					•
Poor contingency planning					•	•	•
Poor communications protection	•						
Poor procedures							•
Susceptible to hazards		•	•				•
Countermeasures	*Alarms *Guards *I.D. badges *Locks on doors	*Smoke/heat detectors *Sprinkler & alarm systems *Area clear of combustible material *"No smoking" rule	*Water sensors *Anti-static carpet *Uninterruptible power supply *Lightning arrestors *Grounding	*Dedicated communication lines *Monitored access points to computers, networks *Shielded computer enclosure	*Assign computer passwords *Monitor employee's use vs. access rights *Install rigorous modification & verification routines when data is manipulated *Install rigorous handshaking (agreement) terms between computers before a session is initiated	*Ensure different security control at different levels of access *Install operating system kernel dedicated to security *Remotely monitor status of both physical and logical machines on system	*Peer review of software *Better documentation *Better training *Institute standard operating procedures

Figure 25.7 Computer vulnerabilities and countermeasures.

Computer Crime Prevention

In the fight against unauthorized access from external sources, or attempts from inside by personnel to exceed their authorized access levels, three distinct computer-crime prevention tools have emerged: firewalls, auditing, and risk assessments. For instance, firewalls are software programs specifically designed as a security interface between the Internet and a local host. When correctly installed and maintained, they safeguard against unauthorized access from the Internet, and can control access from within a company network to the Internet. Usually placed upon a secure workstation dedicated only to hosting security software, most firewalls use Unix as their native operating system. An effective firewall also has mandatory file and virus checking to reduce the likelihood of importing malicious computer worms or code.

Auditing encrypted data transactions with logs (keeping track of who accesses what and with which processes) is a natural byproduct of a firewall. In addition to providing security, this data record can reveal historical patterns in both internal and external attempts to break into computer systems and unencrypted data. A critically important security job is timely reviews of the log and user activity, by both trusted human administrators and automatic procedures that take action upon evidence of anomalies.

Risk analysis—establishing a plan for the security and privacy of each computer system—balances the cost of various types of protection against the costs of doing without them. Periodic risk assessment is the best proactive weapon against computer crimes. Indeed, Section 6 of the Computer Security Act of 1987 (Public Law 100-235) mandates that U.S. Government computer systems containing sensitive information undergo approved risk analyses.

Techno-Criminal

Computer crime is a grave problem. It threatens national security with opportunities for modern criminals that go far beyond anything previously experienced. Although improvements in security are helping to keep it under control, the criminals are keeping pace with technology.

Surveys, case studies, and observations suggest that major problems will be encountered before the year 2001. Satellite communications and networks will continue to be vulnerable, and financial, medical and credit reporting networks will endure major outages as a result. Political extremists and terrorists targeting critical services will score successes, as will organized crime. Major international high-technology financial thefts involving electronic fund transfers and *Internet commerce* will take place. To top it all off, court-qualified investigators and laboratory evidence technicians will be in short supply.

But let us put all this as personally as possible. If you are a manager or owner of a business, computer crime can undermine everything you have worked so hard to accomplish within your organization. Computer criminals, masquerading as authorized users, may be able to figure out how to access and steal the business plans you have labored over. Trade secrets about the product on the verge of being released may help a competitor beat you to market. Disclosure of confidential material may also lead to a loss of credibility with your vendors and put your company at risk of not receiving government contracts.

If you are involved in law enforcement, whether as an investigator or a prosecutor, you may have to deal with either a computer crime investigation or a case where computers have been used by those responsible for other crimes. You may have to assist in the preparation of a subpoena for computer crime evidence, participate in the collection of computers and computer media during an arrest or during the execution of a search warrant, or, be called upon to conduct a major investigation of a computer crime.

As the victim of a computer crime, you may be asked by law enforcement to assist in tracking a computer trespasser, or in putting together data that will later serve as evidence in the investigation and prosecution of a suspected computer criminal. If you are an ordinary computer user, realize that you, too, are vulnerable. If you fail to protect your log-in account password, files, disks and tapes, and other computer equipment and data with encryption, they might be subject to attack. Even if what you have is not confidential in any way, having to reconstruct what has been lost could cost hours, days, or longer in productivity and annoyance.

Finally, in this era of networked computers and satellite communications, even if your own data are not a worry, you have a responsibility to protect others. Someone who breaks into your account could use that account to become a privileged user at your site. If you are connected to other machines, the intruder could then use your system's networking facilities to connect to other machines that may contain even more vulnerable information.

The word responsibility, at all levels, sums it up. By working together responsibly, far more often than not the good guys can outmatch their adversaries.

Well, sometimes. It all depends on whether the good guys are on the same experience level as their adversaries. Case in point, what about the computer geeks who can take down the satellite communications and networks of military sites, nuclear-weapons labs, Fortune 100 companies, and scores of other institutions? It might be partly the good guys' fault if they are not up to par with these electronic burglars.

The Electronic Prince of Thieves

To computer wizards, the term *hacker* is reserved for unusually clever software programmers. To them, the electronic burglars who break into computers are not hackers but *crackers*. They convene to discuss their targets and techniques on underground Internet chat channels—where each member's typed comments appear simultaneously on everyone else's screen.

In 1992, the hottest chat channel in the world was called #hack— *pound-hack.* One frequent visitor was an apprentice cracker who called himself Phantom Dialer or, more often, Phantomd. He watched in fascination as expert crackers from the United States, Europe, Israel, and Australia swapped cryptic words of advice. When he tried to chime in, many hackers gave him a hard time. They called him a lamer, a loser, a poser. Because his keyboarding was slow, they said he must be stupid.

A cracker known as Grok was more sympathetic. He treated Phantomd as a younger, less experienced version of himself, an open vessel in which to pour knowledge. He provided Phantomd with programs that were the electronic equivalent of a burglar's power tools.

Some of Grok's tools were Trojan horses, called that because of the nasty surprises they hid. Replicas of legitimate programs, they could let crackers enter networks covertly and hide their tracks before they left. Phantomd picked up other programs, too, including a cracker favorite, Crack. It guesses passwords by systematically trying every word in a dictionary, along with common names, geek slang words, and other favored keyboard patterns. It lets crackers enter computer realms, such as private files, where they are forbidden to go.

Tools in hand, Phantomd accomplished the biggest invasion of supposedly secure computers since the Internet was created. No one knows how many systems he broke into—it was at least several hundred and possibly thousands. Many were among the most sensitive (and supposedly secure) in the nation, including those of classified military sites, nuclear weapons labs, bank ATM systems, Fortune 100 companies, even dam control systems. When the FBI, assisted by computer administrators, caught Phantomd, his identity was so stunning that the case was kept confidential and discussed nowhere in print until just recently.

Today, hundreds of thousands of marginally skilled crackers could do what he did. Most computer experts agree that during this decade, Internet security has gotten worse. Personal correspondence, bank accounts, business transactions, local emergency systems, national defense—all have been jeopardized. Much of society seems to be rushing onto the Internet, but the long-secret escapade of Phantomd demonstrates how easily people can roam unconstrained on the information superhighway and, if they have a mind to, do overwhelming damage.

Perhaps most surprising, however, many crackers can be contained by relatively simple precautions. Security experts have long urged such safeguards on the people who run computer networks and encrypted satellite communications as well as on those who use them. An Internet crime wave may be avoidable, but only if these cautions are finally taken seriously.[13]

Intruder Alert!

The first time anyone at the Portland Center for Advanced Technology (the high-tech center at Portland State University in Oregon) noticed the intruder in the computer network was in April of 1992. By July, computer system administrators (sysadmins, in the jargon) realized that a cracker named Phantomd was using the Portland network to hop through the Internet to the Massachusetts Institute of Technology (MIT), across the country.

There, he read faculty e-mail and learned that professors often wrote their friends or students something like, *"I'm going to be away for two weeks, so if you want to use my MIT account the password is NNXXN."* People frequently used the same password for all their accounts, it turned out, not wanting to memorize multiple ones. With this simple insight, Phantomd realized, he could go far.

Another trick of his was looking for .rhosts files. Sysadmins set them up to allow computer users who work on several networks to pass from one to another without the bother of typing passwords. Such shortcuts are a typical feature of Unix, the software that underlies most large networks and the whole of the Internet.

By looking for reused passwords and carelessly placed .rhosts files, Phantomd was able to leap from MIT to many other networks, and from those to dozens more. In one of them, he found a passwordless *root* (or control) account that let him comb through a network until he turned up a machine with an .rhosts file that led to a network at the National Aeronautics and Space Administration (NASA).

Visiting the NASA site, he learned that the agency had an enormous computer system, a network of networks. Phantomd could copy password files from these networks to other computers, including a little-used workstation at MIT. As he had with the Portland State password files, he ran the NASA files through Crack, the password-guessing software. It usually decoded a few passwords, enough to let him tunnel farther into NASA.

Phantomd reported his exploits to his mentor, Grok. By June 1992, he was in deep enough that Grok told him to be careful. *Don't worry*, Phantomd replied. No way was he going to crash the space shuttle. Grok was not amused. Presumably, NASA took precautions against damage from intruders, but the agency was supposed to keep people like Phantomd from getting in at all. Because NASA was involved in so many fields, from medicine to microwaves, the potential consequences of Phantomd's security breaks were huge.

Phantomd had no intention of stopping. He hoped that NASA's supercomputers could help him overcome Crack's limitations. The program guessed hundreds of passwords a second but needed weeks to generate the millions of possible matches required to break a password. But if Phantomd could gain access to NASA's supercomputers, he could grind through all combinations in a few hours. That would give him access to places beyond NASA—Places nobody in #hack had ever reached.[14]

The Hunt Is On

One of the sysadmins at the Portland Center for Advanced Technology was a Sri Lankan émigré named Janaka Jayawardene. After he was alerted to Phantomd's break-in, he decided to eavesdrop on the cracker. Jayawardene collected a 3-inch stack of computer paper recording the intruder's movements on the Net. He concluded that Phantomd was part of an underground cabal that felt no compunction about cracking.

Alarmed by this activity, Jayawardene asked for help in tracking down Phantomd, first from the university phone company, eventually from the FBI. He struck out everywhere, at half a dozen places. After several more months of monitoring Phantomd's break-ins, Jayawardene *observed* him entering Los Alamos and Lawrence Livermore national laboratories. Because of the gravity of the harm he could do there (the labs develop nuclear weapons) the sysadmin tried the FBI's Portland office again. This time, he connected with Special Agent E. Brent Rasmussen, one of the few agents who had ever worked on a cracker case—though, oblivious to the insult to computer good guys, the bureau invariably referred to the perpetrators as *hackers*.

For several years, sysadmins all over the country had been hounding the FBI to do something about electronic crime. The agency scoffed at such requests. But after agitation by a few of their agents, the FBI agreed to set up a National Computer Crime Squad. It opened in March 1993, in the Washington, DC, metropolitan region. At first, most of the calls that came into it were from sysadmins who had observed crackers knocking on their doors—unsuccessfully trying to break in. Successful break-ins were rarely observed.

But the 11-member squad had been hearing from agents about organizations hit by Phantomd, including Intel, the world's largest manufacturer of computer chips. Then the squad learned from Rasmussen that something solid had popped out of the haze of cracker incidents. The intruder might turn out to be a thrill-seeking adolescent equipped with a modem. Yet the description (of a single individual who worked with others and had been running through networks across the nation) corresponded to a major worry for the FBI.

Criminals could spend months trolling through the global satellite network for systems worth cracking. After accumulating a list of targets, the crackers could in theory commit a rash of crimes in two minutes—and leave the country. If they were terrorists, they could black out a power system or open the sluice of a dam and, from a hotel room abroad, watch the chaos or destruction on CNN.

Because Phantomd was sometimes entering the Portland State system through the account of an undergraduate named Patrick Humphreys, Jayawardene asked a student sysadmin to put Humphreys on notice. If he was involved with Phantomd, his computer privileges would be canceled—he might even be kicked out of school. While Humphreys was not involved, he said he had a friend named S. Singer who might know something. Humphreys and Singer had met in high school through Dragonfire, a computer bulletin board in Portland. The two had hit it off, online and in the real world. Humphreys had visited Singer at his house outside Portland. There, he had met S. Singer's younger brother.

M. Singer (S. Singer's younger brother) was a strange kid with poor eyesight and other frailties who all but lived through his computer and who sometimes broke into systems to get to online bulletin boards. When the student sysadmin confronted Humphreys, Humphreys called S. Singer to ask if his brother was breaking into systems again. "Oh, yeah," S. Singer said, "my brother has gotten into all this stuff." M. Singer picked up the phone and said exactly the same thing. Humphreys told Jayawardene and Jayawardene told the FBI.

The agents assumed the worst, because the consequences of ignoring a criminal could be horrendous. Even if it were only a teenager who was doing the cracking, he could be helping criminals. Without understanding what he was doing, he still could be scoping out sites for them to hit.

M. Singer was the prime suspect, but the logs from Portland State and MIT, as well as from other sysadmins who had been tracking him, would not be enough to get an arrest warrant. Consulting with the computer-crime squad, Rasmussen realized that the only way to establish that M. Singer's computer was the source of the attacks was to eavesdrop on all data transmissions to and from his home phone line — to wiretap his computer.

To FBI agents who monitored transmissions in Washington, the amount of time that M. Singer spent at the keyboard was amazing. Eleven, even 13 hours a day was standard; he often did more. He sometimes stayed online through three shifts of observers. The datatappers gave up observing him live after he ended an epic 40-hour session with a four-hour string of H's. Apparently, he had fallen asleep, his face flopped on the keyboard. When he woke up, he resumed work immediately.

Phantomd would sometimes switch to a modem that the FBI's equipment could not decode. As a result, the FBI agents feared that they were not following the worst of what the cracker was doing.

The agents were right. Singer and another cracker (a more knowledgeable one whose online name was Jsz) had given up their dream of running Crack on a supercomputer. They had discovered that the program could not take advantage of a supercomputer's speed unless the software was rewritten. But they were working on a more dangerous plan: tapping into the Internet backbones.

Initially put together by the National Science Foundation, the backbones (a network of superfast communication lines) are the interstate highway system of the Internet. Bank transactions, stock market information, credit reports, corporate strategies, top-secret research, private e-mail, marketing numbers, data on utilities and waste dumps and building construction — all pass through the backbones.

To exploit the river of information, Jsz had written a new kind of program that read all the data that passed in or out of a computer. Called a sniffer and now common, it can be set to strain for signals of any sort — key words such as merger or proxy, say, suggesting financial transactions, if he had wanted to make a killing in the market.

Phantomd and Jsz were not much interested in moneymaking schemes. They wanted something more fundamental: access. It meant belonging to an exclusive club of people who never had doors closed in their faces and who roamed effortlessly through the spaces of the global satellite communication network. To gain access, Jsz had designed the sniffer to filter streams of data at the start of a session on the Internet. The program would record the first few keystrokes, which contained the items of interest: the account name and the password.

After testing the sniffer on machines throughout the Internet, Phantomd decided to install the software on the backbones. But the network was bristling with security measures, so Phantomd and Jsz needed something special to get in. They had it — in the form of Sendmail, the ubiquitous program that sends and receives electronic mail on Unix machines.

Because Sendmail must be able to put messages into every user's electronic mailbox, the program must have access to every part of the network. It is also notoriously buggy. In his Internet wanderings, Phantomd penetrated several networks used by security experts, and he managed to read their e-mail. He was rewarded with an early peek at a newly discovered Sendmail bug.

Alarms on the backbone network were designed to go off if anyone seized access. But Sendmail already had access, so by using it Phantomd would not trip the alarms. By typing in some relatively simple commands, he could use Sendmail to e-mail himself control of a computer network.

He slipped into subnetworks of the backbone. Working diligently through the fall of 1993, Phantomd was able to install sniffers on the major backbone networks and leave the system.

In each system, Phantomd monitored a tiny slice of backbone traffic for a short interval. That was enough to let him accumulate about 70 megabytes of sniffer logs with hundreds of thousands (perhaps even millions) of passwords and account names. Military networks, government systems, commercial satellite communications — these fed data through the backbone, where it became his prey.[15]

Classified

From the datatap, the FBI agents saw that Phantomd was slipping into one system after another. He moved so quickly from one network to the next that it was as if he already had passwords for the systems. Many of the networks belonged to universities and the like, but they were not important enough for the FBI agents to step in and protect right away. Their instructions were to wait for something big.

Singer soon provided it. The datataps showed the suspect breaking into the computers of the Naval Research Laboratory, a weapons-research center just outside Washington, DC's beltway, and the Ballistic Research Laboratory, a facility operated by the Department of Defense in Aberdeen, MD. In both places, he rifled files, established phony accounts, and replaced software with Trojan horses.

The FBI agents called the sysadmin at Naval Research, who denied adamantly that any break-in could have taken place. The FBI agents drove to the laboratory with a transcript to prove otherwise. When they finally left, the Navy sysadmin was satisfyingly wheyfaced.

At the ballistics center, the two agents met with an easygoing computer manager. The manager spread the printout of the transcript on the table. Working his jaw, he scanned silently through the exchanges. Then he said: *Oooooh.* The FBI agents looked at one another. Oooooh?

The manager explained: The cracker had gotten right next to some black systems—computers that contained classified secrets. Researchers were not supposed to keep classified data on computers accessible from the Internet, but many people broke the rules. If the ballistics center was lucky, everyone had been punctilious. But it made no one happy to think that defense security depended on luck.

The FBI finally drew the line at letting the suspect sneak into military systems while they were watching. The FBI consulted with the Justice Department. The bureau was finally going to visit this M. Singer.

Early morning. Two days before Christmas 1993. M. Singer was upstairs and online. Agent Rasmussen, parked a block away, confirmed the suspect's status by radio. He and five other agents moved toward the house and took up positions at the front and back doors. Agent Rasmussen knocked sharply on the front door.

It was opened by S. Singer, in his underwear. He grimaced at the people in suits and ties on his front stoop. What were the Jehovah's Witnesses doing at the door this early? Then Steve saw the guns and badges. Agent Rasmussen was already charging past him.

At the top of the stairs, Agent Rasmussen spotted a closed door and rapped. Open up! It's the FBI! A shrill, irritated voice responded: Shut up! Agent Rasmussen had not waited for the answer, though; he shouldered his way in, intending to yank Singer from his computer before he could destroy evidence. But he was momentarily stopped by the room's darkness and a terrible smell. When the details emerged from the gloom, they were paralyzing. The room was squalid, heaped almost knee-high with mounds of paper, clothing, food, and electronic components. The clothing was filthy and mildewed; moldy pizza slices slid out of greasy boxes; and pancakes swam in fetid butter.

Approaching Agent Rasmussen from the center of the mess was a bent figure whose face and body were wrapped in what looked like an Air Force parka. His hands were up in surprise, the fingers bent as if gnarled. He wore glasses but Agent Rasmussen could not make out eyes behind the lenses.

As Agent Rasmussen took M. Singer downstairs, other FBI agents searched his room, packing up the computer. They asked S. Singer if his brother had access to his computer, too. S. Singer said yes, so the agents took it as well. Meanwhile, Agent Rasmussen was in the kitchen, still baffled by the scene upstairs. It was like nothing he had ever seen before.[16]

Wakeup Call

Agent Rasmussen sat in the interrogation room while other FBI agents tried to extract information from M. Singer. The dialogue soared over the agent's head into the realm of baud, bits, and bytes. One look at M. Singer twisted in his chair, however, and it was clear to Agent Rasmussen that no jury would vote to convict him. Maybe that was justice — prison would be a disaster for him; he would never survive it.

On the other hand, Agent Rasmussen was not sure whether the kid grasped that he was in trouble. M. Singer spoke at length only when the discussion veered toward the technical. Just one thing had sunk in — the FBI had taken away his computer. When it was brought in for him to demonstrate, his posture transformed. Sitting up straight, he cradled the keyboard like a baby in his lap. He typed with an awkward fluency, his hands seeming to bat at the keys in short, rapid up-and-down motions. Almost 21, he seemed unable to understand the consequences of his actions or why anyone would be upset by them. The FBI agents asked him if he had ever found a system he could not get into. M. Singer paused a long time before responding: No.

In Washington, DC, the Justice Department did not seek an indictment. Agent Rasmussen was furious as was the head of the computer-crime squad. Both blame the collapse of the case on Joshua Silverman, the prosecutor in charge. But for the government, the advantages of his decision are clear. The case remains open — and out of the public record.

The FBI has not returned M. Singer's computer. Shuttling among the homes of family members, the former cracker has been trying to put together a small company that would provide Internet access to people's homes — Quicknet, he calls it. He borrowed money from his mother to obtain the equipment; it litters his bedroom. Today, the world's most successful known cracker lives on Social Security disability payments.[17]

Lack of Security

How could someone of such limited capacity break in wherever he wanted? Could he have been a front for a more purposeful, wicked party? Agent Rasmussen (now retired) believes that to be possible. But there is another, unnerving explanation: skulking around the Internet is easy. When Phantomd was operating, anyone with enough time, patience, and motivation could do it. Since then, many security experts believe, the situation has gotten worse. In your network, just when you think everything is safe, an intruder could be hovering in the virtual shadows.

The Internet's lack of security has many causes, beginning with the reluctance of computer users to employ the simplest safeguards, such as un-Crackable passwords. People find it harder to type a random password like *tn%xlrZ@* than an easily guessed one like *cool* or *hello*.

Systems administrators are guilty of security lapses, too. In many cases they are inexperienced and untrained, because the supply of top-notch administrators has not kept pace with the explosive growth of the Internet. Worse, many sysadmins who know better are unwilling to practice safe computing. Given the pressure of their jobs (how many people wait patiently when their computers crash?), sysadmins are reluctant to make their own lives more difficult by enforcing security measures. They do not like the grief they get when they ask users to change their passwords frequently.

Hardware and software companies could solve many of these problems by building into their products security precautions that are difficult to disable. But the vendors know that while their customers claim to want security, they dislike paying any price.

Consider the practice of selling new network computers with prein- stalled accounts that have common names like sys (for system) or bin (binary). When purchasers get stuck setting up their machines, the preinstalled accounts let technicians dial in from the factory to fix the problem. The accounts make life easier for the customer—but are just as helpful to crackers like Phantomd, who check for the presence of such accounts.

User resistance, sysadmin inexpertise, vendor foot-dragging—they all combine to make the cracker's job easy. At the end of 1997, Dan Farmer, a security expert at Sun Microsystems and one of the authors of *Satan*, tested almost 3000 computer networks with obvious security needs. Almost 76% had security holes so large that, in less than a minute, intruders could have entered and seized control, possibly obtaining data from a bank, altering an article at a magazine, or outing a sex-club habitué.

Crackers can also penetrate networks in more sophisticated ways. Typical intruders remain as amateurish as ever but are increasingly equipped with powerful cracker tools, as Phantomd was. In 1996, networks across the globe were hit by RootKit, a cracker's version of a software suite. Put together by an unknown cracker and distributed via the World Wide Web, RootKit automatically grabs control of a computer, installs Trojan horses, and sets up a descendant of Jsz's original sniffer. To run it, a cracker simply types *make all install.*

Computer-security experts often complain that the media glamorize crackers, portraying them as wünderkinder who outwit grownups using laptops and encrypted code. In fact, the rise of ever simpler cracking programs like RootKit means that patient people with modest computer skills but excellent tools pose as grave a danger. Phantomd/Matt Singer is a harbinger of the future.

Many incidents continue to be adolescent pranks, but security experts universally agree that the political and financial risks of break-ins will rise. The U.S. Department of Defense has more than 3.2 million computers linked to some 20,000 local networks, which are tied to 200 long-distance networks. These networks, the Pentagon says, are under siege.

According to data from the Defense Information Systems Agency (DISA), the military may have experienced 330 million attacks in 1996, the most recent year for which figures are available. More than likely, DISA believes, about 75% of the attempted intrusions were successful; typically, fewer than 2% were detected. Many attacks on these military systems are trivial pranks, like the pornographic defacing of the Air Force's main Web page just recently. But some are not—like the case of Dutch crackers who stole information about troop movements and missile capabilities from 45 U.S. military sites during the Gulf war and offered it to the Iraqis. If, as DISA suggests, the number of incidents is tripling every year, the Department of Defense network will be attacked 6 million times in 1999. Computer crime is a major threat to national security.

It is also an economic threat. Estimates of the financial impact of computer crime have risen with the number of cracker incidents. In 1996, the FBI estimated that digital criminals cost U.S. businesses $8.6 billion a year, with the losses ranging from outright industrial espionage and willful destruction of files and data to the simple cost of forcing sysadmins to plug holes. Also in 1996, according to an aval survey conducted by the Computer Security Institute in San Francisco, about 2 out of every 6 networks on the Internet were penetrated.

In mid-1997, the Computer Security Institute and the FBI announced the results of a joint survey of 539 U.S. corporations, government agencies, financial institutions, and universities. Fifty-three percent had discovered electronic malfeasance on site in the previous year, with many of the companies believing that the attacks were performed by crackers for foreign competitors. More than two-thirds of the incidents came from outside attacks (the rest, from insiders who abused their access). Fearing negative publicity, fewer than 18% of the institutions reported those attacks to law enforcement. The surveys suggest that computer criminals are costing private enterprise hundreds of billions of dollars.[18]

Can anything make computers safer from intrusion? Any organization can take steps to help secure its systems.

Strict Password Policy Enforcement

A password should never be a real word in any language, frontwards or backwards, nor should it be a proper name or a plain number. Instead, it should be made like this: Take an ordinary word and replace a few of the letters with numbers; or take the first letter of each word in a short, easily memorized sentence. A password should never be sent by e-mail or given out over the phone. And a password should be changed often.[19]

Proper Employment of Firewalls

A firewall is a specialized computer or program that sits between an organization's network and the outside world. Outsiders who try to tie into the network via phone or satellite must satisfy this device before entering the network. Firewalls typically check the origin and type of all incoming data to block suspicious or high-risk activity—messages directed at the system's controls or e-mail from unknown addresses. Firewalls are valuable tools, but the software can be confusing to set up. By some estimates, more than half of all firewalls are not set up properly and crackers can blow right by them.[20]

Use Encryption

Encryption software scrambles data so that even if crackers get to e-mail or files, they will not be able to read them without a *key*—a longish password that unscrambles the data. As previously stated, encryption is so effective that the federal government has tried (mostly unsuccessfully) to limit its unrestricted use, fearing that criminals will encrypt their own computer data, stymieing law enforcement agents and prosecutors. But encrypted data can be inconvenient: If you lose or forget your key, you cannot get to the data. And encryption is not foolproof. Even if crackers cannot unscramble files, they can delete them. But by any measure, encryption is a big step toward lowering risks.[21]

Report All Attacks

The main reason crackers break into computers easily is that most organizations conceal their problems, fearing that publicity would make them look hapless or attract more attacks. So crackers do not get caught, other organizations are not alerted to dangers, and the public remains unaware of the magnitude of the problem.[22]

If all attacks were reported, major institutions would feel pressure they do not today—software vendors, to improve the built-in security of networks; systems administrators, to take appropriate precautions; the government, to beef up enforcement; and, the rest of us, to engage in safer computing. Until that happens, as hackers and crackers put it, *Internet security* will remain an oxymoron. The more important the Internet grows, the more vulnerable it becomes.

From Here

Chapter 25 continued the discussion started in Chapter 24 with an examination of how to collar domestic and international cybercrooks. The chapter also explored crackers who can take down encrypted military sites, nuclear weapons labs, Fortune 100 companies, and scores of other institutions.

Chapter 26 discusses a variety of options for the satellite encryption policy, some of which have been suggested or mentioned in different forums (in public and/or private input received by the committee or by various members of the U.S. Congressional Committee of Privacy and Technology). These policy options include alternative export control regimes for cryptography and alternatives for providing third-party access capabilities when necessary. In addition, Chapter 26 addresses several issues related to or affected by cryptography that will appear on the horizon in the foreseeable future.

Endnotes

1   David J. Icove, "Collaring The Cybercrook: An Investigator's View," Manager, Special Projects and Technical Investigations, U.S. Tennessee Valley Authority Police, 400 West Summit Hill Drive, WT CP, Knoxville, TN 37902-1499, 1997, p. 31.

2   Julian Harrison, Liaison International & The Gamma Liaison Network, 6606 Sunset Blvd., #201, Hollywood, CA 90028, 1997.

3   Federal Bureau of Investigation, FBI.

4   Department of Defense.

5   David J. Icove, "Collaring The Cybercrook: An Investigator's View," p.33.

6   Ibid.

7   FDIC's Public Information Center, 801 17th Street, N.W., Room 100, Washington, DC 20434.

8   Corporate and Trust Services (Caribbean) Limited, P. O. Box 990, Ryan's Place, High Street, St. John's, Antigua, W.I., 1998.

9   State Department Bureau for International Narcotics & Law Enforcement Affairs, Washington, DC, 1998.

10   Ibid.

11   David J. Icove, "Collaring The Cybercrook: An Investigator's View, p. 35.

12   Ibid., pp. 34–35.

13   David H. Freedman and Charles C. Mann, *At Large: The Strange Case of the World's Biggest Internet Invasion,* Simon & Schuster, One Lake Street, Upper Saddle River, NJ 07458, 1997.

14   Ibid.

15   Ibid.

16   Ibid.

17   Ibid.

18   Ibid.

19   Ibid.

20   Ibid.

21   Ibid.

22   Ibid.

Part Five

Results and Future Directions

26

Satellite Encryption Technology Policy Options, Analysis, and Forecasts

The current national satellite encryption policy defines only one point in the arena of possible policy options. A major difficulty in the public debate over the satellite encryption policy has been incomplete explanation as to why the government has rejected certain policy options. This chapter explores a number of possible alternatives to current national satellite encryption policy, selected by Congress either because they address an important dimension of the national satellite encryption policy or because they have been raised by a particular set of stakeholders. Although in the U.S. Congressional Committee of Privacy and Technology's judgment these alternatives deserve analysis, it does not follow that they necessarily deserve consideration for adoption. The committee's judgments about appropriate policy options are discussed in Chapter 30, "Summary, Conclusions and Recommendations."

Export Control Options for Satellite Encryption Technology

An export control-regime (a set of laws and regulations governing what may or may not be exported under any specified set of circumstances) has many dimensions that can be considered independently. These dimensions include:

- The type of export license granted.
- The strength of a product's encryption capabilities.
- The default encryption settings on the delivered product.
- The type of product.
- The extent and nature of features that allow exceptional access.
- The ultimate destination or intended use of the delivered product.[1]

The Type of Export License Granted

Three types of export licenses are available:

- A general license, under which export of an item does not in general require prior government approval but nonetheless is tracked under an export declaration.
- A special license, under which prior government approval is required but which allows multiple and continuing transactions under one license validation.
- An individual license, under which prior government approval is required for each and every transaction.[2]

As a general rule, only individual licenses are granted for the export of items on the U.S. Munitions List, which includes *strong* encryption. However, as noted in Chapter 5, the current export control regime for encryption involves a number of categorical exemptions as well as some uncodified *in-practice* exemptions.

The Strength of a Product's Encryption Capabilities

Current policy recognizes the difference between RC2/RC4 algorithms using 40-bit keys and other types of encryption. It also places fewer and less severe restrictions on the former.

The Default Encryption Settings on the Delivered Product

Encryption can be tacitly discouraged, but not forbidden, by the use of appropriate settings. Software, and even software-driven devices, commonly have operational parameters that can be selected or set by a user. An example is the fax machine that allows many user choices to be selected by keyboard actions. The parameters chosen by a manufacturer before it ships a product are referred to as the *defaults* or *default condition*. Users are generally able to alter such parameters at will.

The Type of Product

Many different types of products can incorporate encryption capabilities. Products can be distinguished by medium (hardware versus software) and/or intended function (computer versus satellite communications).

The Extent and Nature of Features that Allow Exceptional Access

The Administration has suggested that it would permit the export of encryption software with key lengths of 64 bits or less if the keys were *properly* escrowed. Thus, inclusion in a product of a feature for exceptional access could be made one condition for allowing the export of that product. In addition, the existence of specific institutional arrangements (which specific parties would hold the information needed to implement exceptional access) might be made a condition for the export of these products.

Note: At the time of this writing, the precise definition of properly escrowed is still under debate and review in the Administration. The most recent language on this definition is provided in Chapter 13.

The Ultimate Destination or Intended Use of the Delivered Product

United States' export controls have long distinguished between exports to *friendly* and *hostile* nations. In addition, licenses have been granted for the sale of certain controlled products only when a particular benign use (financial transactions) could be certified. A related consideration is the extent to which nations cooperate with respect to reexport of a controlled product and/or export of their own products. For example, CoCom member nations in principle agreed to joint controls on the export of certain products to the Eastern bloc; as a result, certain products could be exported to CoCom member nations much more easily than to other nations.

Note: CoCom refers to the Coordinating Committee, a group of Western nations (and Japan) that agreed to a common set of export control practices during the Cold War to control the export of militarily useful technologies to Eastern bloc nations. CoCom was disbanded in March 1994, and a successor regime known as the New Forum replaced it.

At present, there are few clear guidelines that enable vendors to design a product that will have a high degree of assurance of being exportable (see Chapters 5 and 14). The remainder of this part of the chapter describes a number of options for controlling the export of encryption, ranging from the sweeping to the detailed.

Complete Elimination of Export Controls on Encryption

The complete elimination of export controls (both the USML and the Commerce Control List controls) on encryption is a proposal that goes beyond most made to date, although certainly such a position has advocates. If export controls on encryption were completely elimin-ated, it is possible that within a short time, most information technol-ogy products exported from the United States would have encryption capabilities. It would be difficult for the U.S. government to influence the capabilities of these products, or even to monitor their deployment and use worldwide, because numerous vendors would most probably be involved.

> **Note:** The simple elimination of U.S. export controls on encryption does not address the fact that other nations may have import controls and/or restrictions on the use of cryptography internally. Furthermore, it takes time to incorporate products into existing infrastructures, and slow market growth may encourage some vendors to take their time in developing new products. Thus, simply eliminating U.S. export controls on cryptography would not ensure markets abroad for U.S. products with encryption capabilities. In-deed, the elimination of U.S. export controls could in itself stimulate foreign nations to impose more stringent import controls.

The worldwide removal of all controls on the export, import, and use of products with encryption capabilities would likely result in greater standardization of encryption techniques. Standardization brought about in this manner would result in:

- Higher degrees of international interoperability of these products.
- Broader use, or at least more rapid spread, of encryption capabilities as a result of the strong distribution capabilities of U.S. firms.
- Higher levels of confidentiality, as a result of greater ease in adopting more powerful algorithms and longer keys as stan-dards.
- Greater use of cryptography by hostile, criminal, and un-friendly parties as they, too, begin to use commercial prod-ucts with strong encryption capabilities.[3]

On the other hand, rapid, large-scale standardization would be unlike-ly unless a few integrated software products with encryption capabili-ties were able to achieve worldwide usage very quickly. Consider, for example, that although there are no restrictions on domestic use of encryption in the United States, interoperability is still difficult, in many cases owing to variability in the systems in which the cryptography is embedded. Likewise, many algorithms stronger than DES are well known, and there are no restrictions in place on the domestic use of such algorithms, and yet only DES even remotely approaches common usage (and not all DES-based applications are interoperable).

For reasons well articulated by the national security and law enforcement communities (see Chapter 4) and accepted by the Administration, the complete elimination of export controls on products with encryption capabilities does not seem reasonable in the short term. Whether export controls will remain feasible and efficacious in the long-term has yet to be seen, although clearly, maintaining even their current level of effectiveness will become increasingly difficult.

Transfer of all Cryptography Products to the Commerce Control List

As discussed in Chapter 5, the Commerce Control List (CCL) complements the U.S. Munitions List (USML) in controlling the export of cryptography (see the Sidebar, "Important Differences Between The U.S. Munitions List And The Commodity Control List" in Chapter 5, which describes the primary difference between the USML and the CCL.). In 1994, Representative Maria Cantwell (D-WA) introduced legislation to transfer all mass market software products involving cryptographic functions to the CCL. Although this legislation never passed, it resulted in the promise and subsequent delivery of an executive branch report on the international market for computer software with encryption.

The Cantwell bill was strongly supported by the software industry because of the liberal consideration afforded products controlled for export by the CCL. Many of the bill's advocates believed that a transfer of jurisdiction to the Commerce Department would reflect an explicit recognition of encryption as a commercial technology that should be administered under a dual-use export control regime. Compared to the USML, they argued that the CCL is a more balanced regime that still has considerable effectiveness in limiting exports to target destinations and end users.

On the other hand, national security officials regard the broad authorities of the Arms Export Control Act (AECA) as essential to the effective control of encryption exports. The AECA provides authority for case-by-case regulation of exports of encryption to all destinations, based on national security considerations. In particular, licensing decisions are not governed by factors such as the country of destination, end users, end uses, or the existence of bilateral or multilateral agreements that often limit the range of discretionary action possible in controlling exports pursuant to the Export Administration Act. Further, the national security provisions of the AECA provide a basis for classifying the specific rationale for any particular export licensing decision made under its authority, thus protecting what may be very sensitive information about the particular circumstances surrounding that decision.

Although sympathetic to the Cantwell bill's underlying rationale, the Administration believes that the Cantwell bill does not address the basic dilemma of the satellite encryption policy. As acknowledged by some of the bill's supporters, transfer of a product's jurisdiction to the CCL does not mean automatic decontrol of the product, and national security authorities could still have considerable input into how exports are actually licensed. In general, the committee believes that the idea of split jurisdiction, in which some types of cryptography are controlled under the CCL and others under the USML, makes considerable sense given the various national security implications of widespread use of encryption. However, where the split should be made is a matter of discussion; the committee expresses its own judgments on this point in Chapter 30.

End-Use Certification

Explicitly exempted under the current International Traffic in Arms Regulations (ITAR) is the export of encryption for ensuring the confidentiality of financial transactions, specifically for encryption equipment and software that are *specially designed, developed or modified for use in machines for banking or money transactions, and restricted to use only in such transactions.* In addition, according to senior National Security Agency (NSA) officials, encryption systems, equipment, and software are in general freely exportable for use by U.S.-controlled foreign companies and to banking and financial institutions for purposes other than financial transactions, although NSA regards these approvals as part of the case-by-case review associated with equipment and products that do not enjoy an explicit exemption in the ITAR.

In principle, the ITAR could explicitly exempt products with encryption capabilities for use by foreign subsidiaries of U.S. companies, foreign companies that are U.S. controlled, and banking and financial institutions. Explicit *vertical* exemptions for these categories could do much to alleviate confusion among users, many of whom are currently uncertain about what cryptographic protection they may be able to use in their international satellite communications, and could enable vendors to make better informed judgments about the size of a given market.

Specific vertical exemptions could also be made for different industries (health care or manufacturing) and perhaps for large foreign-owned companies that would be both the largest potential customers and the parties most likely to be responsible corporate citizens. Inhibiting the diversion to other uses of products with encryption capabilities sold to these companies could be the focus of explicit contractual language binding the recipient to abide by certain terms that would be required of any vendor as a condition of sale to a foreign company, as it is today under USML procedures under the ITAR. Enforcement of end-use restrictions is discussed in Chapter 5.

Nation-by-Nation Relaxation of Controls and Harmonization of U.S. Export Control Policy on Satellite Encryption with Export/Import Policies of Other Nations

The United States could give liberal export consideration to products with satellite encryption capabilities intended for sale to recipients in a select set of nations; exports to nations outside this set would be restricted. Nations in the select set would be expected to have a more or less uniform set of regulations to control the export of encryption, resulting in a more level playing field for U.S. vendors. In addition, agreements would be needed to control the reexport of products with encryption capabilities outside this set of nations.

Note: For example, products with satellite encryption capabilities can be exported freely to Canada without the need of a USML export license if intended for domestic Canadian use.

Nation-by-nation relaxation of controls is consistent with the fact that different countries generally receive different treatment under the U.S. export control regime for military hardware. For example, exports of U.S. military hardware have been forbidden to some countries because they were terrorist nations, and to others because they failed to sign the nuclear nonproliferation treaty. A harmonization of export control regimes for satellite encryption would more closely resemble the former CoCom approach to control dual-use items than the approach reflected in the unilateral controls on exports imposed by the USML.

From the standpoint of U.S. national security and foreign policy, a serious problem with harmonization is the fact that the relationship between the United States and almost all other nations has elements of both competition and cooperation that may change over time. The widespread use of U.S. products with strong encryption capabilities under some circumstances could compromise U.S. positions with respect to these competitive elements, although many of these nations are unlikely to use U.S. products with encryption capabilities for their most sensitive satellite communications.

Finally, as is true for other proposals to liberalize U.S. export controls on satellite encryption, greater liberalization may well cause some other nations to impose import controls where they do not otherwise exist. Such an outcome would shift the onus for impeding vendor interests away from the U.S. government. However, depending on the nature of the resulting import controls, U.S. vendors of information technology products with encryption capabilities might be faced with the need to conform to a multiplicity of import control regimes established by different nations.

Liberal Export for Strong Satellite Encryption with Weak Defaults

An export control regime could grant liberal export consideration to products with satellite encryption capabilities designed in such a way that the defaults for usage result in weak or nonexistent encryption (see the Sidebar that follows). As well, users could invoke options for stronger encryption through an affirmative action.

Possible Examples of Weak Satellite Encryption Defaults

The product does not specify a minimum password length. Many users will generate short, and thus poor or weak, passwords.

The product does not perform link encryption automatically. The user on either side of the satellite communication link must select an option explicitly to encrypt the communications before encryption happens.

The product requires user key generation rather than simple passwords and retains a user key or generates a record of one. Users might well accidentally compromise it and make it available, even if they had the option to delete it.

The product generates a key and instructs the user to register it. E-mail encryption is not automatic. The sender must explicitly select an encryption option to encrypt messages.[4]

For example, such a product might be a telephone designed for end-to-end security. The default mode of operation could be set in two different ways. One way would be for the telephone to establish a secure connection if the called party has a comparable unit. The second way would be for the telephone always to establish an insecure connection. Establishing a secure connection would require an explicit action by the user. All experience suggests that the second way would result in far fewer secure calls than the first way.

Note: Of course, other techniques can be used to further discourage the use of secure modes. For example, the telephone could be designed to force the user to wait several seconds for establishment of the secure mode.

An export policy favoring the export of satellite encryption products with weak defaults benefits the information-gathering needs of law enforcement and signals intelligence efforts because of user psychology. Many people, criminals and foreign government workers included, often make mistakes by using products *out of the box* without any particular attempt to configure them properly. Such a policy could also take advantage of the distribution mechanisms of the U.S. software industry to spread weaker defaults.

Experience to date suggests that good implementations of encryption for confidentiality are transparent and automatic and thus do not require positive user action. Such implementations are likely to be chosen by organizations that are most concerned about confidentiality and that have a staff dedicated to ensuring confidentiality (by resetting weak vendor-supplied defaults). End users that obtain their products with satellite encryption capabilities on the retail store market are the most likely to be affected by this proposal, but such users constitute a relatively small part of the overall market.

Liberal Export for Cryptographic Applications Programming Interfaces

A cryptographic applications programming interface is a well-defined boundary between a baseline product (such as an operating system, a database management program, or a word-processing program) and a cryptography module that provides a secure set of cryptographic services such as authentication, digital signature generation, random number generation, and stream or block mode encryption. The use of a CAPI allows vendors to support cryptographic functions in their products without actually providing them at distribution.

Even though such products have no cryptographic functionality per se and are therefore not specifically included in Category XIII of the ITAR, license applications for the export of products incorporating CAPIs have in general been denied. The reason is that strong cryptographic capabilities could be deployed on a vast scale if U.S. vendors exported applications supporting a common CAPI and a foreign vendor then marketed an add-in module with strong encryption capabilities.

Note: This discussion refers only to *documented* or *open* CAPIs (CAPIs that are accessible to the end user). Another kind of CAPI is *undocumented* and *closed*, that is, it is inaccessible to the end user, though it is used by system developers for their own convenience. While a history of export licensing decisions and practices supports the conclusion that most products implementing *open* CAPIs will not receive export licenses, history provides no consistent guidance with respect to products implementing CAPIs that are inaccessible to the end user.

To meet the goals of less restrictive export controls, liberal export consideration could be given to products that incorporate a CAPI designed so that only *certified* encryption modules could be incorporated into and used by the application. That is, the application with the CAPI would have to ensure that the CAPI would work only with certified encryption modules. This could be accomplished by incorporating into the application a check for a digital signature whose presence would indicate that the add-on encryption module was indeed certified; if and only if such a signature were detected by the CAPI would the product allow use of the module.

One instantiation of a CAPI is the CAPI built into applications that use the Fortezza card (discussed in Chapter 13). The CAPI software for Fortezza is available for a variety of operating systems and PC-card reader types; such software incorporates a check to ensure that the device being used is itself a Fortezza card. The Fortezza card contains a private Digital Signature Standard (DSS) key that can be used to sign a challenge from the workstation. The corresponding DSS public key is made available in the CAPI, and thus the CAPI is able to verify the authenticity of the Fortezza card.

A second approach to the use of a CAPI has been proposed by Microsoft and is now eligible for liberal export consideration by the State Department (see the Sidebar that follows). The Microsoft approach involves three components: an operating system with a CAPI embedded within it, modules providing encryption services through the CAPI, and applications that can call on the modules through the CAPI provided by the operating system. In principle, each of these components is the responsibility of different parties: Microsoft is responsible for the operating system, cryptography vendors are responsible for the modules, and independent applications vendors are responsible for the applications that run on the operating system.

The Microsoft CryptoAPI

In June 1995, Microsoft received commodity jurisdiction (CJ) to the Commerce Control List (CCL) for Windows NT with CryptoAPI (a Microsoft trademark) plus a *base* cryptomodule that qualifies for CCL jurisdiction under present regulations (it uses a 40-bit RC4 algorithm for confidentiality). A similar CJ application for Windows 95 and 98 is pending. The *base* cryptomodule can be supplemented by a cryptomodule provided by some other vendor of encryption, but the cryptographic applications programming interface within the operating system will function only with cryptomodules that have been digitally signed by Microsoft, which will provide a digital signature for a cryptomodule only if the cryptomodule vendor certifies that it (the module vendor) will comply with all relevant U.S. export control regulations. In the case of a cryptomodule for sale in the United States only, Microsoft will provide a digital signature upon the module vendor's statement to that effect.

Responsibility for complying with export control regulations on encryption is as follows:

Windows NT (and Windows 95 and 98, should the pending application be successful) qualify for CCL jurisdiction on the basis of a State Department export licensing decision.

Individual cryptomodules are subject to a case-by-case licensing analysis, and the encryption vendor is responsible for compliance.

Applications that use Windows NT or Windows 95 and 98 for encryption services should not be subject to export control regulations on cryptography. Microsoft is seeking an advisory opinion to this effect so that applications vendors do not need to submit a request for a CJ cryptography licensing decision.[5]

From the standpoint of national security authorities, the effectiveness of an approach based on the use of a certified CAPI/module combination depends on a number of factors. For example, the product incorporating the CAPI should be known to be implemented in a manner that enforces the appropriate constraints on cryptomodules that it calls; furthermore, the code that provides such enforcement should not be trivially bypassed. The party certifying the cryptomodule should protect the private signature key used to sign it. Vendors would still be required to support domestic and exportable versions of an application if the domestic version was allowed to use any module while the export version was restricted in the set of modules that would be accepted, although the amount of effort required to develop these two different versions would be quite small.

The use of CAPIs that check for appropriate digital signatures would shift the burden for export control from the applications or systems vendors to the vendors of the cryptographic modules. This shift could benefit both the government and vendors, because of the potential to reduce the number of players engaged in the process. For example, all of the hundreds of e-mail applications on the market could quickly support encrypted e-mail by supporting a CAPI developed by a handful of software and/or hardware encryption vendors. The encryption vendors would be responsible for dealing with the export and import controls of various countries, leaving e-mail application vendors to export freely anywhere in the world. Capabilities such as escrowed satellite encryption could be supported within the encryption module itself, freeing the applications or system vendor from most technical, operational and political issues related to export control.

A trustworthy CAPI would also help to support satellite encryption policies that might differ among nations. In particular, a given nation might specify certain performance requirements for all encryption modules used or purchased within its borders. International interoperability problems resulting from conflicting national satellite encryption policies would still remain.

Note: An approach to this effect is the thrust of a proposal from Hewlett-Packard. The Hewlett-Packard International Cryptography Framework (ICF) proposal includes a stamp size *policy card* (smart card) that would be inserted into a cryptographic unit that is a part of a host system. Cryptographic functions provided within the cryptographic unit could be executed only with the presence of a valid policy card. The policy card could be configured to enable only those cryptographic functions that are consistent with government export and local policies. The *policy card* allows for managing the use of the integrated cryptography down to the application specific level. By obtaining a new policy card, customers could be upgraded to take advantage of varying cryptographic capabilities as government policies or organizational needs change. As part of an

ICF solution, a network security server could be implemented to provide a range of different security services including verification of the other three service elements (the card, the host system, the cryptographic unit.

Liberal Export for Escrowable Products with Encryption Capabilities

As discussed in Chapter 13, the Administration's proposal of August 17, 1995 would allow liberal export consideration for software products with encryption capabilities whose keys are *properly escrowed*. In other words, strong cryptography would be enabled for these products only when the keys were escrowed with appropriate escrow agents.

An escrowed satellite encryption product differs from what might be called an *escrowable* product. Specifically, an escrowed satellite encryption product is one whose key must be escrowed with a registered, approved agent before the use of (strong) cryptography can be enabled, whereas an escrowable product is one that provides full cryptographic functionality that includes optional escrow features for the user. The user of an escrowable product can choose whether or not to escrow the relevant keys, but regardless of the choice, the product still provides its full suite of encryption capabilities.

Note: For example, an escrowable product would not enable the user to encrypt files with passwords. Rather, the installation of the product would require the user to create a key or set of named keys, and these keys would be used when encrypting files. The installation would also generate a protected *safe copy* of the keys with instructions to the user that they should register the key *somewhere*. It would be up to the user to decide where or whether to register the key.

Liberal export consideration for escrowable products could be granted and incentives promulgated to encourage the use of escrow features. While the short-term disadvantage of this approach from the standpoint of U.S. national security is that it allows encryption stronger than the current 40-bit RC2/RC4 encryption allowed under present regulations to diffuse into foreign hands, it has the long-term advantage of providing foreign governments with a tool for influencing or regulating the use of cryptography as they see fit. Currently, most products with encryption capabilities do not have built-in features to support escrow built into them. However, if products were designed and exported with such features, governments would have a hook for exercising some influence. Some governments might choose to require the escrowing of keys, while others might simply provide incentives to encourage escrowing. In any event, the diffusion of escrowable products abroad would raise the awareness of foreign governments, businesses, and individuals about encryption and thus lay a foundation for international cooperation on the formulation of the national satellite encryption policies.

Alternatives to Government Certification of Escrow Agents Abroad

As discussed in Chapter 13, the Administration's August 1995 proposal focuses on an implementation of escrowed satellite encryption that involves the use of *escrow agents certified by the U.S. government or by foreign governments with which the U.S. government has formal agreements consistent with U.S. law enforcement and national security requirements* (see the Sidebar, "Administration's Draft Software Key Escrow Export Criteria in Chapter 13). This approach requires foreign customers of U.S. escrowed satellite encryption products to use U.S. escrow agents until formal agreements can be negotiated that specify the responsibilities of foreign escrow agents to the United States for law enforcement and national security purposes.

Skeptics ask what incentives the U.S. government would have to conclude the formal agreements described in the August 1995 proposal if U.S. escrow agents would, by default, be the escrow agents for foreign consumers. They believe that the most likely result of adopting the Administration's proposal would be U.S. foot-dragging and inordinate delays in the consummation of formal agreements for certifying foreign escrow agents. The following approaches address problems raised by certifying foreign escrow agents: (1) informal arrangements for cooperation; and (2) contractual key escrow.

Informal Arrangements for Cooperation

One alternative is based on the fact that the United States enjoys strong cooperative law enforcement relationships with many nations with which it does not have formal agreements regarding cooperation. Negotiation of a formal agreement between the United States and another nation could be replaced by presidential certification that strong cooperative law enforcement relationships exist between the United States and that nation. Subsequent cooperation would be undertaken on the same basis that cooperation is offered today.

Contractual Key Escrow

A second alternative is based on the idea that formal agreements between nations governing exchange of escrowed key information might be replaced by private contractual arrangements. A user that escrows key information with an escrow agent, wherever that agent is located, would agree contractually that the U.S. government would have access to that information under a certain set of carefully specified circumstances. A suitably designed exportable product would provide strong encryption only upon receipt of affirmative confirmation that the relevant key information had been deposited with escrow agents requiring such contracts with users. Alternatively, as a condition of sale, end users could be required to deposit keys with escrow agents subject to such a contractual requirement.

Use of Differential Work Factors in Satellite Encryption

Differential work factor satellite encryption is an approach to cryptography that presents different work factors to different parties attempting to cryptanalyze a given piece of encrypted information. Iris Associates, the creator of Notes, proposed such an approach for Lotus Notes Version 4 and higher to facilitate its export, and the U.S. government has accepted it. Specifically, the international edition of Lotus Notes Version 4 and higher is designed to present a 40-bit work factor to the U.S. government and a 64-bit work factor to all other parties. It implements this differential work factor by encrypting 24 bits of the 64-bit key with the public-key portion of an RSA key pair held by the U.S. government. Because the U.S. government can easily decrypt these 24 bits, it faces only a 40-bit work factor when it needs access to a communications stream overseas encrypted by the international edition. All other parties attempting to cryptanalyze a message face a 64-bit work factor.

Note: Recall from Chapter 3 that a work factor is a measure of the amount of work that it takes to undertake a brute-force exhaustive cryptanalytic search.

Differential work factor satellite encryption is similar to partial key escrow (described in Chapter 13) in that both provide very strong protection against most attackers but are vulnerable to attack by some specifically chosen authority. However, they are different in that differential work factor satellite encryption does not require user interaction with an escrow agent, and so it can offer strong cryptography *out of the box.* Partial key escrow offers all of the strengths and weaknesses of escrowed satellite encryption, including the requirement that the enabling of strong cryptography does require interaction with an escrow agent.

Separation of Cryptography from Other Items on the U.S. Munitions List

As noted in Chapter 5, the inclusion of products with encryption capabilities on the USML puts them on a par with products intended for strictly military purposes (tanks, missiles). An export control regime that authorized the U.S. government to separate encryption (a true dual-use technology) from strictly military items would provide much needed flexibility in dealing with nations on which the United States wishes to place sanctions.

Alternatives for Providing Government Exceptional Government Access to Encrypted Satellite Data

Providing government exceptional access to encrypted satellite data is an issue with a number of dimensions. Some of these relate directly to encryption.

A Prohibition of the Use and Sale of Satellite Encryption Lacking Features for Exceptional Access

One obvious approach to ensuring government exceptional access to satellite encrypted information is to pass legislation that forbids the use of cryptography lacking features for such access, presumably with criminal penalties attached for violation. Given that escrowed satellite encryption appears to be the most plausible approach to providing government exceptional access, the term *unescrowed encryption* is used here as a synonym for cryptography without features for exceptional access. Indeed, opponents of the Escrowed Encryption Standard (EES) and the Clipper chip have argued repeatedly that the EES approach would succeed only if alternatives were banned. Many concerns have been raised about the prospect of a mandatory prohibition on the use of unescrowed encryption.

From a law enforcement standpoint, a legislative prohibition on the use of unescrowed encryption would have clear advantages. Its primary impact would be to eliminate the commercial supply of unescrowed products with encryption capabilities—vendors without a market would most likely not produce or distribute such products, thus limiting access of criminals to unescrowed encryption and increasing the inconvenience of evading a prohibition on use of unescrowed encryption. At the same time, such a prohibition would leave law-abiding users with strong concerns about the confidentiality of their information being subject to procedures beyond their control. A legislative prohibition of the use of unescrowed encryption also raises specific technical, economic and legal issues.

Concerns about Personal Freedom

The Clinton Administration has stated that it has no intention of outlawing unescrowed encryption, and it has repeatedly and explicitly disavowed any intent to regulate the domestic use of cryptography. However, no Administration can bind future administrations (a fact freely acknowledged by administration officials). Thus, some critics of the Administration position believe that the dynamics of the encryption problem may well drive the government (sooner or later) to prohibit the use of encryption without government access.

Note: For example, Senator Charles Grassley (R-IA) introduced legislation (The Anti-Electronic Racketeering Act of 1995) on June 27, 1995, to *prohibit certain acts involving the use of computers in the furtherance of crimes.* The proposed legislation makes it unlawful *to distribute computer software that encodes or encrypts electronic or digital satellite communications to computer networks that the person distributing the software knows or reasonably should know, is accessible to foreign nationals and foreign governments, regardless of whether such software has been designated as nonexportable, except for software that uses a universal decoding device or program that was provided to the Department of Justice prior to the distribution.*

The result is that the Administration is simply not believed when it forswears any intent to regulate encryption used in the United States. Two related concerns are raised: the slippery slope and the misuse of deployed infrastructure for encryption.

The Slippery Slope   Many skeptics fear that the current satellite encryption policy is the first step down a slippery slope toward a more restrictive policy regime under which government may not continue to respect limits in place at the outset. An oft-cited example is current use of the Social Security Number, which was not originally intended to serve as a universal identifier when the Social Security Act was passed in 1935 but has, over the last 50 years, come to serve exactly that role by default, simply because it was there to be exploited for purposes not originally intended by the enabling legislation.

Misuse of Deployed Infrastructure for Cryptography   Many skeptics are concerned that a widely deployed infrastructure for encryption could be used by a future administration or Congress to promulgate and/or enforce restrictive policies regarding the use of cryptography. With such an infrastructure in place, critics argue that a simple policy change might be able to transform a comparatively benign deployment of technology into an oppressive one. For example, critics of the Clipper proposal were concerned about the possibility that a secure telephone system with government exceptional access capabilities could, under a strictly voluntary program to encourage its purchase and use, achieve moderate market penetration. Such market penetration could then facilitate legislation outlawing all other cryptographically secure telephones.

Note: By contrast, a deployed infrastructure could have characteristics that would make it quite difficult to implement policy changes on a short time scale. For example, it would be very difficult to implement a policy change that would change the nature of the way in which people use today's telephone system. Not surprisingly, policy makers would prefer to work with infrastructures that are quickly responsive to their policy preferences.

Adding to these concerns are suggestions such as those made by a responsible and senior government official that even research in encryption conducted in the civilian sector should be controlled in a legal regime similar to that which governs research with relevance to nuclear weapons design (see the Sidebar, "Bobby Inman on the Classification of Cryptologic Research"). Ironically, former NSA Director Bobby Inman's comments on scientific research appeared in an article that called for greater cooperation between academic scientists and national security authorities and used as a model of cooperation an arrangement, recommended by the Public Cryptography Study Group, that has worked generally well in balancing the needs of academic science and those of national security.

Note: The arrangement recommended by the Public Cryptography Study Group called for voluntary prepublication review of all cryptography research undertaken in the private sector.

Nevertheless, Inman's words are often cited as reflecting a national security mind-set that could lead to a serious loss of intellectual freedom and discourse. More recently, FBI Director Louis Freeh stated to the Administration that *other approaches may be necessary* if technology vendors do not adopt escrowed satellite encryption on their own. Moreover, the current Administration has explicitly rejected the premise that *every American, as a matter of right, is entitled to an unbreakable encryption product*

Bobby Inman on the Classification of Cryptologic Research

In 1982, then-Deputy Director of the Central Intelligence Agency Bobby R. Inman wrote that a source of tension arises when scientists, completely separate from the federal government, conduct research in areas where the federal government has an obvious and preeminent role for society as a whole. One example is the design of advanced weapons, especially nuclear ones. Another is cryptography. While nuclear weapons and cryptography are heavily dependent on theoretical mathematics, there is no public business market for nuclear weapons. Such a market, however, does exist for cryptographic concepts and gear to protect certain types of business communications.

However, cryptologic research in the business and academic arenas, no matter how useful, remains redundant to the necessary efforts of the federal government to protect its own communications. He was concerned that indiscriminate publication of the results of that research will come to the attention of foreign governments and entities and, thereby, could cause irreversible and unnecessary harm to U.S. national security interests.

While key features of science (unfettered research, and the publication of the results for validation by others and for use by all mankind) are essential to the growth and development of science, nowhere in the scientific ethos is there any requirement that restrictions cannot or should not, when necessary, be placed on science. Scientists do not immunize themselves from social responsibility simply because they are engaged in a scientific pursuit. Society has recognized over time that certain kinds of scientific inquiry can endanger society as a whole and has applied either directly, or through scientific/ethical constraints, restrictions on the kind and amount of research that can be done in those areas.[6]

Given concerns about possible compromises of personal and civil liberties, many skeptics of government in this area believe that the safest approach is for government to stay out of the satellite encryption policy entirely. They argue that any steps in this area, no matter how well intentioned or plausible or reasonable, must be resisted strongly, because such steps will inevitably be the first poking of the camel's nose under the tent.

Technical Issues

Even if a legislative prohibition on the use of unescrowed encryption were enacted, it would be technically easy for parties with special needs for security to circumvent such a ban. In some cases, circumvention would be explicitly illegal, while in others it might well be entirely legal. For example, software for unescrowed encryption can be downloaded from the Internet. Such software is available even today. Even if posting such software in the United States were to be illegal under a prohibition, it would nonetheless be impossible to prevent U.S. Internet users from downloading software that had been posted on sites abroad.

Superencryption can be used. Superencryption (sometimes also known as double encryption) is encryption of traffic before it is given to an escrowed encryption device or system. For technical reasons, superencryption is impossible to detect without monitoring and attempting to decrypt all escrow-encrypted traffic, and such large-scale monitoring would be seriously at odds with the selected and limited nature of wiretaps today.

An additional difficulty with superencryption is that it is not technically possible to obtain escrow information for all layers simultaneously, because the fact of double and triple encryption cannot be known in advance. Even if the second (or third or fourth) layers of encryption were escrowed, law enforcement authorities would have to approach separately and sequentially the escrow agents holding key information for those layers.

Talent for hire is easy to obtain. A criminal party could easily hire a knowledgeable person to develop needed software. For example, an out-of-work or underemployed scientist or mathematician from the former Soviet Union would find a retainer fee of $800 per month to be a king's ransom.

> **Note:** Many high-technology jobs are moving overseas in general, not just to the former Soviet Union.

Information can be stored remotely. An obvious noncryptographic circumvention is to store data on a remote computer whose Internet address is known only to the user. Such a computer could be physically located anywhere in the world (and might even automatically encrypt files that were stored there). But even if it were not encrypted, data stored on a remote computer would be impossible for law enforcement officials to access without the cooperation of the data's owner. Such remote storage could occur quite legally even with a ban on the use of unescrowed encryption.

Demonstrating that a given satellite communication or data file is *encrypted* is fraught with ambiguities arising from the many different possibilities for sending information. For example, an individual might use an obscure data format. While ASCII is the most common representation of alphanumeric characters today, Unicode (a proposed 16-bit representation) and EBCDIC (a more-or-less obsolete 8-bit representation) are equally good for sending plain English text.

Another possibility might be an individual talking to another individual speaking in a language such as Navajo. Or, an individual talking to another individual might speak in code phrases.

Also, an individual might send compressed digital data that could easily be confused with encrypted data despite having no purpose related to encryption. If, for example, an individual develops his or her own good compression algorithm and does not share it with anyone, that compressed bit stream may prove as difficult to decipher as an encrypted bit stream.

Note: One problem in using compression schemes as a technique for ensuring confidentiality is that almost any practical compression scheme has the characteristic that closely similar plaintexts would generate similar ciphertexts, thereby providing a cryptanalyst with a valuable advantage not available if a strong encryption algorithm is used.

In addition, an individual might deposit fragments of a text or image that he or she wished to conceal or protect in a number of different Internet-accessible computers. The plaintext (the reassembled version) would be reassembled into a coherent whole only when downloaded into the computer of the user.

Additionally, an individual might use steganography. This is the name given to techniques for hiding a message within another message. For example, the first letter of each word in a sentence or a paragraph can be used to spell out a message, or a photograph can be constructed so as to conceal information. Specifically, most black-and-white pictures rendered in digital form use at most 2^{16} (65,536) shades of gray, because the human eye is incapable of distinguishing any more shades. Each element of a digitized black-and-white photo would then be associated with 16 bits of information about what shade of gray should be used. If a picture were digitized with 24 bits of gray scale, the last 8 bits could be used to convey a concealed message that would never appear except for someone who knew to look for it. The digital size of the picture would be 50% larger than it would ordinarily be, but no one but the creator of the image would know.

None of the preceding alternative coding schemes provides confidentiality as strong as would be provided by good encryption, but their extensive use could well complicate attempts by the government to obtain plaintext information. Given so many different ways to subvert a ban on the use of unescrowed encryption, emergence of a dedicated subculture is likely in which the nonconformists would use coding schemes or unescrowed encryption impenetrable to all outsiders.

Economic Concerns

An important economic issue that would arise with a legislative prohibition on the use of unescrowed encryption would involve the political difficulty of mandating abandonment of existing user investments in products with encryption capabilities. These investments, considerable even today, are growing rapidly, and the expense to users of immediately having to replace unescrowed encryption products with escrowed ones could be enormous.

Note: Existing unescrowed encryption products could be kept in place if end users could be made to comply with a prohibition of the use of such products. In some cases, a small technical fix might suffice to disable the encryption features of a system. Such fixes would be most relevant in a computing environment in which the software used by end users is centrally administered (as in the case of many corporations) and provides system administrators with the capability for turning off encryption. In other cases, users (typically individual users who had purchased their products from retail store outlets) would have to be trusted to refrain from using encryption.

A further expense would be the labor cost involved in decrypting existing encrypted archives and reencrypting them using escrowed satellite encryption products. One potential mitigating factor for cost is the short product cycle of information technology products. Whether users would abandon nonconforming products in favor of new products with escrowing features (knowing that they were specifically designed to facilitate exceptional access) is open to question.

Legal and Constitutional Issues

Even apart from the issues previously described, which in the Administration's view are quite significant, a legislative ban on the domestic use of unescrowed encryption would raise constitutional issues. Insofar as a prohibition on unescrowed encryption were treated for constitutional purposes as a limitation on the content of communications, the government would have to come forward with a compelling state interest to justify the ban. To some, a prohibition on the use of unescrowed encryption would be the equivalent of a law proscribing use of a language (Spanish), which would almost certainly be unconstitutional. On the other hand, if such a ban were regarded as tantamount to eliminating a method of satellite communication (were regarded as content-neutral), then the courts would employ a simple balancing test to determine its constitutionality. The government would have to show that the public interests were jeopardized by a world of unrestrained availability of satellite encryption, and these interests would have to be weighed against the free speech interests sacrificed by the ban. It would also be significant to know what alternative forms of methods of anonymous communication would remain available with a ban and how freedom of speech would be affected by the specific system of escrow chosen by the government. These various considerations are difficult and, in some cases, impossible to estimate in advance of particular legislation and a particular case; but, the First Amendment issues likely to arise with a total prohibition on the use of unescrowed encryption are not trivial.

A step likely to raise fewer constitutional problems, but not eliminate them, is one that would impose restrictions on the commercial sale of unescrowed products with encryption capabilities. Under such a regime, products with encryption capabilities eligible for sale would have to conform to certain restrictions intended to ensure public safety, in much the same way that other products such as drugs, automobiles, and meat must satisfy particular government regulations. *Freeware* or home-grown products with satellite encryption capabilities would be exempt from such regulations as long as they were used privately. The problem of already-deployed products would remain, but in a different form. New products would either interoperate or not interoperate with existing already deployed products. If noninteroperability were required, users attempting to maintain and use two noninteroperating systems would be faced with enormous expenses. If interoperability were allowed, the intent of the ban would be thwarted.

Finally, any national policy whose stated purpose is to prevent the use of unescrowed encryption preempts decision making that the Administration believes properly belongs to users. As noted in Chapter 13, escrowed satellite encryption reduces the level of assured confidentiality in exchange for allowing controlled exceptional access to parties that may need to retrieve encrypted data. Only in a policy regime of voluntary compliance can users decide how to make that trade-off. A legislative prohibition of the use or sale of unescrowed encryption would be a clear statement that law enforcement needs for exceptional access to information clearly outweigh user interests in having maximum possible protection for their information, a position that has yet to be defended or even publicly argued by any player in the debate.

Criminalization of the Use of Satellite Encryption in the Commission of a Crime

Proposals to criminalize the use of satellite encryption in the commission of a crime have the advantage that they focus the weight of the criminal justice system on the *bad* guy without placing restrictions on the use of cryptography by *good guys.* Further, deliberate use of encryption in the commission of a crime could result in considerable damage, either to society as a whole or to particular individuals (in circumstances suggesting premeditated wrongdoing)—an act that society tends to view as worthy of greater punishment than a crime committed in the heat of the moment.

Two approaches could be taken to criminalize the use of satellite encryption in the commission of a crime: First of all, construct a specific list of crimes in which the use of satellite encryption would subject the criminal to additional penalties. For example, using a deadly weapon in committing a robbery or causing the death of someone during the commission of a crime is each a crime that leads to additional penalties. Second, develop a blanket provision stating that the use of satellite encryption for illegal purposes (or for purposes contrary to law) is itself a felony.

In either event, additional penalties for the use of satellite encryption could be triggered by a conviction for a primary crime, or they could be imposed independently of such a conviction. Precedents include the laws criminalizing mail fraud (fraud is a crime, generally a state crime, but mail fraud—use of the mails to commit fraud—is an additional federal crime) and the use of a gun during the commission of a felony.

Intentional use of satellite encryption in the concealment of a crime could also be criminalized. Because the use of satellite encryption is a *prima facie* act of concealment, such an expansion would reduce the burden of proof on law enforcement officials, who would have to prove only that cryptography was used intentionally to conceal a crime. Providers of satellite encryption would be criminally liable only if they had knowingly provided cryptography for use in criminal activity. On the other hand, a law of more expansive scope might well impose additional burdens on businesses and raise civil liberties concerns.

In considering legal penalties for misuse of satellite encryption, the question of what it means to *use* cryptography must be addressed. For example, if and when encryption capabilities are integrated seamlessly into applications and are invoked automatically without effort on the part of a user, should the use of these applications for criminal purposes lead to additional penalties or to a charge for an additional offense? Answering yes to this question provides another avenue for prosecuting a criminal (recall that Al Capone was convicted for income tax evasion rather than bank robbery). A second question is what counts as *satellite encryption*. As previously noted in the discussion of prohibiting unescrowed encryption, a number of mathematical coding schemes can serve to obscure the meaning of plaintext even if they are not satellite encryption schemes in the technical sense of the word. These and related questions must be addressed in any serious consideration of the option for criminalizing the use of satellite encryption in the commission of a crime.

Technical Nonescrow Approaches for Obtaining Access to Information

Escrowed satellite encryption is not the only means by which law enforcement can gain access to encrypted data. For example, as advised by Department of Justice guidelines for searching and seizing computers, law enforcement officials can approach the software vendor or the Justice Department computer crime laboratory for assistance in cryptanalyzing encrypted files. These guidelines also advise that *clues to the password (may be found) in the other evidence seized—stray notes on hardware or desks; scribble in the margins of manuals or on the jackets of disks. Agents should consider whether the suspect or someone else will provide the password if requested.* Moreover, product designs intended to facilitate exceptional access can include alternatives with different strengths and weaknesses such as link encryption, weak encryption, hidden back doors, and translucent cryptography.

Link Encryption

With link encryption, which applies only to satellite communications and stands in contrast to end-to-end encryption (see the Sidebar that follows, and Table 26.1), a plaintext message enters a satellite com-

Table 26.1 Comparison of End-to-End and Link Encryption

	End-to-end encryption	Link encryption
Controlling Party	User	Link provider
Suitable traffic	Most suitable for encryption of individul messages	Facilities bulk encryption of data
Potential leaks of plaintext	Only at transmitting and receiving stations	At either end of the link, which may or may not be within the user's security perimeter
Point of responsibility	User must take responsibility	Link provider takes responsibility

munications link, is encrypted for transmission through the link, and is decrypted upon exiting the link.[7] In a satellite communication that may involve many links, sensitive information can be found in plaintext form at the ends of each link (but not during transit). Thus, for purposes of protecting sensitive information on an open network accessible to anyone (the Internet is a good example), link encryption is more vulnerable than end-to-end encryption, which protects sensitive information from the moment it leaves party A to the moment it arrives at party B. However, from the standpoint of law enforcement, link encryption facilitates legally authorized intercepts, because the traffic of interest can always be obtained from one of the nodes in which the traffic is unencrypted.

Link Versus End-to-End Encryption of Satellite Communications

End-to-end encryption involves a stream of data traffic (in one or both directions) that is encrypted by the end users involved before it is fed into the satellite communications link (see Table 26.1). Traffic in between the end users is never seen in plaintext, and the traffic is decrypted only upon receipt by an end user. Link encryption is encryption performed on data traffic after it leaves one of the end users; the traffic enters one end of the link, is encrypted and transmitted, and then is decrypted upon exit from that link.[8]

On a relatively closed network or one that is used to transmit data securely and without direct user action, link encryption may be cost-effective and desirable. A good example is encryption of the wireless radio link between a GSM cellular telephone and its ground station; the cellular handset encrypts the voice signal and transmits it to the ground station, at which point it is decrypted and fed into the land-based network. Thus, the land based network carries only unencrypted voice traffic, even though it was transmitted by an encrypted cellular telephone. A second example is the *bulk* encryption of multiple channels (each individually unencrypted) over a multiplexed fiber-optic link. In both of these instances of link encryption, only those with access to carrier facilities (presumably law enforcement officials acting under proper legal authorization) would have the opportunity to tap such traffic.

Weak Encryption

Weak encryption allowing exceptional access would have to be strong enough to resist brute-force attack by unauthorized parties (business competitors via a satellite link) but weak enough to be cracked by authorized parties (law enforcement agencies). However, weak encryption is a moving target. The difference between cracking strong and weak encryption by brute-force attack is the level of computational resources that can be brought to such an attack, and those resources are ever increasing. In fact, the cost of brute-force attacks on cryptography drops exponentially over time, in accordance with Moore's law.

> **Note:** Moore's law is an empirical observation that the cost of computation drops by a factor of two approximately every 18 months.

Widely available technologies now enable multiple distributed work-stations to work collectively on a computational problem at the behest of only a few people. The Sidebar, "Successful Challenges To 40-Bit Encryption," in Chapter 5 discusses the brute-force cryptanalysis of messages encrypted with the 40-bit RC4 algorithm, and it is not clear that the computational resources of unauthorized parties can be limited in any meaningful way. In today's environment, unauthorized parties will almost always be able to assemble the resources needed to mount successful brute-force attacks against weak encryption, to the detriment of those using such cryptography. Thus, any technical dividing line between authorized and unauthorized decryption would change rather quickly.

Hidden Back Doors

A *back door* is an entry point to an application that permits access or use by other than the normal or usual means. Obviously, a back door known to government can be used to obtain exceptional access. Back doors may be open or hidden. An open back door is one whose existence is announced publicly. An example is an escrowed satellite encryption system, which everyone knows is designed to allow exceptional access. By its nature, an open back door is explicit; it must be deliberately and intentionally created by a designer or implementer.

> **Note:** Of course, the fact that a particular product is escrowed may not necessarily be known to any given user. Many users learn about the features of a product through reading advertisments and operating manuals for the product; if these printed materials do not mention the escrowing features, and no one tells the user, he or she may well remain ignorant of them, even though the fact of escrow is *public knowledge.*

A hidden back-door is one whose existence is not widely known, at least upon initial deployment. It can be created deliberately (by a designer who insists on retaining access to a system that he or she may have created) or accidentally (as the result of a design flaw). Often, a user wishing access through a deliberately created hidden back door must pass through special system-provided authorization services. Almost by definition, an accidentally created hidden back door requires no special authorization for its exploitation, although finding it may require special knowledge. In either case, the existence of hidden back doors may or may not be documented; frequently, it is not.

Particularly harmful hidden back doors can appear when *secure* applications are implemented using insecure operating systems; more generally, *secure* applications layered on top of insecure systems may not be secure in practice. Cryptographic algorithms implemented on weak operating systems present another large class of back doors that can be used to undermine the integrity and the confidentiality that cryptographic implementations are intended to provide. For example, a database application that provides strong access control and requires authorization for access to its data files but is implemented on an operating system that allows users to view those files without going through the database application, does not provide strong confidentiality. Such an application may well have its data files encrypted for confidentiality.

The existence of back doors can pose high-level risks. The shutdown or malfunction of life-critical systems, loss of financial stability in electronic commerce, and compromise of private information in database systems can all have serious consequences. Even if back doors are undocumented, they can be discovered and misused by insiders or outsiders. Reliance on *security by obscurity* is always dangerous, because trying to suppress knowledge of a design fault is generally very difficult. If a back door exists, it will eventually be discovered, and its discoverer can post that knowledge worldwide. If systems containing a discovered back door were on the Internet or were accessible via satellite or by modem, massive exploitation could occur almost instantaneously, worldwide. If back doors lack a capability for adequate authentication and accountability, then it can be very difficult to detect exploitation and to identify the culprit.

Translucent Encryption

Translucent encryption has been proposed by RSA as an alternative to escrowed satellite encryption. The proposed technical scheme, which involves no escrow of unit keys, would ensure that any given message or file could be decrypted by the government with probability p. The value of p $(0 < p < 1)$ would be determined by the United States Congress. In other words, on average, the government would be able to decrypt a fraction p of all messages or files to which it was given legal access. Today (without encryption), $p = 1$. In a world of strong (unescrowed) encryption, $p = 0$. A large value of p favors law enforcement, while a small value of p favors libertarian privacy. The RSA proposes that some value of p will balance the interests on both sides.

It is not necessary that the value of p be fixed for all time or be made uniform for all devices. The value of p could be set differently for cellular telephones and for e-mail, or it could be raised or lowered as circumstances dictated. The value of p would be built into any given encryption device or program.

Note: In contrast to escrowed satellite encryption, translucent encryption requires no permanent escrowing of unit keys, although it renders access indeterminate and probabilistic.

Intranet-Based Encryption

In principle, secure telephony can be made the responsibility of telephone service providers. Under the current regulatory regime, tariffs often distinguish between data and voice. Circuits designated as carrying ordinary voice (also to include fax and modem traffic) could be protected by encryption supplied by the service provider, perhaps as an extra security option that users could purchase. Common carriers (service providers in this context) that provide encryption services are required by the Communications Assistance for Law Enforcement Act to decrypt for law enforcement authorities upon legal request.

> **Note:** The *trusted third party* (TTP) concept discussed in Europe is similar in the sense that TTPs are responsible for providing key management services for secure satellite communications. In particular, TTPs provide session keys over secure channels to end users that they can then use to encrypt satellite communications with parties of interest; these keys are made available to law enforcement officials upon authorized request.

The simplest version of network- or intranet-based encryption would provide for link satellite encryption (encrypting the voice traffic only between switches via a satellite). Link satellite encryption would leave the user vulnerable to eavesdropping at a point between the end-user device and the first switching office. In principle, a secure end-user device could be used to secure this *last mile* link.

> **Note:** The *last mile* is a term describing that part of a local telephone network between the premises of an individual subscriber and the central-office switch from which service is received. The vulnerability of the *last mile* is increased because it is easier to obtain access to the physical connections and because the volume of traffic is small enough to permit the relevant traffic to be isolated easily. On the other hand, the vulnerability of the switch is increased because it is often accessible remotely through dial-in ports.

Whether telecommunications service providers will move ahead on their own with intranet-based encryption for voice traffic is uncertain for a number of reasons. Because most people today either believe that their calls are reasonably secure or are not particularly concerned about the security of their calls, the extent of demand for such a service within the United States is highly uncertain. Furthermore, by moving ahead with voice encryption, telephone companies would be admitting that calls carried on their network are today not as secure as they could be. Such an acknowledgment might undermine their other business interests. Finally, making intranet-based encryption work internationally would remain a problem, although any scheme for ensuring secure international satellite communications will have drawbacks.

More narrowly focused intranet-based encryption could be used with that part of the network traffic that is widely acknowledged to be vulnerable to interception—namely, wireless voice communications. Wireless communications can be tapped *in the ether* on an entirely passive basis, without the knowledge of either the sending or receiving party. Of particular interest is the cellular telephone network. All of the current standards make some provisions for encryption. Encryption of the wireless link is also provided by the GSM, a European standard for mobile communications. In general, satellite communication is encrypted from the mobile handset to the cell, but not end to end. Structured in this manner, encryption would not block the ability of law enforcement to obtain the contents of a call, because access could always be obtained by tapping the ground station.

At present, satellite transmission of most wireless communications is analog. Unless special measures are taken to prevent surveillance, analog transmissions are relatively easy to intercept. However, it is widely expected that wireless communications will become increasingly digital in the future, with two salutary benefits for security. One is that compared to analog signals, even unencrypted digital communications are difficult for the casual eavesdropper to decipher or interpret, simply because they are transmitted in digital form. The second is that digital satellite communications are relatively easy to encrypt.

Security for Satellite Data Communications

The body responsible for determining technical standards for Internet communications, the Internet Engineering Task Force[9], has developed standards for the Internet Protocol (version 6 or higher, also known as IPv6 or higher) that require conforming implementations to have the ability to encrypt data packets, with the default method of encryption being DES as shown in Table 26.2. However, IPv6 standards are silent with respect to key management, and so leave open the possibility that escrow features might or might not be included at the vendor's option.

> **Note:** The Network Working Group has described protocols that define standards for encryption, authentication, and integrity in the Internet Protocol. These protocols are described in the following documents contained in Table 26.2.

Table 26.2 Internet Protocols[10]

RFC	Title
1825	Security Architecture for the Internet Protocols; describes the security mechanisms for IP version 4 (IPv4) and IP version 6 (IPv6).
1826	IP Authentication Header (AH; describes a mechanism for providing cryptographic authentication for IPv4 and IPv6 datagrams).
1827	IP Encapsulating Security Payload (ESP; describes a mechanism that works in both IPv4 and IPv6 for providing integrity and confidentiality to IP datagrams).
1828	IP Authentication using keyed MD5; describes the use of a particular authentication technique with IP-AH.
1829	The ESP DES-CBC Transform; describes the use of a particular encryption technique with the IP Encapsulating Security Payload (ESP).

If the proposed standards are finalized, vendors may well face a Hobson's choice: to export Internet routing products that do not conform to the IPv6 standard (to obtain favorable treatment under the current ITAR, which do not allow exceptions for encryption stronger than 40-bit with RC2 or RC4), or to develop products that are fully compliant with IPv6 (a strong selling point), but only for the domestic market. Still, escrowed implementations of IPv6 would be consistent with the proposed standard and might be granted commodities jurisdiction to the Commerce Control List under regulations proposed by the Administration for escrowed satellite encryption products.

Distinguishing between Encrypted Voice and Data Communications Services for Exceptional Access

For purposes of allowing exceptional access, it may be possible to distinguish between encrypted voice and satellite data communications, at least in the short run. Specifically, a proposal by the JASON study group (a DOD Advisory panel consisting of scientists, communication specialists, and weapons designers) suggests that efforts to install features for exceptional access should focus on secure voice communications, while leaving to market forces the evolution of secure data communications and storage. This proposal rests on the following propositions.

First of all, telephony, as it is experienced by the end user, is a relatively mature and stable technology, compared to satellite data communications services that evolve much more rapidly. Many people (perhaps the majority of the population) will continue to use devices that closely resemble the telephones of today, and many more people are familiar with telephones than are familiar with computers or the Internet.

An important corollary is that regulation of rapidly changing technologies is fraught with more danger than is the regulation of mature technologies, simply because regulatory regimes are inherently slow to react and may well pose significant barriers to the development of new technologies. This is especially true in a field moving as rapidly as information technology.

Second, telephony has a long-standing regulatory and technical infrastructure associated with it, backed by considerable historical precedent, such as that for law enforcement officials obtaining wiretaps on telephonic communications under court order. By contrast, satellite data communications services are comparatively unregulated (see the Sidebar that follows).

Two Primary Rate and Service Models for Telecommunications Today

Regulated Common Carrier Telephony Services: Regulated common carrier telephony services are usually associated with voice telephony, including fax and low-speed modem data communications. If a *common carrier* provision applies to a given service provider, the provider must provide service to anyone who asks at a rate that is determined by a public utilities commission. Common carriers often own their own transport facilities (fiber-optic cables, telephone wires, and so on), and thus the service provider exerts considerable control over the routing of a particular communication. Pricing of service for the end user is often determined on the basis of actual usage. The carrier also provides value-added services (call waiting) to enhance the value of the basic service to the customer. Administratively, the carrier is usually highly centralized.

Bulk Data Transport: Bulk services are usually associated with data transport (data sent from one computer to another) or with *private* telephony (a privately owned or operated branch exchange for telephone service within a company). Pricing for bulk services is usually a matter of negotiation between provider and customer and may be based on statistical usage, actual usage, reliability of transport, regional coverage, or other considerations. Policy for use is set by the party that pays for the bulk service, and thus, taken over the multitude of organizations that use bulk services, is administratively decentralized. In general, the customer provides value-added services. Routing paths are often not known in advance, but instead may be determined dynamically.[10]

In remarks to the committee, FBI Director Louis Freeh pointed out that it was voice communications that drove the FBI's desire for passage of the Communications Assistance for Law Enforcement Act (CALEA). He acknowledged that other mechanisms for communication might be relevant to law enforcement investigations but nonlegislative approaches have been used to deal with those mechanisms.

Demand for secure telephone communications, at least domestically, is relatively small, if only because most users consider today's telephone system to be relatively secure. A similar perception of Internet security does not obtain today, and thus the demand for highly secure satellite data communications is likely to be relatively greater and should not be the subject of government interference.

Under the JASON proposal, attempts to influence the inclusion of escrow features could affect only the hardware devices that characterize telephony today (a dedicated fax device, an ordinary telephone). In general, these devices do now allow user programming or additions, and in particular, lack the capability enabling the user to provide encryption easily.

The JASON study also recognized that technical trends in telecommunications are such that telephony will be increasingly indistinguishable from satellite data communications. One reason is that satellite communications are becoming increasingly digital. A bit is a bit, whether it was originally part of a voice communication or part of a data communication, and the purpose of a satellite communications infrastructure is to transport bits from Point A to Point B, regardless of the underlying information content. Reconstituting the transported bits into their original form will be a task left to the parties at Point A and Point B. Increasingly, digitized signals for voice, data, images, and video will be transported in similar ways over the same satellite or network facilities, and often they will be combined into single multiplexed streams of bits as they are carried along.

Note: However, the difficulty of searching for a given piece of information does depend on whether it is voice or text. It is quite straightforward to search a given digital stream for a sequence of bits that represents a particular word as text, but quite difficult to search a digital stream for a sequence of bits that represents that particular word as voice.

For example, a voice-generated analog sound wave that enters a telephone may be transmitted to a central switching office, at which point it generally is converted into a digital bit stream and merged with other digital traffic that may originally have been voices, television signals, and high-speed streams of data from a computer. The network transports all of this traffic across the country by a fiber-optic cable and converts the bits representing voice back into an analog signal only when it reaches the switching office that serves the telephone of the called party. To a contemporary user of the telephone, the conversation proceeds just as it might have done 40 years ago (although probably with greater fidelity), but the technology used to handle the call is entirely different.

Alternatively, a computer connected to a data network can be converted into the functional equivalent of a telephone. Some on-line service providers will be offering encrypted satellite voice communications capability in the near future, and the Internet itself can be used today to transport real-time voice and even encrypted satellite video communications, albeit with relatively low fidelity and reliability but also at very low cost. Before these modalities become acceptable for mainstream purposes, the Internet (or its successor) will have to implement on a wide scale new protocols and switching services to eliminate current constraints that involve time delays and bandwidth limitations.

> **Note:** For example, an IBM catalog offers for general purchase a *DSP Modem and Audio Card* with *Telephony Enhancement* that provides a full-duplex speaker telephone for $365. The card is advertised as being able to make the purchaser's PC into *a telephone communications center with telephone voice mail, caller ID, and full duplex speakerphone capability (for true simultaneous, two-way communications)*.

A second influence that will blur the distinction between voice and data is that the owners of the devices and lines that transport bits today are typically the common carriers—firms originally formed to carry long-distance telephone calls and today subject to all of the legal requirements imposed on common carriers (see the Sidebar "Two Primary Rate and Service Models for Telecommunications Today"). However, these firms sell transport capacity to parties connecting data networks, and much of today's bulk data traffic is carried over communications links that are owned by the common carriers. The Telecommunications Reform Act of 1996 will further blur the lines among service providers.

The lack of a technical boundary between telephony and satellite data communications is the result of the way in which today's networks are constructed. Networks are built upon a protocol *stack* that embodies protocols at different layers of abstraction. At the very bottom are the protocols for the physical layer that define the voltages and other physical parameters that represent ones and zeros. On top of the physical layer are other protocols that provide higher-level services by making use of the physical layer. Because the bulk of network traffic is carried over a physical infrastructure that was designed for voice communications (the public switched telecommunications network), interactions at the physical layer can be quite naturally regarded as being in the domain of *voice*. But interactions at higher layers in the stack are more commonly associated with *data*.

Acknowledging these difficulties, the JASON study concluded that limiting efforts to promote escrowed satellite encryption products to those associated with voice communications had two important virtues. First, it would help to preserve law enforcement needs for access to a communications mode (namely telephony) that is widely regarded as important to law enforcement. Second, it would avoid premature government regulation in the data services area (an area that is less important historically to criminal investigation and prosecution than is telephony), thus avoiding the damage that could be done to a strong and rapidly evolving U.S. information technology industry. It would take (several years to a decade) for the technical *loopholes* described previously to become significant, thus giving law enforcement time to adapt to a new technical reality.

A Centralized Satellite Decryption Facility for Government Exceptional Access

Proposed procedures to implement the retrieval of keys escrowed under the Clipper initiative call for the escrowed key to be released by the escrow agencies to the requesting law enforcement authorities upon presentation of proper legal authorization, such as a court order. Critics have objected to this arrangement because it potentially compromises keys for all time—that is, once the key to a specific telephone has been divulged, it is in principle possible to eavesdrop forever on conversations using that telephone, despite the fact that court-ordered wiretaps must have a finite duration.

To counter this criticism, administration officials have designed a plan that calls for keys to be transmitted electronically to EES-decryption devices in such a way that the decryption device will erase the key at the time specified in the court order. However, acceptance of this plan relies on assurances that the decryption device would indeed work in this manner. In addition, this proposal is relevant only to the final plan—the interim procedures specify manual key handling.

Another way to counter the objection to potential long-lasting compromise of keys involves the use of a centralized government-operated satellite decryption facility. Such a facility would receive EES-encrypted satellite traffic forwarded by law enforcement authorities and accompanied by appropriate legal authorization. Keys would be made available by the escrow agents to the facility rather than to the law enforcement authorities themselves, and the plaintext would be returned to the requesting authorities. Thus, keys could never be kept in the hands of the requesting authorities, and concern about illicit retention of keys by law enforcement authorities could be reduced. Of course, concerns about retention by the satellite decryption facility would remain, but since the number of decryption facilities would be small compared to the number of possible requesting law enforcement authorities, the problem would be more manageable. As the satellite decryption facilities would likely be under centralized control as well, it would be easier to promulgate and enforce policies intended to prevent abuse.

> **Note:** The Administration suspects that the likelihood of abusive exercise of wiretap authority is greater for parties that are farther removed from higher levels of government, although the consequences may well be more severe when parties closer to the top levels of government are involved. A single *bad apple* near the top of government can set a corrupt and abusive tone for an entire government, but at least *bad apples* tend to be politically accountable. By contrast, the number of parties tends to increase as those parties are farther and farther removed from the top, and the likelihood that at least some of these parties will be abusive seems higher. Put differently, the Administration believes that state/local authorities are more likely to be abusive in their exercise of wiretapping authority simply because they do the majority of the wiretaps. While Title III calls for a report to be filed on every federal and state wiretap order, the majority of missing reports are mostly from state wiretap orders rather than federal orders.

One important aspect of this proposal is that the particular number of facilities constructed and the capacity of each could limit the number of simultaneous wiretaps possible at any given time. Such a constraint would force law enforcement authorities to exercise great care in choosing targets for interception, just as they must when they are faced with constraints on resources in prosecuting cases. A result could be greater public confidence that wiretaps were being used only in important cases. On the other hand, a limit on the number of simultaneous wiretaps possible is also a potential disadvantage from the standpoint of the law enforcement official, who may not wish to make resource-driven choices about how and whom to prosecute or investigate. Making encryption keys directly available to law enforcement authorities allows them to conduct wiretaps unconstrained by financial and personnel limitations.

A centralized satellite decryption facility would also present problems of its own. For example, many people would regard it as more threatening to give a centralized entity the capability to acquire and decrypt all traffic than to have such capabilities distributed among local law enforcement agencies. In addition, centralizing all wiretaps and getting the communications out into the field in real time could require a complex infrastructure. The failure of a centralized facility would have more far-reaching effects than a local failure, crippling a much larger number of wiretaps at once.

Looming Issues

Two looming issues have direct significance for a national satellite encryption policy: determining the level of encryption needed to protect against high-quality attacks, and organizing the U.S. government for a society that will need better information security.

The Adequacy of Various Levels of Satellite Encryption against High-Quality Attack

What level of encryption strength is needed to protect information against high quality attack? For purposes of analysis, this discussion considers only perfect implementations of encryption for confidentiality (implementations without hidden *trap doors*, installed on secure operating systems, and so on). Thus, the only issue of significance for this discussion is the size of the key and the algorithm used to encrypt the original plaintext.

Any cryptanalysis problem can be solved by brute force given enough computers and time. The question is whether it is possible to assemble enough computational resources to allow a brute-force cryptanalysis on a time scale and cost reasonable for practical purposes.

Note: As discussed in Chapter 5, a message encoded with a 40-bit RC4 algorithm was recently broken in 8 days by a brute-force search through the use of a single workstation optimized for speed in graphics processing.

Even so, such a key size is adequate for many purposes (credit card purchases). It is also sufficient to deny access to parties with few technical skills, or to those with access to limited computing resources. But if the data being protected is valuable (if it refers to critical proprietary information), 40-bit keys are inadequate from an information security perspective. The reason is that for logistical and administrative reasons, it does not make sense to require a user to decide what information is or is not critical—the simplest approach is to protect both critical and noncritical information alike at the level required for protecting critical information. If this approach is adopted, the user does not run the risk of inadequately protecting sensitive information. Furthermore, the compromise of a single piece of information can be catastrophic. And, as it is generally impossible to know if a particular piece of information has been compromised, those with a high degree of concern for the confidentiality of information must be concerned about protecting all information at a level higher than the thresholds offered by the 8-day cryptanalysis time previously described.

From an interceptor's point of view, the cryptanalysis times provided by such demonstrations are quite daunting, because they refer to the time needed to cryptanalyze a single message. A specific encrypted message cryptanalyzed in this time may be useful when it is known with high probability to be useful. However, such times are highly burdensome when many messages must be collected and processed to yield one useful message. An eavesdropper could well have considerable difficulty in finding the ciphertext corresponding to critical information, but the information security manager cannot take the chance that a critical piece of information might be compromised anyway.

Note: In general, information security managers must develop a model of the threat and respond to that threat, rather than simply assuming the worst (for which the only possible response would be to do *everything*). However, in the case of encryption and in the absence of governmental controls on technology, strong encryption costs about the same as weak encryption. Under such circumstances, it makes no sense at all for the information security manager to choose weak encryption.

A larger key size increases the difficulty of a brute-force search. For symmetric algorithms, a 56-bit key entails a work factor that is 2^{16} (65,536) times larger than that of a 40-bit key, and implies a search time of about 1430 years to accomplish (assuming that the algorithm using that key would take about the same time to execute as the RC4 algorithm). Using more computers could decrease the time proportionally.

> **Note:** A discussion of key lengths for asymmetric algorithms is contained in Chapter 3.

Large speed-up factors for search time would be possible through the use of special-purpose hardware, which can be optimized to perform specific tasks. Estimates have been made regarding the amount of money and time needed to conduct an exhaustive key search against a message encrypted using the DES algorithm. Recent work suggests the feasibility of using special-purpose processors costing a few million dollars working in parallel or in a distributed fashion to enable a brute-force solution of a single 56-bit DES cipher on a time scale of hours. When the costs of design, operation, and maintenance are included (and these costs are generally much larger than the cost of the hardware itself), the economic burden of building and using such a machine would be significant for most individuals and organizations. Criminal organizations would have to support an infrastructure for cracking DES through brute-force search clandestinely to avoid being targeted and infiltrated by law enforcement officials. As a result, developing and sustaining such an infrastructure would be even more difficult for criminals attempting to take that approach.

Such estimates suggest that brute-force attack against 56-bit algorithms such as DES would require the significant effort of a well-funded adversary with access to considerable resources. Such attacks would be far more likely from foreign intelligence services or organized criminal cartels with access to considerable resources and expertise, for whom the plaintext information sought would have considerable value, than from the casual snoop or hacker who is merely curious or nosy.

Thus, for routine information of relatively low or moderate sensitivity or value, 56-bit protection probably suffices at this time. But for information of high value, especially information that would be valuable to foreign intelligence services or major competitors, the adequacy in a decade of 56-bit encryption against a determined and rich attacker is open to question.

Organizing the U.S. Government for Better Information Security on a National Basis

As discussed in Chapter 14, no organization or entity within the federal government has the responsibility for promoting information security in the private sector or for coordinating information security efforts between government and nongovernment parties. The national institute of Standards and Technology (NIST) is responsible for setting Federal Information Processing Standards, and from time to time the private sector adopts these standards, but NIST has authority for information security only in unclassified government information systems. Given the growing importance of the private nongovernment sector technologically and the dependence of government on the private information infrastructure, security practices of the private information infrastructure may have a profound effect on government activities, both civilian and military.

How can coordination be pursued? Coherent policy regarding information assurance, information security, and the operation of the information infrastructure itself is needed. Business interests and the private sector need to be represented at the policymaking table, and a forum for resolving policy issues is needed. And, as the details of implementation are often critical to the success of any given policy, policy implementation and policy formulation must go hand in hand.

Information security functions that may call for coordinated national action vary in scale from large to small. One example would be assisting individual companies in key commercial sectors at their own request to secure their corporate information infrastructures by providing advice, techniques, and analysis that can be adopted at the judgment and discretion of the company involved. In some key sectors (banking and telecommunications), conduits and connections for such assistance already exist as the result of government regulation of firms in those sectors. At present, the U.S. government will provide advice regarding information security threats, vulnerabilities, and solutions only to government contractors (and federal agencies).

A second example is educating users both inside and outside government about various aspects of better information security. For instance, many product vendors and potential users are unaware of the fact that there are no legal barriers to the use of encryption domestically. Outreach efforts could also help in publicizing the information security threat.

Certifying appropriate entities that perform some cryptographic service is a third example. For instance, a public-key infrastructure for authentication requires trusted certification authorities. Validating the *bona fides* of these authorities (through a licensing procedure) will be an essential aspect of such an infrastructure. In the event that private escrow agents become part of an infrastructure for the wide use of encryption, such agents will need to be approved or certified to give the public confidence in using them.

A fourth example is setting *de jure* standards for information security. As previously discussed, the NIST charter prevents it from giving much weight to commercial or private sector needs in the formulation of Federal Information Processing Standards if those needs conflict with those of the federal government, even when such standards affect practice in the private sector. Standards of technology and of practice that guide the private sector should be based on private sector needs, both to promote *best practices* for information security and to provide a legitimate defense in liability cases involving breaches of information security.

How such functions should be implemented is another major question. The committee does not wish to suggest that the creation of a new organization is the only possible mechanism for performing these functions. Some existing organization or entity could well be retooled to service these purposes. But it is clear that whatever entity assumes these functions must be highly insulated from political pressure (arguing for a high degree of independence from the executive branch), broadly representative (arguing for the involvement of individuals who have genuine policy-making authority drawn from a broad range of constituencies, not just government), and fully capable of hearing and evaluating classified arguments if necessary (arguing the need for security clearances).

Note: The Administration concluded that the broad outlines of the national satellite encryption policy can be argued on an unclassified basis. Nevertheless, it is a reality of decision making in the U.S. government on these matters that classified information may nevertheless be invoked in such discussions and uncleared participants asked to leave the room. To preclude this possibility, participating members should have the clearances necessary to engage as full participants in order to promote an effective interchange of views and perspectives.

One proposal that has been discussed regarding these responsibilities involves the Federal Reserve Board. The Federal Reserve Board oversees the Federal Reserve System (FRS), the nation's central bank. The FRS is responsible for setting monetary policy (setting the discount rate), the supervision of banking organizations and open market operations, and providing services to financial institutions. The Board of Governors is the FRS's central coordinating body. Its seven members are appointed by the President of the United States and confirmed by the Senate for 14-year terms. These terms are staggered to insulate the governors from day-to-day political pressure. Its primary function is the formulation of monetary policy, but the board of governors also has supervisory and regulatory responsibilities over the activities of banking organizations and the Federal Reserve Banks.

A second proposal has been made by the Cross-Industry Working Team (XIWT) of the Corporation for National Research Initiatives for the U.S. government to establish a new Joint Security Technology Policy Board as an independent agency of the government. Under this proposal, the board would be an authoritative agency and coordination body officially chartered by statute or executive order *responsible and answerable* for federal performance across all of its agencies, and for promotion of secure information technology environments for the public. In addition, the board would solicit input, analysis, and recommendations about security technology policy concerns from private sector groups and government agencies; represent these groups and agencies within the board; disseminate requests, inquiries, and information back to these groups and agencies; review draft legislation in cognizant areas and make recommendations about the legislation; and, represent the U.S. government in international forums and other activities in the domain of international security technology policy.

The board would be chaired by the Vice President of the United States and would include an equal number of members appointed from the private sector and the federal government.

A third proposal, perhaps more in keeping with the objective of minimal government, could be to utilize existing agencies and organizational structures. The key element of the proposal would be to create an explicit function in the government, that of domestic information security. Because information policy intersects with the interests and responsibilities of several agencies and cabinet departments, the policy role should arguably reside in the Executive Office of the President. Placing the policy function there would also give it the importance and visibility it requires. It might also be desirable to give specific responsibility for the initiation and coordination of policy to a Counselor to the President for Domestic Information Security (DIS). This individual could chair an interagency committee consisting of agencies and departments with a direct interest in and responsibilities for information security matters, including the operating agency, economic policy agencies (Departments of Treasury and Commerce), law enforcement agencies and bureaus (FBI, DEA, ATF), and international affairs and intelligence agencies (Departments of State, Defense, CIA and NSA).

Operationally, a single agency could have responsibility for standards setting, certification of escrow agents, approval of certificate holders for authentication purposes, public education on information security, definition of *best practices*, management of encryption on the Commerce Control List, and so on. The operating agency could be one with an economic policy orientation, such as the Department of Commerce. An alternative point of responsibility might be the Treasury Department, although its law enforcement responsibilities could detract from the objective of raising the economic policy profile of the information security function.

The public advisory committee, which is an essential element of this structure, could be made up of representatives of the computing, telecommunications and banking industries, as well as *public* members from academia, law, and so on. This committee could be organized along the lines of the President's Foreign Intelligence Advisory Board and could report to the Counselor for DIS.

Summary

This chapter described a number of possible policy options, but did not attempt to pull together how these options might fit together in a coherent policy framework. This is the function of Chapter 30. In any event, this chapter described various mechanisms that might be used to manage the export of products with satellite encryption capabilities as described in detail in the Sidebar that follows.

Mechanisms of Export Management

Type: Total Embargo

Description: All or most exports of encryption to target country prohibited (this would be more restrictive than today's regime).

Hypothetical example: No products with encryption capabilities can be exported to Vietnam, Libya, Iraq, Iran.

When appropriate: Appropriate during wartime or other acute national emergency or when imposed pursuant to United Nations or other broad international effort.

Type: Selective export prohibitions

Description: Certain products with encryption capabilities barred for export to target country.

Hypothetical example: Nothing cryptographically stronger than 40-bit RC4 can be exported to South Korea, Taiwan.

When appropriate: Appropriate when supplier countries agree on items for denial and cooperate on restrictions.

Type: Selective activity prohibitions

Description: Exports of encryption for use in particular activities in target country prohibited.

Hypothetical example: PGP allowed for export to pro-democracy groups in People's Republic of China but not for government use.

When appropriate: Appropriate when supplier countries identify proscribed operations and agree to cooperate on restrictions.

Type: Transactional licensing

Description: Products with encryption capabilities require government agency licensing for export to a particular country or country group.

Hypothetical example: State Department individual validated license for a DES encryption product. Licensing actions may be conditioned on end-use verification or post-export verification.

When appropriate: Appropriate when product is inherently sensitive for export to any destination, or when items have both acceptable and undesired potential applications. Also requires an effective multilateral control regime.

Type: Bulk licensing

Description: Exporter obtains government authority to export categories of products with encryption capabilities to particular consignees for a specified time period.

Hypothetical examples: Commerce Department distribution license, ITAR foreign manufacturing license. Note that categories can be determined with considerable freedom. Enforcement may rely on after-the-fact audits.

When appropriate: Same as preceding circumstances, but when specific transaction facts are not critical to effective export control.

Type: Preexport notification

Description: Exporter must prenotify shipment; government agency may prohibit, impose conditions, or exercise persuasion.

Hypothetical example: Requirement imposed on vendors of products with encryption capabilities to notify the U.S. government prior to shipping product overseas.

When appropriate: Generally regarded as an inappropriate export control measure because exporter cannot accept last-minute uncertainty.

Type: Conditions on general authority or right to export

Description: Exporter not required to obtain government agency license but must meet regulatory conditions that preclude high-risk exports. In general, 40-bit RC2/RC4 encryption falls into this category once the Commodity Jurisdiction procedure has determined that a particular product with encryption capabilities may be governed by the CCL.

Hypothetical example: Commerce Department general licenses.

When appropriate: Appropriate when risk of diversion or undesired use is low.

Type: Postexport record keeping

Description: While no license may be necessary, exporter must keep recordkeeping records of particulars of exports for specified period and submit or make available to government agency.

Hypothetical example: Vendor is required to keep records of foreign sales of 40-bit RC2/RC4 encryption products under a Shippers Export Declaration.

When appropriate: Appropriate when it is possible to monitor exports of weak encryption for possible diversion.[11]

From Here

Chapter 26 discussed a variety of options for satellite encryption policy, some of which have been suggested or mentioned in different forums (in public and/or private input received by the committee or by various members of the committee). These policy options include alternative export control regimes for cryptography and alternatives for providing third-party access capabilities when necessary. In addition, Chapter 26 addressed several issues related to or affected by cryptography that will appear on the horizon in the foreseeable future. Chapter 27 covers the following: tightly-coupled secure GPS/INS Systems; military leads in secure GPS/INS Integration; an optimized secure GPS carrier phase ambiguity search method focusing on high speed and reliability; secure intelligent vehicle highway systems; and, preoperational testing of encrypted data link-based air traffic management systems in Magadan, Far East Russia.

Endnotes

1   "Information Security and Privacy in Network Environments," National Security Agency, April 1994 (as printed in U.S. Congress, Office of Technology Assessment, OTA-TCT-606, U.S. Government Printing Office, Washington DC, September 1994, pp. 166-206).

2   Ibid., p. 167.

3   Ibid., p. 170.

4   Ibid., p. 203.

5   Ibid., p. 204.

6   Bobby Inman, "Classifying Science: A Government Proposal," *Aviation Week and Space Technology,* February 8, 1982.

7   "Information Security and Privacy in Network Environments," National Security Agency, p. 205.

8   Ibid.

9   Internet Engineering Task Force, Network Working Group as Requests for Comments (RFCs) in August 1995:

10   © 1996 by the National Academy of Sciences. Courtesy of the National Academy Press, 2101 Constitution Avenue, NW, Washington D.C., 20418.

11   "Information Security and Privacy in Network Environments," National Security Agency, p. 206.

12   Adapted from National Research Council, *Finding Common Ground: U.S. Export Controls in a Changed Global Environment*, National Academy Press, Washington, DC, 1990, p. 109.

27

Future Trends in Mobile Technology for the Human Race: Secure GPSs for Everyone!

The rapid growth of pagers, cellular radio, cellular data, personal communications systems (PCS), wireless LANs, etc., has been nothing short of phenomenal. We stand on the brink of a massive global explosion in wireless systems and services. With a variety of secure commercial satellite systems going operational in the next five years, this is a dynamic and exciting time in mobile technology and/or global positioning systems (GPS). Future implementations are certain to bring advances in encrypted speech compression, adaptive modulation/coding and mitigation, plus smart antennas, and all digital implementation of software multimode radios. These fast-moving developments mandate teamwork and continual interfacing between systems and circuit designers, as well as between RF and signal processing technologists.

The expansion of the secure wireless mobile market is in direct conflict with limitations in the frequency spectrum. Most of today's systems use frequency division multiple access (FDMA), or time division multiple access (TDMA) in the case of digital systems. Another multiple access scheme, code division multiple access (CDMA), which includes extensive use of spread spectrum technology, has been applied in military encrypted satellite communications for some time. There have been numerous proponents for introducing CDMA into civil systems; the first, the USA standard for cellular mobile radio systems, IS-95, has been established.

The CDMA format offers frequency and interference advantages that can enhance spectral efficiency and capacity. The advantages and disadvantages of CDMA over the established FDMA/TDMA are still being debated. The answer lies in parameters such as propagation conditions, frequency ranges, cell sizes, as well as transmission quality, system loads, market and regulatory aspects. To add to the puzzle, there is more than one CDMA approach. International interest and participation indicates that CDMA will play a role in future personal and public radio communications. To support the forthcoming personal communications systems (PCS) revolution, satellites are on the drawing board for launch into four different earth orbits; low, medium, high, and geostationary. Encrypted satellite communications will probably be designed to coexist with cellular in a dual mode that should augment user coverage. A major performance hurdle is that encrypted satellite systems are limited by thermal noise, whereas cellular tends to be limited by co-channel interference. To promote synergy between satellite and cellular systems, satellites will be driven to use identical signaling protocols, and handsets are expected to be size-wise compatible with cellular. In addition, satellite system power constraints require positive antenna gain and low noise front ends. The desire of operators for a broad network with as few cells as possible conflicts with large concentrations of users and limited bandwidth that forces frequency reuse. This presents a variety of technical challenges in antennas as well as frequency assignment, channelization and signal processing. Developing and integrating a net of satellite cells requires careful allocation and management of resources to be efficient.

With that in mind, the first part of this chapter asks the burning question: *With regards to accelerating advances in mobile technology, why aren't tightly coupled global positioning systems/inertial navigation systems (GPS/INS) everywhere, on aircraft, ships and land vehicles?* Good question! Why?

Tightly Coupled Secure GPS/INS Systems

It has been recognized for decades that tightly coupled GPS/INS (TCGI) performance is superior to loosely coupled, but for years the challenge of integrating components was found to be so difficult that, today, most GPS/INS systems are loosely coupled simply because they are easier to build. The tightly coupled systems combine the disparate raw data of GPS and INS sensors in a single, optimum Kalman filter, rather than cascading two filters, one for each sensor. Also, in some TCGI systems the inertial information is fed back into the receiver so that the receiver can track through increased noise and dynamics. Owing to greater synergism of elements, the tightly coupled systems are significantly more accurate, robust, jam-resistant, and less costly, but tend to be complex, inflexible, and hard to build. That difficulty, combined with the high cost of an inertial navigation system (INS) or inertial measurement unit (IMU), has dampened the widespread use of TCGI systems.

To address that problem, an approach has been worked out by Knight Systems of Rancho Palos Verdes, CA. The approach consists of a means to speed up and simplify the integration.

> **Disclaimer:** This author does not in any way endorse any of Knight Systems' products or methodologies. They are mentioned here simply for information and comparison purposes.

A TCGI prototype has been built using the idea, and then van tested. The resulting proof of concept was initially reported at the Institute of Navigation and the Central Inertial Guidance Test Facility in 1993. The new development at that time was the introduction of a software package called GINI that implements the approach, and makes it easy for systems integrators to build secure TCGI systems. The GINI package was first made generally known in September, 1995. It consists of an object library of C-language functions that can be built into GPS/INS products and information technology (IT) utilities. The library offers standardization of the TCGI integration process, ease of use, rapid turnaround, flexibility, and optimal performance. Real-time, near-real-time, and postprocessing applications using GINI are under development now, although a demonstration involving receiver aiding has not been done yet, and none is in progress as of this writing. The rest of this part of the chapter describes the GINI approach to integration, reviews the proof-of-concept demonstration, describes the GINI library, and explains what a system integrator has to do to use it. The essence of the approach is finding practical ways to introduce modularity into secure systems that seemed either hopelessly interwoven or else impossible to integrate at all (see the Sidebar that follows).

Military Leads in Secure GPS/INS Integration

The fundamental question is whether GPS and its derivatives can totally eliminate on-board Inertial Navigation Systems (INS). The military has been integrating a secure GPS with their on-board systems for some time and the combination of GPS and INS has been shown to improve navigation accuracy substantially. As a result the push for improved INS performance has subsided over the last 10 years. An additional benefit of an integrated system appears to be that the INS can coast, allowing GPS to withstand RF threats, and ultimately recover from any resulting navigation errors. Coasting by INS can also provide the advantage of integrity monitoring in the case of unanticipated faults. However, relying on an inertial system under critical Cat III landing conditions, where the aircraft is committed, is still questionable. Attitude determination tests have also been conducted where antennas were strategically placed on the aircraft in an attempt to derive attitude. Results have shown that an attitude reference is still required, and it does not appear that GPS can justify instrument elimination, especially for airframe stabilization.

In addition to the military, large commercial transport aircraft are beginning to utilize both INS and GPS systems. By not being restricted to ground navigation aids they will have the flexibility to fly direct routes, achieving substantial fuel savings, while preserving safe separation. This is a major step toward the end-state *free flight* mode that the airlines are seeking. There is a question on how to overlay GPS on current civil aviation avionics, and show a favorable benefit to cost. The use of lower accuracy instruments (Fiber-Optic Gyros) can substantially reduce overall avionics cost, but much of the current onboard processing capability is already maxed out with a variety of flight management functions.

Approach and Purpose

The reasons for modularity are compelling. If it were possible to develop a TCGI system by mixing and matching modules, as if choosing receivers, inertial systems, and navigation algorithms from a menu, the task of integration would be drastically simplified. Once a system was built it could be changed by substituting other modules. The system integration would be simplified by dividing technical challenges into smaller pieces that are easier to manage, by hiding information within modules so that the way a module does its job is transparent to the rest of the system, and by localizing the effects of mode transitions so that a state change in one subsystem does not cause a transient in another. Finally, it would reduce the cost of integration both directly by shortening the time required to get a product to market, and indirectly by allowing an evolutionary approach to system development.

The goals of modularity and tight coupling are in conflict. Modularity is generally facilitated if the exchanges between subsystems can be boiled down to a few simple connections, where a low data rate can be used, round-off errors and small timing errors are acceptable, and the data are allowed to be half a second stale going across an interface. Those are the advantages of loose coupling. Tight coupling requires precise, high-rate encrypted data exchanges, and exacting synchronization of subsystems.

Nevertheless, it was concluded in 1992 that modularity was possible *within families* of products that adhere to a shared interface protocol. Since that time, the concept has been broadened and generalized. Instead of *families of products*, the concept is more nearly described by a set of requirements placed on the sensors to be compatible. Those requirements are:

- ◆ GPS Receiver
  - ◆ Able to navigate autonomously.
  - ◆ Measure the ranges and carrier phase simultaneously at regular epochs, typically 1 Hz.
  - ◆ Output ranges, carrier phase, raw ephemeris, position, velocity, accuracy statistics, and status.
  - ◆ Time tag the receiver's data, and support time tagging of INS data, with GPS time.
  - ◆ Allow the receiver clock to float.

- ♦ If the receiver will be INS-aided, then make provision for applying GINI-computed signal Doppler, and for adjusting receiver bandwidth to accommodate INS error.
- ♦ INS or IMU
  - ♦ Self-moding and control after application of power
  - ♦ Generate and output conventional $\Delta V, \Delta \theta$.
  - ♦ Sample $\Delta V, \Delta \theta$ simultaneously at regular epochs, typically 100 Hz or higher.
  - ♦ Compensate $\Delta V, \Delta \theta$ as necessary to achieve the advertised accuracy.
  - ♦ Support time tagging of $\Delta V, \Delta \theta$ accumulation intervals with GPS time, generally to within a small fraction of a millisecond.
  - ♦ Additionally, if a navigation-grade INS is used:
    - ♦ Provide attitude output.
    - ♦ Navigate autonomously.[1]

As originally conceived, the INS $\Delta V, \Delta \theta$ accumulation intervals were to be controlled from the GPS receiver's clock, so that the INS divides a receiver 1-second interval into a hundred 10 ms intervals. That remains a preferred option, but the concept has been generalized to allow the two sensors to be asynchronous, as long as INS $\Delta V, \Delta \theta$ accumulation intervals can be accurately tagged with GPS time.

Figure 27.1 shows a functional block diagram of the TCGI system, with inertial, GPS, and software elements.[2] Notice that the figure is a *functional* diagram, meaning that nothing constrains where or how the functions are implemented. The GPS receiver might be embedded in the INS physical housing, and GINI might very well be hosted in either the receiver or INS processor.

The approach is not a panacea. Drawbacks of the approach are, reduced fault tolerance, requirements placed on sensors to be compatible, and adaptation.

Fault Tolerance

Tightly coupled systems are inherently less fault tolerant than loosely coupled ones because the tight coupling means that a failure in one element is more likely to affect other elements. To address that concern, the receiver and INS are required to function as autonomous

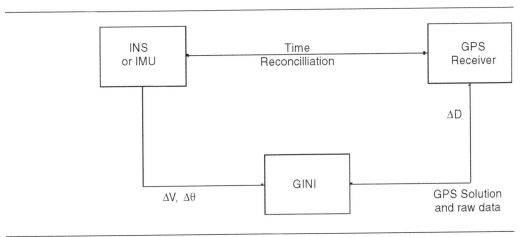

Figure 27.1 Functional elements.

navigators, and then their raw data are optimally combined in a third navigation solution by GINI. There is then no single-point failure that can hobble all three navigation solutions, although an INS failure would leave the receiver unaided. The requirement to navigate autonomously also means that the sensor manufacturers can apply a great deal of testing independently of any integration, and can release a product that is well understood within their own realm of expertise. A receiver that can navigate autonomously is obviously working and doing most of the things needed to support a TCGI integration.

Sensors' Requirements Placement

The toughest requirements involve timing and making the sensors' raw data available for output. Clearly, it is not possible to connect just any INS with just any receiver, especially among older sensors that were only intended to operate as stand-alone elements of federated systems, or present-day inexpensive sensors that simply lack the outputs. If the GPS receiver's raw data are decrypted and cannot be allowed outside the receiver's processor, then GINI moves into the receiver's processor (where it remains functionally distinct from receiver software). Memory and throughput requirements are described later.

The requirement for the receiver's clock to float, rather than be driven toward GPS time, enables the navigation Kalman filter to predict the

clock's behavior. That predictability is key to a major TCGI performance advantage. It is possible for the TCGI to operate in a poor signal
environment where GPS signals are only momentarily and infrequently
available, and yet perform almost as well as in an environment where
the signals are fully and continuously available. As for adaptation, the
aspects of TCGI that are most resistant to modularity are, achieving
optimal performance with varying INS and IMU types and adjusting
the finite state logic.

Performance

There can be a wide diversity among receiver types, but to GINI the
differences all boil down to data volume and statistics of performance.
That makes it easy to achieve commonality among receiver types.
Achieving commonality among INS and IMU types, though, is not as
easy. One approach would be to dump enough (modeled) process noise
into the navigation filter that the distinction between sensor types
would be blurred. That would achieve commonality at the expense of
performance. The optimal approach used by GINI is to take the same
INS or IMU error model that would be used in a Monte Carlo
simulation, and *then probe* that model so that GINI designs its own
Kalman Q algorithm during a startup procedure. It is much easier for
a user to supply an error model than to tune a Kalman filter. Historically, the biggest problem that has been encountered in maintaining
filter software has been that the filter has to be retuned for each new
application. With GINI's approach, it is only necessary to provide a
new sensor error model, and the filter tunes itself.

Logic

If one system requires transfer alignment and another does not, then
something has to be different between those two systems, and there is
no point around which to form commonality. The GINI solution is to
provide a library of functions, and allow users to build systems by
plugging functions together according to need. That means that *the
face* put on the system is primarily up to the system integrator. For
example, the integrator may develop some special augmentation techniques, work out a way to resolve carrier cycle ambiguities in real time,
and blend that with an integrity monitoring algorithm, resulting in an
overall system that exhibits a new level of synergism and unique,
distinguishing characteristics. But as far as GINI is concerned, the

range measurements simply became more accurate. GINI does not need to know *how* that improvement was achieved. In keeping with modularity, GINI is itself a module.

Proof of Concept

Van testing of an early GINI prototype was accomplished in a collaboration with Allen Osborne Associates in Westlake Village, CA, and Inertial Science, Inc., in Newbury Park, CA. An INS and a receiver were placed inside a van, and then driven onto rural highways. A single wire was all that was necessary to connect the receiver and INS, demonstrating the simplicity of the GINI integration approach. Contributors to the proof of concept each spent a few days on the effort, demonstrating the reduction in TCGI development time. The GPS and INS data were recorded on floppy disks, and then postprocessed. It was determined that the equipment and integration software worked perfectly through tests of functionality, open-loop tests, tests of measurement residuals, tests of filter corrections to the INS, and tests of convergence of the GPS data on the TCGI solution. All the tests passed as expected.

Description of GINI

Major GINI functions are, strapdown navigation, satellite orbital calculations, Kalman filtering, and receiver aiding. The GINI will interface with a strapdown inertial system, but not a gimbaled one. Gyro quality can be anything from 0.001°/hr to 30°/hr. The GINI forms the optimal solution with whatever sensors it has to use.

Navigation

The TCGI solution for position and velocity can be read out at rates up to 10–40 Hz. Attitude can be accessed at rates up to the INS $\Delta V, \Delta\theta$ rate. Output formats include latitude/longitude, earth-centered coordinates, range XYZ, speed/track/climb, east/north/up velocity, roll/pitch/heading, and attitude direction cosines. All output data can be made available at the last data capture event, receiver measurement, or INS $\Delta V, \Delta\theta$, although currently, there is no ability (within GINI) to translate to an arbitrary point on the vehicle. The equations of motion use the exact, ellipsoidal, rotating earth model of the World Geodetic System, 1984. A wander-azimuth mechanization is used for numerical accuracy. Sculling compensation is applied if not already done in the INS, and an 8th-order Bortz attitude formulation is used.

Kalman Filter

Nominally, the filter tracks 20 states (or elements of error) in position, velocity, attitude, gyro drift, accelerometer bias, clock effects, and lever arm. The filter can process the encrypted pseudo-range and carrier phase data of up to 12 satellites at once, using a Bierman factorization for numerical stability. The filter adapts to processor load, meaning that if the processor has not finished with the previous filter cycle by the time a new cycle would normally begin, then the filter can hold off for another second or two. A full state transition model is used to describe the evolution of navigation effects, including the Schuler 84-minute loop, Coriolis terms, earth 24-hour loop, and higher-order effects. GINI is tolerant of nonlinearities and poor *a priori* assumptions, and will converge under dynamics with initial roll, pitch, and azimuth errors of up to 60°. Additional sensors, such as barometric altimeters and Doppler radars, are not present in GINI, as of this writing, but the hooks are present for quickly adding them.

Satellite Orbital Calculations

The GINI computes satellite positions and ranges. The range computations include satellite clock error, equipment group delay, relativistic effects, and tropospheric delay and GINI maintains a library of packed ephemeris, both the current and preceding issue, for all 24 satellites. Unpacked ephemeris is maintained for up to 12 satellites at once. Satellite ephemeris is provided to GINI in the form of packed data obtained directly from receiver demodulation of satellite broadcasts, with parity and checksum bits removed. And GINI will accept the information in binary or ASCII, with 8 or 10 words per subframe, and in natural or permuted byte order. In van testing, the differences between calculated and measured ranges were 23 cm RMS for (augmented P-code) pseudo-ranges, and 1.5 mm for differenced carrier phase, with a base-station separation of 10 miles, validating the satellite orbital software.

Receiver Aiding

One of the primary goals of modularity is to isolate the effect of mode transitions, so that a state change in one subsystem does not cause a hiccup in another. The GINI approach to receiver aiding achieves that goal by mapping INS $\Delta V, \Delta \theta$ to receiver frequency increments, AD. All the receiver then has to do is add up a long string of (D to have GPS signal Doppler at any given time. The mapping is insensitive to system events, so in keeping with modularity, the receiver does not even notice the cutover to a new issue of ephemeris. The computed range, range rate, and (D, for up to 12 satellites at once and with multiple GPS antennas, are all provided to the receiver, typically at 100 Hz. When required, the receiver can use the aiding information to continue tracking the GPS code signal under conditions that are too noisy to track the GPS carrier signal, or to know exactly where to look in phase and frequency for a signal that has temporarily vanished.

Written in C Code

A number cruncher without complicated system requirements, GINI is written in a subset of plain C, compatible with ANSI, Microsoft, Borland, and Kernighan & Ritchie. The GINI deliverable consists of a static library that is linked to the user's other software. It requires about 130 Kb of computer memory. Throughput varies, but a typical configuration with a 20-state Kalman filter requires 1.3% of computing capacity of a 120 MHz Pentium desktop computer in order to keep up with real time. At that rate, GINI could process a 2-hour field recording in about 90 seconds.

User Responsibilities

Briefly, the user has to bring GPS and INS data into a processor where GINI will reside, assemble the information into data structures defined in GINI header files, and then call GINI functions in a time-ordered sequence. The user then calls other GINI functions to access the TCGI navigation results and monitor performance. Receiver measurements of range and carrier phase have to be edited to remove erroneous information and GINI functions are provided to support the user's editing policy. Also, if the receiver has not already augmented its measurements (applied differential corrections), then the system integrator will have to do that. It does not matter to GINI whether the receiver is augmented, but civilian users should plan on augmenting the receiver's measurements at least in postprocessing if not in real time.

Probably the toughest single challenge the system integrator will face is time tagging INS $\Delta V, \Delta \theta$ accumulation intervals with GPS time, accurate to a small fraction of a millisecond. The receiver can easily time tag its own data to microsecond accuracy. If a camera or geographic data collection system is involved, then the data capture events have to be tagged with GPS time and flagged in GINI input buffers. If a full real-time system is envisioned, then a real-time executive is required that allocates up to two simultaneous task levels to GINI, typically 100 Hz and 0.5 Hz. The inertial information and data capture times and flags have to be moved to GINI input buffers within milliseconds, although half-second delays are acceptable in getting receiver data to GINI. If a near-real-time or postprocessing mode will be used instead, then a real-time executive is not required. The GINI functions all merge into a single task level and latency requirements are relaxed.

Most users will not need receiver aiding, but if it is needed, then the receiver has to be modified to apply the aiding information. The system integrator will be involved with the data transport, receiver moding, receiver bandwidth selection, and system stability issues.

The significance of this work is that tightly coupled GPS/INS systems are becoming much easier to design, assemble, test, and modify. An approach that speeds up the integration has been worked through, published, proved in field trials, and now made generally available to systems integrators in the form of a software library.

The Global Positioning System (GPS) has been used in many ways that were unimagined by its original planners and implementers. This part of the chapter discusses one such application—the surveillance of commercial aircraft to aid in the development of a secure airspace environment.

Monitoring Encrypted GPS Position Reports

The GPS provides system users with the ability to determine their own position with an accuracy, reliability, and cost that is unprecedented. Voice procedures augmented by radar have been the primary tools for air traffic surveillance since the end of World War II. But for some countries the equipage of aircraft with GPS and an encrypted data link capable of carrying position reports to the ATC authorities is providing a viable alternative to long-range secondary radar systems.

In the summer and fall of 1995, ARINC Inc. (communications and information processing systems company based in Annapolis, MD) installed and demonstrated equipment in Magadan, Russia, which permits air traffic controllers of MagadanAero-Control to monitor GPS position reports generated by aircraft as far away as Canada and the South Pacific. The position reports were displayed against maps and flight tracks. This equipment has clearly demonstrated an alternative technology for the upgrading of the ATC system in Siberia and other remote areas.

Surveillance of Air Traffic Currently Enroute

Voice procedures, whereby a controller talks to an aircrew using HF or VHF radios, are the predominant means of controlling air traffic in remote areas. This process works well when flight density is low. When traffic volume increases, a track system is often used to increase the capacity of the airspace while maintaining safety. Secondary radar (which positively identifies each target) provides a means to identify and separate traffic at the highest densities, consistent with a high level of safety. There are two important characteristics of aircraft position information that differentiate radar from voice procedures — interval between reports and method of visualization.

Radar data is collected for each target on each revolution of the antenna, typically 5–20 seconds. Originally radar data was displayed on a plan position indicator that showed a sweep radiating from the center of a circular display, which was synchronized to the antenna. Each target illuminated by the radar caused an intensification of the sweep at the corresponding range. Primary radars still use this method of display.

More modern, secondary surveillance radars depend upon a transponder onboard the aircraft to send positive identification and other information back to the antenna when the aircraft is illuminated by the radar beam. These response messages are processed and displayed against a map of the area, without the distraction of a displayed sweep. The identification information and other data, such as altitude, speed, etc., are displayed along with an icon showing the position of the aircraft.

Voice reports are typically issued by the pilot at either fixed intervals of time or when passing named waypoints along the flight path previously approved by the civil aviation authority. This results in an interval of 15 minutes to an hour between reports. The controller uses some form of written tracking, either handwritten or printed flight progress strips, to visualize the aircraft's position.

A modern radar situation display shows all aircraft in the airspace, each with identification, against reference maps of coastlines, flight routes, fixed waypoints, either alone or in combination. Unexpected situations, such as aircraft headed toward the same point in space, are graphically illustrated.

Voice reports must be incorporated into a controller's internal awareness of the current situation in his/her airspace. While this is routinely done by hundreds of people at any time of the day or night, it is easy to see the limitations of this method. The GPS and and encrypted data link can provide an alternative to internal visualization, which equals secondary radar in the clarity and immediacy of information presentation.

The GPS and Encrypted Data Link

The GPS provides current three-dimensional position information to within roughly 100 meters. Aircraft can and do use this information for navigation, as has no doubt been reported. If that information could be shared with the ATC authorities automatically (without increased workload on the part of pilots or controllers), it would provide an unprecedented means of surveillance.

Aircraft identification, position, and other status information is routinely sent via an encrypted data link to ground destinations. The Aircraft Communications And Reporting System (ACARS) has provided a means to send these encrypted messages since 1978. Commercial aircraft the world over routinely send and receive airline operational control (AOC) messages over ACARS. More than 8 million of these messages are exchanged each month with the more than 5000 ACARS-equipped aircraft in the world.

Originally, ACARS was implemented as a VHF data link and was limited to line-of-sight communications (120–150 miles maximum). However, in 1992, the aeronautical mobile satellite system provided by Inmarsat was approved by the FAA for use in exchanging routine aircraft position reports, clearance requests, and ATC clearances in the Pacific Ocean. This encrypted data link message information was used by the controllers in the same way as voice messages (as printed position reports or verbal requests/clearances).

Magadan Installation

Figure 27.2 shows the overall architecture of the Magadan experiment.[3] Both the air/ground voice and data networks are shown providing communication to the aircraft.

The preoperational system installed in August and September of 1995 consists of the elements shown in Figure 27.3.[4] Those system elements which are installed in Russia are included within the highlighted area. The Orbital satellite GES was an existing installation, as was the air traffic controller console, which is identified as the voice controlled aircraft position in Figure 27.3. All other system elements within the hatched area were installed in Magadan as part of this demonstration. All of the system elements outside of the hatched area, with the exception of thc multiplexer and modem, predate this project.

The Gateway is the central communication element of the installation. Not only does it interface to the wide area network, identified as ACARS/ADNS, but it performs message conversion and routing functions that permit the various ATC application modules (CPDLC, ADS and voice position report transcription) to receive information in one consistent data format, regardless of the source of the information. The Gateway function is necessary in order to support the mixed fleet of

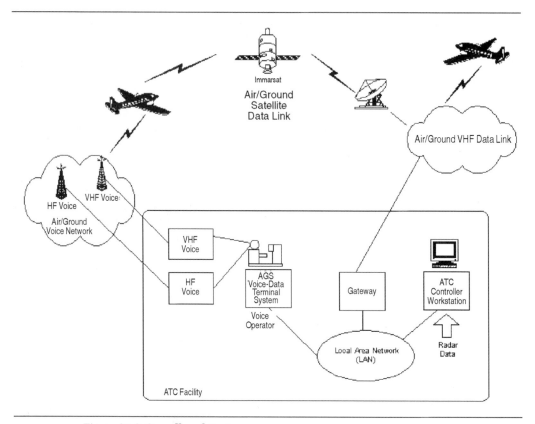

Figure 27.2 Overall architecture.

aircraft that will be found in any realistic airspace support concept. The demonstration equipment accommodates many categories of air-craft:

- ♦ 747-400 FANS l aircraft.
- ♦ Package A or other character-oriented application aircraft.
- ♦ Aeronautical mobile satellite and/or VHF data link equipped aircraft.
- ♦ Russian domestic aircraft or other aircraft with only VHF and/or HF voice communications equipment.[5]

A primary goal of the demonstration is to integrate the display of variously equipped aircraft into a single situation display. A complete diagram of Magadan installation is shown in Figure 27.4.[6]

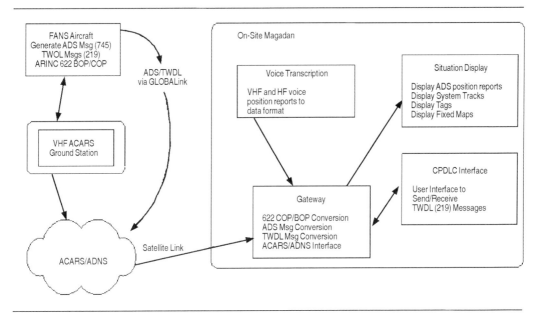

Figure 27.3 Magadan block diagram.

FANS 1 Aircraft

All twenty-four (24) United Airlines 747-400 aircraft have been up-
graded with the FANS 1 avionics package since June 1995. Presently,
Cathay Pacific, Qantas, and Air New Zealand also are operating FANS
1 avionics. The FANS 1 upgrade package in the Flight Management
System (FMS) provides (among other features) for Automatic Depend-
ent Surveillance (ADS) and Controller Pilot Datalink Communications
(CPDLC) message communication via encrypted data link. All United
FANS 1 aircraft have both VHF and aeronautical mobile satellite (via
Inmarsat) encrypted data link capability.

Non-FANS Aircraft

Numerous non-FANS aircraft presently operate near or within the
Magadan flight information region. Some of these carry the Package *A*
avionics. Others may upgrade to FANS 1 capability during the period
of the demonstration. Most U.S. aircraft and many foreign aircraft are

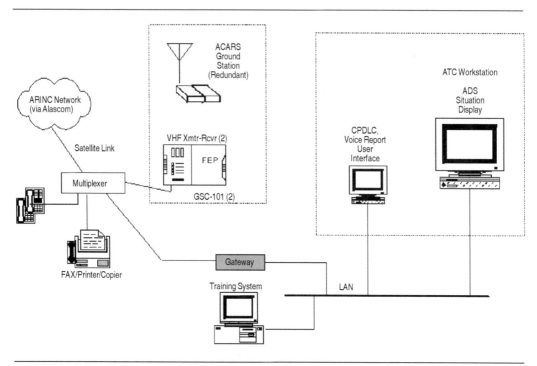

Figure 27.4 Magadan installation.

equipped with a VHF encrypted data link as a minimum and may have character-oriented message-based applications for position reporting.

Russian Domestic Aircraft

Russian domestic aircraft typically have VHF or HF voice communications, with no FANS or VHF data link equipment. They are able to maintain voice communications using VHF and HF voice communications stations. The position of these aircraft is determined by onboard navigation instruments and is communicated by voice to the ACC where the ATC operator manually enters the position report into the tracking system. Aircraft positions will be displayed on the same situation display as data-linked position reports. Between manual position report entries, the system will predict future aircraft positions. This is based upon information about the next waypoint and estimated time over that waypoint that is given in the voice report.

Operational Testing

The first flight to communicate with Magadan was United flight 863 SFO-SYD on August 24, 1995. The ADS messages were received at 5-minute intervals and the complete Boeing CPDLC script was transacted in approximately 1.5 hours. A test script was used to assure coordination between pilot and controller.

On September 2–3, 1995, United flight 881 ORD-NRT became the first flight over Far East Russia to be tracked by ADS and CPDLC. During this flight all communications were performed by a Russian controller. oth ADS and CPDLC were exercised, although a number of free text messages were used instead of their equivalent CPDLC messages. The flight was initially tracked on the ground at Chicago's O'Hare International Airport by the ADSATC system in Annapolis, MD. Both ADS and CPDLC were tested from Annapolis for about 6 hours prior to a handoff to Magadan.

On October 19, 1995, United flight 1854 PEK-SFO became the first FANS 1 flight to overfly Magadan. The connection was established as the flight entered Russian airspace in the Khabarovsk region. The default ADS interval of 15 minutes was used throughout this flight. Again, a Russian controller handled the flight. As the flight entered the Magadan sector, all directions given by voice were also given by CPDLC. The flight was tracked until it neared Alaskan airspace, at which time a handoff was initiated to Annapolis. Annapolis established communications with the flight and tracked it to SFO. Table 27.1 summarizes the first three FANS 1 test flights tracked in Magadan.[7]

Global Positioning System technology also offers highly accurate and secure satellite positioning services around the world, at no charge. As

Table 27.1 FANS 1 Test Flights Tracked in Magadan

Flight	Date	Departure	Destination	# AFN Msga	# ADS Msgs	# CPDLC Msgs
UAL863	24Aug95	SFO	SYD	7	20 (est)	40
UAL881	2–3Sep95	ORD	NRT	4	85	51
UAL1854	19Oct95	PEK	SFO	7	10 (est)	57

previously mentioned, the GPS system makes use of spread spectrum signaling and a triangulation process to calculate the location of an object within 16 m. Naturally, the positioning information obtained through GPS can be fed back to a central data collection area for Intelligent Vehicle Highway Systems (IVHS) applications—the focus of the last part of this chapter.

Secure Intelligent Vehicle Highway Systems

Intelligent Vehicle Highway Systems (IVHS), now renamed Intelligent Transportation Systems (ITS), describes a collective approach to the problem of enhancing mobility and traffic handling capacity, improving the conditions of travel and reducing the adverse environmental effect, and eventually automating the existing surface transportation systems through the application of a range of advanced technologies. Once a significant portion of IVHS infrastructure is in place, the automated vehicle operation phase can be implemented. It is well known that due to a number of economical, social and environmental reasons, the conventional approach of additional highway construction is no longer the best solution to the rapidly growing demand for higher level secure and safer traffic in the highway system. As such, IVHS presents the next major breakthrough in the modernization of the surface transportation systems. An additional benefit of IVHS is the increase in productivity due to saving of the time otherwise spent in traveling. This has primarily resulted as a consequence of congestion reduction envisioned by ITS. Moreover, enhancing the roadway system efficiency and improving the flow of traffic not only reduces the energy consumption but also diminishes the environmental pollution. Finally, IVHS improves safety by maximizing the flow of secure route information to the driver.

Satellite communications requirements are widely regarded as pivotal in any IVHS network. In particular, many encrypted satellite communications links in an IVHS system are required to be wireless. The communications between the vehicle and IVHS infrastructure, which is one of the most essential links, is required to be wireless, mobile, and for the most part, interactive and very secure. Several wireless communication technologies have been considered, tested, and implemented for various IVHS functions. This ranges from RF and IR beacon technology to mobile encrypted satellite communications. However, as the IVHS wireless system architecture is not yet fully developed and defined, the question of which wireless technology is most suitable is still, to a large extent, an open one. A survey of these technologies is presented in this part of chapter. An important issue regarding wireless communications links in IVHS is whether a single wireless technology may be used to satisfy all of IVHS wireless communications needs, or economic and technical restrictions would rule that various satellite communications functions of IVHS be implemented using a suitable media for each function. As IVHS passes through the *initiative* stage and enters the R&D and implementation platform, it appears that the latter is the case. Currently many operational wireless communication systems are in place to realize certain secure IVHS communications functions. This part of the chapter explores a number of secure wireless communications technologies that have been applied in currently operational or planned IVHS projects and/or considered to be suitable for secure IVHS applications.

IHVS Major Technologies

The transportation technologies that will be developed under the umbrella of the IVHS program consists of five major components:

- ♦ Advanced Traffic Management System (ATMS).
- ♦ Advanced Traveler Information Systems (ATIS).
- ♦ Commercial Vehicle Operations (CVO).
- ♦ Advanced Vehicle Control Systems (AVCS).
- ♦ Advanced Public Transportation Systems (APTS).[8]

Advanced Traffic Management System (ATMS)

This is the backbone and the most Important component of IVHS. The ATMS collects and uses real-time traffic data on major arteries, expressways, and transit networks, and initiates proper and secure feedback signals to the surface transportation system to control the flow of traffic. Rapid detection of incidents that impede traffic flow, provides input into a decision response system that can accelerate recovery of roadway congestion and minimize probability of accidents.

The ATMS consists of three main parts. The first is the surveillance system that monitors the operation status of a roadway network. The second is a real-time traffic-adaptive control system that receives traffic data from the surveillance system and adapts network control variables such as traffic signals, freeway ramp meters, and changeable electronic messages for optimal performance. In addition, there are also system operator support systems, such as simulation models and expert systems that facilitates real-time control and management of the network. An important feature of ATMS is that substantial portions of ATMS functionality can be added to the highway system without requirement for additional electronic systems in the vehicle. The Los Angeles Smart Corridor is an ATMS project.

Advanced Traveler Information Systems (ATIS)

The ATIS is a framework through which a wide variety of information is made available to the driver, as well as to the general traveler, to help them efficiently and safely reach their intended destination. It essentially consists of three components. The first, is the in-vehicle route guidance system, which includes electronic maps, route information, and in-vehicle display system, and enables the driver to select the best route to the designation. The ATIS second component is model development to optimize network routine and usage. The last, the mechanism that disseminates traffic, weather, and route information, allows pretrip and enroute planning as well as alternate route selection. The Chicago ADVANCE project and the Southern California PATH FINDER are examples of ATIS projects.

Commercial Vehicle Operations (CVO)

This component of IVHS addresses the special needs of fleets of commercial and government vehicles. It applies many aspects of ATIS technology to dynamic fleet management problems. Among the functions that CVO technology can provide are vehicle locations and identification, two-way communication with dispatcher. While CVO clearly requires the addition of equipment to the vehicles and outside of the vehicle, the perceived cost benefits are substantial. A successful example of a CVO project is Qualcomm's OmniTRACS, which is a two-way secure mobile satellite communications and tracking system, developed for the trucking industry.

Advanced Vehicle Control Systems (AVCS)

This technology aims to assist the driver to make safer trips. The AVCS elements can span from sensing impending collisions to intervening by braking or steering to avoid collisions. The ultimate goal of AVCS is to develop an automated highway technology in which the driver no longer drives, and cars could journey from one place to another on designated highways without human intervention. This would require a substantial secure (encrypted) IVHS infrastructure and significant availability of in-vehicle equipment. Extensions of AVCS include support for High Occupancy Vehicle (HOV) traffic in specially monitored lanes. California PATH is an AVCS project.

Advanced Public Transportation Systems (APTS)

This component of IVHS applies advanced navigation and encrypted satellite communication technologies to all aspects of public transportation systems. As such, APTS addresses issues such as the optimal utilization of mass transportation systems, for example, HOVs.

IVHS-Required Encrypted Satellite Communications Links

A complete IVHS system consists of four primary components. In order to perform various IVHS functions, these components need to be interconnected through communications links. Figure 27.5 illustrates a general block diagram of an IVHS network.[9] The four primary components are identified as: traffic management system; vehicles; road surveillance network; and roadside system. The commercial truck is a subset of the vehicle component; however, a separate commercial trucks block is included in the diagram to highlight some important CVO functions of an IVHS network. The required encrypted satellite communications links between various components of an IVHS network are shown using arrows. To distinguish between wireline and wireless communications links, wireline connections are designated using solid black arrows. It should be noted that some IVHS wireless links, such as vehicle-to-vehicle (for AVCS applications) that are not shown in the block diagram, will not be discussed in this chapter.

Prior to the deployment of an encrypted satellite communications infrastructure for an IVHS network, several important requirements

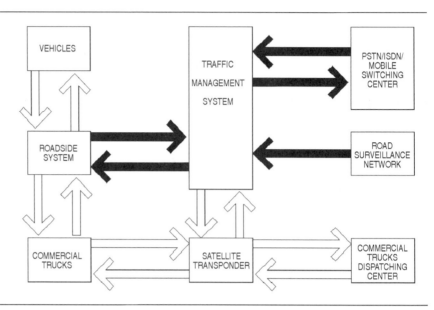

Figure 27.5 A generic block diagram for an IVHS network.

should be contemplated. Primarily, as neither the architecture nor all desired functions of IVHS networks are fully defined yet, extensive room for growth in any encrypted satellite communications system contemplated for IVHS applications should be provided. Secondly, the system has to be highly reliable, and like any other communications system, it needs to be cost effective in using the transmission media and equipment. Connectivity with other components of ITS and public telecommunications networks such as public switched telephone network (PSTN), integrated services digital network (ISDN), and Mobile Cellular/PCN (personal communications network), is a requirement. Finally, compliance with national and international communications standards is another important issue.

The encrypted satellite communications link that connects the traffic management system to the surveillance/roadside system, normally a wireline communications network, forms the backbone of ATMS, which would later support other IVHS functions as well. This link will not be described any further in this chapter. A key IVHS wireless interconnection is the vehicle to infrastructure communications link. This includes vehicle/roadside communications (VRC) for ATMS,

ATIS, APTS, and AVCS applications, as well as commercial trucks to communication satellites link for CVO applications. As ITS networks grow to incorporate more functions, the demand for VRC grows.

Wireless Technologies for Secure IVHS Applications

A number of wireless technologies have been or currently are being considered and studied for IVHS applications. The important issues in selecting a wireless technology for ITS purposes are capability of growth, scope of application, cost of deployment and maintenance, data rate and capacity, mode of communications, and area coverage. In what follows, a brief survey of wireless technologies that have been applied, tested, or considered for this purpose is presented.

The IR and RF Beacon Technology

The IR and RF beacon technology supports satellite communications up to 1 Mb/s, over short distances with very limited coverage. This technology cannot support functions that require spontaneous communication. The Japanese road/automobile communications system (RACS) that takes advantage of RF beacon technology for ATIS applications, is an example. This system supports noncontinuous one-way and/or two-way data communications in a limited zone (about 65 m length around intermittently installed RF beacons). Encrypted data transmission is achieved during the short period of time in which the vehicle passes through the zone. The RACS carries out several ATIS, ATMS and CVO functions such as navigation, dissemination of traffic information, automatic vehicle identification (AVl), information collection regarding traffic and road incidents, automatic toll collection, and, on-request information services such as weather report and sight-seeing information.

RF Data Network

The RF data network offers a good solution for ATIS-type applications if it is to be implemented locally. It is relatively inexpensive, provides complete coverage (in a locale), is interactive and supports up to a data rate of 19,200 bits/s or higher. A dedicated RF data network infrastructure for IVHS would provide flexibility, cost effectiveness, and room for future growth.

Analog Cellular Phone Technology

Analog cellular phone technology provides an inexpensive coverage if the vehicle is already equipped with a mobile phone. As of 1987, Analog Cellular System reached its maximum capacity with only about 2% of vehicles using this service in the U.S. Nevertheless, many IVHS projects (under development or operational) are at least partially using the present analog cellular networks. The Chicago ADVANCE and the San Francisco/Bay Area Travel Info projects are examples of this sort. The feasibility of integrating some of IVHS wireless needs into cellular networks will improve as this technology is gradually replaced with its digital counterpart.

Digital Cellular Phone Technology

Digital cellular phone technology comes with the promise of cellular CDMA to increase the capacity of cellular networks by more than 20-fold (relative to the analog AMPS system). Therefore, it is only natural to think that part of IVHS wireless communication needs should be integrated into future digital cellular networks.

Personal Communications Services (PCS)

Personal communications services (PCS) refer to low-power digital cellular voice-data communications systems, to be used predominantly by pedestrians in major metropolitan areas. The PCS is already providing services in some of the big cities. Thus, PCS technology has some potential for carrying out secure IVHS wireless communications functions within the metropolitan areas.

Mobile Satellite Services (MSS)

Mobile satellite services (MSS) provide very wide area coverage for short interactive vehicle to dispatcher and/or to traffic management center message communications. A successful application of MSS technology in IVHS can be found in the Qualcomm's OmniTRACS project. It operates on a secondary basis on Ku band (12-14 GHz), which has been allocated for fixed satellite services (FSS) on a primary basis. OmniTRACS provides a two-way messaging and positioning service to up to 90,000 users. In order to mitigate interference to satellite primary application users, a direct sequence spread-spectrum signaling is used to spread the power spectral density of each mobile user over a bandwidth of 1 MHz. Additionally, a frequency hopping-spread spectrum signaling is applied to distribute the power spectral density generated by the combination of all active mobile users over a flexible bandwidth of 54 MHz. This has provided an excellent service for the trucking industry. However, if data from other mobile users in addition to trucks for some IVHS function is to be handled by MSS, it would soon put a strain on the primary users of MSS.

Cellular Digital Packet Data (CDPD)

Cellular digital packet data (CDPD) is a novel technology that provides packet-switched data transmission using the existing analog cellular networks. The packets of data are transmitted during the time slots in which the cellular radio channel is idle (between telephone calls). This overlay implementation not only makes use of the existing analog cellular infrastructure, but also enables encrypted data transmission using the same device. Mobility, ease of use, and wide coverage make CDPD technology suitable for providing wide-area distribution and collection of IVHS data.

Global Positioning System

As previously mentioned, global positioning system technology offers highly accurate satellite positioning services around the world at no charge. The GPS system makes use of spread-spectrum signaling and a triangulation process to calculate the location of an object within 16 meters. Naturally, the positioning information obtained through GPS can be fed back to a central data collection area for IVHS applications. In fact, many IVHS projects, such as the Chicago ADVANCE and the Minnesota Travlink take advantage of GPS technology.

Other wireless technologies considered for IVHS applications include *FM Sideband Broadcast*, and *Meteor Burst Communications*. The FM sideband is the least expensive technology in which the data is superimposed on an FM broadcast radio. The data rate is around 1100 bits/s. The link is one way and the coverage is limited to areas covered by FM radio broadcast stations.

Technical feasibility and economic necessity has determined that various secure IVHS wireless communications needs should be satisfied using different wireless technologies. In most operational IVHS projects, a multitude of wireless technologies are applied. The use of various cellular technologies to meet the wireless communications requirement for many IVHS functions has proven feasible and cost effective.

Summary

The first part of this chapter addressed the question: *Why are tightly coupled GPS/INS systems not found everywhere, on aircraft, ships, and land vehicles?* Two barriers to widespread use are cited. On e is the high cost of the INS, and the other is the cost and complexity of tightly coupled GPS/lNS integration. One of those two barriers has recently been diminished drastically with the development of a standardized software package for secure tightly coupled integration. In the past, only the largest corporations have been able to pay the initial development cost for secure tightly coupled GPS/ INS integration, usually with funding from a large defense program. Using the new software package, integration and van test can be accomplished in a matter of days, and this has been demonstrated with field trials. The package is intended primarily for small companies that otherwise would not be able to build tightly coupled GPS/INS systems at all. What would have been a prohibitive 4- or 5-man year development effort is reduced to a few man weeks. To accomplish an integration, the system integrator has to find a way, through serial interfaces or by some other means, to get the INS measurements of acceleration (accumulated velocity change ΔV) and attitude rate (accumulated angle change $A\theta$) into a processor, along with the raw data of a GPS receiver. He or she also has to find a way to time tag the INS ΔV, $A\theta$ with GPS time. The rest of tightly coupled GPS/INS integration is predominantly accomplished in the standardized software package. That leaves the cost of the INS as the only remaining barrier to the very widespread use of GPS/lNS, and invites new development of low-cost inertial sensors. The focus of this part of the chapter was on the software package, and how it achieves standardization and ease of use while retaining the flexibility to produce optimal results with a variety of secure INS and GPS receiver types.

Next, the chapter pointed out that since August of 1995, ARINC and MagadanAeroControl have been testing the encrypted communication and display of GPS-based ADS reports from aircraft flying over Far East Russia. In addition, two other important capabilities have been tested:

♦ Display of aircraft reporting by voice.
♦ Controller-pilot encrypted data link communications.

The Magadan experiment has tested the core elements of the international agreement for future air traffic management as promulgated by the International Civil Aviation Organization (ICAO) as a satellite communications, navigation and surveillance/air traffic management (CNS/ATM) concept.

Finally, intelligent vehicle highway systems (IVHS) have generated a new challenge for communications industries. This is particularly true for secure mobile satellite communications, as most essential secure IVHS communications links require wireless and encrypted mobile transmission. In this chapter IVHS and its major components were briefly described. A generic block diagram for an IVHS network was presented. This block diagram defines the communications requirements for the IVHS network. The complete characterization of encrypted satellite communication between the vehicle and IVHS infrastructure is determined by the system architecture; however, by its very nature, this encrypted link is required to be wireless, mobile, and for the most part, interactive. A survey of wireless technologies that have been tested, or implemented, and/or considered suitable for IVHS applications were discussed in some detail. Examples of operational, and/or under development IVHS projects that apply the corresponding wireless technology were also provided.

From Here

Chapter 27 covered the following: Tightly coupled secure GPS/INS Systems, military leads in secure GPS/INS Integration, an optimized secure GPS carrier phase ambiguity search method focusing on high speed and reliability, secure intelligent vehicle highway systems, and, preoperational testing of encrypted data link-based air traffic management systems in Magadan, Far East Russia. Chapter 28 examines why ones and zeros, not bombs and bullets, may win tomorrow's battles; why, within 25 years, accurate, up-to-date encrypted information from satellites about what is happening on a battlefield will provide more of an advantage than any weapon system; why a satellite computer picture will show where all U.S. forces are and where many of the enemy forces are; and how satellites collect information. The chapter also looks at the soldier of the future; naval forces; air and space-based forces; and, battles without soldiers.

Endnotes

1 Donald T. Knight, "Rapid Development of Tightly Coupled GPS/INS Systems," Knight Systems, 6810 Los Verdes Drive #3, Rancho Palos Verdes, CA 90275, IEEE Aerospace and the Electronic Systems Magazine, IEEE Aerospace and Electronic Systems Society, 3 Park Avenue, New York, NY 10016-5997, Volume 12, Number 2, February 1997, p. 15.

2 Ibid.

3 Alexander P. Shuvae and Roy T. Oishi, "Preoperational Testing Of Data Link-Based Air Traffic Management Systems In Magadan, Far East Russia," ARINC, 2551 Riva Road, Annapolis, MD 21401-7465, IEEE Aerospace and the Electronic Systems Magazine, IEEE Aerospace and Electronic Systems Society, 3 Park Avenue, New York, NY 10016-5997, Volume 12, Number 2, December 1996, p. 10..

4 Ibid., p. 11.

5 Ibid.

6 Ibid., p.12

7 Ibid.

8 Behnam, Kamali, "Some Applications Of Wireless Communications In Intelligent Vehicle Highway Systems," Department of Electrical and Computer Engineering, Mercer University School of Engineering, 1400 Coleman Avenue, Macon, GA 31207-0001, IEEE Aerospace and the Electronic Systems Magazine, IEEE Aerospace and Electronic Systems Society, 3 Park Avenue, New York, NY 10016-5997, Volume 12, Number 2, November 1996, p. 9.

9 Ibid., p. 10.

28

Satellite Encryption for the Electronic Battlefield of Tomorrow: Warfare 2030

It is 2030 and Iraq is *going north*—invading Turkey. The United States decides to help outgunned Turkey to repel the attackers, much as it liberated Kuwait during the Persian Gulf war. Yet in this series of Pentagon war games, concocted to illustrate how a future war might be fought, the American plan of attack bears little resemblance to Operation Desert Storm.

One big difference: Iraq's arsenal of stealthy cruise missiles forces the Americans to operate without the ports, airfields, base camps, and *iron mountains* of supplies that provided a huge edge in the Mideast desert. Iraq can spot such sprawling facilities with its intelligence satellites and blast them with its huge arsenal of cruise missiles, despite American missile defenses. Even tents housing command posts are vulnerable. As a result, U.S. fighter jets remain sheltered on airfields and aircraft carriers outside the combat theater, and U.S. ground troops are unable to land in large numbers. Grand maneuvers, like the massive armored attack into southern Iraq in 1991, are out of the question.

The United States responds with tricks of its own. A fleet of remotely controlled unmanned aerial vehicles, or UAVs, is launched over the entire theater. Some linger for days, continuously observing enemy troops; others carry weapons that are fired as needed. Several stealthy, highly automated *arsenal ships* sneak close to shore and deposit hundreds of missiles each on the Iraqi targets. Many of the missiles release *brilliant* submunitions that hover like hawks until they spot a tank or other target to home in on and destroy.

The most potent new U.S. weapon, however, is not a bomb, but a ganglion of electronic ones and zeroes. The American *eyes and ears*— from satellites and UAVs to air-dropped sensors that detect vehicle movements—are linked together to generate a real-time picture of nearly everything happening on the battlefield. Small squads of ground troops carry displays showing where they are, where the other U.S. squads are, and where the enemy is. Missile and artillery batteries use the same pictures to target enemy positions. The God's-eye view lets commanders make rapid decisions, giving American forces a devastating advantage: the ability to act so fast that the enemy cannot respond (see the Sidebar that follows).

Battlefield Reaction

Lifting the Fog of War

Within 85 years, accurate, up-to-date information about what is happening on a battlefield will provide more of an advantage than any weapon system. An encrypted computer picture will show where all U.S. forces are and where many of the enemy forces are.

Collecting Information

Ground troops: Small squads of widely dispersed ground troops may scout out enemy targets and punch their locations into an information system shared by all troops. Computing devices worn on the wrist or helmet could show where every squad is and let units communicate with each other and with commanders.

Ships: Powerful radars from numerous ships will be stitched together into one enormous radar sweep, which will become part of an even larger picture of the whole theater.

Unmanned aerial vehicles: Dozens of UAVs could form a *web* over the battlefield relaying pictures of a given area at a moment's notice. Some will be stealthy, while others will fly above the range of fighter jets that could shoot them down. Improved cameras will be able to see in bad weather, day or night.

Reconnaissance satellites: Satellites will be able to pinpoint objects with better than 1-meter resolution, and sweep areas of 100,000 square miles or more.

Remote sensors: Devices airdropped onto a battlefield should be able to detect heat, movement, sound or even diesel fumes produced by enemy forces and relay data on troop movements.

The Big Picture

Linking it all together: Information from all sources would be integrated to form a common picture of the changing situation on the battlefield. An exponential increase in the bandwidth needed to transmit encrypted data by satellite would let troops at different locations throughout the theater receive continual updates, so that all U.S. forces are looking at the same picture at the same time.

Mobile operations center: Commanders will stay linked to the common picture when they are on the move.

Acting on Information

Long-range strike: Stealthy aircraft, ships and ground batteries far to the rear will provide *massed effects* (concentrated barrages of bombs and missiles) without the need to mass troops in large formations vulnerable to enemy attack. Computers will likely be able to recognize targets automatically from the encrypted data provided by sensors, though humans will probably retain control over what gets hit.

Real time: The battlefield picture could be automatically updated as often as every 19 seconds. Clicking on an icon would pull up details like UAV aerial footage.[1]

Many military strategists think computing power and encrypted satellite communications may dramatically reduce friction and produce smaller but decidedly deadlier military forces. New technology *will provide an order of magnitude improvement in lethality.* In other words, if you can see every armored concentration, every surface-to-air missile shooter, every Scud launcher, maybe even where the leadership is, then you win.

Military Revolution

There are competing visions, however, of how technology could (or should) reshape the U.S. military. Some long-range thinkers believe a military revolution is at hand: Future battles will turn on long-distance artillery and missile shootouts instead of titanic armored clashes. Small squads of troops, linked by computers, could carry as much punch as a division of 18,000 soldiers.

Note: Combat information centers aboard today's ships are an early glimpse of the future. Eventually, real-time digital pictures should replace the grease pencils and acetate maps in today's command centers.

Military pragmatists, however, worry that radical change could create new vulnerabilities as proven tenets of warfare are cast aside. Some resistance comes from fear that big weapons platforms such as tanks, aircraft carriers, and fighter jets (which account for hundreds of thousands of military careers and defense-industry jobs) may be supplanted by smaller, stealthier and sometimes unmanned systems. And the history of warfare is replete with reminders that better technology is not always triumphant. Future Bosnias, where a heavy presence is the key to success, could militate against change, perpetuating the need to put *boots on the ground*.

The Electronic Revolution

For now, though, high technology remains an American edge. United States intelligence estimates say it will be at least 12 years before another nation, such as China or a resurgent Russia, could challenge American military dominance. But like other breakthroughs that revolutionized warfare, from the stirrup to the steam engine, microprocessors and satellite dishes are mostly civilian technologies that rivals can readily copy, adapt, or improve. If we correctly capture the promise of the electronic revolution, we can assure we remain the preeminent military power. If we miss the boat, all bets are off.

If the United States does not reshape its military to take advantage of new capabilities, America could repeat the mistake of the French in the years preceding World War II. While the Germans developed the blitzkrieg concept around their new tanks, the French tried to meld the tank with the trench-warfare tactics of World War I and invested heavily in building the Maginot line of defensive forts. In 1940, German panzer divisions raced around the Maginot line and seized Paris within a month.

Just as the French planned for the next war by studying the last one, so some strategists worry that the U.S. military remains fixated on America's last war—Desert Storm. Indeed, the stunning American success in the desert could stymie innovation. When you have a leading capability, it is very hard to come along with something better.

Nor was the gulf war the high-tech showcase it appeared to be at the time. A 4-year study recently released by Congress's General Accounting Office found that the vaunted *smart bombs* aimed at air shafts and entranceways were less effective than military leaders and manufacturers claimed. And U.S. strategies and tactics in the Persian Gulf in many ways recalled earlier, conventional wars. The colossal maneuvers of the coalition armies may in retrospect appear like the final charges of cavalry in the nineteenth century, an anomaly in the face of modern firepower.

Faster and better encrypted information will be the key to dominating the high-tech battlefields of the future, say many military planners. Tracking and managing battles electronically may seem unremarkable, yet events in Desert Storm underscore just how difficult it is to see and plot fast-moving developments on a battlefield, even against a weak foe.

The 35 U.S. friendly-fire deaths (one-fourth of all Americans killed) occurred largely because tank gunners and fighter pilots could not always keep track of changing battle lines and moving vehicles. And the lack of accurate, timely surveillance toward the end of the war led General H. Norman Schwarzkopf, the allied commander-in-chief, to think that most Iraqi troops were trapped in Kuwait, a leading factor in his call for a cease-fire. In fact, thousands of Iraqis were fleeing with their equipment back into Iraq, where they reorganized and later helped suppress Shiite and Kurdish uprisings against Saddam Hussein.

The Techno Battlefield

In future conflicts, high-bandwidth satellite transmitters will unleash a torrent of encrypted data to troops. And a *system of systems* will help mold scraps of discrete information (traditionally processed through different pipelines) into a single mosaic that will present a clearer picture of the battlefield. United States forces in Bosnia already are glimpsing the possibilities. Recently, they gained access to a new *secret Internet* run from the Pentagon, where they can get maps, logistics updates, aerial photos and other intelligence information. It is a giant step in the right direction.

A new Air Force *situational-awareness system* (hurriedly fielded in southern Europe after Capt. Scott O'Grady's F-16 fighter was shot down over Bosnia in 1995) may someday give air warriors as great a combat edge as stealth aircraft, precision-guided munitions, or any other technological advance of the past 23 years. By linking numerous surveillance sources at the speed of light, the system shows air commanders every plane, missile battery, and relevant activity on a single computer screen and lets them rapidly relay warnings or targeting tasks to pilots.

Just a year ago, it took several minutes for analysts to match data from computers in different rooms and figure out that O'Grady's F-16 was headed into a Bosnian Serb surface-to-air missile *trap*. A decade from now, commanders using the system may be able to direct attacks with decisive swiftness simply by touching target icons nominated by software.

Continuous Surveillance

Bosnia also has been a proving ground for unmanned aerial vehicles, which fill a major gap in the intelligence net — the ability to loiter over an area and provide continuous surveillance instead of making periodic passes as a satellite or spy plane would. The Predator UAVs that started flying missions over Bosnia in 1995 shoot still or moving pictures *clear enough to see people hiding among the trees.* Future UAVs will likely be stealthy and roam at high altitudes for days. Some technologists foresee front-line troops using palm-size UAVs to relay pictures from several hundred yards forward, although a 30-mile-per-hour crosswind could disrupt the tiny spies.

Military leaders are beginning to grasp what better and faster encrypted satellite information can mean. A new Navy system just installed, for instance, combines radar sweeps from numerous ships, planes, and even ground stations into one enormous picture shared by all. The radars are not new, but the ability to rapidly swap voluminous radar data is. Instead of being able to see only as far as the horizon (the limit of a single radar's range), a ship will be able to see what all the other ships in its network can see.

Navy worries about ships' vulnerabilities have been mounting since an Iraqi Exocet missile killed 37 sailors on the frigate USS Stark in 1987. Recently a Navy spokesperson said current defenses *will be stressed well beyond capabilities* by emerging missile and radar-jamming technologies. An expanded radar network should help offset those threats, letting ships fire at missiles or airplanes three to four times farther away, according to one admiral. Some ships also could *hide* by turning off their radars, which signal their location to enemy attackers. Such *Silent Sams*, lurking ahead of the fleet, could then use other ships' radar to ambush attacking planes.

> **Note:** The Joint Surveillance Target Attack Radar System (JSTARS) aircraft provides a huge advantage by defining where enemy forces are located. Future versions may have to be stealthy, though, to evade a foe's own advanced weapons.

At the Army's Command and General Staff College at Fort Leavenworth, KS, planners are as excited about *encrypted data fusion* and *digitization* as their predecessors were about the first tanks and machine guns. During the Army's Prairie Warrior war game in 1996, five separate information technology systems were meshed into a common battlefield picture. That let the up-and-coming junior officers playing the role of a division staff plan to act much faster than usual. The impact on the typically formidable opposition force: Whole regiments disappeared.

During Prairie Warrior, a group of battle planners also tapped into the system to generate *flag plans* for use when something unexpected happened. At any given time, the team had 9 or 10 detailed alternatives ready to go, compared with two or three in a conventional planning cell. That increases the likelihood of always having the initiative and never having to react. It gives average guys like the battle planners a Napoleonic vision.

Nobody disputes the virtues of enhanced encrypted information. Elbows fly, however, over how it will change the role of sea, air and ground troops. Even though soldiers will be equipped with advanced weapons and satellite communications gear, warfare will basically be the same in 7 to 12 years.

Yet the Army has been criticized even within its own ranks for an interim new division structure (to be the backbone of the *next Army*) that relies on the same number of tanks as the heavy divisions equipped to confront the Soviets in the cold war. The Army has to get on with some obvious changes.

Senior Air Force battle planners predict, not surprisingly, that precision weapons, many dropped from the air, will destroy enemy forces with little or no need for close-in fighting. If the Air Force does this right, the term *battlefield* will become archaic — a place where soldiers go to die. See the Sidebar that follows for additional information on the new battlefield technology and tactics.

Superiority in the Battlefield

New Technology and Tactics

In future conflicts, rivals may also have advanced technology, which means U.S. troop formations or bases may be easy targets.

Ground Forces

Small squads linked electronically may be able to spread out over a large area in order to stay hidden. Platoons in a rifle company could be several miles apart, in contrast with the 2000-meter separation typical today.

The soldier of the future: Links to data from numerous sources would give individual soldiers much more information about what is going on around them, letting them act much more quickly.

Smart helmet: A *heads-up display* would project target information and images of terrain from other sources, while live video from a helmet — mounted camera would be fed back into the network.

Personal weapon: Sensors on the gun would project targets onto the heads-up display for more accurate fire against hard-to-see objects.

Clothing and gear: Protection from chemical and biological agents would be incorporated into the uniform, and soldiers will carry a *combat identification* device meant to preempt fire by friendly forces.

Naval Forces

Arsenal ships: New Navy arsenal ships would be packed with 600 or more missiles each, which could be fired remotely by a commander on another ship-or on land. The arsenal ship would be low floating or submersible, allowing it to sneak close to shore without being detected. Its firepower would be called upon mostly at the start of a conflict.

Air and Space-Based Forces

Directed energy weapons: A laser fired from a 747 would target ballistic missiles launched at U.S. forces. The lumbering 747 would still have to avoid enemy airspace.

Space weapons: Within 40 years, satellites may provide the widest coverage for laser beams able to shoot down missiles.

Precision-guided munitions (PGMs): By taking out targets on the first or second try, accurate *smart bombs* would dramatically reduce the amount of material troops need to bring battle.

Brilliant submunitions: Some missiles may release dozens of submunitions that hover like a hawk over enemy territory until they find a tank or other object to home in on and destroy.[2]

Hidden Opponents

Marine Corps futurists come down somewhere in between. They foresee troops still trained for traditional amphibious missions but able to operate in small, widely dispersed squads as well. Instead of doing a lot of shooting themselves, the squads would seek out enemy targets and relay coordinates to artillery or missile systems far to the rear, or even to ships. Then they would quickly move out. The battered enemy would never see its opponents and would have no target for return fire. The risk of massing troops would probably lead to *islands* of fighting, rather than discernible front lines.

Future war-fighting experiments, including a recent Army exercise featured 6000 soldiers—mostly outfitted with helmet-mounted video displays, computer devices, and nearly 200 other futuristic systems— began to paint tomorrow more clearly. But whatever the possibilities, an Internet of fighters carries risks that should be familiar to anybody whose PC has ever crashed. If the network goes down, the soldiers could be in worse shape than if they had not had it.

Army planners in particular are ever wary of a foe like the Viet Cong, able to defeat technology with rudimentary tactics and a willingness to sacrifice soldiers. As a hedge, even ambitious future plans envision a high proportion of troops trained to operate conventionally. Still, one must accept an increased risk at present.

At the same time, the services will need to address the new demands of the digital world. Computers processing warehouses' worth of new data, for instance, can overload analysts. There is a case for less imagery, not more. One hope is that new software will be able to sort through millions of pictures, say, and discard those that are not useful.

The Online General

Silicon warfare also will demand more training. The handful of young officers selected to run the keyboards during Prairie Warrior were atypically wired, complaining for example that their terminals lacked *video feed routers* and *collaborative planning tools*. But for many soldiers, Windows 98 is some mystical thing out there. And, as in the business world, the most mystified soldiers may be the highest ranking.

To coax its leaders online, the Army now issues every new one-star general a laptop along with the customary pistol. The Marine Chief of Staff sends periodic e-mail messages to make sure the machines get a workout. The Marines are trying to expose their troops to computers early in their careers by handing out free copies of *Marine Doom,* a leatherneck version of the well-known computer game.

But the latest technology is a fast-moving target, and paying for it may be the toughest problem of all. Military leaders say they are already $40 billion short of what they need just for routine modernization of trucks, jets, and other equipment. And Pentagon budgets are more likely to shrink than to expand.

Note: Already in action, tank crews will see pictures on their computer screens showing where other U.S. troops are (to reduce friendly fire casualties) and where many of the enemy forces are. The *sensors* contributing images to such an information network will include unmanned aerial vehicles like the Predator, now being used to shoot still and moving pictures over Bosnia. The B-2 bomber controversial for its high cost and cold-war heritage, may be well suited for future conflicts. Its stealth will let it penetrate hostile airspace, while its long range means it will not have to fly from air bases close to the theater and vulnerable to enemy attack.

The costliest items on the Pentagon's shopping list — a new version of the Navy's F/A-18 jet, the stealthy F-22 fighter for the Air Force and a *joint-strike fighter* to be used by the Navy, Air Force and Marine Corps — may be a large target for savings. Most analysts agree such weapons — estimated to cost about $550 billion — would be marvelous in a Desert Storm II. But in many futuristic war games, they figure marginally against foes with plenty of missiles to blast airfields and aircraft carriers. Improved Patriot and other missile-defense systems — including an *airborne laser* on a 747 — could provide some protection.

Battlefields under the Sea

Representatives for The Center for Strategic and Budgetary Assessments worry that spending on such expensive conventional weapons could crowd out investment in new systems. In a May, 1997 report on the Navy, they argued for reducing the number of aircraft carriers — each of which costs about $6 billion to purchase — from 14 to as few as 10. That would liberate money to convert retired Trident submarines into stealthy troop transports or submersible *battleships* carrying precision munitions, and to experiment with different versions of the missile-packing arsenal ship. The Navy plans to build an arsenal ship, but it may simply retrofit old cruiser hulls instead of trying out submersibles or other stealthy designs. Meanwhile, the Navy also is working up plans for a costly new supercarrier.

Critics argue that more military bases need to be closed and more work turned over to private contractors, to free up money for new investments. Republican Senator John McCain (AZ) disagrees with that approach. He indicates that *you're not going to get significant money out of base closings and privatizations.*

Earlier in 1997, McCain proposed a more drastic plan for cutting back on training and other war preparations for certain troops in order to divert money to new technology. Yet other critics argue that it is dangerous to tamper with readiness because advanced technology may have *little impact on the outcomes of combat* when clashing armies are both highly skilled. In that case, the force with better training and smarter military doctrine wins.

A major Pentagon study could realign spending priorities. Even if it does not, the dominance of information in warfare may be inexorable. Platforms such as jets, carriers, and tanks are no longer the centerpiece. They're just reusable containers. The relation of information from the sensor to the shooter is the centerpiece. Warriors, man your terminals.

Nevertheless, military planners have some futuristic weapons in mind. Let us see what kind.

Soldierless Battles

What is an Air Force without pilots? The idea may bomb with jet jocks who have a need for speed. But unmanned fighters and bombers should be feasible by 2035, according to *New World Vistas*, a recent Air Force assessment of future technologies. And they would be some rather sporty numbers.

Small attack planes launched from a *mother ship* (a long-range air-borne aircraft carrier) could perform aerobatic maneuvers generating up to 20 times the force of gravity, more than twice what a human pilot can withstand. An unmanned bomber might travel as fast as Mach 20 — more than 15,000 miles per hour, or more than seven times the speed of the fastest plane today. Humans may still be manning the controls — but they would be in a ground station looking through virtual-reality goggles, not in a cockpit at 70,000 feet.

Sci-fi Toys

While the Pentagon is turning much more frequently these days to off-the-shelf technology found in homes and offices, the military's sci-fi set is still holding its own. The most promising far-out ideas are those, like remote-controlled combat planes, that take friendly troops out of harm's way and put enemy troops in it.

Other ideas may end up better suited for Hollywood. Some futurists, for instance, think a powered *exoskeleton* could enable an infantryman to walk 200 miles a day and lift hundreds of pounds (see the Sidebar that follows). But by the time such a device materializes (if it ever does), robotic vehicles might do the job just as well. The notion of jet packs that would let individual soldiers zip through the sky still enthralls some long-range thinkers.

Sci-fi Gizmos

Will They Fly?

Increasing use of off-the-shelf technology has not suppressed the Pentagon's appetite for futuristic gizmos. The following are some of the ideas that might catch on in 35 to 40 years.

Micro unmanned aerial vehicle: A flying version of this model could be launched from a soldier in a foxhole to see what is a mile or two up ahead. Images could then be transmitted to a small satellite dish and viewed on a laptop—as long as wind does not sweep away the tiny plane.

Unmanned combat aircraft: Unmanned fighters could go much faster and withstand up to twice as much gravitational force as jets built to accommodate a human. The *fotofighter* conceived by the Air Force would have another handy feature: lasers able to shoot down several attacking missiles or planes at once, from any angle.

Unmanned ground vehicles: Instead of venturing onto extremely dangerous turf, soldiers could send robotic vehicles. They would carry cameras able to see what is going on at ground level and be armored against mines or other explosives.

Exoskeletons: Exoskeletons could make supermen of soldiers, with an external mechanical structure that enhances their ability to carry a heavy load or travel by foot. Then again, soldiers may fall down under the increased pressure.

> ***New materials:*** Stealthy tents and *covert parafoils*, for parachuting into enemy territory unobserved, are two possible uses for advanced materials.
>
> ***Vaccines and performance drugs:*** Genetically engineered vaccines could make soldiers immune to chemical and biological agents. Other drugs could keep them awake for days at a time and maybe alert, too.[3]

Some weapons of the future are already familiar. Lasers have been around for over 30 years—yet their potential still sounds impressive. The Air Force foresees a *fotofighter* that by 2035 will be able to fire lasers in all directions simultaneously; it would be *invulnerable* from any angle, *unless* the enemy has lasers, too. Then new counter-measures will be needed. High-power microwaves could be beamed up to remotely piloted aircraft to renew their power sources, letting them stay aloft indefinitely.

Directed Energy Attack

The first use of lasers or other forms of *directed energy* as battlefield weapons will probably involve destroying the bad guy's satellite communications and the electronics at the heart of sophisticated weapons systems. Lasers today look as if they would be less effective as replacements for tank guns and artillery, but there are strong incentives to keep trying: Any technology able to reduce the tons of heavy ammunition typically dragged into battle would provide a dramatic improvement in mobility, one key to survival. The holy grail for logisticians would be a nonoil-based power source that would free gas-guzzling tanks and other vehicles from the umbilical cord of supply trucks that trails after them in battle. Some sort of electric drive system holds the most hope, though so far most assessments have been less than positive.

Of course, a battlefield illuminated with blinding flashes from lasers means soldiers will need laser-protection visors on their *smart helmets*. As envisioned today, the helmet would have a *heads-up* display mounted in front of one eye, showing footage of surrounding terrain from other sensors in the satellite communications network. A mouse-type clicker hanging near a shirt pocket would let the soldier call up more detail or switch from picture to picture. Special sensors could also let soldiers aim their weapons more accurately just by looking at an object. A laser range finder and satellite navigation system would quickly pinpoint the target's location and relay it to the weapon.

Images on the heads-up display, transmitted from cameras attached to robotic vehicles, could give soldiers *telepresence* (as opposed to physical presence) in perilous terrain. Some areas of the battlefield might be devoid of soldiers but filled with remote-controlled missile launchers, optical and acoustic sensors searching for targets, and mines that pop into the air and chase moving objects. Tracking devices could clamp onto enemy vehicles as they pass by to keep tabs on their whereabouts or put them in a missile's cross hairs. Cameras mounted on a remote-controlled, pizza-size *ducted fan* could hover several feet off the ground to explore buildings and see what is inside. An endoscopic device could even poke through walls.

Space could be an expanding frontier for warring forces in 30 to 40 years as they try to blind each other's reconnaissance satellites with lasers fired from the ground or air. Satellites eventually could be armed with their own lasers, in which case they would light up the heavens while they shot at each other. *Space mines* would be a menace, too. The small satellites would be launched toward larger ones, then destroy themselves along with the target. Satellites that survive could unleash a new rain of terror using technology left over from the Reagan-era Strategic Defense Initiative: Satellite bombs could be hurled toward Earth at 5000 miles per hour, arriving with complete surprise and reaching targets buried hundreds of feet deep.

Hiding from all these new threats sounds exhausting—which may be one reason researchers are beginning to explore chemical or biological ways to make troops stay awake longer and perform better. A scientific mix of new stimulants and sedatives, along with precisely timed naps, could let troops stay alert for two to three straight days, by one estimate. Memory-enhancing drugs could make human computers out of some troops. And by 2030, pilots and others may operate machines simply by thinking. Such *brain-activated control* would rely on neurosensors in a skullcap to convert brain waves into electronic commands. Not even the Vulcans on *Star Trek* had that in their bag of tricks.

The first part of the chapter discussed weapons that will zap you to death and reduce your surroundings to ashes. But what about weapons that do not? What about weapons that will zap you, fry you, stun you, but not kill you? The Pentagon's quest for nonlethal arms is amazing. But is it smart?

Nonlethal Weapons

Tucked away in the corner of a drab industrial park in Huntington Beach, CA, is a windowless, nondescript building. Inside, under extremely tight security, engineers and scientists are working on devices whose ordinary appearance masks the oddity of their function. One is cone-shaped, about the size of a fire hydrant. Another is a 4-foot long metal tube, mounted on a tripod, with some black boxes at the operator's end. These are the newest weapons of war.

For hundreds of years, sci-fi writers have imagined weapons that might use energy waves or pulses to knock out, knock down, or otherwise disable enemies—without necessarily killing them. And for a good 50 years, the U.S. military has quietly been pursuing weapons of this sort. Much of this work is still secret, and it has yet to produce a usable *nonlethal* weapon. But now that the cold war has ended and the United States is engaged in more humanitarian and peacekeeping missions, the search for weapons that could incapacitate people without inflicting lethal injuries has intensified. Police, too, are keenly interested. Scores of new contracts have been let, and scientists, aided by government research on the *bioeffects* of beamed energy, are searching the electromagnetic and sonic spectra for wavelengths that can affect human behavior. Recent advancements in miniaturized electronics, power generation, and beam aiming may finally have put such pulse and beam weapons on the cusp of practicality, some experts say.

Weapons already exist that use lasers, which can temporarily or permanently blind enemy soldiers. So-called acoustic or sonic weapons, like the ones in the forementioned lab, can vibrate the insides of humans to stun them, nauseate them, or even *liquefy their bowels and reduce them to quivering diarrheic messes*, according to a Pentagon briefing. Prototypes of such weapons were recently considered for tryout when U.S. troops intervened in Somalia. Other, stranger effects also have been explored, such as using electromagnetic waves to put human targets to sleep or to heat them up, on the microwave-oven principle. Scientists are also trying to make a sonic cannon that throws a shock wave with enough force to knock down a person.

While this and similar weapons may seem far-fetched, scientists say they are natural successors to projects already underway—beams that disable the electronic systems of aircraft, computers, or missiles, for instance. Once you are into these antimaterial weapons, it is a short jump to antipersonnel weapons. That is because the human body is essentially an electrochemical system, and devices that disrupt the electrical impulses of the nervous system can affect behavior and body functions. But these programs (particularly those involving antipersonnel research) are so well guarded that details are scarce. People in the military go silent on this issue more than on any other issue. People just do not want to talk about this.

Ongoing Projects

The effort to develop exotic weapons is surprising in its range. Scores of projects are underway, most with funding of several hundred thousand dollars each. One Air Force lab plans to spend more than $300 million by 2005 to research the *bioeffects* of such weaponry.

The benefits of bloodless battles for soldiers and law enforcement are obvious. But the search for new weapons (cloaked as they are in secrecy) faces hurdles. One is the acute skepticism of many conventional-weapons experts. It is interesting technology but it will not end bloodshed and wars. Some so-called nonlethal weapons could end up killing rather than just disabling victims if used at the wrong range. Others may easily be thwarted by shielding.

Sterner warnings come from ethicists. Years ago the world drafted conventions and treaties to attempt to set rules for the use of bullets and bombs in war. But no treaties govern the use of unconventional weapons. And no one knows what will happen to people exposed to them over the long term.

Moreover, medical researchers worry that their work on such things as the use of electromagnetic waves to stimulate hearing in the deaf or to halt seizures in epileptics might be used to develop weaponry. In fact, the military routinely has approached the National Institutes of Health (NIH) for research information. The Defense Advanced Projects Agency (DARPA) has come to NIH every few years to see if there are ways to incapacitate the central nervous system remotely. But nothing has ever come of it. That is too science fiction and far-fetched. Still, the Pentagon plans to conduct human testing with lasers and acoustics in the future. The testing will be constrained and highly ethical. It may not be far off. The U.S. Air Force expects to have microwave weapons by the year 2025 and other nonlethal weaponry sooner. When that does happen there will be a public uproar. There needs to be an open debate on them now.

Dazzling Lasers

What happened with U.S. forces in Somalia foreshadows the impending ethical dilemmas. In early 1995, some U.S. marines were supplied with so-called dazzling lasers. The idea was to inflict as little harm as possible if Somalis turned hostile. But the marines' commander then decided that the lasers should be *detuned* to prevent the chance of their blinding citizens. With their intensity thus diminished, they could be used only for designating or illuminating targets (see the Sidebar that follows).

Laser Weapons

Light beams affect mind and body: Lasers emit a high-intensity light, which can force an individual to turn away or can cause blinding. Those that result in permanent blinding have been banned by international treaty. But dazzling lasers might be used in hostage situations, prison riots, and special operations.

Dazzling effects: Lasers can force the pupil to close, or they can burn the light-sensitive retina or cornea, depending on the intensity of the beam.

Laser guns: Lasers have been developed to be mounted on existing weapons, such as the M-16 rifle.[4]

On March 1, 1995, commandos of U.S. Navy SEAL Team 5 were positioned at the south end of Mogadishu airport. At 7 A.M., a technician from the Air Force's Phillips Laboratory, developer of the lasers, used one to illuminate a Somali man armed with a rocket-propelled grenade. A SEAL sniper shot and killed the Somali. There was no question the Somali was aiming at the SEALs. But the decision not to use the laser to dazzle or temporarily blind the man irks some of the nonlethal-team members. They were not allowed to disable these guys because that was considered inhumane. Putting a bullet in their head is somehow more humane?

Despite such arguments, the International Red Cross and Human Rights Watch have since led a fight against antipersonnel lasers. In the fall of 1995, the United States signed a treaty that prohibits the development of lasers designed *to cause permanent blindness*. Still, laser weapons are known to have been developed by the Russians, and proliferation is a big concern. Also, the treaty does not forbid dazzling or *glare* lasers, whose effects are temporary. United States military labs are continuing work in this area, and commercial contractors are marketing such lasers to police.

Sonic Weapons

The next debate may well focus on acoustic or sonic weapons. Benign sonic effects are certainly familiar, ranging from the sonic boom from an airplane to the ultrasound instrument that *sees* a baby in the uterus. The military is looking for something less benign—an acoustic weapon with frequencies tunable all the way up to lethal. Indeed, Huntington Beach-based Scientific Applications & Research Associates Inc. (SARA) has built a device that will make internal organs resonate: the effects can run from discomfort to damage or death. If used to protect an area, its beams would make intruders increasingly uncomfortable the closer they get. SARA has built several prototypes. Such acoustic fences could be deployed today. Researchers at SARA estimate that 6 to 11 years will be needed to develop acoustic rifles and other more exotic weapons, but there have been rumors that the technology is only 2 to 3 years away. The military also envisions acoustic fields being used to control riots or to clear paths for convoys (see the following Sidebar).

Nonlethal Acoustic Weapons

Arms for crowd control and invisible fencing: Acoustic frequencies could be used to guard sensitive facilities, rescue hostages, clear paths for military convoys, disperse crowds, or target individuals.

Acoustic fencing: An array of acoustic devices can keep people away. The closer they get, the worse they feel.

Acoustic effects on the body: Sonic frequencies can cause tiny hair cells in the inner ear to vibrate, creating sensations like motion sickness, vertigo, and nausea. They can also make internal organs resonate, resulting in pain, spasms, or even death.[5]

These acoustic devices have already been tested at the Camp Pendleton Marine Corps Base, near the company's Huntington Beach office. And they were considered for Somalia. The military asked for acoustics. But the Department of Defense said, no because they were still untested. The Pentagon feared they could have caused permanent injury to pregnant women, the old, or the sick. Acoustics is seen as just one more tool for the military and law enforcement. Like any tool, it can be abused. But like any tool, it can be used in a humane and ethical way.

Toward the end of World War II, the Germans were reported to have made a different type of acoustic device. It looked like a large cannon and sent out a sonic boomlike shock wave that in theory could have felled a B-17 bomber. In the mid-1940s, the U.S. Navy created a program called Project Squid to study the German vortex technology. The results are unknown. But an American inventor says he replicated the Nazi device in his laboratory in 1949. Against hard objects the effect was astounding, he recalls: It could snap a board like a twig. Against soft targets like people, it had a different effect. Experimental human subjects felt as if they had been hit by a thick rubber blanket. The idea seemed to flounder for years until recently, when the military was intrigued by its nonlethal possibilities. The Army and Navy now have vortex projects underway. The SARA lab has tested its prototype device at Camp Pendleton (see the following Sidebar).

Vortex Technology

These arms can knock down people or even aircraft: The vortex gun expels a doughnut-shaped shock wave that could knock people down. The gun could also be filled with gases or chemical agents. A vortex ring of pepper spray, for instance, would stun its victims with both a physical blow and a chemical irritant. For example:

- They may be hand held or vehicle mounted.
- Explosive charge creates vortex in shock tube.
- Shock wave hits body.
- The vortex ring would travel at hundreds of miles per hour.
- The vortex ring must spin at Mach 1 or faster to create shock wave.
- They may also contain chemical agents such as pepper spray.

Vortex weapons: The vortex gun fires a doughnut-shaped wave with a powerful center. Lab tests show vortices can break wooden boards across a room. When they strike a person, the effect is like being hit with a heavy blanket.[6]

Microwave It

The Soviets were known to have potent blinding lasers. They were also feared to have developed acoustic and radiowave weapons. The 1987 issue of Soviet Military Power, a cold war Pentagon publication, warned that the Soviets might be close to a *prototype short-range tactical RF (radio frequency) weapon*. That year it was also reported that the Soviets had used such weapons to kill goats at a 1-km range. The Pentagon, it turns out, has been pursuing similar devices since the 1960s.

Typical of some of the more exotic proposals are those from the Health Sciences Research Division of Oak Ridge National Laboratory, which also does research on crime control. One of the projects the laboratory is working on is an electromagnetic gun that would *induce epileptic-like seizures* for a short duration of time. Another was a *thermal gun that would have the operational effect of heating the body to 107 to 109 °F* for a very short duration (5 minutes). Such effects would bring on discomfort, fevers, possible brain damage, or even death.

But, unlike the work on blinding lasers and acoustic weapons, progress here has been slow. The biggest problem is power. High-powered microwaves intended to heat someone standing 300 yards away to 107 °F may kill someone standing 20 yards away. On the other hand, electromagnetic fields weaken quickly with distance from the source. And beams of such energy are difficult to direct to their target. Mission Research Corp. of Albuquerque, NM, has used a computer model to study the ability of microwaves to stimulate the body's peripheral nervous system (PNS). If sufficient peripheral nerves fire, then the body shuts down to further stimuli, producing the so-called stun effect. But, the ranges at which this can be done are only a few meters (see the following Sidebar).

Electromagnetic Heat

A tunable *weapon that can discomfort or cook the enemy:* As antipersonnel weapons, microwaves could be used as *barriers,* causing short term pain or burns to those who enter their path. Phaser-like microwave *stun guns* have also been contemplated, but major technical hurdles still need to be overcome before their successful development.

STATUS: Research is classified. Prototypes reportedly exist and are ready for testing.

Short term microwave effects on the body: Microwaves have a wide range of biological consequences. A heating effect is produced by excitation of water molecules. Army experiments with animals in nonweapons programs show that short term microwave exposure can lead to memory impairment, possible cardiac arrest, a *stun* effect, and seizures.[7] Long term exposure to microwaves can lead to certain memory impairment, loss of memory altogether, and death.

Nonetheless, government laboratories and private contractors are pursuing numerous similar programs. A 1996 Air Force Scientific Advisory Board report on future weapons, for instance, includes a classified section on a radio frequency or *RF Gunship*. Other military documents confirm that radio-frequency antipersonnel weapons programs are underway. And the Air Force's Armstrong Laboratory at Brooks Air

Force Base in Texas is heavily engaged in such research. According to budget documents, the lab intends to spend more than $220 million over the next seven years to *exploit less-than-lethal biological effects of electromagnetic radiation for Air Force security, peacekeeping, and war-fighting operations.*

Low-Frequency Electrical Activity in the Brain

By using very low-frequency electromagnetic radiation (the waves way below radio frequencies on the electromagnetic spectrum) researchers at the Armed Forces Radiobiology Research Institute in Bethesda, MD found they could induce the brain to release behavior-regulating chemicals. They could put animals into a stupor by hitting them with these frequencies. They got chick brains (*in vitro*) to dump 90% of the natural opioid substances (e.g., endorphin) in their brains. The researchers even ran a small project that used magnetic fields to cause certain brain cells in rats to release a very small amount of histamine. In humans, this would cause short duration instant flulike symptoms and produce nausea. These fields were extremely weak. They were undetectable. The effects were nonlethal and reversible. You could disable a person temporarily. It would have been like a stun gun.

Note: These are exotic weapons that do *not* produce traditional medical symptons that one would expect. They are designer type weapons that have been modified to produce these symptoms.

The researchers never tested any of the hardware in the field. The program was scheduled for four years, apparently was closed down after two. The work was really outstanding. They could have had a weapon in one year. The researchers were told their work would be unclassified, *unless it works.* Because it worked, there is suspicion that the program became a *black ops* project. Other scientists tell similar tales of research on electromagnetic radiation turning top secret once successful results were achieved. There are clues that such work is continuing. In 1995, the annual meeting of four-star U.S. Air Force generals (called CORONA) reviewed more than 2000 potential projects. One was called *Put the Enemy to Sleep/Keep the Enemy From Sleeping.* It called for exploring *acoustics, microwaves,* and *brain-wave manipulation* to alter sleep patterns. It was one of only four projects approved for initial investigation.

Note: Black ops are secretly funded DOD, CIA, and NSA operations and/or projects whose known existence by the general public would pose a threat to national security.

First Contact

As the military continues its search for nonlethal weapons, one device that works on contact has already hit the streets. It is called the *Pulse Wave Myotron*. A sales video shows it in action. A big, thuggish-looking *criminal* approaches a well-dressed woman. As he tries to choke her, she touches him with a white device about the size of a pack of cigarettes. He falls to the floor in a fetal position, seemingly paralyzed but with eyes open, and he does not recover for minutes.

Contact with the Myotron feels like millions of tiny needles are sent racing through the body. This is a result of scrambling the signals from the motor cortex region of the brain. It is horrible. It is no toy. The Myotron overrides voluntary (but not involuntary) muscle movements, so the victim's vital functions are maintained. Sales are targeted at women, but law enforcement officers and agencies (including the Arizona state police and bailiffs with the New York Supreme Court) have purchased the device. A special model built for law enforcement, called the Black Widow, is being tested by the FBI. The Russian government just ordered 200,000 of them.

The U.S. military also has shown interest in the Myotron. About the time of the Gulf War, the company that developed Myotron got calls from people in the military. They asked the company about bonding the Myotron's pulsewave to a laser beam so that everyone in the path of the laser would collapse. While it could not be done, the company was nonetheless warned to keep quiet. Company officials were told that these calls were totally confidential, and that they would completely deny it if anyone ever mentioned it.

Some say such secrecy is necessary in new-weapons development. But others think it is a mistake. Because the programs are secret, the sponsorship is low level, and the technology is unconventional. The military has not done any of the things necessary to determine if the money is being well spent or the programs are a good idea. It should not be long before the evidence is in.

From Here

Chapter 28 examined why ones and zeros, not bombs and bullets, may win tomorrow's battles; why within 25 years, accurate, up-to-date encrypted information from satellites about what is happening on a battlefield will provide more of an advantage than any weapon system; why a satellite computer picture will show where all U.S. forces are and where many of the enemy forces are; and discussed the role satellites will play in collecting information. The chapter also looked at the soldier of the future; Naval Forces; Air and Space-Based Forces; and, battles without soldiers. Chapter 29 concludes with a discussion on how attacks on satellites could threaten national security as we approach the year 2000; and, how DOD and businesses face significant challenges in countering attacks on satellites.

Endnotes

1 Defense Advanced Research Projects Agency, Department of Defense (DoD), 3701 North Fairfax Drive, Arlington, VA 22203-1714, 1998.

2 Ibid.

3 Ibid.

4 Ibid.

5 Ibid.

6 Ibid.

7 Ibid.

29

Armageddon: Year 2000 Satellite Encryption Crises!

What got the Department of Defense (DOD), the IRS, and the Social Security Administration gathering in Washington for a strategy session recently? Budget deliberations? Government downsizing? Nope. The federal agencies are teaming up with the private sector to combat one of the most potentially crippling forces known to modern computing: the Year 2000 Problem.

After midnight, December 31, 1999, computer systems and satellite communications throughout the world are at risk of failing; and they are open to hacker attacks from unscrupulous individuals who would take advantage of the fact that computers may confuse the year 2000 with the year 1900 on January 1, 2000, and go backward in time instead of forward when the new century begins. The severity of the problem was raised when Congress was told that if businesses and governments continue to ignore this issue, disruption of routine business operations and the inability of the Federal Government to deliver services to the American public could result.

In case your favorite info-tech professional has not cornered you to lament what is variously known as the Millennium Bug, the Year 2000 Problem, or simply Y2K, a bit of background is in order. During the 1960s, computer memory was extremely limited and extraordinarily expensive. As a result, programmers were forced to economize when writing software, allocating only a two-digit number for each year's date. So, the year 1998, to a computer, is simply '98' (the '19' is assumed). It does not take a computer science degree to figure out the pitfall in this penny-pinching scheme: As midnight revelers celebrate the new year on January 1, 2000, some 90% of the world's computer hardware and software will "think" it is the first day of 1900.

What sounds like a simple computer glitch has enormous business ramifications. Although the bulk of Y2K foul-ups will occur in 2000, consumers will not have to wait for the new year to witness the effects. Most corporations and government entities refuse to talk openly about Y2K's effect for fear of public embarrassment and mass hysteria. But Y2K-induced snafus are already becoming the stuff of folklore. According to the Gartner Group, a Stamford, CT-based technology consulting firm, a state correctional facility (Gartner will not reveal which one) recently released several prisoners by mistake. The inmates' sentences extended well into the next century, but the prison's computer calculations showed they were long overdue for release. Some diligent human caught the glitch, and the prisoners were rounded up without incident. According to Gartner, mixups like this will consume tens if not hundreds of work-years within the typical organization. All told, Gartner estimates that corporations will spend between $400 billion and $700 billion grappling with Y2K over the next three years.

According to a Congressional Research Service memorandum dated April 12, 1996, many people initially doubted the seriousness of this problem, assuming that a technical fix will be developed. Others suspect that the software services industry may be attempting to overstate the problem to sell their products and services. Most agencies and businesses, however, have come to believe that the problem is real, that it will cost billions of dollars to fix, and that it must be fixed by January 1, 2000, to avoid a flood of erroneous automatic transactions. The memorandum further suggests that it may already be too late to correct the problem in all of the Nation's computers, and that large corporations and Government agencies should focus on only their highest priority systems.

The Committee on Government Reform and Oversight is deeply concerned that many Federal Government departments, agencies and businesses are not moving with necessary dispatch to address the year 2000 computer problem (see Sidebar in this chapter, "Countdown to Y2K: FAA Hardly Reacting"). Without greater urgency, those agencies risk being unable to provide services or perform functions that they are charged by law with performing. In addition, they are opening themselves up to hacker attacks that could pose a threat to national security. Senior agency management and corporate chief information officers (CIOs) must take aggressive action if these problems are to be avoided.

Attacks Could Threaten National Security

Because so few incidents are actually detected and reported, no one knows the full extent of the damage that could be caused by computer attacks as the year 2000 approaches. However, according to many DOD and private sector experts, the potential for catastrophic damage is great given: (1) the known vulnerabilities of the Department of Defense's (DOD's) command and control, military research, logistics, and other systems; (2) weaknesses in national information infrastructure systems, such as public intranets which DOD depends upon; and (3) the threat of terrorists or foreign nationals using sophisticated offensive information warfare techniques. They believe that attackers could disrupt military operations and threaten national security by successfully compromising DOD and private sector information and systems or denying service from vital commercial satellite communications backbones or power systems.

The National Security Agency (NSA) has acknowledged that potential adversaries are developing a body of knowledge about the DOD's and other U.S. satellite communication systems, and about methods to attack these systems. According to NSA, these methods, which include sophisticated computer viruses and automated attack routines, allow adversaries to launch untraceable attacks from anywhere in the world. In some extreme scenarios, experts state that terrorists or other adversaries could seize control of DOD information technology satellite communication systems and seriously degrade the nation's ability to deploy and sustain military forces. The Department of Energy (DOE) and NSA estimate that more than 120 countries have established computer and satellite communication attack capabilities. In addition, most countries are believed to be planning some degree of information warfare as part of their overall encryption security strategy.

At the request of the Office of the Secretary of Defense for Command, Control, Communications and Intelligence, the Rand Corporation conducted exercises known as *The Day After* between January and June 1995 to simulate an information warfare attack. Senior members of the national security community and representatives from national security-related telecommunications and information systems industries participated in evaluating and responding to a hypothetical conflict between an adversary and the United States and its allies in the year 2000.

In the scenario, an adversary attacks computers and satellite communication systems throughout the United States and allied countries, causing accidents, crashing systems, blocking encrypted satellite communications, and inciting panic (see the Sidebar that follows). For example, in the scenario, automatic tellers at two of Georgia's largest banks are attacked. The attacks create confusion and panic when the automatic tellers wrongfully add and debit hundreds of thousands of dollars from customers' accounts. A freight train is misrouted when a logic bomb is inserted into a railroad computer system, causing a major accident involving a high-speed passenger train in Pennsylvania. Meanwhile, telephone service is sabotaged in New York, a major airplane crash is caused in France (see Sidebar in this chapter "Countdown to Y2K: FAA Hardly Reacting"); and Rome, Italy, loses all power service. An all-out attack is launched on computers and encrypted satellite communications at most military installations, slowing down, disconnecting, or crashing the systems. Weapons systems designed to pinpoint enemy tanks and troop formations begin to malfunction due to electronic infections (see Sidebar in this chapter, "Countdown to Armageddon: How to Avert an Accidental War").

Cyberwarfare 2000

With just a few keystrokes, a terrorist's strike may be just a modem or satellite away. Click, click. That's all it might take, and then:

Wall Street takes a dive and throws financial markets into chaos. Banks lose their electronic records and explosive international

tensions ignite. Airplane traffic plunges disastrously into mountainsides and oceans. Municipal water districts shut down, leaving
entire cities dry.

High-tech mayhem possible: Such incidents regularly provide
plot lines for Hollywood movies. But a presidential commission now
formally predicts such incidents could be all too real in the near
future. That is because as networked computers and satellite communications expand their control over the nation's energy, power,
water, finance, communications and emergency systems, the possibility of electronic attack on encrypted systems and catastrophic
terrorism becomes increasingly possible.

A serious threat is sure to evolve if we do not take steps now to
protect these systems in the future according to the Commission on
Critical Infrastructure Protection. Made up of prominent federal law
officials and business leaders, the commission has been researching
such possibilities of electronic terrorism for more than a two years.
Recently, the commission delivered a lengthy classified report to the
White House making its confidential recommendations to the president on strategies to avoid such calamities.

A brand new battlefield: Nobody around the world today would
attempt to defeat the U.S. on the battlefield. Instead, they will be
seeking means to find vulnerabilities in America's satellite communication systems without having to confront the U.S. with the
conventional armed way of the past.

A recent study by the commission advises greater cooperation and
sharing of information between the government and private industry
to avert terrorism. The commission also identified a need for accelerated research and a nationwide program to educate people on the
scope of the problem.

Such policies are long overdue. The nation has a lot of experience
and knowledge from the Cold War that is not being utilized as much
as it could be.

Misdirected doomsaying: Not everyone is convinced of impending doom. Although federal law authorities have steadily been
working to expand their regulatory control of the Internet, their
efforts are frequently misdirected and based more on fear than on a
sound understanding of technology.

Trying to control the Internet is like trying to control the weather.
The specter of electronic terrorism and even nuclear war could best

be avoided by governments and organizations simply not using interconnected computer lines and unencrypted satellite communications for critical information. Certainly there are some in the military who have been making up these horror stories, but it is hard to give them a whole lot of credence. Most of the corporate computer folks understand that the Internet is not all that secure—let alone satellite communications.

Hackers crossing the line: People have also made *stupid mistakes* in the past, including leaving the Tennessee Valley Authority (TVA) vulnerable to computer hackers who could have accessed control of flood gates to major dams through a modem. But that was a long time ago.

But other more recent events provide lawmakers cause for alarm. According to one 1996 report, hackers have broken into Defense Department computer systems more than 300,000 times. Other incidents have involved computer thefts from banks and the tampering of emergency 911 systems in Florida via unencrypted satellite communication transmissions.

These are the warning signs of a new kind of danger. Major industries, ranging from utility companies to food processors, airlines to telephone companies, are all tempting targets.

Nonphysical attacks: The threats historically have been physical attacks, but now there is the attraction of cyberterrorism, which can be perpetrated 9000 miles away on a computer or from a satellite thousands of miles above the earth. And every major system has been hacked into at some time.

The commission's report does not dwell much on the dangers posed by the networked computer technology, but on dangers where the *physical and virtual worlds* converge. That is what is worrisome.[1]

Countdown to Y2K: FAA Hardly Reacting

With January 1, 2000 rapidly approaching, and concern growing over the year 2000 programming problem, the U.S. Federal Aviation Administration (FAA) is still in an *awareness* phase, which means it has not developed a course of action to make its air traffic control infrastructure Y2K compliant. Maybe it is not time to panic, but concern is definitely in order.

According to a report by the U.S. General Accounting Office (GAO), the FAA has failed to designate problem areas and has no way of knowing how serious its Y2K problem is or what it will do to address it. Making the skies safe for January 1, 2000 is no small task: With 24 million lines of code, 60 computer languages and more than 360 computer systems, the FAA has a difficult road ahead. The National Air Traffic Controllers Association, (NATCA) has been following the FAA's progress—or lack thereof—and things are not pretty.

NATCA has been watching to see if time lines are being met—and they are not. Controllers *are not optimistic* about the situation. They are very aware of it and do not believe that the FAA can do it on time.

Jane F. Garvey, FAA Administrator, testifying before Congress recently in response to the GAO's report, assured the members of the House Committee on Science that the FAA will be ready. Aviation safety will not be compromised. Ensuring that the FAA meets this challenge is a top priority for them. The FAA is aware they are behind schedule.

Shutdown: The biggest problem will be the possible shutdown of the host computer, which ties together the 30 air traffic control centers in the U.S. that control aircraft after they leave the airports. Among the specific functions that can be lost are the ability to transfer control from one controller to another; systems that warn controllers when two aircraft are too close to one another and when an aircraft is too close to the ground; the loss of route lines that display the path of an aircraft; halos, which display a 5-mile ring around an aircraft on the radar scope; and the quick look function, which allows a controller to view aircraft on an adjacent controller's scope in the event that his or her scope fails.

Another big problem is the lack of a contingency plan. The controllers need to be trained to handle the problems that will arise if the systems fail. After all, this does not happen every day. There is a backup system called direct access radar channel (DARC), but the FAA is not sure if DARC is Y2K compliant either.

Imagine what January 1, 2000 will be like. The problems, if they do occur, will be immediate. Aircraft will have to be put in holding patterns at different altitudes causing delays. The aircraft are supposed to be kept 5 miles apart, and 5 miles is less than 1 minute away. In addition to safety, it would have a definite economic impact. Amtrak might make some money. Most people would be better off at home watching the ball drop in Times Square than traveling that day.

The GAO report says the systems in use are unique to the FAA and not off-the-shelf systems that can be easily maintained by system vendors. But that is just one part of the problem. The enormous challenge involved in correcting these systems is not primarily technical, it is managerial.

Across the globe: Worldwide, the airline industry carries almost 2.3 billion passengers a year and nearly 30 million tons of freight. Every 10 years or so, these numbers double according to the International Air Transport Association (IATA), a worldwide organization that represents and serves the airline industry.

The IATA Information Management Committee has addressed the Y2K issue by establishing a Y2K group composed of representatives from 40 airlines around the world. All of the representatives agree that without proper planning and management, this problem could significantly disrupt airline operations—and businesses around the world.[2]

Countdown to Armageddon: How to Avert an Accidental War

At the height of the cold war, fears of an accidental nuclear exchange loomed large both in Hollywood and in Washington, DC. Movies like *Fail Safe* and *Dr. Strangelove* played on the risks posed by communication glitches and rogue commanders. Some policymakers at the White House, the Pentagon, and on Capitol Hill also were concerned about the dangers of maintaining a huge nuclear arsenal on hair-trigger alert. But most accepted the risks as the unavoidable price of deterring the Soviets from launching a massive, surprise attack.

Today, the cold war is over, and under the framework of the Strategic Arms Reduction Treaty, or START I, the U.S. and Russian strategic nuclear arsenals have been cut by about one-third. Both countries have mothballed thousands of short-range tactical weapons and taken their long-range bombers off standby alert. In 1994, Presidents Clinton and Yeltsin further agreed to stop targeting strategic missiles at each other's country—although this step had little strategic significance because it takes only a few seconds to reenter targeting coordinates into a missile's guidance system.

Many Americans feel a lot safer. But in fact the danger of nuclear annihilation has scarcely diminished. In their continuing preoccupation with the now unlikely possibility of a first-strike attack, both the United States and Russia have kept thousands of strategic nuclear warheads on hair-trigger alert. Accordingly, both countries together remain ready to fire more than 5000 warheads within a half-hour's notice.

It is an anachronism. The biggest concern is the instability of Russian forces today. There are still a significant number of nuclear weapons that could be fired at the United States under mistaken circumstances or during an all-out attack on computers and unencrypted satellite communications by hackers as we approach the countdown to the year 2000—the *Eve Of Destruction*. A growing chorus of experts, such as former Senator Sam Nunn, want Moscow and Washington to lower the risk of accidental or unauthorized attacks by *dealerting* their nuclear forces—that is, increasing the amount of time needed to launch missiles from a few pressure-packed minutes to hours or even days.

Near-fatal catastrophe: Russian President Boris Yeltsin knows all about the potential for a catastrophic mistake. In January 1995, a meteorological missile fired from Norway to study the northern lights activated Russia's early-warning system. Although the Norwegians had notified Russian officials well before the launch, the message had not been passed along to Russia's high command. Yeltsin and his top generals agonized for several minutes over whether the rocket was part of a surprise U.S. attack. Fortunately, a few minutes short of the deadline for firing a retaliatory salvo, the scientific rocket headed out toward the ocean and away from Russia.

Note: It makes one wonder though, if the cold war is over, why would Yeltsin and his top generals agonize over whether the rocket was part of a surprise U.S. attack?

The risk of a horrible mistake is only growing with the widening political and economic turmoil in Russia. The loss of former Soviet radar installations that are outside Russia's current borders has left significant gaps in the country's early-warning network, making Russian fingers on the button ever more jittery. Again, the question is asked, why would the Russians think the U.S. would launch a surprise nuclear attack on its own widening financial and aerospace business ventures within Russia? Oh, I get it! Maybe the Russian military thinks they are playing an interactive video game. Or, maybe they think they are on an episode of *Sliders*, a TV show about a parallel world where the cold war is still on. Vodka talking here I suspect!

Anyway, stories abound in Russia of utility companies' shutting off power to nuclear weapons facilities because the military has not paid the bills, and of nuclear control equipment failing because thieves have stolen communications cables for their copper. Morale in the military also has plummeted as a result of chronic food and housing shortages. Indeed, in February, 1997, Russia's then defense minister, Igor Rodionov, warned that cuts in defense spending were pushing Russia toward a point *"beyond which its missiles and nuclear systems become uncontrollable."*

The Clinton administration is for the first time formally reviewing options for taking U.S. and Russian nuclear arsenals off hair-trigger alert. The possibilities include taking warheads off missiles and storing the warheads up to hundreds of miles away from the silos, removing missiles' guidance systems (reinstalling them would take expert technicians at least a few hours), and keeping ballistic-missile submarines either in port or at patrol locations well out of range of their targets. Compliance with these measures would be possible to verify through on-site inspections or encrypted satellite images.

Deliberately increasing the time needed to launch nuclear weapons would have an added benefit: At a time when the START approach of shrinking the overall arsenal is under increasing attack in the Russian Parliament, this would introduce a new approach to arms control. A sensible first step for President Clinton is to immediately dealert long-range missiles slated for retirement under the stalled START II agreement. If President Yeltsin responded in kind, the two presidents could engage in a virtual disarmament process that, over time, might eventually succeed in dismantling the nuclear balance of terror without going to the controversial extreme of completely eliminating nuclear weapons.

Dealerting missiles would not only make the world safer but could also provide President Clinton with the historic achievement he has long sought. The limited test ban treaty is still considered one of the great achievements of the Kennedy presidency. And in a survey conducted in September, 1997, reducing the risk of nuclear war ran second only to improving U.S. education as the most important legacy President Clinton could leave his country[3] (aside from the Monica Lewinsky debacle).

The Day After exercises were designed to assess the plausibility of information warfare scenarios as part of the Y2K problem and to help define key issues to be addressed in this area. The exercises highlighted some defining features of information warfare, including the fact that attack mechanisms and techniques can be acquired with relatively modest investment. The exercises also revealed that no adequate tactical warning system exists for distinguishing between information warfare attacks and accidents. Perhaps most importantly, the study demonstrated that because the U.S. economy, society, and military rely increasingly on a high-performance networked information infrastructure, this infrastructure presents a set of attractive strategic targets for opponents who possess information warfare capabilities.

The Defense Science Board, a Federal Advisory Committee established to provide independent advice to the Secretary of Defense, acknowledged the threat of an information warfare attack and the damage that could be done in its October 1994 report, titled Information Architecture for the Battlefield. The report states that,

> there is mounting evidence that there is a threat that goes beyond hackers and criminal elements. This threat arises from terrorist groups or nation states, and is far more subtle and difficult to counter than the more unstructured but growing problem caused by hackers. The threat causes concern over the specter of military readiness problems caused by attacks on DOD computer systems, but it goes well beyond the DOD. Every aspect of modern life is tied to a computer system or satellite at some point, and most of these systems are relatively unprotected. This is especially so for those tied to the NII (National Information Infrastructure).

The report added that a large structured attack with strategic intent against the United States could be prepared and exercised under the guise of unstructured activities and that such an attack could ''cripple U.S. operational readiness and military effectiveness.'' These studies demonstrate the growing potential threat to national security posed by computer attacks in conjunction with the Y2K problem. Information warfare will increasingly become an inexpensive but highly effective tactic for disrupting military operations. Successfully protecting information and detecting and reacting to computer attacks presents DOD and our nation with significant challenges.

The Department of Defense and Businesses Face Significant Challenges in Countering Attacks on Satellites

The task of precluding unauthorized users from compromising the confidentiality, integrity, or availability of information is increasingly difficult given the complexity of DOD and the private sector's information infrastructure, growth of and reliance on outside intranets including the Internet, and, the increasing sophistication of the attackers and their tools on unencrypted satellite communications. Absolute protection of DOD and private sector information is neither practical nor affordable. Instead, the DOD and the private sector must turn to risk management to ensure intranet and satellite communications security. In doing so, however, they must make tradeoffs that consider the magnitude of the threat, the value and sensitivity of the information to be protected, and the cost of protecting it.

Elements of a Good Intranet and Satellite Communications Security Program

According to DOD and private sector security experts, certain core elements have emerged as critical to effective intranet and satellite communications security. A good computer and satellite communications security program begins with top management's understanding of the risks associated with networked computers, and a commitment that intranet and satellite communications security will be given a high priority. At DOD, management attention to intranet and satellite communications security has been uneven. The DOD information infrastructure has evolved into a set of individual computer systems and interconnected intranets, many of which were developed without sufficient attention to the entire infrastructure. While some local area networks and DOD installations have excellent intranet and satellite communications security programs, others do not. However, the overall infrastructure is only as secure as the weakest link. Therefore, all components of the DOD infrastructure must be considered when making investment decisions.

In addition, policies and procedures must also reflect this philosophy and guide implementation of the DOD's overall intranet and satellite communications security program as well as the security plans for individual DOD installations. The policies should set minimum standards and requirements for key intranet and satellite communications security activities and clearly assign responsibility and accountability for ensuring that they are carried out. Further, sufficient personnel, training, and resources must be provided to implement these policies.

While not intended to be a comprehensive list, the following are intranet and satellite communications security activities that all of the intranet and satellite communications security studies and industry experts agreed were important:

1. Clear and consistent intranet and satellite communications security policies and procedures.
2. Vulnerability assessments to identify intranet and satellite communications security weaknesses at individual DOD installations.
3. Mandatory correction of identified intranet/system and satellite communications security weaknesses.
4. Mandatory reporting of attacks to help better identify and communicate vulnerabilities and needed corrective actions.
5. Damage assessments to reestablish the integrity of the information compromised by an attacker.
6. Awareness training to ensure that computer users understand the security risks associated with networked computers and practice good intranet and satellite communications security.
7. Assurance that intranet managers and system administrators have sufficient time and training to do their jobs.
8. Prudent use of firewalls, smart cards, and other technical solutions.
9. An incident response capability to aggressively detect and react to attacks and track and prosecute attackers.

The DOD has recognized the importance of good intranet and satellite communications security. In other words, they have recognized that the vulnerability to systems and intranets and satellite communications is increasing. Also, the ability of individuals to penetrate intranets and satellite communications, and deny, damage, or destroy data has been demonstrated on many occasions. As DOD's warfighters become more and more dependent on our information technology systems, the potential for disaster is obvious.

In addition, as part of DOD's Federal Managers' Financial Integrity Act requirements, they have identified intranet and satellite communications security as a system weakness in its Fiscal Year 1998 Annual Statement of Assurance, a report documenting high-risk areas requiring management attention. The DOD has acknowledged a significant increase in attacks on its information technology systems and its dependence on these systems.

Also, DOD has implemented a formal defensive information warfare program. This program was started in December 1992 through Defense Directive 3600.1. The directive broadly states that measures will be taken as part of this program to "protect friendly information technology systems by preserving the availability, integrity, and confidentiality of the systems and the information contained within those systems." The Defense Information Systems Agency (DISA), in cooperation with the military services and defense agencies, is responsible for implementing the program. The DOD's December 1995 Defensive Information Warfare Management Plan defines a three-pronged approach to protect against, detect, and react to threats to the DOD information infrastructure. The plan states that DOD must monitor and detect intrusions or hostile actions as they occur, react quickly to isolate the systems under attack, correct the intranet and satellite communications security breaches, restore service to authorized users, and improve intranet and satellite communications security.

The DISA has also taken a number of actions to implement its plan, the most significant being the establishment of its Global Control Center at DISA headquarters. The center provides the facilities, equipment, and personnel for directing the defensive information warfare program, including detecting and responding to computer and satellite communications attacks. And DISA has also established its Automated Systems Security Incident Support Team (ASSIST) to provide a centrally coordinated around-the-clock Defense response to attacks. The ageny also performs other services to help secure DOD's information infrastructure, including conducting assessments of DOD organizations' vulnerability to computer and satellite communications attacks. The Air Foce Informaton Warfare Center (AFIWC) has developed a computer emergency response capability and performs functions similar to those of DISA. The Navy and Army have just established similar capabilities through the Fleet Information Warfare Center (FIWC) and Land Information Warfare Activity (LIWA), respectively.

The DOD and the private sector are incorporating some of the elements described previously as necessary for strengthening intranet and satellite communications security and countering computer attacks, but there are still areas where improvement is needed. Even though the technology environment has changed dramatically in recent years, and the risk of attacks has increased, top management at many organizations do not consider intranet and satellite communications security to be a priority. As a result, when resources are allocated, funding for important protective measures, such as training or the purchase of protection technology, takes a back seat.

As discussed in the remainder of this chapter, DOD and businesses need to establish a more comprehensive intranet and satellite communications security program. A program that ensures that sufficient resources are directed at protecting information technology (IT). Specifically: (1) Most DOD and private sector policies for protecting, detecting, and reacting to computer and satellite communications attacks are outdated and incomplete; (2) computer users are often unaware of system vulnerabilities and weak intranet and satellite communications security practices; (3) system and intranet administrators are not adequately trained and do not have sufficient time to perform their duties; (4) technical solutions to intranet and satellite communications security problems show promise, but these alone cannot guarantee protection; and (5) while DOD's incident response capability is improving, it is not sufficient to handle the increasing threat.

Department of Defense and Corporate Policies on Intranet and Satellite Communications Security are Outdated and Incomplete

The military services, DOD agencies, and corporations have issued a number of intranet and satellite communications security policies, but they are dated, inconsistent, and incomplete. At least 67 separate DOD policy documents address various computer and intranet security issues. The most significant DOD policy documents include Defense Directive 3600.1, discussed earlier, and Defense Directive 5200.28, entitled "Security Requirements for Automated Information Systems," dated March 21, 1988, which provides mandatory minimum intranet and satellite communications security requirements. In addition, Defense Directive 8000.1, entitled "Defense Information Management Program," dated October 27, 1992, requires DISA and the military services to provide technology and services to ensure the availability, reliability, maintainability, integrity, and security of Defense information. However, these and other policies relating to computer and satellite communications attacks are outdated and inconsistent. They do not set standards, mandate specific actions, or clearly assign accountability for important intranet and satellite communications security activities such as vulnerability assessments, internal reporting of attacks, correction of vulnerabilities, or damage assessments.

Shortcomings in DOD's computer security policy have been reported previously. The Joint Security Commission found similar problems in 1994, and noted that DOD's policies in this area were developed when computers were physically and electronically isolated. Consequently, the Commission reported that DOD intranet and satellite communications security policies were not suitable for today's highly networked environment. The Commission also found that DOD policy was based on a philosophy of complete risk avoidance, rather than a more realistic and balanced approach of risk reduction. In addition, the Commission found a profusion of policy formulation authorities within DOD. This has led to policies being developed that create inefficiencies and implementation problems when organizations attempt to coordinate and interconnect their computer systems. Nevertheless, DOD policies do not specifically require the following important intranet and satellite communications security activities.

Vulnerability Assessments

The DISA established a Vulnerability Analysis and Assessment Program in 1992 to identify vulnerabilities in DOD information technology systems. The Air Force and Navy have similar programs, and the Army has just started assessing its systems. Under its program, DISA attempts to penetrate selected DOD information technology systems using various techniques, all of which are widely available on the Internet. The DISA personnel attack vulnerabilities which have been widely publicized in their alerts to the military services and defense agencies. Assessment is performed at the request of the targeted DOD installation and, upon completion, systems and security personnel are given a detailed briefing. Typically, DISA and the installation develop a plan to strengthen the site's defenses, more effectively detect intrusions, and determine whether systems administrators and intranet and satellite communications security personnel are adequately experienced and trained. Air Force and Navy on-line assessments are similar to DISA vulnerability assessments.

However, there is no specific DOD-wide policy requiring vulnerability assessments or criteria for prioritizing who should be targeted first. This has led to uneven application of this valuable risk assessment mechanism. Some installations have been tested multiple times while others have never been tested. As of March 1996, vulnerability assessments had been performed on less than 1% of the thousands of defense systems around the world. The DISA and military services recognize this shortcoming, but state that they do not have sufficient resources to do more. This is a concern because vulnerabilities in one part of DOD's information infrastructure make the entire infrastructure vulnerable.

Correction of Vulnerabilities The DOD does not have any policy requirement for correcting identified deficiencies and vulnerabilities. The DOD's computer emergency response teams—ASSIST, AFIWC, FIWC, and LIWA—as well as the national computer emergency response team at the Software Engineering Institute routinely identify and broadcast to DOD intranet administrators system vulnerabilities and suggested fixes. However, the lack of specific requirements for correcting known vulnerabilities has led to no action or inconsistent action on the part of some DOD organizations and installations.

Reporting Attacks   The DOD also has no policy requiring internal reporting of attacks or guidance on how to respond to attacks. System and intranet administrators need to know when and to whom attacks should be reported and what response is appropriate for reacting to attacks and ensuring systems availability, confidentiality, and integrity. Reporting attacks is important for DOD to identify and understand the threat (size, scale, and type of attack), as well as to measure the magnitude of the problem for appropriate corrective action and resource allocation. Further, because a computer and satellite communications attack on a federal facility is a crime, it should be reported.

Damage Assessments   Finally, there is no policy for DOD organizations to assess damage to their systems once an attack has been detected. As a result, these assessments are not usually done. For example, Air Force officials have reported that the Rome Air Force Base Laboratory computer security incident was the exception rather than the rule. They reported that system and intranet administrators, due to lack of time and money, often simply 'patch' their systems, restore service, and hope for the best. However, these assessments are essential to ensure the integrity of the data in those systems and to make sure that no malicious code was inserted that could cause severe problems later.

The Department of Defense and Corporate Personnel Lack Sufficient Awareness and Technical Training

The Software Engineering Institute's Computer Emergency Response Team estimates that at least 80% of the intranet and satellite communications security problems it addresses involve poorly chosen or poorly protected passwords and lack of encryption by computer users. According to the Institute, many computer users do not understand the technology they are using, the vulnerabilities in the intranet and satellite communications environment they are working in, and the responsibilities they have for protecting critical information. They also often do not understand the importance of knowing and implementing good intranet and satellite communications security policies, procedures, and techniques. The DOD and corporate management officials generally agreed that user awareness training was needed, but stated that installation commanders and their private sector counterparts do not always understand intranet and satellite communications security risk and, thus, do not always devote sufficient resources to the problem. Top officials for both government and the private sector have reported that they are trying to overcome the lack of resources by low cost alternatives such as banners that warn individuals of their intranet and satellite communications security responsibilities when they turn on their computers.

In addition, intranet and system administrators often do not know what their responsibilities are for protecting their systems, and for detecting and reacting to intrusions. Critical intranet and satellite communications security responsibilities are often assigned to personnel as additional or ancillary duties.

Recent surveys of 35 individuals responsible for managing and securing systems at five military installations were conducted by the Government Accounting Office (GAO). Seventeen stated that they did not have enough time, experience, or training to do their jobs properly. In addition, 18 stated that system administration was not their full-time job, but rather an ancillary duty. The GAO findings were confirmed by an Air Force survey of system administrators. It found that 436 of 810 respondents were unaware of procedures for reporting vulnerabilities and incidents; 350 of 626 respondents had not received any intranet and satellite communications security training and 488 of 817 respondents reported that their intranet and satellite communications security responsibilities were ancillary duties.

In addition, DOD officials stated that it is not uncommon for installations to lack a full-time, trained, experienced intranet and satellite communications security officer. Intranet and satellite communications security officers generally develop and update the site's security plan; enforce security statutes and policy; aggregate and report all security incidents and changes in the site's security status; and evaluate security threats and vulnerabilities. They also coordinate intranet and satellite communications security with physical and personnel security; develop back-up and contingency plans; manage access to all information technology systems with sound password, encryption and user identification procedures; ensure that audit trails of log-ins to systems are maintained and analyzed; and perform a host of other duties necessary to secure the location's computer and satellite communications systems. Without a full-time intranet and satellite communications security official, these important security activities are usually done in an ad hoc manner or not done at all. The DOD officials again cited the low priority installation commanders give intranet and satellite communications security duties as the reason for the lack of full-time, trained, experienced security officers.

The DOD has developed training courses and curricula that focus on the secure operation of computer and satellite communications systems and the need to protect information. For example, DISA's Center for Information Systems Security offers courses on the vulnerability of intranets, satellite communications, and computer systems security. Each of the military services also provides training in this area. While the GAO did not assess the quality of the training, it is clear that not enough training is done; DOD officials cite resource constraints as the reason for this limitation. To illustrate, in its August 1995 Command and Control Protect Program Management Plan, the Army noted that it had approximately 5000 systems administrators, but few of these had received formal intranet and satellite communications security training. The plan stated that the systems administrators have not been taught intranet and satellite communications security basics such as how to detect and monitor an active intrusion, establish countermeasures with encryption, or respond to an intrusion. The plan added that a single course is being developed to train systems administrators, but that no funds are available to conduct the training. This again demonstrates the low priority top DOD management officials often give intranet and satellite communications security.

In its February 1994 report, Redefining Security, the Joint Security Commission had similar concerns, stating:

> Because of a lack of qualified personnel and a failure to provide adequate resources, many intranet and satellite communications security tasks are not performed adequately. Too often critical intranet and satellite communications security responsibilities are assigned as additional or ancillary duties.

The report added that the DOD lacks comprehensive, consistent training for intranet and satellite communications security officers, and that DOD's current intranet and satellite communications security training efforts produce inconsistent training quality and, in some cases, a duplication of effort. The report concluded that, despite the importance of intranet and satellite communications security awareness, training and education programs these programs tend to be frequent and ready targets for budget cuts.

According to DOD officials, installation commanders may not understand the risks associated with networked computers, and thus may
not have devoted sufficient priority or resources to address these
problems. These officials also cite the lack of a professional job series
for intranet officials as a contributing factor to poor security practices
at DOD installations. Until intranet and satellite communications
security is supported by the personnel system (including potential for
advancement, financial reward, and professional training) it will not
be a full-time duty. As a result, intranet and satellite communications
security will continue to be the purview of part-time, inadequately
trained personnel.

Technical Solutions Show Promise, but Cannot Alone Provide Adequate Protection

As described here, DOD and the private sector are developing a variety
of technical solutions that should assist the DOD in preventing,
detecting, and reacting to attacks on its computer and satellite communications systems in the Year 2000. However, knowledgeable attackers with the right tools and encryption know-how can defeat these
technologies. Therefore, these should not be an entity's sole means of
defense. Rather, they should be prudently used in conjunction with
other intranet and satellite communications security measures discussed in this chapter. Investment in these technologies should also be
based on a comprehensive assessment of the value and sensitivity of
the information to be protected.

One important technology is a smart card called Fortezza. The card and
its supporting equipment, including card readers and software, were
developed by the NSA. The card is based on the Personal Computer
Memory Card International Association industry standard and is a
credit card size electronic module that stores digital information that
can be recognized by an intranet or and satellite communications
system. The card will be used by DOD and other government agencies
(like NASA) to provide data encryption and authentication services.
The DOD plans to use the card in its Defense Message System and
other systems around the world.

Another technology that DOD is implementing is firewalls. Firewalls
are hardware and software components that protect one set of system

resources from attack by outside intranet users by blocking and checking all incoming intranet traffic. Firewalls permit authorized users to access and transmit privileged information and deny access to unauthorized users. Several large commercial vendors have developed firewall applications which DOD is using and tailoring for specific organization's computing and satellite communications needs and environments. Like any technology, firewalls are not perfect; hackers have successfully circumvented them in the past. They should not be an installation's sole means of defense, but should be used in conjunction with the other technical, physical, and administrative solutions discussed in this chapter.

Many other technologies exist and are being developed today that DISA, NSA, and the military services are using and considering for future use. These include automated biometrics systems that examine an individual's physiological or behavioral traits and use that information to identify an individual. Biometrics systems are available today, and are being refined for future applications, that examine fingerprints, retina patterns, voice patterns, signatures, and keystroke patterns. In addition, a technology in development called location-based authentication may help thwart attackers by pinpointing their location. This technology determines the actual geographic location of a user attempting to access a system. For example, if developed and implemented as planned, it could prevent a hacker in a foreign country, pretending to come from a military installation in the United States, from logging into a DOD system in the Year 2000.

These technical products show promise in protecting DOD systems and information from unauthorized users. However, they are expensive — firewalls can cost from $6000 to $50,000 for each Internet access point, and Fortezza cards and related support could cost about $400 for each computer. They also require consistent and department-wide implementation to be successful, continued development to enhance their utility, and usage by personnel who have the requisite skills and training to appropriately use them. Once again, no single technical solution is foolproof and, thus, combinations of protective mechanisms should be used. Decisions on which mechanisms to use should be based on an assessment of threat, the sensitivity of the information to be protected, and the cost of protection.

The Department of Defense and Corporate Incident Response Capability Is Limited

Because absolute intranet and satellite communications security is not possible and some attacks will succeed in the Year 2000, an aggressive incident response capability is a key element of a good security program. The DOD has several organizations whose primary mission is incident response (the ability to quickly detect and react to computer attacks). These organizations (DISA's Center for Information Systems Security, ASSIST, and the military service teams) as discussed previously in this chapter, provide intranet and satellite communications monitoring and incident response services to military installations. The AFIWC, with its Computer Emergency Response Team and Countermeasures Engineering Team, was established in 1993 and has considerably greater experience and capability than the other military services. Recognizing the need for more incident response capability, the Navy established the FIWC in 1995, and the Army established its LIWA in 1997. However, these organizations are not all fully staffed and do not have the capability to respond to all reported incidents, much less the incidents not reported. For example, when the FIWC was established, 40 personnel slots were requested, but only 4 were granted. Similarly, the LIWA is just beginning to build its capability.

Rapid detection and reaction capabilities are essential to effective incident response. The DOD is installing devices at numerous military sites to automatically monitor attacks on its computer systems. For example, the Air Force has a project underway called Automated Security Incident Measurement (ASIM), which is designed to measure the level of unauthorized activity against its systems. Under this project, several automated tools are used to examine intranet and satellite communications activity and detect and identify unusual intranet and satellite communications events, for example, Internet addresses not normally expected to access DOD computers. These tools have been installed at only 47 of the 219 Air Force installations around the world. Selection of these installations was based on the sensitivity of the information, known system vulnerabilities, and past hacker activity. Data from the ASIM is analyzed by personnel responsible for securing the installation's intranet and satellite communications system. Data is also centrally analyzed at the AFIWC in San Antonio, TX.

Air Force officials at AFIWC and at Rome Laboratory told GAO that

ASIM has been extremely useful in detecting attacks on Air Force systems. They added, however, that as currently configured, ASIM information is only accumulated and automatically analyzed nightly. As a result, a delay occurs between the time an incident occurs and the time when ASIM provides information on the incident. They also stated that ASIM is currently configured for selected operating systems and, therefore, cannot detect activity on all Air Force computer and satellite communications systems. They added that they plan to continue refining the ASIM to broaden its use for other Air Force operating systems and enhance its ability to provide data on unauthorized activity more quickly. The AFIWC officials believe that a well-publicized detection and reaction capability can be a successful deterrent to would-be attackers.

The Army and Navy are also developing similar devices, but they have been implemented in only a few locations. The Army's system, known as Automated Intrusion Monitoring System (AIMS), has been in development since June 1995, and is intended to provide both a local and theater-level monitoring of computer and satellite communications attacks. Currently, AIMS is installed at the Army's 5th Signal Command in Worms, Germany and will be used to monitor Army computers and satellite communications systems scattered throughout Europe.

The DISA officials told GAO that although the services' automated detection devices are good tools, they need to be refined to allow DOD to detect unauthorized activity as it is occurring. DISA's Defensive Information Warfare Management Plan provides information on new or improved technology and programs planned for the next 1 to 5 years. These efforts included a more powerful intrusion detection and monitoring program, a malicious code detection and eradication program, and a program for protecting DOD's vast information infrastructure. These programs, if developed and implemented as planned, should enhance DOD's ability to protect and react to attacks on its computer systems in the Year 2000.

Summary

Electronic terrorists may soon wreak havoc via satellite and computer on the nation's energy, power, water, finance, encrypted communications and emergency systems. Law officials and business leaders think the government should do something about it. But, what is the government doing to protect itself?

Unknown and unauthorized individuals are increasingly attacking and gaining access to highly sensitive unclassified information on DOD computer and satellite communications systems. Given the threats the attacks pose to military operations and national security, GAO was asked to report on the extent to which DOD systems are being attacked, the potential for further damage in the Year 2000 to information and systems, and the challenges DOD faces in securing sensitive information.

Attacks on DOD computer and satellite communications systems as we approach the Year 2000 are a serious and growing threat. The exact number of attacks cannot be readily determined because only a small portion are actually detected and reported. However, Defense Information Systems Agency (DISA) data implies that DOD may have experienced as many as 470,000 attacks in 1996. The DISA information also shows that attacks are successful 76% of the time, and that the number of attacks is doubling each year, as Internet use increases along with the sophistication of hackers and their tools.

At a minimum, these attacks are a multimillion dollar nuisance to DOD. At their worst, they are a serious threat to national security as the Year 2000 crisis approaches. Attackers have seized control of entire DOD systems, many of which support critical functions, such as weapons systems research and development, logistics, and finance. Attackers have also stolen, modified, and destroyed data and software. In a well-publicized attack on Rome Laboratory, the Air Force's premier command and control research facility, two hackers took control of laboratory support systems, established links to foreign Internet sites, and stole tactical and artificial intelligence research data.

The potential for catastrophic damage as we approach the year 2000 is great. Organized foreign nationals or terrorists could use *information warfare* techniques to disrupt military operations by harming command and control systems, the public switch network, and other systems or intranets and satellite communications DOD and the private sector relies on.

The DOD is taking action to address the growing year 2000 problem, but faces significant challenges in controlling unauthorized access to its computer and satellite communications systems. Currently, DOD is attempting to react to successful attacks as it learns of them, but it has no uniform policy for assessing risks, protecting its systems, responding to incidents, or assessing damage. Training of users and system and intranet administrators is inconsistent and constrained by limited resources. Technical solutions being developed, including firewalls, sophisticated encryption, smart cards and intranet monitoring systems, will improve protection of DOD and private sector information. However, the success of these measures depends on whether DOD and the private sector implement them in tandem with better policy and personnel solutions.

From Here

Chapter 29 concluded with a discussion on how attacks on satellites could threaten national security as we approach the year 2000; and, how DOD and businesses face significant challenges in countering attacks on satellites. Chapter 30 wraps up with a discussion on the findings and recommendations with regards to: the problem of information vulnerability; cryptographic solutions to satellite information vulnerabilities; the policy dilemma posed by satellite encryption; national satellite encryption policy for the information age; and, how satellite encryption is essential to freedom.

Endnotes

1   The President's Commission on Critical Infrastructure Protection, P. O. Box 46258, Washington, DC 20050-6258, 1998.

2   Ibid.

3   Ibid.

30

Summary, Recommendations and Conclusions

In an age of explosive worldwide growth of electronic data storage and satellite communications, many vital national interests require the effective protection of information. Especially when used in coordination with other tools for information security, encryption in all of its applications, including data confidentiality, data integrity, and user authentication, is a most powerful tool for protecting information.

Summary

Because digital representations of large volumes of information are increasingly pervasive, both the benefits and the risks of digital representation have increased. The benefits are generally apparent to users of information technology—larger amounts of information, used more effectively and acquired more quickly, can increase the efficiency with which businesses operate, open up entirely new business opportunities, and play an important role in the quality of life for individuals.

The risks are far less obvious. As discussed in Chapter 2, one of the most significant risks of a digital information age is the potential vulnerability of important information as it is communicated and stored. When information is transmitted via satellite in computer-readable form, it is highly vulnerable to unauthorized disclosure or alteration:

Many communications are carried over channels (satellites, cellular telephones, and local area networks) that are easily tapped. Tapping wireless channels is almost impossible to detect and to stop, and tapping local area networks may be very hard to detect or stop as well. Other electronic communications are conducted through data networks that can be easily penetrated (the Internet).

Approximately 20 billion words of information in computer-readable form can be scanned for 1 dollar today (as discussed in Chapter 2), allowing intruders, the malicious, or spies to separate the wheat from the chaff very inexpensively. For example, a skilled person with criminal intentions can easily develop a program that recognizes and records all credit card numbers in a stream of unencrypted data traffic. The decreasing cost of computation will reduce even further the costs involved in such search.

Note: The feasibility of designing a program to recognize text strings that represent credit card numbers has been demonstrated most recently by the First Virtual Corporation, San Diego, CA.

Many users do not know about their vulnerabilities to the theft or compromise of information; in some instances, they are ignorant of or even complacent about them. Indeed, the insecurity of computer networks today is much more the result of poor operational practices on the part of users and poor implementations of technology on the part of product developers than of an inadequate technology base or a poor scientific understanding.

In the early days of computing, the problems caused by information vulnerability were primarily the result of relatively innocent trespasses of amateur computer hackers who were motivated mostly by technical curiosity. But this is true no longer, and has not been true for some time. The fact that the nation is moving into an information age on a large scale means that a much larger number of people are likely to have strong financial, political, or economic motivations to exploit information vulnerabilities that still exist. For example, electronic interceptions and other technical operations account for the largest portion of economic and industrial information lost by U.S. corporations to foreign parties, as noted in Chapter 2.

Today, the consequences of large-scale information vulnerability are potentially quite serious. For example, U.S. business, governmental, and individual satellite communications are targets or potential targets for intelligence organizations of foreign governments, competitors, vandals, suppliers, customers, and organized crime. Businesses send through electronic channels considerable amounts of confidential information, including items such as project and merger proposals, trade secrets, bidding information, corporate strategies for expansion in critical markets, research and development information relevant to cost reduction or new products, product specifications, and expected delivery dates. Most importantly, U.S. businesses must compete on a worldwide basis. International exposure increases the vulnerability to compromise sensitive information. Helping to defend U.S. business interests against such compromises of information is an important function of law enforcement.

American values such as personal rights to privacy are at stake. Private citizens may conduct sensitive financial transactions electronically or by telephone. Data on their medical histories, including mental illnesses, addictions, sexually transmitted diseases and personal health habits, are compiled in the course of providing medical care. Driving records, spending patterns, credit histories, and other financial information are available from multiple sources. All such information warrants protection.

The ability of private citizens to function in an information economy is at risk. Even today, individuals suffer as criminals take over their identities and run up huge credit card bills in their name. Toll fraud on cellular telephones is so large that some cellular providers have simply terminated international connections in the areas that they serve. Inaccuracies as the result of incorrectly posted information ruin the credit records of some individuals. Protecting individuals against such problems warrants public concern and is again an area in which law enforcement and other government authorities have a role to play. The federal government has an important stake in assuring that its important and sensitive political, economic, law enforcement and military information, both classified and unclassified, is protected from misuse by foreign governments or other parties whose interests are hostile to those of the United States.

Elements of the U.S. civilian infrastructure such as the banking system, the electric power grid, the public switched telecommunications network (PSTN), and the air traffic control system are central to so many dimensions of modern life that protecting these elements must have a high priority. Defending these assets against information warfare and crimes of theft, misappropriation, and misuse potentially conducted by hostile nations, terrorists, criminals, and electronic vandals is a matter of national security and will require high levels of information protection and strong security safeguards.

Encryption Solutions to Information Vulnerabilities

Encryption does not solve all problems of information security; for example, cryptography cannot prevent a party authorized to view information from improperly disclosing that information. Although it is not a silver bullet that can stand by itself, encryption is a powerful tool that can be used to protect information stored and communicated in digital form via satellite. Encryption can help to assure confidentiality of data; to detect unauthorized alterations in data and thereby help to maintain its integrity; and, to authenticate the asserted identity of an individual or a computer system (see Chapter 3). Used in conjunction with other information security measures, encryption has considerable value in helping law-abiding citizens, businesses, and the nation as a whole defend their legitimate interests against information crimes and threats such as fraud, electronic vandalism, the improper disclosure of national security information, or information warfare.

Modern cryptographic techniques used for confidentiality make it possible to develop and implement ciphers that are for all practical purposes impossible for unauthorized parties to penetrate but that still make good economic sense to use. Strong encryption is economically feasible today. For example, many integrated circuit chips that would be used in a computer or a satellite communications device can inexpensively accommodate the extra elements needed to implement the DES encryption algorithm. If implemented in software, the cost is equally low, or even lower.

Public-key encryption can help to eliminate the expense of using couriers, registered mail, or other secure means for exchanging keys. Compared to a physical infrastructure for key exchange, an electronic infrastructure based on public-key encryption to exchange keys will be faster and more able to facilitate secure satellite communications between parties that have never interacted directly with each other prior to the first communication. Public-key encryption also enables the implementation of the digital equivalent of a written signature, enabling safer electronic commerce.

Encryption can be integrated by vendors into end-user applications and hardware for the benefit of the large majority of users who do not have the technical skill to perform their own integration. Encryption can also be made automatic and transparent in ways that require no extra action on the part of the user, thus ensuring that cryptographic protection will be present regardless of user complacency or ignorance.

The Policy Dilemma Posed by Encryption

The confidentiality of information that encryption can provide is useful not only for the legitimate purposes of preventing information crimes (the theft of trade secrets or unauthorized disclosure of sensitive medical records) but also for illegitimate purposes (shielding from law enforcement officials a conversation between two terrorists planning to bomb a building). Although strong, automatic encryption implemented as an integral part of information technology and satellite communications provides confidentiality for *good guys* against *bad guys* (U.S. business protecting information against economic intelligence efforts of foreign nations), it unfortunately also protects *bad guys* against *good guys* (terrorists evading law enforcement agencies). Under appropriate legal authorization such as a court order, law enforcement authorities may gain access to *bad guy* information for the purpose of investigating and prosecuting criminal activity. Similarly, intelligence gathering for national security and foreign policy purposes depends on having access to information of foreign governments and other foreign entities, (See Chapter 4.) Because such activities benefit our society as a whole (by limiting organized crime and terrorist activities), *bad guy* use of encryption used for confidentiality poses a problem for society as a whole, not just for law enforcement and national security personnel.

Considered in these terms, it is clear that the development and widespread deployment of encryption that can be used to deny government access to information represents a challenge to the balance of power between the government and the individual. Historically, all governments, under circumstances that further the common good, have asserted the right to compromise the privacy of individuals (through opening mail, tapping telephone calls, inspecting bank records). Unbreakable encryption for confidentiality provides the individual with the ability to frustrate assertions of that right.

The confidentiality that encryption can provide thus creates conflicts. Nevertheless, all of the issues previously described—privacy for individuals, protection of sensitive or proprietary information for businesses and other organizations in the prevention of information crimes, ensuring the continuing reliability and integrity of nationally critical information systems and networks, law enforcement access to stored and communicated information for purposes of investigating and prosecuting crime, and national security access to information stored or communicated by foreign powers or other entities and organizations whose interests and intentions are relevant to the national security and the foreign policy interests of the United States—are legitimate. Informed public discussion of the issues must begin by acknowledging the legitimacy of both information security for law-abiding individuals and businesses and information gathering for law enforcement and national security purposes.

A major difficulty clouding the public policy debate regarding encryption has been that certain elements have been removed from public view due to security classification. Although many of the details relevant to policy makers are necessarily classified, these details are not central to making policy arguments one way or the other. Classified material, while important to operational matters in specific cases, is not essential to the big picture of why policy has the shape and texture that it does today nor to the general outline of how technology will, and policy should, evolve in the future.

To manage the policy dilemma created by encryption, the United States has used a number of tools to balance the interests previously described. For many years, concern over foreign threats to national security has been the primary driver of a national encryption policy that has sought to maximize the protection of U.S. military and diplomatic communications while denying the confidentiality benefits of cryptography to foreign adversaries through the use of controls on the export of cryptographic technologies, products, and related technical information (see Chapter 5). More recently, the U.S. government has aggressively promoted escrowed encryption as the technical foundation for national cryptography policy, both to serve domestic interests in providing strong protection for legitimate uses while enabling legally authorized access by law enforcement officials when warranted and also as the basis for more liberal export controls on cryptography (see Chapter 13).

Both escrowed encryption and export controls have generated considerable controversy. Escrowed encryption has been controversial because its promotion by the U.S. government appears to some important constituencies to assert the primacy of information access needs of law enforcement and national security over the information security needs of businesses and individuals. Export controls on encryption have been controversial because they pit the interests of U.S. vendors and some U.S multinational corporations against some of the needs of national security.

National Encryption Policy for the Information Age

In a world of ubiquitous computing and satellite communications, a concerted effort to protect the information assets of the United States is critical. While encryption is only one element of a comprehensive approach to information security, it is nevertheless an essential element.

Basic Principle

United States national policy should be changed to support the broad use of encryption in ways that take into account competing U.S. needs and desires for individual privacy, international economic competitiveness, law enforcement, national security, and world leadership.

In practice, this principle suggests three basic objectives for national cryptography policy:

1. Broad availability of encryption to all legitimate elements of U.S. society.
2. Continued economic growth and leadership of key U.S. industries and businesses in an increasingly global economy, including but not limited to U.S. computer, software and communications companies.
3. Public safety and protection against foreign and domestic threats.[1]

Broad Availability of Encryption   Encryption supports the confidentiality and integrity of digitally represented information (computer data, software, video) and the authentication of individuals and computer systems communicating with other computer systems. These capabilities are important in varying degrees to protecting the information security interests of many different private and public stakeholders, including law enforcement and national security. Furthermore, encryption can help to support law enforcement objectives in preventing information crimes such as economic espionage.

Continued Economic Growth And Leadership   Such leadership is an integral element of national security. United States companies in information technology today have undeniable strengths in foreign markets, but current national encryption policy threatens to erode these advantages. The largest economic opportunities for U.S. firms in all industries lie in using encryption to support their critical domestic and international business activities, including international, intrafirm and interfirm satellite communications with strategic partners, cooperative efforts with foreign collaborators and researchers in joint business ventures, and real-time connections to suppliers and customers, rather than in selling information technology (see Chapter 5).

Public Safety and Protection   Insofar as possible, satellite communications and stored information of foreign parties whose interests are hostile to those of the United States should be accessible to U.S. intelligence agencies. Similarly, the satellite communications and stored information of criminal elements that are a part of U.S. and global society should be available to law enforcement authorities as authorized by law (see Chapter 4).

Basic Principle Conclusions

Objectives 1 and 2 argue for a policy that actively promotes the use of strong encryption on a broad front and that places few restrictions on the use of cryptography. Objective 3 argues that some kind of government role in the deployment and use of encryption may continue to be necessary for public safety and national security reasons. These three objectives can be met within a framework recognizing that on balance, the advantages of more widespread use of encryption outweigh the disadvantages.

Encryption is one important tool for protecting information and it is very difficult for governments to control. Thus, there is a general belief that the widespread nongovernment use of encryption in the United States and abroad is inevitable in the long run. Encryption is important because when it is combined with other measures to enhance information security, it gives end users significant control over their information destinies. Even though export controls have had a nontrivial impact on the worldwide spread of encryption in previous years, over the long term cryptography is difficult to control because the relevant technology diffuses readily through national boundaries. Export controls can inhibit the diffusion of products with encryption capabilities but cannot contain the diffusion of knowledge (see Chapter 5). The spread of encryption is inevitable because in the information age the security of information will be as important in all countries as other attributes valued today, such as the reliability and ubiquity of information.

Given the inevitability that encryption will become widely available, policy that manages how cryptography becomes available can help to mitigate the deleterious consequences of such availability. Indeed, governments often impose regulations on various types of technology that have an impact on the public safety and welfare, and encryption may well fall into this category. National policy can have an important effect on the rate and nature of the transition from today's world to that of the long-term future. Still, given the importance of encryption to a more secure information future and its consequent importance to various dimensions of economic prosperity, policy actions that inhibit the use of cryptography should be scrutinized with special care.

The Clinton Administration's policy recommendations are intended to facilitate a judicious transition between today's world of high information vulnerability and a future world of greater information security, while to the extent possible meeting government's legitimate needs for information gathering for law enforcement, national security, and foreign policy purposes. The national encryption policy should be expected to evolve over time in response to events driven by an era of rapid political, technological, and economic change.

The Administration recognizes that the national encryption policy is intended to address only certain aspects of a much larger information security problem faced by citizens, businesses, and government. Nevertheless, the Administration found that current national policy is not adequate to support the information security requirements of an information society. Encryption is an important dimension of information security, but current policy discourages the use of this important tool in both intentional and unintentional ways, as described in Chapters 5 and 14. For example, through the use of export controls, national policy has explicitly sought to limit the use of encryption abroad but has also had the effect of reducing the domestic availability of products with strong encryption capabilities to businesses and other users. Furthermore, government action that discourages the use of encryption contrasts sharply with national policy and technological and commercial trends in other aspects of information technology. Amidst enormous changes in the technological environment in the past 30 years, today the federal government actively pursues its vision of a national information infrastructure, and the use of computer and satellite communications technology by private parties is growing rapidly.

A mismatch between the speed at which the policy process moves and the speed with which new products develop has had a profound impact on the development of the consensus necessary with respect to the encryption policy (Chapters 5 and 14). This mismatch has a negative impact on both users and vendors. For example, both are affected by an export control regime that sometimes requires many months or even years to make case-by-case decisions on export licensing, while high-value sales to these users involving integrated products with encryption capabilities can be negotiated and consummated on a time scale of days or weeks. Because the basic knowledge underlying encryption is well known, cryptographic functionality can be implemented into new products on the time scale of new releases of products (several months to a year). Both users and vendors are affected by the fact that significant changes in the export control regulations governing encryption have not occurred since 1992, at a time when needs for information security are growing, a period that could have accommodated several product cycles. Promulgation of encryption standards not based on commercial acceptability (the Escrowed Encryption Standard (FIPS 185), the Digital Signature Standard (FIPS 180-1)) raised significant industry opposition (from both vendors and users) and led to controversy and significant delays in or outright resistance to commercial adoption of these standards.

These examples suggest that the time scales on which encryption policy is made and is operationally implemented are incompatible with the time scales of the marketplace. A more rapid and market-responsive decisionmaking process would leverage the strengths of U.S. businesses in the international marketplace before significant foreign competition develops. As is illustrated by the shift in market position from IBM to Microsoft in the 1980s, the time scale on which significant competition can arise is short indeed.

Attempts to promote a policy regime that runs against prevailing commercial needs, practice, and preference may ultimately result in a degree of harm to law enforcement and national security interests far greater than what would have occurred if a more moderate regime had been promoted in the first place. The reason is that proposed policy regimes that attempt to impose market-unfriendly solutions will inevitably lead to resistance and delay; whether desirable or not, this is a political reality. Responsible domestic businesses, vendors, and end users are willing to make some accommodations to U.S. national interests in law enforcement and national security, but cannot be expected to do so willingly when those accommodations are far out of line with the needs of the market. Such vendors and users are likely to try to move ahead on their own (and quickly so) if they believe that government requirements are not reasonable. Moreover, foreign vendors may well attempt to step into the vacuum. The bottom line is that the U.S. government may have only a relatively small window of time in which to influence the deployment of encryption worldwide.

Public debate has tended to draw lines that divide the policy issues in an overly simplistic manner (setting the privacy of individuals and businesses against the needs of national security and law enforcement). As observed in the previous discussion, such a dichotomy does have a kernel of truth. But viewed in the large, the dichotomy as posed is misleading. If encryption can protect the trade secrets and proprietary information of businesses and thereby reduce economic espionage (which it can), it also supports in a most important manner the job of law enforcement. If encryption can help protect nationally critical information systems and networks against unauthorized penetration (which it can), it also supports the national security of the United States. Framing the national encryption policy in this larger context would help to reduce some of the polarization among the relevant stakeholders.

Finally, the national encryption policy of the United States is situated in an international context, and the formulation and implementation of U.S. policy must take into account international dimensions of the problem if U.S. policy is to be successful. These international dimensions, discussed in Chapter 14 include the international scope of business today; the possibility of significant foreign competition in information technology; an array of foreign controls on the export, import, and use of cryptography; important similarities in the interests of the United States and other nations in areas such as law enforcement and antiterrorist activities; and, important differences in other areas such as the relationship between the government and the governed.

Recommendations

The following recommendations address several critical policy areas. Each recommendation is cast in broad terms, with specifically actionable items identified for each when appropriate. The broad picture of the encryption policy can be understood on an unclassified basis, no findings or recommendations were held back on the basis of classification.

First Recommendation

No law should bar the manufacture, sale or use of any form of encryption within the United States. This recommendation is consistent with the position of the Clinton Administration that legal prohibitions on the domestic use of any kind of encryption are inappropriate.

For technical reasons described in Chapter 26, a legislative ban on the use of unescrowed encryption would be largely unenforceable. Products using unescrowed encryption are in use today by millions of users, and such products are available from many difficult-to-censor Internet sites abroad. Users could preencrypt their data, using whatever means were available, before their data were accepted by an escrowed encryption device or system. Users could store their data on remote computers, accessible through the click of a mouse but otherwise unknown to anyone but the data owner; such practices could occur quite legally even with a ban on the use of unescrowed encryption. Knowledge of strong encryption techniques is available from official U.S. government publications and other sources worldwide, and experts who understand how to use such knowledge might well be in high demand from criminal elements. Even demonstrating that a given satellite communication or data file is *encrypted* may be difficult to prove, as algorithms for data compression illustrate. Such potential technical circumventions suggest that even with a legislative ban on the use of unescrowed encryption, determined users could easily evade the enforcement of such a law.

In addition, a number of constitutional issues, especially those related to free speech, would be almost certain to arise. Insofar as a ban on the use of unescrowed encryption would be treated (for constitutional purposes) as a limitation on the *content* of satellite communications, the government would have to come forward with a compelling state interest to justify the ban. These various considerations are difficult, and in some cases impossible to estimate in advance of a particular legislation as applied to a specific case. But, the First Amendment issues likely to arise with a ban on the use of unescrowed encryption are not trivial. In addition, many people believe with considerable passion that government restrictions on the domestic use of encryption would threaten basic American values such as the right to privacy and free speech. Even if the constitutional issues could be resolved in favor of some type of ban on the use of unescrowed encryption, these passions would surely result in a political controversy that could divide the nation and at the very least impede progress on the way to the full use of the nation's information infrastructure.

Finally, a ban on the use of any form of encryption would directly challenge the principle that users should be responsible for assessing and determining their own approaches to meeting their security needs. This principle is explored later in greater detail in the third Recommendation.

Second Recommendation

The national encryption policy should be developed by the executive and legislative branches on the basis of open public discussion and governed by the rule of law. In policy areas that have a direct impact on large segments of the population, history demonstrates that the invocation of official government secrecy often leads to public distrust and resistance. Such a result is even more likely where many members of society are deeply skeptical about government.

The encryption policy set in the current social climate is a case in point. When encryption was relevant mostly to government interests in diplomacy and national security, government secrecy was both necessary and appropriate. But in an era in which encryption plays an important role in protecting information in all walks of life, public consensus and government secrecy related to information security in the private sector are largely incompatible. If a broadly acceptable social consensus that satisfies the interests of all legitimate stakeholders is to be found regarding the nation's cryptographic future, a national discussion of the issue must occur.

The nation's best forum for considering multiple views across the entire spectrum is the U.S. congress, and only comprehensive congressional deliberation and discussion conducted in the open can generate the public acceptance that is necessary for policy in this area to succeed. In turn, a consensus derived from such deliberations, backed by explicit legislation when necessary, will lead to greater degrees of public acceptance and trust, a more certain planning environment, and better connections between policy-makers and the private sector on which the nation's economy and social fabric rest. For these reasons, congressional involvement in the debate over the encryption policy is an asset rather than a liability. Moreover, some aspects of the encryption policy will require legislation if they are to be properly implemented as discussed later in the chapter.

This argument does not suggest that there are no legitimate secrets in this area. However, in accordance with the Administration's conclusion that the broad outlines of the national encryption policy can be analyzed on an unclassified basis, the Administration believes that the U.S. Congress can also debate the fundamental issues in the open. Nor is the Administration arguing that *all* aspects of policy should be handled in Congress. The executive branch is necessarily an important player in the formulation of the national encryption policy, and of course it must *implement* policy. Moreover, while working with the Congress, the executive branch must develop a coherent voice on the matter of the encryption policy (one that it does not currently have) and establish a process that is efficient, comprehensive, and decisive in bringing together and rationalizing many disparate agency views and interests. Instances in which legislation may be needed are found later in the chapter.

Third Recommendation

The national encryption policy affecting the development and use of commercial cryptography should be more closely aligned with market forces. As encryption has assumed greater importance to nongovernment interests, the national cryptography policy has become increasingly disconnected from market reality and the needs of parties in the private sector. As in many other areas, the national policy on encryption that runs counter to user needs and against market forces is unlikely to be successful over the long term. User needs will determine the large-scale demand for information security, and policy should seek to exploit the advantages of market forces whenever and wherever possible. Indeed, many decades of experience with technology deployment suggest that reliance on user choices and market forces is generally the most rapid and effective way to promote the widespread utilization of any new and useful technology.

Considerations of public safety and national security make it undesirable to maintain an entirely laissez-faire approach to the national encryption policy. Government intervention in the market should be carefully tailored to specific circumstances.

A national encryption policy that is aligned with market forces would emphasize the freedom of domestic users to determine cryptographic functionality, protection, and implementations according to their security needs as they see fit. Innovation in technologies such as escrowed encryption would be examined by customers for their business fitness of purpose. Diverse user needs would be accommodated; some users will find it useful to adopt some form of escrowed encryption to ensure their access to encrypted data, while others will find that the risks of escrowed encryption (the dangers of compromising sensitive information through a failure of the escrowing system) are not worth the benefits (the ability to access encrypted data to which keys have been lost or corrupted). As no single cryptographic solution or approach will fit the business needs of all users, users will be free to make their own assessments and judgments about the products they wish to use. Such a policy would permit, indeed encourage, vendors to implement and customers to use products that have been developed within an already-existing framework of generally accepted encryption methods and to choose key sizes and management techniques without restriction.

Standards are another dimension of the national encryption policy with a significant impact on commercial cryptography and the market (see Chapter 14). Cryptographic standards that are inconsistent with prevailing or emerging industry practice are likely to encounter significant market resistance. Thus, to the maximum extent possible, the national encryption policy that is more closely aligned with market forces should encourage adoption by the federal government and private parties of cryptographic standards that are consistent with prevailing industry practice.

Finally, users in the private sector need confidence that products with cryptographic functionality will indeed perform as advertised. To the maximum degree possible, the national encryption policy should support the use of algorithms, product designs, and product implementations that are open to public scrutiny. Information security mechanisms for widespread use that depend on a secret algorithm or a secret implementation invite a loss of public confidence, because they do not allow open testing of the security, they increase the cost of hardware implementations, and they may prevent the use of software implementations as described later in the chapter. Technical work in encryption conducted in the open can expose flaws through peer review and assure the private sector user community about the quality and integrity of the work underlying its cryptographic protection (see Chapter 13).

Government classification of algorithms and product implementations clearly inhibits public scrutiny, and for the nongovernment sector, government classification in encryption is incompatible with most commercial and business interests in information security. Moreover, the use of classified algorithms largely precludes the use of software solutions, as it is impossible to prevent a determined and technically sophisticated opponent from reverse-engineering an algorithm implemented in software. A similar argument applies to unclassified company-proprietary algorithms and product designs. However, the concerns that arise with classified algorithms and implementations are mitigated somewhat by the fact that it is often easier for individuals to enter into the nondisclosure agreements necessary to inspect proprietary algorithms and product designs than it is to obtain U.S. government security clearances. Legally mandated security requirements to protect classified information also add to costs in a way that protection of company-proprietary information does not.

Fourth Recommendation

Export controls on encryption should be progressively relaxed but not eliminated. For many years, the United States has controlled the export of cryptographic technologies, products, and related technical information as munitions (on the U.S. Munitions List (USML) administered by the State Department). These controls have been used to deny potential adversaries access to U.S. encryption technology that might reveal important characteristics of U.S. information security products and/or be used to thwart U.S. attempts at collecting signals intelligence information. To date, these controls have been reasonably effective in containing the export of U.S. hardware-based products with encryption capabilities (see Chapter 5). However, software-based products with encryption capabilities and cryptographic algorithms present a more difficult challenge because they can more easily bypass controls and be transmitted across national borders. In the long term, as the use of encryption grows worldwide, it is probable that national capability to conduct traditional signals intelligence against foreign parties will be diminished (as discussed in Chapter 4).

The current export control regime on strong encryption is an increasing impediment to the information security efforts of U.S. firms competing and operating in world markets, developing strategic alliances internationally, and forming closer ties with foreign customers and suppliers. Some businesses rely on global networks to tie together branch offices and service centers across international boundaries. Other businesses are moving from a concept of operations that relies on high degrees of vertical integration to one that relies on the *outsourcing* of many business functions and activities. Consistent with rising emphasis on the international dimensions of business (for both business operations and markets), many U.S. companies must exchange important and sensitive information with an often-changing array of foreign partners, customers, and suppliers. Under such circumstances, the stronger level of cryptographic protection available in the United States is not meaningful when an adversary can simply attack the protected information through foreign channels.

Export controls also have had the effect of reducing the domestic availability of products with strong encryption capabilities. As noted in Chapter 5, the need for U.S. vendors (especially software vendors) to market their products to an international audience leads many of them to weaken the encryption capabilities of products available to the domestic market, even though no statutory restrictions are imposed on that market. Thus, domestic users face a more limited range of options for strong encryption than they would in the absence of export controls.

Looking to the future, both U.S. and foreign companies have the technical capability to integrate high-quality cryptographic features into their products and services. As demand for products with encryption capabilities grows worldwide, foreign competition could emerge at a level significant enough to damage the present U.S. world leadership in this critical industry. Today, U.S. information technology products are widely used in foreign markets because foreign customers find the package of features offered by those products to be superior to packages available from other non-U.S. vendors, even though encryption capabilities of U.S. products sold abroad are known to be relatively weak. However, for growing numbers of foreign customers with high security needs, the incremental advantage of superior nonencryption features offered by U.S. products may not be adequate to offset perceived deficiencies in encryption capability. Under such circumstances, foreign customers may well turn to non-U.S. sources that offer significantly better encryption capabilities in their products.

Overly restrictive export controls thus increase the likelihood that significant foreign competition will step into a vacuum left by the inability of U.S. vendors to fill a demand for stronger encryption capabilities integrated into general-purpose products. The emergence of significant foreign competition for the U.S. information technology industry has a number of possible long-term negative effects on U.S.

national and economic security that policymakers would have to weigh against the contribution these controls have made to date in facilitating the collection of signals intelligence in support of U.S. national security interests (a contribution that will probably decline over time). Stimulating the growth of important foreign competitors would undermine a number of important national interests.

First of all, the national economic interest is supported by continuing and even expanding U.S. world leadership in information technology. Today, U.S. information technology vendors have a window of opportunity to set important standards and deploy an installed base of technology worldwide, an opportunity that should be exploited to the maximum degree possible. Conversely, strong foreign competition would not be in the U.S. economic self-interest.

Second, traditional national security interests are supported by U.S. vendors in supplying products with encryption capabilities to the world market. For example, it is desirable for the U.S. government to keep abreast of the current state of commercially deployed encryption technology, a task that is much more difficult to accomplish when the primary suppliers of such technology are foreign vendors rather than U.S. vendors.

Third, U.S. business needs for trustworthy information protection are supported by U.S. encryption products. Foreign vendors could be influenced by their governments to offer for sale to U.S. firms products with weak or poorly implemented encryption. If these vendors were to gain significant market share, the information security of U.S. firms could be adversely affected.

Finally, influence over the deployment of encryption abroad is supported by the significant impact of U.S. export controls on encryption as the result of the strength of the U.S. information technology industry abroad. To the extent that the products of foreign competitors are available on the world market, the United States loses influence over encryption deployments worldwide.

The importance of the U.S. information technology industry to U.S. economic interests and national security is large enough that some prudent risks can be taken to hedge against the potential damage to that industry, and some relaxation of export controls on encryption is warranted. In the long term, U.S. signals intelligence capability is likely to decrease in any case. Consequently, the benefits of relaxation (namely helping to promote better information security for U.S. companies operating internationally and to extend U.S. leadership in this critical industry) are worth the short-term risk that the greater availability of U.S. products with stronger encryption capabilities will further impede U.S. signals intelligence capability.

Relaxation of export controls on encryption is consistent with the basic principle of encouraging the use of encryption in an information society for several reasons. First, relaxation would encourage the use of encryption by creating an environment in which U.S. and multinational firms and users are able to use the same security products in the United States and abroad and thus to help promote better information security for U.S. firms operating internationally. Second, it would increase the availability of good encryption products in the United States. Third, it would expand U.S. business opportunities overseas for information technology sales incorporating stronger encryption for confidentiality by allowing U.S. vendors to compete with foreign vendors on a more equal footing, thereby helping to maintain U.S. leadership in fields critical to national security and economic competitiveness (as described in Chapter 5).

Some of these thoughts are not new. For example, in referring to a decision to relax export controls on computer exports, then-Deputy Secretary of Defense William Perry said that *however much we want to control (computers) that are likely to be available on retail mass markets, it will be impractical to control them, and that we have to recognize we don't have any ability to control computers which are available on the mass retail market from non-COCOM countries. He further noted that the U.S. government can no longer set the standards and specifications of computers. They're going to be set in the commercial industry, and our job is to adapt to those if we want to stay current in the latest computer technology.* Exports of information technology products with encryption capabilities are not qualitatively different.

At the same time, encryption is inherently dual-use in character (more so than most other items on the USML), with important applications to both civilian and military purposes. While this fact suggests to some that the export of all encryption should be regulated under the Commerce Control List (CCL), the fact remains that encryption is a particularly critical military application for which few technical alternatives are available. The USML is designed to regulate technologies with such applications for reasons of national security (as described in Chapters 4 and 5), and thus the current export control regime on encryption should be relaxed but not eliminated. This action would have two major consequences. First of all, relaxation will achieve a better balance between U.S. economic needs and the needs of law enforcement and national security. Second, retention of some controls will mitigate the loss to U.S. national security interests in the short term, allow the United States to evaluate the impact of relaxation on national security interests before making further changes, and *buy time* for U.S. national security authorities to adjust to a new technical reality.

Consistent with the Third Recommendation, the export control regime for encryption should be better aligned with technological and market trends worldwide. Recommendations later in the chapter reflect how the present export control regime should be relaxed expeditiously. However, it should be noted that some explicit relaxations in the export control regime have occurred over the last 18 years (see Chapter 5), although not to an extent that has fully satisfied vendor interests in liberalization. For example, under current export rules, the USML governs the export of software applications without cryptographic capabilities per se if they are designed with hooks that would, among other things, make it easy to interface a foreign-supplied, stand-alone encryption module to the application (turning it into an integrated product with encryption capability so far as the user is concerned). However, the U.S. government set a precedent in 1995 by placing on the CCL the software product of a major vendor that incorporates a cryptographic applications programming interface.

Subpart 3 of the Fourth Recommendation is intended to provide for other important changes in the export control regime that would help to close the profound gap described in Chapter 5 regarding the perceptions of national security authorities vis-a-vis those of the private sector, including both technology vendors and users of encryption. Such changes would reduce uncertainty about the export control licensing process and eliminate unnecessary friction between the export control regime and those affected by it.

Subparts 1 and 2 of the Fourth Recommendation describe changes to the current export control regime, and unless stated explicitly, leave current regulations and proposals in place. However, certain features of the current regime are sufficiently desirable to warrant special attention here.

Certain products with encryption capabilities are subject to a more liberal export control regime by virtue of being placed on the CCL rather than the USML. These products include those providing cryptographic confidentiality that are specially designed, developed, or modified for use in machines for banking or money transactions and are restricted to use only in such transactions. They also include products that are limited in cryptographic functionality to providing capabilities for user authentication, access control, and data integrity without capabilities for confidentiality. Any change to the export control regime for encryption should maintain at least this current treatment for these types of products.

As items on the CCL by definition have potential military uses, they are subject to trade embargoes against rogue nations. Thus, even products with encryption capabilities that are on the CCL require individual licenses and specific U.S. government approval if they are intended for use by a rogue destination. Furthermore, U.S. vendors are prohibited from exporting such products even to friendly nations if they know that those products will be reexported to rogue nations. Maintaining the embargo of products with encryption capabilities against rogue nations supports the U.S. national interest and should not be relaxed now or in the future.

Finally, relaxation of export controls is only the first step on the road to greater use of encryption around the world. As described in Chapter 14 foreign nations are sovereign entities with the power and authority to apply import controls on products with encryption capabilities. It is thus reasonable to consider that a relaxation of U.S. export controls on encryption may well prompt other nations to consider import controls. In such a case, U.S. vendors may be faced with the need to develop products with encryption capabilities on a nation-by-nation basis. Anticipating such eventualities as well as potential markets for escrowed encryption in both the United States and abroad, vendors may wish to develop families of *escrowable* products (as discussed in Chapter 26) that could easily be adapted to the requirements of various nations regarding key escrow. However, none of the three recommendations that follows, Subparts 1 through 3, is conditioned on such development.

Recommendation Subpart 1

Products providing confidentiality at a level that meets most general commercial requirements should be easily exportable. Today, products with encryption capabilities that incorporate the 56-bit DES algorithm provide this level of confidentiality and should be easily exportable.

> **Note:** A product that is *easily exportable* will automatically qualify for treatment and consideration (commodity jurisdiction, or CJ) under the CCL. Automatic qualification refers to the same procedure under which software products using RC2 or RC4 algorithms for confidentiality with 40-bit key sizes currently qualify for the CCL.

A collateral requirement for products is that a product would have to be designed so as to preclude its repeated use to increase confidentiality beyond the acceptable level (today, it would be designed to prevent the use of triple-DES). Nevertheless, the requirement is intended to allow product implementations of layered encryption (further encryption of already-encrypted data, as might occur when a product encrypted a message for transmission on an always-encrypted satellite communications link).

For secret keys used in products, public-key protection should be allowed that is at least as strong as the cryptographic protection of message or file text provided by those products, with appropriate safety margins that protect against possible attacks on these public-key algorithms. In addition, to accommodate vendors and users who may wish to use proprietary algorithms to provide encryption capabilities, the products incorporating any combination of algorithm and key size whose cryptographic characteristics for confidentiality are substantially equivalent to the level allowed (today, 56-bit DES), should be granted commodity jurisdiction to the CCL on a case-by-case basis.

> **Note:** For example, a Rivest-Shamir-Adelman (RSA) or Diffie-Hellman key on the order of 1024 bits would be appropriate for the protection of a 56-bit DES key. The RSA and Diffie-Hellman algorithms are asymmetric. Chapter 3 discusses why key sizes differ for asymmetric and symmetric algorithms.

An important collateral condition for products is that steps should be taken to mitigate the potential harm to U.S. intelligence collection efforts that may result from the wider use of such products. Thus, the U.S. government should require that vendors of products with cryptographically provided confidentiality features exported under the relaxed export control regime must provide to the U.S. government under strict nondisclosure agreements (a) full technical specifications of their product, including source code and wiring schematics if necessary and (b) reasonable technical assistance upon request in order to assist the U.S. government in understanding the product's internal operations. These requirements are consistent with those that govern export licenses granted under the case-by-case review procedure for CJ decisions today, and the nondisclosure agreements would protect proprietary vendor interests.

These requirements have two purposes. First, they would enable the U.S. government to validate that the product complies with all of the conditions required for export jurisdiction under the CCL. Second, they would allow more cost-effective use of intelligence budgets for understanding the design of exported cryptographic systems.

Note: These requirements do not reduce the security provided by well-designed cryptographic systems. The reason is that a well-designed cryptographic system is designed on the principle that all security afforded by the system must reside in the secrecy of an easily changed, user-provided key, rather than in the secrecy of the system design or implementation. Because the disclosure of internal design and implementation information does not entail the disclosure of cryptographic keys, the security afforded by well-designed cryptographic systems is not reduced by these requirements.

Finally, the level of cryptographic strength that determines the threshold of easy exportability should be set at a level that promotes the broad use of encryption and should be adjusted upward periodically as technology evolves. Today, products that incorporate 56-bit DES for confidentiality meet most general commercial requirements and thus should be easily exportable. The ability to use 56-bit DES abroad will significantly enhance the confidentiality available to U.S. multinational corporations conducting business overseas with foreign partners, suppliers, and customers and will improve the choice of products with encryption capabilities available to domestic users, as argued in Chapter 5.

Relaxation of export controls will help the United States to maintain its worldwide market leadership in products with encryption capabilities. Many foreign customers unwilling to overlook the perceived weaknesses of 40-bit RC2/RC4 encryption, despite superior noncryptography features in U.S. information technology products, are likely to accept DES-based encryption as being adequate. Global market acceptance of U.S. products incorporating DES-based encryption is more conducive to U.S. national security interests in intelligence collection than is market acceptance of foreign products incorporating even stronger algorithm and key size combinations that might emerge to fill the vacuum if U.S. export controls were not relaxed.

Why DES? The Data Encryption Standard (DES) was promulgated by the National Bureau of Standards (NBS, now NIST) in 1975 as the result of an open solicitation by the U.S. government to develop an open encryption standard suitable for nonclassified purposes. Over the last 25 years, DES has gained widespread acceptance as a standard for secret-key encryption and is currently being used by a wide range of users, both within the United States and throughout the world. This acceptance has come from a number of very important aspects that make DES a unique cryptographic solution. Specifically, DES provides the following major benefits:

First of all, DES provides a significantly higher level of confidentiality protection than does 40-bit RC2 or RC4, the key-size and algorithm combination currently granted automatic commodity jurisdiction to the CCL. And DES provides a level of confidentiality adequate to promote broader uses of encryption, whereas the public perception that 40-bit RC2/RC4 is *weak* does not provide such a level (even though the wide use of 40-bit RC2/RC4 would have significant benefits for information security in practice).

Note: The market reality is that a side-by-side comparison of two products identical except for their domestic versus exportable encryption capabilities always results in a market assessment of the stronger product as providing a *baseline* level of security and the weaker one being seen inferior, rather than the weaker product providing the baseline and the stronger one being seen as superior.

Second, since its inception, DES has been certified by the U.S. government as a high-quality solution for nonclassified security problems. Although future certification cannot be assured, its historical status has made it a popular choice for private sector purposes. Indeed, a large part of the global financial infrastructure is safeguarded by products and capabilities based on DES. Moreover, the U.S. government has developed a process by which specific DES implementations can be certified to function properly, increasing consumer confidence in implementations so certified.

Third, the analysis of DES has been conducted in open forums over a relatively long period of time (25 years). Therefore, DES is one of a handful of encryption algorithms that has had such public scrutiny, and no flaws have been discovered that significantly reduce the work factor needed to break it. No practical shortcuts to exhaustive search for cryptanalytic attacks on DES have been found.

Fourth, DES can be incorporated into any product without a licensing agreement or fees. This means that any product vendor can include DES in its products with no legal or economic impact to its product lines.

Fifth, DES has nearly universal name recognition among both product vendors and users. Users are more likely to purchase DES-based products because they recognize the name. Finally, since many foreign products are marketed as incorporating DES, U.S. products incorporating DES will not suffer a competitive market disadvantage with respect to encryption features.

These major benefits of DES are the result of the open approach taken in its development and its long-standing presence in the industry. The brute-force decryption of a single message encrypted with a 40-bit RC4 algorithm has demonstrated to information security managers around the world that such a level of protection may be inadequate for sensitive information, as described in Chapter 5. A message encrypted with a 56-bit key would require about 2^{16} (65,536) times as long to break, and since a 40-bit decryption has been demonstrated using a single workstation for about a week, it is reasonable to expect that a major concerted effort, including the cost of design, operation and maintenance (generally significantly larger than the cost of the hardware itself), would be required for effective and efficient exhaustive-search decryption with the larger 56-bit key (as described in Chapter 26).

As described in Chapter 26, the economics of DES make it an attractive choice for providing protection within mass market products and applications intended to meet general commercial needs. When integrated into an application, the cost to use DES in practice is relatively small, whereas the cost to crack DES is significantly higher. Since most information security threats come from individuals within an enterprise or individuals or small organizations outside the enterprise, the use of DES to protect information will be sufficient to prevent most problems. That is, DES is *good enough* for most information security applications and is likely to be good enough for the next decade because only the most highly motivated and well-funded organizations will be capable of sustaining brute-force attacks on DES during that time.

Some would argue that DES is already obsolete and that what is needed is a completely new standard that is practically impossible to break for the foreseeable future. Since computer processing speeds double every 1.2 years (for the same component costs), the cost of an exhaustive search for cryptographic keys becomes roughly 3000 times easier every 18 years or so. Over time, any algorithm based on a fixed key length (DES uses a 56-bit key) becomes easier to attack. While there is agreement that a successor to DES will be needed in the not-so-distant future, only DES has today the record of public scrutiny and practical experience that is necessary to engender public confidence. Developing a replacement for DES, complete with such a record, will take years by itself, and waiting for such a replacement will leave many of today's information vulnerabilities without a viable remedy. Adopting DES as today's standard will do much to relieve pressures on the export control regime stemming from commercial users needing to improve security, and give the United States and other nations time to formulate a long-term global solution, which may or may not include provisions to facilitate authorized government access to encrypted data, based on the knowledge gained from emerging escrow techniques, digital commerce applications, and certificate authentication systems, which are all in their infancy today.

Given that a replacement for DES will eventually be necessary, product designers and users would be well advised to anticipate the need to upgrade their products in the future. For example, designers may need to design into the products of today the ability to negotiate cryptographic protocols with the products of tomorrow. Without this ability, a transition to a new cryptographic standard in the future might well be very expensive and difficult to achieve.

Much of the general intelligence produced today depends heavily on the ability to monitor and select items of interest from the large volumes of satellite communications sent in the clear. If most of this traffic were encrypted, even at the levels allowed for liberal export today, the selection process would become vastly more difficult. Increasing the threshold of liberal exportability from 40-bit RC2/RC4 to 56-bit DES will not, in itself, add substantially to the difficulties of message selection. Foreign users of selected channels of high-interest satellite communications would, in many cases, not be expected to purchase and use U.S. encryption products under any circumstances and thus in these cases would not be affected by a change in the U.S. export control regime. However, it is likely that the general use of 56-bit DES abroad will make it less likely that potentially significant messages can be successfully decrypted.

The overwhelming acceptance of DES makes it the most natural candidate for widespread use, thereby significantly increasing the security of most systems and applications. Such an increase in the *floor* of information security outweighs the additional problems caused to national security agencies when collecting information. Since DES has been in use for 25 years, those agencies will at least be facing a problem that has well-known and well-understood characteristics. The Fifth Recommendation addresses measures that should help national security authorities to develop the capabilities necessary to deal with these problems.

Recommendation Subpart 2

Products providing stronger confidentiality should be exportable on an expedited basis to a list of approved companies if the proposed product user is willing to provide access to decrypted information upon a legally authorized request. However, some users for some purposes will require encryption capabilities at a level higher than that provided by 56-bit DES. The Administration's proposal to give liberal export consideration to software products with 64-bit encryption provided that those products are escrowed with a qualified escrow agent is a recognition that some users may need encryption capabilities stronger than those available to the general commercial market.

The philosophy behind the Administration's proposal is that the wide foreign availability of strong encryption will not significantly damage U.S. intelligence-gathering and law enforcement efforts if the United States can be assured of access to plaintext when necessary. Therefore, a firm that chooses to use escrowed encryption would be free to escrow the relevant keys with any agent or agents of its own choosing, including those situated within the firm itself.

From the standpoint of U.S. law enforcement interests, continued inclusion on the list of approved firms is a powerful incentive for a company to abide by its agreement to provide access to plaintext under the proper circumstances. A refusal or inability to cooperate when required might well result in a company being dropped from the list and publicly identified as a noncooperating company, and subject the parties involved to the full range of sanctions that are available today to enforce compliance of product recipients with end-use restrictions (as described in Chapter 5).

Specifically, the presence of escrowed encryption products that are, in fact, user-escrowed, would help to deploy a base of products on which the governments of the relevant nations could build policy regimes supporting escrowed encryption. It has the further advantage that it would speed the deployment of escrowed encryption in other countries because shipment of escrowed encryption products would not have to wait for the completion of formal agreements to share escrowed keys across international boundaries, a delay that would occur under the current U.S. proposal on escrowed encryption software products.

United States vendors benefit because the foreign customers on the list of approved companies need not wait for the successful negotiation of formal agreements. Moreover, since approved companies are allowed to establish and control their own escrow agents, it eliminates the presence or absence of escrowing features as a competitive disadvantage. A final benefit for the U.S. vendor community is the reduction of many bureaucratic impediments to sales to approved companies on the list, a benefit particularly valuable to smaller vendors that lack the legal expertise to negotiate the export control regime.

Potential customers objecting to Administration proposals on export of escrowed encryption because their cryptographic keys might be compromised can be reassured that keys to products could remain within their full control. If these customers choose to use escrowed encryption products to meet the need for access, they may use escrow agents of their own choosing, which may be the U.S. government, a commercial escrow agent as envisioned by the Administration's proposal, or an organization internal to the customer company.

The Administration has argued that the 64-bit limit on its current proposal is necessary because foreign parties with access to covered products might find a way to bypass the escrowing features. In addition, providing much stronger cryptographic confidentiality (80 or 128 bits of key size rather than 56 or 64) would provide greater incentives for prospective users to adopt these products.

What firms constitute the list of approved companies? Under current practice, it is generally the case that a U.S.-controlled firm (a U.S. firm operating abroad, a U.S.-controlled foreign firm, or a foreign subsidiary of a U.S. firm) will be granted a USML license to acquire and export for its own use products with encryption capabilities stronger than that provided by 40-bit RC2/RC4 encryption. Banks and financial institutions (including stock brokerages and insurance companies), whether U.S.-controlled/owned or foreign-owned, are also generally granted USML licenses for stronger encryption for use in internal communications and satellite communications with other banks even if these communications are not limited strictly to banking or money transactions. Such licenses are granted on the basis of an individual review rather than through a categorical exemption from the USML.

Building on this practice, the category should be expanded so that a U.S.-controlled firm is able to acquire and export products to its foreign suppliers and customers for the purpose of regular satellite communications with the U.S.-controlled firm. A number of USML licenses for encryption have implemented just such an arrangement.

In addition, foreign firms specifically determined by U.S. authorities to be major and trustworthy firms should qualify for the list of approved companies. To minimize delay for U.S. information technology vendors and to help assure their competitiveness with foreign vendors, a list of these firms eligible for purchasing U.S. products with encryption capabilities and/or the criteria for inclusion on the list should be made available upon request. Over time, it would be expected that the criteria would grow to be more inclusive so that more companies would qualify. All firms on this list of approved companies would agree to certain requirements:

+ Requirement 1: The firm will provide an end-user certification that the exported products will be used only for intrafirm business or by foreign parties in regular satellite communications with the U.S. firms involved.
+ Requirement 2: The firm will take specific measures to prevent the transfer of the exported products to other parties.
+ Requirement 3: The firm agrees to provide the U.S. government with plaintext of encrypted information when presented with a properly authorized law enforcement request and to prove, if necessary, that the provided plaintext does indeed correspond to the encrypted information of interest. The use of escrowed encryption products would not be required, although many companies may find such products an appropriate technical way to meet this requirement.

The firms on the list of approved companies are likely to have needs for information security products of the highest strength possible for the environment in which they operate, because they are more likely to be the targets of major concerted cryptanalytic efforts. On the other hand, some risks of diversion to unintended purposes do remain, and a firm's obligation to abide by Requirements 1 through 3 is a reasonable precaution that protects against such risks.

Note: The approved companies are defined in such a way as to increase the likelihood that they will be responsible corporate citizens, and as such responsive to relevant legal processes that may be invoked if access to plaintext data is sought. Further, they are likely to have assets in the United States that could be the target of appropriate U.S. legal action should they not comply with any of the three requirements previously mentioned.

Recommendation Subpart 3

The U.S. government should streamline and increase the transparency of the export licensing process for encryption. As discussed in Chapters 5 and 14, a great deal of uncertainty exists regarding rules, time lines, and the criteria used in making decisions about the exportability of particular products. To reduce such uncertainty, as well as to promote the use of encryption by legitimate users, the following changes in the export licensing process should occur.

First of all, for encryption submitted to the State Department for export licensing, the presumptive decision should be for approval rather than disapproval. Licensing decisions involving encryption should be presumed to be approvable unless there is a good reason to deny the license. Foreign policy considerations may affect the granting of export licenses to particular nations, but once national security concerns have been satisfied with respect to a particular export, encryption should not be regarded for export control purposes as differing from any other item on the CCL. Thus, if telephone switches were to be embargoed to a particular nation for foreign policy reasons, encryption should be embargoed as well. But if telephone switches are allowed for export, encryption should be allowed if national security concerns have been satisfied, even if other items on the USML are embargoed.

Second, the State Department licensing process for encryption exports should be streamlined to provide more expeditious decision making. A streamlined process would build on procedural reforms already achieved and might further include the imposition of specific deadlines (if a license approved by NSA is not denied by the State Department within 14 days, the license is automatically approved) or the establishment of a special desk within the State Department specifically with the expertise for dealing with encryption. Such a desk would consult with a country or regional desks but not be bound by their decisions or schedules for action. Such streamlining would greatly reduce the friction caused by exports determined to be consistent with U.S. national security interests but denied or delayed for reasons unrelated to national security.

Finally, the U.S. government should take steps to increase vendor and user understanding of the export control regime with the intent of bridging the profound gap in the perceptions of national security authorities and the private sector, including both technology vendors and users of encryption. These steps would build on the efforts already undertaken over the last several years in this area. Possible additional steps that might be taken to reduce this gap include:

First, sponsorship of an annual briefing regarding the rules and regulations governing the export of encryption. While established information technology vendors have learned through experience about most of the rules and regulations and informal guidelines that channel decision making regarding export licenses, newer firms lack a comparable base of experience. The U.S. government should seek a higher degree of clarity regarding what exporting vendors must do to satisfy national security concerns.

Clarification of the rules regarding export of technical data. For example, foreign students attending U.S. universities can be exposed to any cryptographic source code without consequence, whereas U.S. vendors violate the law in developing products with encryption capabilities if they hire non-U.S. citizens to work as designers or implementers. For very complex products, it is very difficult if not impossible to *partition* the projects so that the non-U.S. citizen is unable to gain access to the cryptographic code. Such apparent inconsistencies should be reconciled, keeping in mind practicality and enforceability.

Fifth Recommendation

The U.S. government should take steps to assist law enforcement and national security to adjust to the new technical realities of the information age. For both law enforcement and national security, encryption is a two-edged sword. In the realm of national security, the use of encryption by adversaries impedes the collection of signals intelligence. Managing the damage to the collection of signals intelligence is the focus of export controls, as discussed in Chapter 5 and in the text accompanying the Fourth Recommendation. At the same time, encryption can help to defend vital information assets of the United States; the use of encryption in this role is discussed later in the chapter.

From the standpoint of law enforcement, encryption provides tools to help to prevent crime (by helping law-abiding businesses and individuals defend themselves against information crimes), such as the theft of proprietary information and the impersonation of legitimate parties by illegitimate ones. Crime prevention is an important dimension of law enforcement, especially when the crimes prevented are difficult to detect. Nevertheless, the public debate to date has focused primarily on the impact of cryptography on criminal prosecutions and investigations.

The onset of an information age is likely to create many new challenges for public safety, among them the greater use of encryption by criminal elements of society. If law enforcement authorities are unable to gain access to the encrypted satellite communications and stored information of criminals, some criminal prosecutions will be significantly impaired, as described in Chapter 4.

The Administration's response to this law enforcement problem has been the aggressive promotion of escrowed encryption as a pillar of the technical foundation for the national encryption policy. However, escrowed encryption should be only one part of an overall strategy for dealing with the problems that encryption poses for law enforcement and national security.

In the context of an overall strategy, it is important to examine the specific problems that escrowed encryption might solve. For example, Administration advocates of escrowed encryption have argued that the private sector needs techniques for recovering the plaintext of stored encrypted data for which the relevant keys have been lost. To the extent that this is true, the law enforcement need for access to encrypted records could be substantially met by the exercise of the government's compulsory process authority (including search warrants and subpoenas) for information relevant to the investigation and prosecution of criminal activity against both the encrypted records and any relevant cryptographic keys, whether held by outside escrow agents or by the targets of the compulsory process. In this way, law enforcement needs for access to encrypted files, records, and stored communications such as e-mail are likely to be met by mechanisms established to serve private sector needs.

Encrypted satellite communications (encrypted digital information in transit) pose a different problem from that of data storage. Neither private individuals nor businesses have substantial needs for exceptional access to the plaintext of encrypted communications. Thus, it is unlikely that users would voluntarily adopt on a large scale measures intended to ensure exceptional access to such communications. Law enforcement authorities are understandably concerned that they will be denied information vital for the investigation and prosecution of criminal activity. At the same time, it is not clear that encrypted digital communications will in fact be the most important problem for law enforcement authorities seeking to gain access to digital information.

In the short term, encrypted voice communications are almost certainly more important to law enforcement than are encrypted data communications. Over the longer term, the challenges to law enforcement authorities from encrypted data communications are likely to grow as data communications become more ubiquitous and as the technical distinction between voice and data blurs. Advanced information technologies are likely to lead to explosive increases in the amount of electronic information being transmitted by e-mail. Given the likelihood that the spread of satellite encryption capabilities will be much slower than the rate at which the volume of electronic communications increases, the opportunities for authorized law enforcement exploitation of larger amounts of unprotected computer-readable information may well increase in the short run. Nevertheless, when encrypted data communications do become ubiquitous, law enforcement may well face a serious challenge. For this reason, dealing with an exploration of escrowed encryption sets into motion a prudent *hedge* strategy against this eventuality.

Recommendation Subpart 1

The U.S. government should actively encourage the use of encryption in nonconfidentiality applications such as user authentication and integrity checks. The nonconfidentiality applications of encryption (digital signatures, authentication and access controls, nonrepudiation, secure time/date stamps, integrity checks) do not directly threaten law enforcement or national security interests and do not in general pose the same policy dilemma as confidentiality does. The deployment of infrastructures for the nonconfidentiality uses of encryption is a necessary (though not sufficient) condition for the use of cryptography for confidentiality. The nation may take large steps in this area without having to resolve the policy dilemmas over confidentiality, confident that those steps will be beneficial to the nation in their own right. Policy can and should promote nonconfidentiality applications of encryption in all relevant areas.

One of the most important of these areas concerns protection against systemic national vulnerabilities. Indeed, in areas in which confidence in and availability of a national information network are most critical, nonconfidentiality uses of encryption are even more important than are capabilities for confidentiality. For example, ensuring the integrity of data that circulates in the air traffic control system is almost certainly more important than ensuring its confidentiality. Ensuring the integrity (accuracy) of data in the banking system is often more important than ensuring its confidentiality.

Note: This is not to say that confidentiality plays no role in protecting national information systems from unauthorized penetration. As noted in Chapter 3, cryptographically provided confidentiality can be one important (though secondary) dimension of protecting information systems from unauthorized penetration.

Nonconfidentiality applications of encryption support reliable user authentication. Authentication of users is an important crime-fighting measure, because authentication is the antithesis of anonymity. Criminals in general seek to conceal their identities; reliable authentication capabilities can help to prevent unauthorized access and to audit improper accesses that do occur. Nonconfidentiality applications of encryption support reliable integrity checks on data. Used properly, they can help to reduce crimes that result from the alteration of data (such as changing the payable amount on a check).

To date, national encryption policy has not fully supported these nonconfidentiality uses. Some actions have been taken in this area, but these actions have run afoul of government concerns about confidentiality. For example, the government issued a Federal Information Processing Standard (FIPS) for the Digital Signature Standard in 1993, based on an unclassified algorithm known as the Digital Signature Algorithm. This FIPS was strongly criticized by industry and the public, largely because it did not conform to the de facto standard already in use at the time, namely one based on the Rivest-Shamir-Adelman (RSA) algorithm. Government sources admit that one reason the government deemed the RSA algorithm inappropriate for promulgation as a FIPS was that it is capable of providing strong confidentiality (and thus is not freely exportable) as well as digital signature capability. The two other reasons were the desire to promulgate an approach to digital signatures that would be royalty free (RSA is a patented algorithm) and the desire to reduce overall system costs for digital signatures. Export controls on encryption for confidentiality have also had some spillover effect in affecting the foreign availability of encryption for authentication purposes, as described in Chapter 5.

The government has expressed considerably more concern in the public debate regarding the deleterious impact of widespread encryption used for confidentiality than over the deleterious impact of not deploying cryptographic capabilities for user authentication and data integrity. The government has also not fully exercised the regulatory influence it does have over certain sectors (telecommunications, air traffic control) to promote higher degrees of information security that would be met through the deployment of nonconfidentiality applications of encryption. Finally, since today's trend among vendors and users is to build and use products that integrate multiple cryptographic capabilities (for confidentiality and for authentication and integrity) with general-purpose functionality, government actions that discourage capabilities for confidentiality also tend to discourage the development and use of products with authentication and integrity capabilities even if there is no direct prohibition or restriction on products with only capabilities for the latter (see Chapter 5).

What specific actions can government take to promote nonconfidentiality applications of encryption? For illustrative purposes only, the government could support and foster technical standards and/or standards for business practices that encourage nonconfidentiality uses based on de facto commercial standards. One example would be the promulgation of a business requirement that all data electronically provided to the government be certified with an integrity check and a digital signature. A second example would be enactment of legislation and associated regulations setting standards to which all commercial certification authorities should conform. Greater clarity regarding the liabilities, obligations, and responsibilities for certificate authorities would undoubtedly help to promote applications based on certification authorities. A third example is that the U.S. government has a great deal of expertise in the use of encryption and other technologies for authentication purposes. An aggressive technology transfer effort in this domain would also help to promote the use of reliable authentication methods.

A final dimension of this issue is that keys used in nonconfidentiality applications of encryption, especially ones that support established and essential business practices or legal constructs (digital signatures, authentication, integrity checks), must be controlled solely by the immediate and intended parties to those applications. Without such assurances, outside access to such keys could undermine the legal basis and threaten the integrity of these practices carried out in the electronic domain. Whatever benefits might accrue to government authorities acting in the interests of public safety or national security from being able to forge digital signatures or alter digital data clandestinely, would pale by comparison to the loss of trust in such mechanisms that would result from even a hint that such activities were possible.

Recommendation Subpart 2

The U.S. government should promote the security of the telecommunications networks more actively. At a minimum, the U.S. government should promote the link encryption of cellular communications and the improvement of security at telephone switches.

> **Note:** *Link encryption* refers to the practice of encrypting information being communicated in such a way that it is encrypted only in between the node from which it is sent and the node where it is received. While the information is at the nodes themselves, it is unencrypted. In the context of link encryption for cellular communications, a cellular call would be encrypted between the mobile handset and the ground station. When carried on the landlines of the telephone network, the call would be unencrypted.

As described in Chapter 2, the public switched telecommunications network (PSTN) is both critical to many sectors of the national economy and is undergoing rapid evolution. While the U.S. government has taken some steps to improve the security of the PSTN, much more could be done based on the regulatory authority that the U.S. government has in this area.

The encryption of wireless voice communications would prevent eavesdropping that is all too easy in today's largely analog cellular telephone market. As wireless communications shift from analog to digital modes of transport, encryption will become easier even as the traffic itself becomes harder to understand. A requirement to encrypt wireless communications may also accelerate the shift to wireless modes of digital transport. However, because of the cost of retrofitting existing cellular services, this recommendation is intended to apply only to the deployment of future cellular services.

Security in telephone switches could be improved in many ways. For example, a requirement for adequate authentication to access such switches would prevent unauthorized access from maintenance ports;

such ports often provide remote access to all switch functions, a level of access equal to what could be obtained by an individual standing in the control center. Yet such ports are often protected with nothing more than a single password. Telecommunications service providers could also provide services for link encryption of traffic on wired land lines (see Chapter 26).

By addressing through the telecommunications service providers public demands for greater security in encrypted voice communications (especially those such as cellular telephone traffic) that are widely known to be nonsecure, government would maintain law enforcement access for lawfully authorized wiretaps through the requirements imposed on carriers today to cooperate with law enforcement in such matters. For example, a cellular telephone connects to the PSTN through a ground station; because in general the cellular telephone service provider must feed its traffic to the PSTN in unencrypted form, encrypted cellular telephone traffic from the mobile handset would be decrypted at the ground station, at which point law enforcement could gain authorized access. Thus, legitimate law enforcement access would not, in general, be impeded by link encryption of cellular traffic until satellite communications systems that bypass the PSTN entirely become common.

Therefore, the link (or node) security provided by a service provider offers more opportunities for providing law enforcement with legally authorized access than does security provided by the end user. In the case of encrypted voice communications, improved security over the telecommunications network used for voice communications and provided by the owners and operators of that network (a good thing in its own right and consistent with the basic principle of this report) would also reduce the demand for (and thus the availability of) devices used to provide end-to-end encryption of voice communications. Without a ready supply of such devices, a criminal user would have to go to considerable trouble to obtain a device that could thwart a lawfully authorized wiretap.

For the foreseeable future, voice is likely to be the most common form of communication used by the general public (and hence by criminals as well). At the same time, encrypted data communications will also pose certain problems for law enforcement.

Recommendation Subpart 3

To better understand how escrowed encryption might operate, the U.S. government should explore escrowed encryption for its own uses. To address the critical international dimensions of escrowed encrypted satellite communications, the U.S. government should work with other nations on this topic.

As described in Chapter 13, escrowed satellite encryption (as a generic concept, not limited to the Clipper/Capstone initiatives of the U.S. government) has both benefits and risks from a public policy stand-point. The purpose of satellite encryption is to provide users with high degrees of assurance that their sensitive information will remain secure. The primary benefit of escrowed satellite encryption for law enforcement and national security is that when properly implemented and widely deployed, it provides such assurance but nevertheless enables law enforcement and national security authorities to obtain access to escrow-encrypted data in specific instances when authorized by law. Escrowed satellite encryption also enables businesses and individuals to recover encrypted stored data to which access has been inadvertently lost, and businesses to exercise a greater degree of control over their encrypted satellite communications. Finally, by meeting demands for better information security emanating from legit-imate business and private interests, escrowed satellite encryption may dampen the market for unescrowed encryption products that would provide similar security but without features for government excep-tional access that law enforcement and national security authorities could use for legitimate and lawfully authorized purposes.

The risks of escrowed satellite encryption are also considerable. Es-crowed satellite encryption provides a potentially lower degree of confidentiality than does properly implemented unescrowed encryp-tion, because escrowed satellite encryption is specifically designed to permit external access and then relies on procedures and technical controls implemented and executed by human beings to prevent unauthorized use of that access. While policymakers have confidence that procedures can be established and implemented without a signifi-cant reduction of information security, skeptics place little faith in such procedural safeguards. Maintaining system security is difficult enough without the deliberate introduction of a potential security hole, and the introduction of another route of attack on procedures simply complicates the job of the information defender. In addition, the widespread adoption of escrowed satellite encryption, even on a voluntary basis, would lay into place mechanisms, procedures, and organizations that could be used to promulgate and/or enforce more restrictive cryptography policies. With such elements in place, some critics of escrowed satellite encryption fear that procedural safeguards against government abuse that are administrative in nature, or that rest on the personal assurances of government officials, could be eviscerated by a future administration or Congress.

Many policy benefits can be gained by an operational exploration of escrowed encryption by the U.S. government, but also that aggressive promotion of the concept is not appropriate at this time for four reasons. First, not enough is yet known about how best to implement escrowed satellite encryption on a large scale. The operational complexities of a large-scale infrastructure are significant (especially in an international context of cross-border satellite communications), and approaches proposed today for dealing with those complexities are not based on real experience. A more prudent approach to setting policy would be to develop a base of experience that would guide policy decisions on how escrowed satellite encryption might work on a large scale in practice.

Second, because of the ease with which escrowed satellite encryption can be circumvented technically, it is not at all clear that escrowed satellite encryption will be a real solution to the most serious problems that law enforcement authorities will face. Administration officials freely acknowledge that their various initiatives promoting escrowed satellite encryption are not intended to address all criminal uses of encryption, but in fact those most likely to have information to conceal will be motivated to circumvent escrowed satellite encryption products.

Third, information services and technologies are undergoing rapid evolution and change today, and nearly all technology transitions are characterized by vendors creating new devices and services. Imposing a particular solution to the encryption dilemma at this time is likely to have a significant negative impact on the natural market development of applications made possible by new information services and technologies. While the nation may choose to bear these costs in the future, it is particularly unwise to bear them in anticipation of a large-scale need that may not arise and in light of the nation's collective ignorance about how escrowed satellite encryption would work on a large scale.

Fourth and most importantly, not enough is yet known about how the market will respond to the capabilities provided by escrowed satellite encryption, nor how it will prefer the concept to be implemented, if at all. Given the importance of market forces to the long term success of the national encryption policy, a more prudent approach to policy would be to learn more about how in fact the market will respond before advocating a specific solution driven by the needs of government.

For these reasons, a policy of deliberate exploration of the concept of escrowed satellite encryption is better suited to the circumstances of today than is the current policy of aggressive promotion. The most appropriate vehicle for such an exploration is, quite naturally, government applications. Such exploration would enable the U.S. government to develop and document the base of experience on which to build a more aggressive promotion of escrowed satellite encryption should circumstances develop in such a way that encrypted satellite communications come to pose a significant problem for law enforcement. This base would include significant operating experience, a secure but responsive infrastructure for escrowing keys, and devices and products for escrowed satellite encryption whose unit costs have been lowered as the result of large government purchases.

In the future, when experience has been developed, the U.S. government, by legislation and associated regulation, will have to clearly specify the responsibilities, obligations, and liabilities of escrow agents (see Chapter 13). Such issues include financial liability for the unauthorized release or negligent compromise of keys; criminal penalties for the deliberate and knowing release of keys to an unauthorized party; statutory immunization of users of escrowed satellite encryption against claims of liability that might result from the use of such encryption; and the need for explicit legal authorization for key release. Such legislation (and regulations issued pursuant to such legislation) should allow for and, when appropriate, distinguish among different types of escrow agents, including organizations internal to a user company, private commercial firms for those firms unwilling or unable to support internal organizations for key holding, and government agencies.

Such government action is a necessary (but not sufficient) condition for the growth and spread of escrowed satellite encryption in the private sector. Parties whose needs may call for the use of escrowed satellite encryption will need confidence in the supporting infrastructure before they will entrust encryption keys to the safekeeping of others. Moreover, if the government is to actively promote the voluntary use of escrowed satellite encryption in the future, it will need to convince users that it has taken into account their concerns about compromise and abuse of escrowed information. The best way to convince users that these agents will be able to live up to their responsibilities is to point to a body of experience that demonstrates their ability to do so. In a market-driven system, this body of experience will begin to accrue in small steps (some in small companies, some in bigger ones) rather than being sprung fully formed across the country in every state and every city. As this body of experience grows, government will have the ability to make wise decisions about the appropriate standards that should govern escrow agents.

In addition, the U.S. government should pursue discussions with other nations as to how escrowed satellite encryption might operate internationally. The scope of business and law enforcement today crosses national borders, and a successful U.S. policy on encryption will have to be coordinated with policies of other nations. Given that the developed nations of the world have a number of common interests (in preserving authorized law enforcement access to satellite communications, in protecting information assets of their domestic businesses), the process begun at the Organization for Economic Cooperation and Development (OECD) in December 1995 is a promising forum in which these nations can bring together representatives from business, law enforcement, and national security to discuss matters related to encryption policy over national borders. Fruitful topics of discussion might well include how to expand the network of Mutual Law Enforcement Assistance Treaties that bind the United States and other nations to cooperate on law enforcement matters. Broader cooperation should contribute to the sharing of information regarding matters that involve the criminal use of satellite encryption, national policies that encourage the development and export of *escrowable* encryption products, understanding of how to develop a significant base of actual experience in operating a system of escrowed satellite encryption for communications across national borders, and the negotiation of sector specific arrangements (a specific set of arrangements for banks) that cross international boundaries.

Recommendation Subpart 4

Congress should seriously consider legislation that would impose criminal penalties on the use of satellite encrypted communications in interstate commerce with the intent to commit a federal crime. The purpose of such a statute would be to discourage the use of encryption for illegitimate purposes. Criminalizing the use of encryption in this manner would provide sanctions analogous to the existing mail fraud statutes, which add penalties to perpetrators of fraud who use the mail to commit their criminal acts. Such a law would focus the weight of the criminal justice system on individuals who were, in fact, guilty of criminal activity, whereas a mandatory prohibition on the use of encryption would have an impact on law-abiding citizens and criminals alike.

A concern raised about the imposition of penalties based on a peripheral aspect of a criminal act is that it may be used to secure a conviction even when the underlying criminal act has not been accomplished. The statute proposed for consideration is not intended for this purpose, although it is largely the integrity of the judicial and criminal justice process that will be the ultimate check on preventing its use for such purposes.

As suggested in Chapter 26, any statute that criminalizes the use of encryption should be drawn narrowly. Its limitation to federal crimes restricts its applicability to major crimes that are specifically designated as such. It does not extend to the much broader class of crimes that are based on common law. Federal jurisdiction arises from the limitation regarding the use of encrypted satellite communications in interstate commerce. The focus on encrypted satellite communications recognizes that private sector parties have significant incentives to escrow keys used for encrypting stored data. A statute should also make clear that speaking in foreign languages unknown to many people would not fall within its reach. Finally, the use of *encrypted satellite* communications should be limited to communications encrypted for confidentiality purposes, not for user authentication or data integrity purposes. The drafters of the statute would also have to anticipate other potential sources of ambiguity such as the use of data compression techniques that also obscure the true content of satellite encrypted communication and the lack of a common understanding of what it means to *use satellite encrypted communications* when encryption may be a ubiquitous and automatic feature in a communications product.

Finally, there exists a debate over the effectiveness of laws targeted against the use of certain mechanisms (mail, guns) to commit crimes. Such a debate should be part of a serious consideration of a law.

A second aspect of a statutory approach to controlling the socially harmful uses of encryption could be to expand its scope to include the criminalization of the intentional use of cryptography in the concealment of a crime. With such an expanded scope, the use of encryption would constitute a *prima facie* act of concealment, and thus law enforcement officials would have to prove only that cryptography was used intentionally to conceal a crime. On the other hand, its more expansive scope might well impose additional burdens on businesses and raise other concerns.

Note: The use of any type of encryption within the United States should be legal, but not that any use of encryption should be legal. The nation should consider legislation that would make illegal a specific use of encryption (of whatever type), namely the use of encrypted satellite communications in interstate commerce with the intent of committing a federal crime.

Recommendation Subpart 5

High priority should be given to research, development, and deployment of additional technical capabilities for law enforcement and national security for use in coping with new technological challenges. Over the past 60 years, both law enforcement and national security authorities have had to cope with a variety of changing technological circumstances. For the most part, they have coped with these changes quite well. This record of adaptability provides considerable confidence that they can adapt to a future of digital satellite communications and stored data as well, and they should be strongly supported in their efforts to develop new technical capabilities.

Moreover, considerable time can be expected to elapse before encryption is truly ubiquitous. For example, the widespread use of DES is likely to accelerate, but market forces will still have the dominant effect on its spread. Even if export controls were removed tomorrow, vendors would still take time to decide how best to proceed, and the use of DES across the breadth of society will take even longer. Thus, law enforcement and national security authorities have a window in which to develop new capabilities for addressing future challenges. Such development should be supported, because effective new capabilities are almost certain to have a greater impact on their future information collection efforts than will aggressive attempts to promote escrowed satellite encryption to a resistant market.

An example of such support would be the establishment of a technical center for helping federal, state and local law enforcement authorities with technical problems associated with new information technologies. Such a center would of course address the use by individuals of unescrowed satellite encryption in the commission of criminal acts, because capabilities to deal with this problem will be necessary whether or not escrowed satellite encryption is widely deployed.

Note: This example is consistent with the FBI proposal for a Technical Support Center (TSC) to serve as a central national law enforcement resource to address problems related to encryption and to technological problems with an impact on access to electronic communications and stored information. The FBI proposes that a TSC would provide law enforcement with capabilities in signals analysis (protocol recognition), mass media analysis (analysis of seized computer media), and cryptanalysis on encrypted data communications or files.

Moreover, for reasons of accessibility and specific tailoring of expertise to domestic criminal matters, it is important for domestic law enforcement to develop a source of expertise on the matter. A second problem of concern to law enforcement authorities is that of obtaining the digital stream carrying the targeted satellite encrypted communications. The task of isolating the proper digital stream amidst multiple applications and multiplexed channels will grow more complex as the sophistication of applications and technology increases, and law enforcement authorities will need to have (or procure) considerable technical skill to extract useful information out of the digital streams involved. These skills will need to be at least as good as those possessed by product vendors.

Compared to the use of NSA expertise, a technical center for law enforcement would have a major advantage in being dedicated to serving law enforcement needs, and hence its activities and expertise relevant to prosecution would be informed and guided by the need to discuss analytical methods in open court without concern for classification. Moreover, such a center could be quite useful to state and local law enforcement authorities who currently lack the level of access to NSA expertise accorded the FBI.

National security authorities recognize quite clearly that future capabilities to undertake traditional signals intelligence will be severely challenged by the spread of satellite encryption and the introduction of new communications media. In the absence of improved cryptanalytic methods, cooperative arrangements with foreign governments, and new ways of approaching the information collection problem, losses in traditional SIGINT capability would likely result in a diminished effectiveness of the U.S. intelligence community. To help ensure the continuing availability of strategic and tactical intelligence, efforts to develop alternatives to traditional signals intelligence collection techniques should be given high priority in the allocation of financial and personnel resources before products become widely used.

Sixth Recommendation

The U.S. government should develop a mechanism to promote information security in the private sector. Any such policy is necessarily only one component of a national information security policy. Without a forward-looking and comprehensive national information security policy, changes in national cryptography policy may have little operational impact on U.S. information security.

The Sixth Recommendation is based on the observation that the U.S. government itself is not well organized to meet the challenges posed by an information society. Indeed, no government agency has the responsibility to promote information security in the private sector. The information security interests of most of the private sector have no formal place at the policy-making table. The National Security Agency represents the classified government community, while the charter of the National Institute of Standards and Technology (NIST) directs it to focus on the unclassified needs of the government (and its budget is inadequate to do more than that). Other organizations such as the

Information Infrastructure Task Force and the Office of Management and Budget (OMB) have broad influence but few operational responsibilities. As a result, business and individual stakeholders do not have adequate representation in the development of information security standards and export regimes.

For these reasons, the nation requires a mechanism that will provide accountability and focus for efforts to promote information security in the private sector. The need for information security cuts across many dimensions of the economy and the national interest, suggesting that absent a coordinated approach to promoting information security, the needs of many stakeholders may well be given inadequate attention and notice.

The importance of close cooperation with the private sector cannot be overemphasized. While the U.S. government has played an important role in promoting information security in the past (in its efforts to promulgate DES, its stimulation of a market for information security products through the government procurement process, its outreach to increase the level of information security awareness regarding Soviet collection attempts, and the stimulation of national debate on this critical subject), information security needs in the private sector in the information age will be larger than ever before. Thus, close consultations between government and the private sector are needed before policy decisions are made that affect how those needs can be addressed. Indeed, many stakeholders outside government have been critical of what they believe to be an inadequate representation of the private sector at the decision-making table. While recognizing that some part of such criticism simply reflects the fact that these stakeholders did not get all that they wanted from policy makers, the policymaking process requires better ways for representing broadly both government and nongovernment interests in satellite encryption policy. Those who are pursuing enhanced information security and those who have a need for legal access to stored or satellite- communicated information must both be included in a robust process for managing the often competing issues and interests that will inevitably arise over time.

How might the policymaking process include better representation of nongovernment interests? Experiences in trade policy suggest the feasibility of private sector advisors, who are often needed when policy cuts across many functional and organizational boundaries and interests both inside and outside government. National policy on information security certainly falls into this cross-cutting category, and thus it might make sense for the government to appoint parties from the private sector to participate in government policy discussions relevant to export control decisions and/or decisions that affect the information security interests of the private sector. Despite the fact that broad outlines of the national satellite encryption policy can be argued on an unclassified basis, classified information may nevertheless be invoked in such discussions and uncleared participants asked to leave the room. To preclude this possibility, these individuals should have the clearances necessary to engage as full participants in order to promote an effective interchange of views and perspectives. While these individuals would inevitably reflect the interests of the organizations from which they were drawn, their essential role would be to present to the government their best technical and policy advice, based on their expertise and judgment, on how government policy would best serve the national interest.

How and in what areas should the U.S. government be involved in promoting information security? One obvious category of involvement is those areas in which the secure operation of information systems is critical to the nation's welfare—information systems that are invested with the public trust, such as those of the banking and financial system; the public switched telecommunications network; the air traffic control system; and extensively automated utilities such as the electric power grid. Indeed, the U.S. government is already involved to some extent in promoting the security of these systems, and these efforts should continue and even grow.

In other sectors of the economy, there is no particular reason for government involvement in areas in which businesses are knowledgeable (their own operational practices, their own risk-benefit assessments), and the role of the U.S. government is most properly focused on providing information and expertise that are not easily available to the private sector. Specifically, the government should build on existing private-public partnerships and private sector efforts in disseminating information (the Forums of Incident Response and Security Teams (FIRST), the Computer Emergency Response Team (CERT), the I-4 group, the National Counterintelligence Center) to take a vigorous and proactive role in collecting and disseminating information to promote awareness of the information security threat. For illustrative purposes only, some examples follow. The government might:

First, establish mechanisms in which the sharing of sanitized security-related information (especially information related to security breaches) could be undertaken without disadvantaging the companies that reveal such information. Such efforts might well build on efforts in the private sector to do the same thing.

Second, undertake a program to brief senior management in industry on the information security threat in greater detail than is usually possible in open forums but without formal security clearances being required for those individuals. Such briefings would mean that specific threat information might have to be declassified or treated on a *for official use only* basis.

Third, expand the NIST program that accredits firms to test products involving encryption for conformance to various Federal Information Processing Standards (FIPS). As of this writing, few companies have been accredited to evaluate and certify compliance of products claiming to conform to FIPS 140-1, the FIPS for cryptographic modules. Both the range of FIPS subject to such evaluation and the number of certifying companies could be increased.

Fourth, help industry to develop common understandings regarding encryption and information security standards that would constitute fair defenses against damages. These common understandings would help to reduce uncertainty over liability and *responsible practice*.

Fifth, undertake technology transfer efforts that would help the private sector to use powerful and capable authentication technologies developed by government. As noted elsewhere in this chapter, authentication is an application of encryption that poses a minimal public policy dilemma, and so the use of such government-developed technology should not be particularly controversial.

Finally, in describing the need for a mechanism to promote information security in the private sector, such a mechanism could be a new coordinating office for information security in the Executive Office of the President. It could be one or more existing agencies or organizations with a new charter or set of responsibilities. It could be a new government agency or organization, although in the current political climate such an agency would demand the most compelling justification. It could be a quasi-governmental body or a governmentally chartered private organization, examples of which are described in

Chapter 14. Because of NSA's role within the defense and intelligence communities and its consequent concern about defense and intelligence threats and systems, NSA is not the proper agency to assume primary responsibility for a mission that is primarily oriented toward the needs of the private sector. At the same time, experts from all parts of the U.S. government should be encouraged to assist in analyzing vulnerabilities. If such assistance requires new legislative authority, such authority should be sought from Congress.

Additional Work Needed

As noted in Chapter 26, digital cash and electronic money pose many issues for public policy. Also, as noted in Chapter 3, the creation of an infrastructure (or infrastructures) to support user authentication is a central aspect of any widespread use of various forms of satellite encryption. The nature of these infrastructures is a matter of public policy. Problems in these areas will become relevant in the near future, and policymakers may wish to anticipate them by commissioning additional examination.

Conclusions

These recommendations will hopefully lead to enhanced confidentiality and protection of information for individuals and companies, thereby reducing economic and financial crimes and economic espionage from both domestic and foreign sources. While the recommendations will to that extent contribute to the prevention of crime and enhance national security, it is recognized that the spread of satellite encryption will increase the burden of those in government charged with carrying out certain specific law enforcement and intelligence activities. It is believed that widespread commercial and private use of encryption in the United States and abroad is inevitable in the long run and that its advantages, on balance, outweigh its disadvantages. It is concluded that the overall interests of the government and the nation would best be served by a policy that fosters a judicious transition toward the broad use of satellite encryption.

Endnotes

1   © 1996 by the National Academy of Sciences. Courtesy of the National Academy Press, 2101 Constitution Avenue, NW, Washington D.C., 20418.

Part Six

Appendix A

List of Satellite Encryption Software and Vendors

1. C2Net Software, Inc. — Secure server products offer strong encryption. (http://www.c2.org/)

2. Cisco Micro Webserver — "Plug-and-play" embedded Web server runs as Windows NT service. (http://www.cisco.com/warp/public/751/webserv/)

3. CodedDrag — Explanation of CodedDrag, a fully-embedded Windows program that encrypts your personal data using strong encryption algorithms like Triple DES and Blowfish. Software download available here. (http://www.fim.uni-linz.ac.at/coded-drag/codedrag.htm)

4. Cypherpunks "brute" key cracking ring — In order to demonstrate that the various export-weakened crypto systems are too insecure to be used for secure commerce and business applications, the SSL Brute project was formed. (http://www.cl.cam.ac.uk/brute/)

5. EFF Privacy, Crypto, ITAR Export Restrictions Archive — From the Electronic Frontier Foundation. Files dealing with crypto export restrictions. (http://www.eff.org/pub/Crypto/ITAR–export/).

6. Encryption Utilities for Windows 95/NT—Information on and links to various encryption tools including PGP, INAS, Security-Plus, Private Idaho and more. (http://www.ccse.net/~tgrant/secure2.htm)

7. Enigma Group: Workflow and Security Architects—Home page for The Enigma Group, a Canadian company dedicated to the development and delivery of secure information systems. (http://www.enigmagroup.com/)

8. Entegrity Solutions: Products—Products for digital certification. (http://www.entegrity.com/Products/products.html)

9. Entrust Technologies—Public key cryptography products that provide encryption. (http://www.entrust.com/)

10. EPIC Online Guide to Practical Privacy Tools—Electronic Privacy Information Center lists anonymity tools, remailers, cookie busters, e-mail and file privacy. (http://www.epic.org/privacy/tools.html)

11. Kryptology—File encryption software, Windows, Windows95, DOS, Unix.

12. New Mexico Software—Encryption software (http://www.new-mexicosoftware.com/).

13. PGP Source Code and Internals Guide—The good news is that they've completely upgraded their Web server with an RDBMS-backed site that can collect your comments, add you to their mailing lists, and process your orders. (http://mitpress.mit.edu/mitp/recent-books/comp/pgp-source.html)

14. RSA Data Security—Provider of encryption and authentication software (http://www.rsa.com/).

15. RSA Products—RSA Data Security offers a wide variety of toolkits that allow software and hardware developers to incorporate encryption technologies into their products. (http://www.rsa.com/rsa/PRODUCTS/)

16. RSA SecurPC—http://www.rsa.com/rsa/PRODUCTS/end–user.html

17. Security Dynamics—Product and acquisitions news from leading enterprise security solution vendor. RSA is a wholly owned subsidiary of Security Dynamics Technologies, Inc. (http://www.securid.com/)

18. SynCrypt—A security program with features ranging from public key e-mail encryption to local file encryption to digital signature. (http://www.syndata.com/)

19. VeriSign, Inc.—Providing public-key cryptography solutions. Issues Digital IDs to support privacy and authentication within a broad range of products and services. (http://www.verisign.com/)

20. WWW Virtual Library: Cryptography, PGP, and Your Privacy— Basics for attaining personal electronic privacy: an examination of cryptography as scientific basis and means for privacy. (http://world.std.com/~franl/crypto.html)

Appendix B

Worldwide Survey of Cryptographic Products

NAI Labs, The Security Research Division of Network Associates, Inc. (formerly the Advanced Research and Engineering Division of Trusted Information Systems (TIS)) has been conducting an encryption survey for over six years. Information about encryption products continues to flow in on a daily basis, and NAI Labs releases a summary of information periodically. The summary statistics as of May, 1999 are reported in sidebar, "Cryptographic Product Survey Results." NAI Labs also has summary listings of foreign and domestic hardware and software products that employ cryptography.

NAI Labs has identified 1773 products worldwide. They also obtained numerous products from abroad and are examining these products to assess their functionality and security. The survey results show that encryption is indeed widespread throughout the world and the quality of foreign and domestic products appears to be comparable.

U.S. export controls have been modified recently to allow export of strong commercial encryption. International customers who need cryptographic-based security for their sensitive information can now turn to U.S. sources as well as foreign sources to fulfill that need.

951

Cryptographic Product Survey Results

(As of May 1999)

Foreign Products:

TIS identified 805 foreign products from 35 countries:

Argentina	Australia	Austria
Belgium	Canada	Czech Republic
Denmark	Estonia	Finland
France	Germany	Greece
Hong Kong	Iceland	India
Iran	Ireland	Isle of Man
Israel	Italy	Japan
Mexico	Netherlands	New Zealand
Norway	Poland	Romania
Russia	South Africa	South Korea
Spain	Sweden	Switzerland
Turkey	UK	

Of these, 167 employ Triple DES, IDEA, Blowfish, RC 5, or CAST-128.

Domestic Products:

TIS identified 968 domestic products, 466 with DES.

Total Products:

Worldwide total of 1773 products produced and distributed by 999 companies (512 foreign, 487 domestic) in at least 68 countries.[1]

Sources of Information

To provide a definitive assessment of the manufacture and distribution of satellite encryption products throughout the world, the survey makes use of numerous sources of information: computer security product guides, including Datapro reports, Elsevier PC Security Guide,

Computer Security Institute (CSI) Computer Security Products Buyer's Guide, and SCMagazine/Infosecurity News Buyer's Guide; various trade press and journal articles; product literature; Internet electronic mailing lists and news groups; the World-Wide Web; and foreign embassies and trade associations.

Confirmation of Products

Information about satellite encryption products is collected into a comprehensive database; and, where possible, verification of product availability through direct inquiries of product manufacturers and distributors. NAI Labs assumes no responsibility for the accuracy or completeness of the information.

Types of Products

The survey includes many types of cryptographic-based security products:

- Hardware, firmware, software, or combinations thereof.
- General-purpose products (word processors, spreadsheets, telephones, or modems), as well as explicit satellite encryption products (a PC file encryption utility).
- Commercial mass-market products, as well as shareware and other products freely available via the World Wide Web.
- Products providing confidentiality, integrity, and/or authentication service using satellite encryption mechanisms.[2]

Obtaining Foreign Products

NAI Labs has obtained a number of products, focusing on software products employing the Data Encryption Standard (DES). The products were purchased via routine channels, either directly from the foreign manufacturer or from a U.S. distributor. The products were shipped to

NAI Labs within a few days or in several cases, overnight. Implementations of DES, RSA, and IDEA were obtained freely over the Internet from sites throughout the world.

Summary

In order to determine how widespread satellite encryption is in the world, NAI Labs has been conducting a survey of domestic and international products that employ cryptography. While some amount of information about specific products here and there has been available, no one has previously assembled a comprehensive database with, where possible, verification of product availability.

Endnotes

1   "Worldwide Survey of Cryptographic Products," NAI Labs, the Security Research Division, Network Associates, Inc., 3060 Washington Road (Rt. 97), Glenwood, Maryland 21738 USA, 1998, pp. 1–2.

2   Ibid., p. 2.

Glossary

- **Algorithm and Key Length** The combination of cryptographic algorithm and its key length(s) often used to establish the strength of an encryption process.

- **AOR** Atlantic Ocean Region. Designation for international satellites that are located over the Atlantic.

- **ARQ** Automatic Repeat ReQuest, method of automatic error correction used for handshake of SITOR terminals in point to point communications to insure accurate communications.

- **Asymmetric Cryptography (also Public-Key Cryptography)** Cryptography based on algorithms that enable the use of one key (a public key) to encrypt a message and a second, different, but mathematically related, key (a private key) to decrypt a message. Asymmetric cryptography can also be used to perform digital signatures and key exchange.

- **Authentication (Of Identity)** An adjunct step to identification that confirms an asserted identity with a specified, or understood, level

955

of confidence. Authentication can be used to provide high assurance that the purported identity is, in fact, the correct identity associated with the entity that provides it. The authentication mechanism can be based on something that the entity knows, has, or is (a password, a smart card that uses some encryption or random number for a challenge-response scheme, or a fingerprint).

- **Authentication Of a Message (Or a File)**   The process of adding one or more additional data elements to commuinications traffic (or files) to ensure the integrity of the traffic (or files). Such additional elements are often called *message authenticator(s)* and would be an example of an integrity lock.

- **Availability**   The property that a given resource will be usable a given time period; for example, that an encrypted file can be decrypted when necessary.

- **Bit Stream (also Digital Stream)**   The running stream of binary symbols representing digitized information; the term is commonly used to refer to digital communications.

- **Capstone/Fortezza Initiative**   A government initiative to promote and support escrowed encryption for data storage and communications.

- **C-Band, Ku-Band, Hybrid satellites**   The type of satellite based on the frequency that the signals operate. C-Band satellites operate at a lower frequency than Ku-Band satellites. Ku-Band also has a smaller bandwidth (about the size of a raindrop) that allow for receive dishes to be smaller than C-Band. Because of the size of the Ku-bandwidth, feeds may experience rain fade with severe storms. Hybrid satellites have both C-Band and Ku-Band capabilities on the same satellite.

- **Channel**   The numerical dial position that a transponder is found on a IRD. The channel number is not always the same as the transponder number.
- **Ciphertext**   Literally, text material that has been encrypted. Also used in a generic sense for the output of any encryption process, no matter what the original digitized input might have been (text, computer files, computer programs, or digitized graphical images).

- **Cleartext (also Plaintext)**   The material entering into an encryption process or emerging from a decryption process. *Text* is used categorically for any digitized material.

- **Clipper Chip** An escrowed encryption chip that implements the Skipjack algorithm to encrypt communications conducted over the public switched network (between telephones, modems, or facsimile equipment).

- **Collateral Cryptography** A collective term meant to include uses of encryption other than for confidentiality; it includes such services as authentication, integrity checks, authoritative date/time stamping, and digital signatures.

- **Cross-strapping** A technology on the new generation satellites (Galaxy IV/Telstar 401) that enables a feed to be transmitted in C-Band and received in Ku-Band and vice versa. Also known as a *turnaround in the air.*

- **Cryptanalysis** The study and practice of various methods to penetrate ciphertext and deduce the contents of the original cleartext message.

- **Crypto** A device to encrypt or decrypt voice or data transmissions.

- **Cryptographic Algorithm** A mathematical procedure, used in conjunction with a closely guarded secret key, that transforms original input into a form that is unintelligible without special knowledge of the secret information and the algorithm. Such algorithms are also the basis for digital signatures and key exchange.

- **Cryptography** Originally, the science and technology of keeping information secret from unauthorized parties by using a code or a cipher. Today, cryptography can be used for many applications that do not involve confidentiality.

- **Data Encryption Standard (Des)** A U.S. government standard (FIPS 46-1) describing a cryptographic algorithm to be used in a symmetric cryptographic application.

- **Date/Time Stamp** The date and time a transaction or document is initiated or submitted to a computer system, or the time at which a transaction is logged or archived. Often it is important that the stamp be certified by some authority to establish legal or other special status. Such a service can be provided by a cryptographic procedure.

- **Decryption** The cryptographic procedure of transforming ciphertext into the original message cleartext.

- **Digital Signature** A digitized analog of a written signature produced by a cryptographic procedure acting (commonly) on a digest of the message to be signed.

- **Digital Signature Standard (Dss)** A U.S. government standard (FIPS 186) describing a cryptographic algorithm for producing a digital signature.

- **D/L or U/L Frequency Downlink or Uplink Frequency** This is, in essence, the street address that you punch into your equipment to either uplink or downlink to a particular transponder.

- **EIRP** Estimated Isotropic Radiated Power. A fancy way of saying what the footprint, or geographic coverage, of the satellite is.

- **Encryption** Also known as scrambling. A system that takes the signal and reconfigures it into something that is unusable without a corresponding decoder. Examples of encryption systems include SA B-MAC, GI Videocipher, and LEITCH.

- **EOL** End of Life. The EOL is based on how much fuel is on board the satellite to keep it in its geosynchronous orbit 22,300 miles above the Equator. Videoconferencing managers planning a future broadcast should take note of EOL dates. Satellites generally have a projected lifetime of about 10 years. New generation satellites being launched (Telstar 4 series) have estimated lifetimes of 12–15 years.

- **Escrowed Encryption Initiative** A voluntary program to improve the security of telephone communications while meeting the legitimate needs of law enforcement.

- **Escrowed Encryption Standard (EES)** A voluntary U.S. government standard for key-escrowed encryption of voice, fax, or computer data transmitted over circuit-switched telephone systems.

- **Exceptional Access** Access to encrypted data granted to a recipient other than the originally intended recipient.

- **FNC: Federal Networking Council** The body responsible for coordinating networking needs among U.S. Federal agencies.

- **GMT** Greenwich Mean Time. The international time standard, originating from Greenwich, England. While U.S. broadcasts are based on Eastern time, all international video transmissions are based on Greenwich Mean Time (55K audio).

- **HPA** High Powered Amplifier. The piece of equipment (also known as transmitter) at the uplink that boosts the signal to the satellite at 186,000 miles per second (the speed of light).

- **IFB** Interrupted Feedback. Here is a fun one—IFB is the earpiece used for communicating between on-air personalities and directors/ producers without being heard by the guest or the viewers.

- **Integrated Product** A product designed to provide the user a capability useful in its own right (word processing) and integrated with encryption capabilities that a user may or may not employ. A product in which the cryptographic capability is fully integrated with the other functionality of the product.

- **Integrity Check** A quantity derived algorithmically from the running digital stream of a message and appended to it for transmission, or from the entire contents of a stored data file and appended to it. Some integrity checks are not cryptographically based (cyclic redundancy checks), but others are.

- **INTELSAT** The international consortium that includes all the countries that own and operate a particular international satellite.

- **Interceptor** A party eavesdropping on communications.

- **IOR** Indian Ocean Region. Designation for international satellites that are located over the Indian Ocean. This term is used extensively in international video traffic, especially to the Middle East.

- **IRD** Integrated Receiver/Decoder. The receiver that converts the downlink signal into something that your TV monitor understands as a picture. The IRD also takes encrypted signals and decodes them into usable video.

- **ITAR** International Traffic in Arms Regulations.

- **Key** A sequence of easily changed symbols that, used with a cryptographic algorithm, provides a cryptographic process.

- **Key Escrow Encryption (also Escrowed Encryption)** An encryption system that enables exceptional access to encrypted data through special data recovery keys held by a trusted party.

- **Key Management** The overall process of generating and distributing cryptographic keys to authorized recipients in a secure manner.

- **LNB** Low Noise Block Converter. The equipment in a downlink that takes the high frequency signal from the satellite and converts it to a lower frequency signal that can be fed through cable to your receiver.

- **Node** A computer system that is connected to a communications network and participants in the routing of messages within that network. Networks are usually described as a collection of nodes that are connected by communications links.

- **Nonrepudiation (Of a Signed Digital Message, Data, or Software)** The status achieved by employing a digital-signature procedure to affirm the identity of the signer of a digital message with extremely high confidence and, hence, to protect against a subsequent attempt to deny authenticity, whether or not there had been an initial authentication.

- **PCMCIA Card** The industry-standard Personal Computer Memory Card Industry Association card and associated electrical interface for various computer components (memory, hard disks, and cryptographic processes). Also known as a PC card.

- **Polarity** The way that the signal is propagated to space in either vertical or horizontal plane. For example, terrestrial TV signals are received by a horizontal antenna, whereas the antenna for your car radio is vertical. Satellite dishes can send and receive in either polarity, depending on which way the feedhorn is rotated.

- **POR** Pacific Ocean Region. Designation for international satellites that are located over the Pacific Ocean. This term is used extensively in international video traffic, especially to Asia and Australia.

- **Private Key** The private (secret) key associated with a given person's public key for a public-key cryptographic system.

- **Public Key** The publicly known key associated with a given person's use of a public-key cryptographic system.

- **RC2/RC4 Algorithms** Two variable-key-length cryptographic algorithms designed by Ronald Rivest of the Massachusetts Institute of Technology. Both are symmetric algorithms.

- **Reliability** The ability of a computer or an information or telecommunications system to perform consistently and precisely according to its specifications and design requirements and to do so with high confidence.

- **RSA Algorithm**   The Rivest-Shamir-Adelman public-key encryption algorithm.

- **Secret-Key Cryptosystem**   A symmetric cryptographic process that uses the same secret key (which both parties have and keep secret) to encrypt and decrypt messages.

- **Security-Specific (Or Stand-Alone) Cryptography Product** An add-on product specifically designed to provide cryptographic capabilities for one or more other software or hardware capabilities.

- **Skipjack**   A classified symmetric key encryption algorithm that uses 80-bit keys; developed by the National Security Agency.

- **STU-III**   A U.S. government secure telephone system using end-to-end encryption.

- **Symmetric Cryptography, Cryptosystem**   A cryptographic system that uses the same key to encrypt and decrypt messages.

- **Third-Party Access**   Eavesdropping on or entry to data communications, telephony, or stored computer data by an unauthorized party.

- **Transponder**   The designation that the carrier gives to the isolated frequency on a satellite. Depending on the satellite, most have 24 transponders or channels.

- **TWTA power**   Traveling Wave Tube Amplifier power. The power at which the signal is delivered from the satellite. Video network managers with sites at the fringe areas of the footprint of a satellite should look at TWTA power for possible weak signals.

- **TVRO**   Television Receive Only. A satellite dish that serves as a downlink only.

- **U.S. DOMSAT**   U.S. Domestic Satellite. Satellites that serve the U.S. Algorithm (a mathematical rule or procedure for solving a problem).

- **Work Factor**   A measure of the difficulty of undertaking a brute-force test of all possible keys against a given ciphertext and known algorithm.

Index